The Dynamics of R

The Dynamics of Rotating Fluids

P. A. DAVIDSON

OXFORD
UNIVERSITY PRESS

OXFORD
UNIVERSITY PRESS

Great Clarendon Street, Oxford, OX2 6DP, United Kingdom

Oxford University Press is a department of the University of Oxford.

It furthers the University's objective of excellence in research, scholarship, and education by publishing worldwide. Oxford is a registered trade mark of Oxford University Press in the UK and in certain other countries.

Published in the United States of America by Oxford University Press
198 Madison Avenue, New York, NY 10016, United States of America

British Library Cataloguing in Publication Data

Data available

Library of Congress Control Number: 2023951737

ISBN 9780198886303
ISBN 9780198886310 (pbk.)

DOI: 10.1093/9780191994272.001.0001

Printed and bound by
CPI Group (UK) Ltd, Croydon, CR0 4YY

Links to third party websites are provided by Oxford in good faith and for information only. Oxford disclaims any responsibility for the materials contained in any third party website referenced in this work.

MIX
Paper | Supporting
responsible forestry
FSC
www.fsc.org
FSC® C013604

For Catherine, Sarah, and James.

Preface

This is neither an undergraduate textbook nor a research monograph. Rather, it sits somewhere in between, aimed at both advanced undergraduates taking specialized courses in fluid dynamics and postgraduate students starting a research career in engineering science, geophysics, or applied mathematics. This raises the interesting question as to what the style and content of such a book should be.

The task of an author writing a mainstream undergraduate text is clear. They should describe as clearly as possible what is taught in their subject, although some brave authors describe what they think *should* be taught. Likewise, an author embarking on a research monograph has the task of documenting the key underlying concepts, as well as their various refinements arising from applications, eventually establishing the broad frontiers of their subject. But how should a text directed at advanced undergraduate and postgraduate students be structured?

Of course, there is no simple answer to this question, and so readers are largely at the mercy of the author's predilections. My own particular prejudice is that such a text should pace itself and not push too hard to get to the frontiers. In particular, such a book should start with a careful survey of the underlying physical concepts, before slowly building up a broad overview of the subject. Of course, I have other prejudices. For example, I believe an author should resist the temptation to indulge in gratuitous mathematics, and focus instead on providing a clear exposition of the underlying physical concepts, using mathematics as and when it is needed, and at an appropriate level. These, then, have been my two guiding principles in planning this book. However, I have lived long enough to know that this will not be to everyone taste, and hope that those readers who find the pace too slow, or the mathematics too pedestrian, will indulge an author too set in his ways to find a new set of prejudices.

In line with these predilections, Part I of this text includes a crash course in fluid dynamics and waves. Students well versed in both of these topics may decide to omit Chapters 2 and 3, and move directly from Chapter 1 to Part II, though the author politely suggests that readers do not take that step lightly.

In any event, Part II represents the centre of mass of the book. Here applications take a back seat, and we explore the fundamental properties of rotating fluids in their simplest environment, free from many of the additional complexities that inevitably accompany naturally occurring rotating flows. The topics covered include Ekman layers and Ekman pumping, progressive inertial waves, inertial modes in closed domains, Rossby waves, rotating shallow-water flow, precession, the stability of rotating fluids, rotating convection, and vortex breakdown. The aim is to explore the essential properties of rotating fluids without the added distractions of, say, stratification or magnetic fields, which so often influence rotating flows in geophysics and astrophysics.

Finally, in Part III, we provide an introduction to some of the more spectacular rotating flows observed in nature, all of which remain active areas of research. This includes tidal vortices, tornadoes, dust devils, tropical cyclones, helical vortices in the liquid core of the earth, zonal (east–west) winds in planetary atmospheres, and, finally, those vast accretion discs that surround young and dying stars. Our understanding of such flows is often somewhat limited, and that is particularly the case for tidal vortices, dust devils, and tornadoes, where relatively little is known about their internal structure. Consequently, some of the discussion in Part III is somewhat qualitative, and that is especially the case in Chapter 15, where tornadoes, dust devils, and tidal vortices are discussed. In any event, although many aspects of these complex phenomena remain poorly understood, it seems appropriate in Part III to provide the reader with a partial overview of these diverse and intriguing flows, in part to show how many of the ideas developed in Part II can be applied in practice, and in part because their fascinating properties warrant our attention. Of course, such an overview can offer only a steppingstone to more serious study, perhaps inspiring further reading.

It only remains to thank Christian Armstrong and John Howe of SAMS, Emmanuel Dormy of ENS, Jerome Noir of ETH, and Rich Rotunno of NCAR for several helpful suggestions, and the team at Oxford University Press, who were a delight to work with. Finally, I must thank Catherine, my long-suffering wife, who patiently endured those long, monastic silences that frequently accompanied the writing of this book.

Peter Davidson,

Cambridge, 2023

Contents

PART II THE THEORY OF ROTATING FLUIDS

PART III ILLUSTRATIVE EXAMPLES OF ROTATING
FLOWS IN NATURE

PART I
AN INTRODUCTION TO FLUID DYNAMICS AND WAVES

Chapter 1
A Qualitative Introduction to Rotating Fluids

1.1 Some Naturally Occurring Rotating Flows

We are all familiar with examples of strongly rotating flows, ranging in scale from the humble bathtub vortex to those vast cyclones that form over warm seas in the tropics. Perhaps the most striking example is the tornado, which can reap so much havoc, and its smaller, weaker cousins, such as waterspouts and dust devils. The oceanic counterpart of a tornado is the tidal vortex, which is generated when strong tidal currents interact with submerged rock formations. The Corryvreckan whirlpool off the coast of Scotland, which almost claimed the life of George Orwell, is one of the more spectacular examples of a tidal vortex (Figure 1.1). Dust devils, on the other hand, form over deserts

Figure 1.1 Corryvreckan whirlpool off the island of Jura. Reproduced from Walter Baxter.

The Dynamics of Rotating Fluids. P. A. Davidson, Oxford University Press. © Peter A Davidson (2024).
DOI: 10.1093/9780191994272.003.0001

Figure 1.2 An illustration of a dust devil on Mars. Courtesy of JPL/MSSS/NASA.

on particularly hot days (Figure 1.2), driven by a combination of unstable convection and wind shear. Tidal vortices and dust devils are typically around 2–20 m across, while tornadoes and waterspouts (tornadoes over water) are around 80–800 m in diameter.

A less familiar class of rotationally dominated flows are the tall, thin, tornado-like vortices which are thought to populate the liquid-metal core of the earth, aligned with the rotation axis. These are, perhaps, around 40 km in diameter and 10^3 km long. Crucially, the helical flow within these vortices maintains the earth's magnetic field against the natural forces of decay by stretching and folding the magnetic field lines. These vortices are driven by buoyancy, but owe their shape and helical nature to the Coriolis force associated with the earth's rapid rotation. Without such vortices there would be no geodynamo, and without its magnetic field the earth would be a barren wasteland, stripped of its atmosphere by the solar wind.

Moving up in scale we have tropical cyclones (Figure 1.3), which are around 10^3 km across. These draw their energy from the warm moisture that evaporates from the sea surface in the equatorial regions, but like the convection cells in the core of the earth, they are shaped by, and owe their existence to, the Coriolis force associated with the earth's rotation.

Farther afield, many astrophysical flows are dominated by rotation. One striking example is the zonal (east–west) winds on the surface of Jupiter and Saturn (Figure 1.4), which are highly turbulent and owe their existence to the

Figure 1.3 Hurricane Isabel as seen from the International Space Station. Courtesy of NASA.

rotation of those planets. The zonal bands shown in Figure 1.4 are around 10^4 km across. At an even larger scale, there are those vast swirling discs of gas (accretion discs) that form around young and dying stars, being around 10^{10} km in diameter. In the case of a young protostar, which grows by accreting mass through its protoplanetary disc, the disc itself often provides an environment conducive to planet formation.

So swirling flows vary in scale from a few centimetres (a bathtub vortex) up to 10^{10} km. It is hard not to be intrigued by these flows, which shape the weather, occasionally reap havoc, maintain the earth's magnetic field and so help preserve our atmosphere, and allow planets to nucleate in distant galaxies. Many of these flows are discussed later in this book (see Table 1.1). First, however, we need to establish the theory of rotating fluids, as this provides a unified framework from which to view these various flows.

1.2 Two Classic Laboratory Experiments: Taylor Columns and Inertial Waves

Rapidly rotating fluids exhibit two surprising properties that underpin many of the phenomena observed in large-scale geophysical flows. On the one hand, it is observed that, in the presence of a strong background rotation, a *slowly*

Figure 1.4 Image of Jupiter taken from NASA's Cassini spacecraft. The zonal flows are evident, as is the Great Red Spot. Courtesy of NASA/JPL/Space Science Station.

changing flow which might normally be expected to be strongly three dimensional is in fact almost two dimensional, with a velocity field which is more or less independent of the coordinate parallel to the rotation axis. This is called the *Taylor–Proudman theorem* after Joseph Proudman, who first predicted the result, and the physicist G.I. Taylor, who subsequently provided a laboratory

Table 1.1 Some familiar (and less familiar) rotating flows arranged, more or less, by size

Class of rotating flow and the chapter where they are discussed	Radius or width
Bathtub vortex (Chapter 2)	~3 cm
Tidal vortices in the oceans and dust devils in deserts (Chapter 15)	1–10 m
Tornadoes and waterspouts (Chapter 15)	40–400 m
Helical convection cells in the molten core of the earth (Chapter 17)	~20 km
Tropical cyclones (Chapter 16)	100–10^3 km
Zonal (east–west) winds in the atmosphere of the gas giants (Chapter 18)	~10^4 km
Protoplanetary accretion discs rotating around young stars (Chapter 19)	~10^{10} km

demonstration of the phenomenon. On the other hand, rapidly rotating fluids which are excited with an angular frequency ϖ in the range $0 < \varpi < 2|\Omega|$, Ω being the background rotation rate, support a form of incompressible, internal wave motion in which the restoring mechanism is the Coriolis force. These are called *inertial waves*, and their characteristics are not unlike those of internal gravity waves in a stratified fluid. So, a rapidly rotating fluid is a wave-bearing system which, if disturbed, will disperse waves. In fact, as we shall see, the ability to sustain internal waves and a predisposition towards two-dimensional motion are both manifestations of the same underlying phenomenon.

The surprising power of the Taylor–Proudman theorem is nicely demonstrated in an experiment first performed by G. I. Taylor (1923) and shown schematically in Figure 1.5. A tank of water is placed on a turntable and a small object is *slowly* traversed across the base of the tank. The entire process is then observed in the rotating frame of reference. One might have expected to see the fluid ahead of the object to move up and over the object to make way for it, as would occur in the non-rotating case. However, it is observed that the resulting motion is in fact two dimensional. In particular, instead of water rising up and over the object, the fluid sitting above the object drifts across the tank keeping pace with the object. It is as if the fluid sitting inside the imaginary cylinder which circumscribes the object is rigidly attached to the object. This cylinder of drifting fluid is known as a *Taylor column*, and the fluid outside the column flows around it as if the column were rigid, thus maintaining a state of two-dimensional motion.

This extraordinary behaviour can be confirmed by placing dye at the points A and B, shown in Figure 1.5(a). The dye at A is seen to drift across the tank, always centred above the object, while that at B splits into two as the

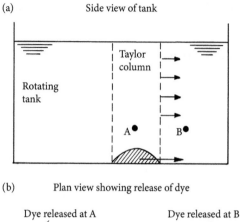

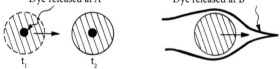

Figure 1.5 A small object is slowly dragged across the base of a rotating tank of water, producing a Taylor column. (a) Side view. (b) Plan view. Reproduced from Davidson (2015).

object passes below it, making way for the Taylor column (Figure 1.5b). Such columnar motion is common in rotating flows.

We might note in passing that the apparent axial rigidity of the fluid exhibited in Taylor's experiment was also observed by Kelvin in 1868, some 55 years earlier. This less well-known experiment is shown in Figure 1.6. Two corks float in a rotating vessel filled with water, one above the other, and when the top cork is pushed downward, the one below also moves down, maintaining the distance between the two. It is as if the axial compression of fluid elements is forbidden in a rapidly rotating fluid.

Returning to Taylor's experiment, one curious feature of the motion is the slow, steady translation of the dye at point A. How does the fluid at A know the instantaneous location of the object below? The answer to this question is more than a little surprising: the information is transmitted upward by inertial waves. In fact, we shall discover that Taylor columns are first established, and then maintained, by low-frequency inertial waves which continually propagate along the rotation axis with a frequency $\varpi \ll 2|\Omega|$. In short, the object at the base of the tank is acting like a radio antenna, continually emitting low-frequency waves which travel upward, carrying the information that the object is moving. To understand this odd behaviour, we first need to describe a second experiment.

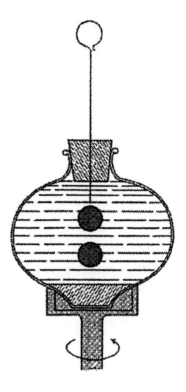

Figure 1.6 Kelvin's experiment of 1868.

Suppose we suspend a disc of diameter d at the centre of a large rotating tank of water and then oscillate that disc vertically with a frequency which lies in the range $0 < \varpi < 2|\Omega|$. It is observed that waves radiate away from the disc in the form of two cones. One cone channels waves upward and the other waves downward, and these conical annuli are terminated by wave fronts that travel away from the disc with the velocity $\mathbf{c}_g$, which is known as the *group velocity*. Within the conical annuli there are wave crests, but surprisingly the crests are not orthogonal to $\mathbf{c}_g$, but rather parallel to $\mathbf{c}_g$. These crests appear out of nowhere on the outer surfaces of the cones, ripple through the cones, and then magically disappear as they reach the inside surfaces of the conical annuli. The crests themselves propagate with the velocity $\mathbf{c}_p$, which is known as the *phase velocity*, and it is clear that $\mathbf{c}_g$ and $\mathbf{c}_p$ are orthogonal. The whole situation is as shown in Figure 1.7, which corresponds to an excitation frequency of around $\varpi \approx \Omega$.

Evidently, inertial waves have some very strange properties, such as the wave energy propagating *along* the wave crests, rather than normal to the crests. We shall have some fun establishing these properties in Chapter 6, but for the moment we shall just take these as being given. Now consider what happens as the frequency of oscillation of the disc is altered. It turns

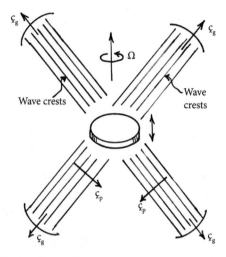

Figure 1.7 An oscillating disc radiates inertial waves at the frequency $\varpi \approx \Omega$. The radiation pattern consists of two conical annuli, one above the disc and one below. The fronts of the wave cones propagate away from the disc with the group velocity c_g, while the wave crests propagate with the phase velocity c_p, which is orthogonal to c_g.

out that, as the frequency increases, the cone angle widens and the magnitude of the group velocity falls, so the propagation of wave energy away from the disc is slowed down and becomes increasingly horizontal. In the limit of $\varpi \to 2|\Omega|$ the group velocity falls to zero, and for $\varpi > 2|\Omega|$ there are no waves at all. Conversely, if we reduce the excitation frequency, then the cone angle reduces and waves are increasingly channelled in the vertical direction. In the limit of $\varpi \to 0$, in which the disc is moving very slowly, the group velocity asymptotes to $c_g \approx \pm \Omega d/\pi$, so the waves travel along the rotation axis at a speed which is very fast by comparison with the movement of the disc.

It turns out that this is more or less what is happening in Taylor's experiment. As the object in Figure 1.5 is *slowly* pulled across the base of the tank it continually emits *low-frequency* waves which travel upward, carrying with them the information that the object is moving, thus maintaining the Taylor column. Of course, we have not provided any evidence to support this claim, but merely suggested an interpretation of the experimental observations. So we shall spend much of Chapter 6 filling in the gaps, providing the theoretical underpinning necessary to establish the picture suggested above.

1.3 Two More Experiments: Ekman Boundary Layers and Stirred Cups of Tea

So far, we have ignored the fluid viscosity, ν, and its role in setting up boundary layers on solid surfaces. In confined, rotating fluids such boundary layers play a particularly important role, establishing weak secondary flows which, over time, dominate the overall behaviour of the fluid. To see how this comes about, let us start with an idealized problem whose nonlinear solution was first given by Kármán (1921).

Consider an infinite disc rotating at the rate Ω in an otherwise stationary fluid, as shown in Figure 1.8. The no-slip condition ensures that the fluid adjacent to the surface of the disc also rotates at the speed Ω, and so a thin boundary layer is established on the disc. Since there is no imposed length scale, the only independent parameters in the problem are Ω and ν, and so dimensional analysis suggests that the boundary-layer thickness scales as $\delta \sim \sqrt{\nu/\Omega}$. Within that boundary layer the fluid is centrifuged radially outward, and so when viewed from above the fluid near the disc spirals outward. It turns out that the radial and azimuthal components of velocity within the boundary layer are similar in magnitude, and so in cylindrical polar coordinates centred on the disc we have $u_r \sim u_\theta \sim \Omega r$. Of course, conservation of mass tells us that the radial outflow within the boundary layer must be replenished by an axial inflow towards the disc, according to $|u_z|\,(\pi r^2) \sim u_r(2\pi r \delta)$, and since $u_r \sim \Omega r$, we conclude that $|u_z| \sim \Omega \delta \sim \sqrt{\Omega \nu}$. In effect, we have a centrifugal fan which entrains remote fluid and then centrifuges that fluid radially outward within a *Kármán layer*. Note that the ratio of the axial inflow to the radial outflow is δ/r, and that this is small, except near $r = 0$. So the axial flow external to the Kármán layer is relatively weak, and indeed it is often difficult to see in an experiment.

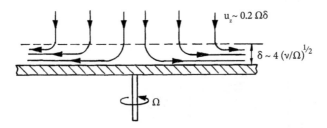

Figure 1.8 The secondary flow in Kármán's problem of flow induced by a rotating disc.

Some 19 years after Kármán's original analysis, Bödewadt (1940) provided the nonlinear solution to the reverse problem, in which fluid rotates at a rate Ω above a stationary disc, as shown in Figure 1.9. Once again, a thin, viscous boundary layer is established on the disc, this time so the fluid velocity can adjust from $u_\theta = \Omega r$ outside the boundary layer to zero at the disc surface. However, the direction of the radial flow is now reversed, for the reasons discussed in Chapter 5, and so the fluid spirals radially *inward* within the boundary layer. As in a Kármán layer, dimensional analysis suggests $\delta \sim \sqrt{\nu/\Omega}$, and we observe that $|u_r| \sim u_\theta \sim \Omega r$. The key difference, however, is that mass conservation now requires that a Bödewadt layer *detrains*, rather that entrains, fluid. Hence, the fluid spirals radially inward along the surface of the disc and then up and out of the boundary layer. A simple mass balance again yields $u_z \sim \Omega \delta \sim \sqrt{\Omega \nu}$, and so the axial flow away from the disc is weak by comparison with the background rotation.

The solutions of Kármán and Bödewadt are particular examples of a broader class of rotating boundary layers called *Ekman layers*, whose defining characteristic is that any differential rotation between a fluid and an adjacent horizontal surface causes mass to be exchanged between the boundary layer and the remote fluid. Although the axial flow outside the boundary layer is relatively weak, and often difficult to see, it has profound consequences in those cases where the flow is confined, such as in an ocean basin or a stirred cup of tea. Let us consider the case of a stirred cup of tea, as shown in Figure 1.10.

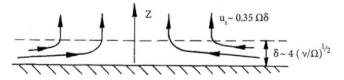

Figure 1.9 The secondary flow induced by a rotating fluid above a stationary surface.

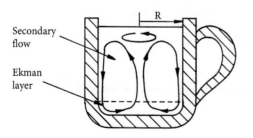

Figure 1.10 Spin-down of a stirred cup of tea.

Suppose that the tea is spun up and the spoon then removed. A natural question to ask is: how long does it take for the tea to stop spinning? To answer this, we note that a Bödewadt layer is established on the bottom of the teacup, inducing a radial inflow along the base of the cup. Mass conservation then requires that the fluid drifts up and out of the boundary layer, after which it is recycled through a layer on the side of the cup. Thus, a secondary flow is established, in which the fluid spirals radially inward through the Bödewadt layer, and then spirals back down through the side-wall boundary layer. This kind of secondary flow is usually called *Ekman pumping*. Crucially, as each fluid particle passes through the Bödewadt layer, it gives up a significant fraction of its kinetic energy, and so the tea comes to rest when the entire contents of the cup have been flushed, once or twice, through the Bödewadt layer. Evidently, the spin-down time is of the order of the turn-over time of the secondary flow, $\tau \sim 2R/u_z \sim 2R/\sqrt{\nu\,\Omega}$, where R is the radius of the cup. This might be compared with spin-down in a long cylinder, where Ekman pumping plays little role. Here the spin-down time is controlled by the time taken for the angular momentum to diffuse to the outer boundary, which turns out to be $\tau \sim R^2/\pi^2\nu$. Suppose, for example, that $R = 4$ cm, $\Omega = 4$ s^{-1}, and $\nu = 10^{-6}$ m^2/s. Then $\tau \sim 40$ s in a teacup, which is about right, whereas the estimate $\tau \sim R^2/\pi^2\nu$ yields $\tau \sim 3$ minutes in a long cylinder.

Evidently, Ekman pumping is a very efficient mechanism for destroying kinetic energy in a confined, swirling flow. We shall explore the consequences of this in Chapter 5.

References

Bödewadt, U.T., 1940, Die Drehstromung uber festem Grunde, *Z. angew Math Mech.*, **20**, 241–53.

Davidson, P.A., 2015, *Turbulence: An Introduction for Scientists and Engineers*, 2nd Ed., Oxford University Press.

Kármán, T., 1921, Uber laminare und turbulente Reibung, *Z. angew Math Mech.*, **1**, 233.

Taylor, G.I., 1923, Experiments on the motion of solid bodies in rotating fluids, *Proc. Roy. Soc., A*, **104**, 213–18.

Chapter 2

A Crash Course on Incompressible Fluid Dynamics

This chapter serves two purposes. First, for those readers who feel the need for a refresher course in fluid dynamics, §2.1–§2.6, which are loosely based on Chapter 2 of Davidson (2021), provide a self-contained introduction to the subject. Second, §2.7–§2.12 introduce a number of more specialized topics which are particularly important in rotating fluids, such as vortex dynamics. Those readers who have a stronger background in fluid dynamics, and who choose to skip §2.1–§2.6, may still wish to consult these later sections.

2.1 An Eulerian Description of Motion and the Convective Derivative

In mechanics, the motion of a particle is usually described by specifying the position of the particle as a function of time, $\mathbf{x}_p(t)$. The velocity and acceleration of that particle are then defined in terms of the time derivatives of $\mathbf{x}_p(t)$. Such an approach is referred to as a *Lagrangian* description of the motion and the resulting time derivatives are known as *Lagrangian time derivatives*. Unfortunately, the Lagrangian formalism is ill-suited to fluid dynamics, where there are an infinite number of particles to track. Rather, in fluid mechanics, the various properties of the flow, such as the pressure distribution, p, or the velocity, $\mathbf{u} = (u_x, u_y, u_z)$, are usually specified as functions of position and time, $\mathbf{x}$ and t. Thus, we talk of the pressure field, $p(\mathbf{x}, t)$, and the velocity field, $\mathbf{u}(\mathbf{x}, t)$, rather like we talk of the magnetic field, $\mathbf{B}(\mathbf{x}, t)$, in electrodynamics. Such an approach, in which $\mathbf{x}$ and t are *independent* variables, is known as an *Eulerian* description of the motion.

A convenient way of visualizing a snapshot of a flow is to use *streamlines*. These are lines drawn in space which are everywhere parallel to $\mathbf{u}$, as illustrated in Figure 2.1. Such lines cannot cross, except where $\mathbf{u} = 0$, and they bunch together where the flow speeds up. Streamlines represent the trajectories of individual fluid particles if the flow is steady, $\mathbf{u} = \mathbf{u}(\mathbf{x})$, but they do

The Dynamics of Rotating Fluids. P. A. Davidson, Oxford University Press. © Peter A Davidson (2024).
DOI: 10.1093/9780191994272.003.0002

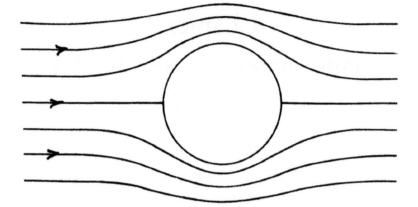

Figure 2.1 Streamlines for inviscid flow over a cylinder.

not coincide with particle trajectories if the flow is unsteady, $\mathbf{u} = \mathbf{u}(\mathbf{x}, t)$, as the streamline pattern then evolves with time. A related concept is that of a *stream-tube*. This is an imaginary tube drawn in space such that $\mathbf{u}$ is every-where parallel to the edges of the tube. Such tubes can be constructed from a collection of streamlines.

Often we want to know the rate of change of some fluid property, say tem-perature, in a given particle as it passes through the flow field. For example, suppose we have a steady flow, $\mathbf{u}(\mathbf{x})$, in which there is a steady temperature field, $T(\mathbf{x})$. Then we might want to know the rate of change of temperature of a particular fluid particle as it slides down a streamline. Note that the tempera-ture of the particle changes with time, despite the fact that the temperature field is steady, because the particle passes through a sequence of different points, each of which is at a different temperature. The rate of change of tempera-ture following a given element of fluid is written as DT/Dt, a notation first introduced by Stokes in 1845. The operator $D(\cdot)/Dt$ is called the *convective derivative*.

Alternatively, we might want to know the rate of change of velocity of a fluid particle as it passes through the flow field, perhaps so that we can apply New-ton's second law to that particle. Note that the acceleration of a fluid element is not $\partial\mathbf{u}/\partial t$, which is the rate of change of $\mathbf{u}$ at a fixed point in space, through which a succession of fluid particles will pass, but rather the rate of change of $\mathbf{u}$ following the fluid element, *i.e.* $D\mathbf{u}/Dt$. So the question now arises as to how to calculate convective derivatives like DT/Dt or $D\mathbf{u}/Dt$ from an Eulerian description of the flow.

Consider the change in temperature of a given fluid particle resulting from small changes in time and position. Then

$$\delta T = (\partial T/\partial t)\,\delta t + (\partial T/\partial x)\,\delta x + \cdots$$

where, because we want to follow a particular fluid particle, δx and δt are related by $\delta x = \mathbf{u}\delta t$. Evidently, if we follow a fluid element, then

$$\delta T = \frac{\partial T}{\partial t}\delta t + (\mathbf{u}\cdot\nabla T)\,\delta t,$$

which yields

$$\frac{DT}{Dt} = \frac{\partial T}{\partial t} + \mathbf{u}\cdot\nabla T. \tag{2.1}$$

Similarly, the acceleration of a fluid particle is

$$\frac{D\mathbf{u}}{Dt} = \frac{\partial \mathbf{u}}{\partial t} + (\mathbf{u}\cdot\nabla)\mathbf{u}. \tag{2.2}$$

When the flow is *steady*, it is often convenient to rewrite the acceleration of a fluid particle in terms of intrinsic coordinates. Let us write $\mathbf{u} = V(s)\hat{\mathbf{e}}_t$, where $V = |\mathbf{u}|$, s is a coordinate measured along the streamline, and $\hat{\mathbf{e}}_t$ is a unit vector tangential to the streamline, as shown in Figure 2.2. We also need to introduce the principal unit normal, $\hat{\mathbf{e}}_n$, which is directed away from the local centre of curvature and is defined in the usual way by $d\hat{\mathbf{e}}_t/ds = -\hat{\mathbf{e}}_n/R$, R being the radius of curvature of the streamline. The acceleration of a fluid particle, written in terms of $V(s)$ and these two unit vectors, is then

$$\frac{D\mathbf{u}}{Dt} = (\mathbf{u}\cdot\nabla)\mathbf{u} = V\frac{\partial}{\partial s}(V\hat{\mathbf{e}}_t) = V\frac{\partial V}{\partial s}\hat{\mathbf{e}}_t - \frac{V^2}{R}\hat{\mathbf{e}}_n. \tag{2.3}$$

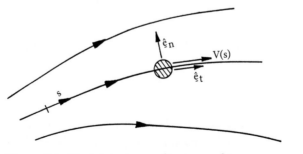

Figure 2.2 The intrinsic coordinates used in equation (2.3).

2.2 Mass Conservation and the Streamfunction

We now consider the consequences of mass conservation. The rate of flow of mass out through a closed surface S is

$$\dot{m} = \oint_S \rho \mathbf{u} \cdot d\mathbf{S}, \tag{2.4}$$

where, by convention, $d\mathbf{S}$ points outward for a closed surface and ρ is density. Mass conservation applied to a control volume V, fixed in space and with bounding surface S, then demands

$$\frac{d}{dt} \int_V \rho dV = \int_V \frac{\partial \rho}{\partial t} dV = -\oint_S \rho \mathbf{u} \cdot d\mathbf{S}. \tag{2.5}$$

Gauss' theorem now yields

$$\int_V \left[\frac{\partial \rho}{\partial t} + \nabla \cdot (\rho \mathbf{u}) \right] dV = 0,$$

and since this holds for *any* volume, V, mass conservation in differential form is

$$\frac{\partial \rho}{\partial t} + \nabla \cdot (\rho \mathbf{u}) = 0, \tag{2.6}$$

which is known as the *continuity equation*. This is frequently rewritten in the more convenient form

$$\frac{D\rho}{Dt} + \rho \nabla \cdot \mathbf{u} = 0. \tag{2.7}$$

We now restrict ourselves to an incompressible fluid, *defined* by the constraint

$$\frac{D\rho}{Dt} = 0.$$

Mass conservation then demands $\nabla \cdot \mathbf{u} = 0$, and so $\mathbf{u}$ is solenoidal in an incompressible fluid. Of course, any solenoidal vector field can be written as the curl of another vector field, say $\mathbf{u} = \nabla \times \mathbf{A}$, where $\mathbf{A}$ is the vector potential for $\mathbf{u}$. It is convenient to choose $\mathbf{A}$ to be solenoidal, so that $\mathbf{A}$ is uniquely defined by the two expressions $\nabla \cdot \mathbf{A} = 0$ and $\nabla \times \mathbf{A} = \mathbf{u}$, along with suitable boundary conditions. This is particularly useful for two-dimensional flows, $\mathbf{u}(x, y) = (u_x, u_y, 0)$, where the simplest choice for $\mathbf{A}$ is $\mathbf{A} = \psi(x, y)\hat{\mathbf{e}}_z$. In such cases,

$$\mathbf{u} = \left(\frac{\partial \psi}{\partial y}, -\frac{\partial \psi}{\partial x}, 0\right), \quad \mathbf{u} \cdot \nabla \psi = 0. \tag{2.8}$$

Evidently, the lines of constant ψ are parallel to $\mathbf{u}$ and so represent streamlines. The function ψ is called the *streamfunction*. It is readily confirmed that the difference in the values of ψ for two adjacent streamlines equals the rate of flow of mass between those streamlines, divided by the fluid density.

The streamlines in an axisymmetric flow, $\mathbf{u}(r, z) = (u_r, 0, u_z)$, expressed in terms of cylindrical polar coordinates (r, θ, z), are given by the stream-surfaces $\Psi = \text{constant}$, where the *Stokes streamfunction*, $\Psi(r, z)$, is defined by

$$\mathbf{u} = \nabla \times \left[\frac{\Psi(r, z)}{r}\hat{\mathbf{e}}_\theta\right] = \left(-\frac{1}{r}\frac{\partial \Psi}{\partial z}, 0, \frac{1}{r}\frac{\partial \Psi}{\partial r}\right). \tag{2.9}$$

2.3 More Kinematics: Characterizing the Spin and Deformation of a Fluid Element

2.3.1 Viscous Shear Stresses and the Need to Distinguish Between Spin and Deformation

Two things happen to a blob of fluid as it slides along a streamline. First, the blob will tend to deform. For example, if it is initially spherical, then it may deform into an ellipsoid. Second, the blob as a whole may spin about its instantaneous centre. Both of these processes are important, but for entirely different reasons. The rate of deformation is crucial because, as we shall see, it sets the level of shear stress in a fluid, while the rate of rotation is important as it leads us to the topic of vortex dynamics and to the role of angular momentum conservation in fluid dynamics.

A hint as to the link between deformation and shear stress is given in Figure 2.3. Consider an initially rectangular fluid element, $\delta x \delta y$, in the parallel shear flow $u_x(y)$. In a short time interval, δt, the line element δy rotates by $\delta \gamma = (du_x/dy)\,\delta t$, and so the angular distortion rate of the fluid element is $d\gamma/dt = du_x/dy$. It turns out that a shear stress, τ_{xy}, is required to sustain this angular

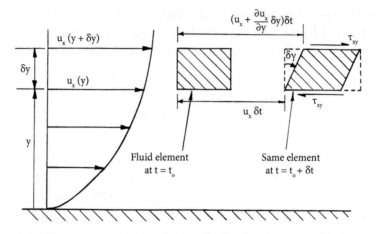

Figure 2.3 The distortion of a rectangular fluid element in a parallel shear flow.

distortion, and *Newton's law of viscosity* asserts that this stress is proportional to the angular distortion rate,

$$\tau_{xy} = \mu\frac{du_x}{dy} = \rho\nu\frac{du_x}{dy}. \tag{2.10}$$

The constant of proportionality, μ, is known as the *dynamic viscosity* of the fluid, while ν is the kinematic viscosity.

Since both deformation and spin are important, albeit for different reasons, let us see if we can unambiguously distinguish between the two. In the interests of simplicity, we start with two-dimensional motion, as shown in Figure 2.4. Consider what happens to a small rectangular element, $\delta x \delta y$, as it is swept along by the flow. In a short time interval δt it deforms as shown. In particular, the anti-clockwise rotation of the short line element δx is $(\partial u_y/\partial x)\,\delta t$, while the clockwise rotation of the line element δy is $(\partial u_x/\partial y)\,\delta t$. We now define the *rate of strain*, or *rate of deformation*, of this fluid element, S_{xy}, to be half of the total angular distortion rate,

$$S_{xy} = \frac{1}{2}\left(\frac{\partial u_y}{\partial x} + \frac{\partial u_x}{\partial y}\right). \tag{2.11}$$

As in Figure 2.3, a shear stress is required to produce this deformation, and the natural generalization of (2.10) is

$$\tau_{xy} = 2\rho\nu S_{xy} = \rho\nu\left(\frac{\partial u_y}{\partial x} + \frac{\partial u_x}{\partial y}\right). \tag{2.12}$$

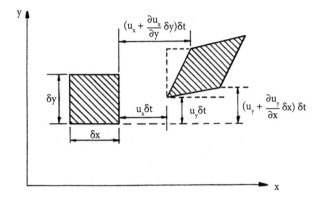

Figure 2.4 The distortion of a rectangular fluid element in a two-dimensional flow.

The *angular velocity* of the element, on the other hand, is defined to be the average rate of rotation of the sides δx and δy. Taking rotation to be positive in the anti-clockwise direction, our angular velocity is then

$$\Omega = \frac{1}{2}\left(\frac{\partial u_y}{\partial x} - \frac{\partial u_x}{\partial y}\right). \tag{2.13}$$

We now introduce the idea of *vorticity*, defined as $\boldsymbol{\omega} = \nabla \times \mathbf{u}$. This allows us to rewrite (2.13) as $\omega_z = 2\Omega$, which tells us that, at least in a two-dimensional flow, the vorticity evaluated at a given location and at a given instant is equal to twice the angular velocity of a fluid particle passing through that point at that instant.

Finally, we generalize these ideas to three dimensions. Consider the relative motion of two adjacent fluid particles, A and B, which are instantaneously located at points $\mathbf{x}$ and $\mathbf{x} + \delta\mathbf{x}$, with velocities $\mathbf{u}$ and $\mathbf{u} + \delta\mathbf{u}$. Clearly, $\delta\mathbf{u}$ is the velocity of particle B as seen by an observer moving with particle A. Now $\delta u_i = (\partial u_i/\partial x_j)\,\delta x_j$ and it is natural to split $\partial u_i/\partial x_j$ into its symmetric and antisymmetric parts, giving

$$\delta u_i = \frac{1}{2}\left[\frac{\partial u_i}{\partial x_j} + \frac{\partial u_j}{\partial x_i}\right]\delta x_j + \frac{1}{2}\left[\frac{\partial u_i}{\partial x_j} - \frac{\partial u_j}{\partial x_i}\right]\delta x_j. \tag{2.14}$$

The rate-of-strain (or rate-of-deformation) tensor, S_{ij}, is now defined as

$$S_{ij} = \frac{1}{2}\left[\frac{\partial u_i}{\partial x_j} + \frac{\partial u_j}{\partial x_i}\right], \tag{2.15}$$

which is an obvious generalization of (2.11). Our expression for $\delta\mathbf{u}$ then becomes

$$\delta\mathbf{u} = \delta\mathbf{u}^{(s)} + \delta\mathbf{u}^{(a)} = S_{ij}\delta x_j + \frac{1}{2}\left(\nabla \times \mathbf{u}\right) \times (\delta\mathbf{x}). \qquad (2.16)$$

We shall now show that, as suggested above, the two terms on the right of (2.16) play very different roles, with S_{ij} associated with deformation, and the vorticity, $\boldsymbol{\omega} = \nabla \times \mathbf{u}$, with the spin of fluid blobs.

2.3.2 The Rate-of-Deformation Tensor

Consider first the rate-of-deformation tensor. The tensor S_{ij} is symmetric and so it may be put into diagonal form through an appropriate orientation of the coordinate system. Let us label these principal axes as 1, 2, and 3 and let a, b, and c be the three principal rates of strain, *i.e.* $a = \partial u_1/\partial x_1$, $b = \partial u_2/\partial x_2$, and $c = \partial u_3/\partial x_3$. In coordinates aligned with the principal axes we then have $\delta\mathbf{u}^{(s)} = (a\delta x_1, b\delta x_2, c\delta x_3)$, where continuity requires $a + b + c = 0$. We conclude that, provided $\boldsymbol{\omega} = 0$, a short material line element oriented parallel to x_1 experiences the relative velocity field $\delta\mathbf{u} = (a\delta x_1, 0, 0)$. This element is therefore stretched or compressed at the rate $a\delta x_1$ while remaining parallel to x_1. Similarly, provided $\boldsymbol{\omega} = 0$, line elements aligned with x_2 or x_3 remain parallel to x_2 or x_3.

In summary, then, provided $\boldsymbol{\omega} = 0$, an initially spherical blob of fluid deforms into an ellipsoid whose principal axes do not rotate. Thus, S_{ij} is associated with the pure deformation of fluid blobs. Such deformations require stresses acting on the fluid, and so a stress tensor, τ_{ij}, is necessarily associated with S_{ij}, as discussed in §2.4.2 below.

2.3.3 Vorticity and the Intrinsic Spin of Fluid Elements

Consider now the second contribution to $\delta\mathbf{u}$ in (2.16), $\delta\mathbf{u}^{(a)} = \frac{1}{2}\boldsymbol{\omega} \times (\delta\mathbf{x})$. This represents rigid-body rotation with angular velocity $\boldsymbol{\omega}/2$ about the point $\mathbf{x}$. We conclude, therefore, that the relative velocity $\delta\mathbf{u}^{(a)}$ rotates fluid elements without causing any deformation of those elements. In short, the vorticity at location $\mathbf{x}$ and at time t is twice the angular velocity, $\boldsymbol{\Omega}$, of a fluid particle passing through that point at that time, the angular velocity being measured about the centre of the particle (Figure 2.5). Evidently, vorticity is all about the intrinsic spin of fluid lumps as they slide along streamlines. It is a crucial concept in fluid mechanics, allowing us to invoke angular momentum conservation in a simple way. Note that the identity $\nabla \cdot (\nabla \times (\sim)) = 0$ ensures that $\nabla \cdot \boldsymbol{\omega} = 0$, and so $\boldsymbol{\omega}$ is solenoidal.

It is important to note that the vorticity, $\boldsymbol{\omega} = \nabla \times \mathbf{u}$, has nothing to do with how much global rotation the flow may possess. For example, the velocity field on the left of Figure 2.6 is $\mathbf{u} = (u_x(y), 0, 0)$. This possesses vorticity, yet the streamlines are straight and parallel. On the other hand, the velocity field on the right is $\mathbf{u}(r) = (0, k/r, 0)$ in (r, θ, z) coordinates. This has no vorticity, except for a singularity at $r = 0$, yet the streamlines are circular. Hence, vorticity measures the intrinsic spin of fluid elements, not global rotation. Flows which are devoid of vorticity are called *irrotational* flows, a term that has carried over to vector calculus in general to mean a vector field whose curl is zero.

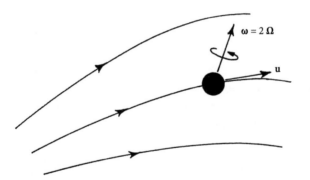

Figure 2.5 The vorticity, $\boldsymbol{\omega} = \nabla \times \mathbf{u}$, at location $\mathbf{x}$ and at time t is twice the angular velocity, $\boldsymbol{\Omega}$, of a fluid element passing through the point $\mathbf{x}$ at time t.

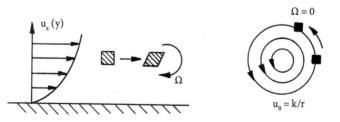

Figure 2.6 The fluid element on the left has vorticity, yet the streamlines are straight, while that on the right has no vorticity, except at $r = 0$, yet the streamlines are circular.

2.4 Newton's Law of Viscosity and the Navier–Stokes Equation

2.4.1 The Stress Tensor and Cauchy's Equation of Motion

Before deriving the equation of motion for a viscous fluid, we first need to say a bit more about the stress tensor, τ_{ij}, which consists of both the *pressure* and *viscous* stresses acting on fluid particles. We start with some simple definitions. Consider a small element of fluid which is instantaneously cubic and aligned with the coordinate axes (x, y, z), as shown in Figure 2.7. It has volume $\delta V = \delta x \delta y \delta z$ and is centred at the location (x_0, y_0, z_0).

Each surface element of our small cube will experience a force which is exerted on it by the surrounding fluid, and the magnitude of that force is necessarily proportional to its area. Hence, the force exerted on the surface element $\delta A = \delta x \delta y$, whose normal is $\hat{\mathbf{e}}_z$, can be written as

$$\delta\mathbf{F}_{(\text{face } +z)} = \left(\tau_{xz}\hat{\mathbf{e}}_x + \tau_{yz}\hat{\mathbf{e}}_y + \tau_{zz}\hat{\mathbf{e}}_z\right)\delta x \delta y, \tag{2.17}$$

which defines τ_{xz}, τ_{yz}, and τ_{zz}. Of course, τ_{xz} and τ_{yz} are the shear stresses acting on this element of area, while τ_{zz} is the normal stress, and these are all evaluated at the centre of that face, $(x_0, y_0, z_0 + \delta z/2)$. Now consider the companion face below. Newton's third law applied across a horizontal plane demands that the stresses on the two faces are in opposite directions, being equal and opposite in the limit of $\delta z \to 0$. So we write

$$\delta\mathbf{F}_{(\text{face } -z)} = -\left(\tau_{xz}\hat{\mathbf{e}}_x + \tau_{yz}\hat{\mathbf{e}}_y + \tau_{zz}\hat{\mathbf{e}}_z\right)\delta x \delta y, \tag{2.18}$$

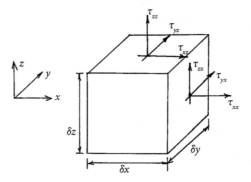

Figure 2.7 The stresses acting on a small rectangular element of fluid.

where the stresses are now evaluated at the location $(x_0, y_0, z_0 - \delta z/2)$. Similarly, the forces acting on the surface elements whose normal vectors are $\pm \hat{\mathbf{e}}_x$ take the form

$$\delta \mathbf{F}_{(\text{faces } \pm x)} = \pm \left(\tau_{xx} \hat{\mathbf{e}}_x + \tau_{yx} \hat{\mathbf{e}}_y + \tau_{zx} \hat{\mathbf{e}}_z \right) \delta y \delta z. \tag{2.19}$$

A torque balance now requires

$$(\tau_{xz} \delta x \delta y) \delta z = (\tau_{zx} \delta y \delta z) \delta x,$$

or $\tau_{xz} = \tau_{zx}$. (Any angular acceleration the cube has does not alter this balance because it leads to a term which is fourth-order in δx.) A consideration of similar pairs of shear stresses shows that, like S_{ij}, the stress tensor, τ_{ij}, is symmetric.

Now suppose that these stresses are allowed to vary slowly in space. Then small differences in stress between companion faces can create a net force on the fluid element. For example, small differences in the stresses top and bottom of the cube lead to the force

$$\delta \mathbf{F}_{(\text{face } +z)} + \delta \mathbf{F}_{(\text{face } -z)} = \left(\frac{\partial \tau_{xz}}{\partial z} \hat{\mathbf{e}}_x + \frac{\partial \tau_{yz}}{\partial z} \hat{\mathbf{e}}_y + \frac{\partial \tau_{zz}}{\partial z} \hat{\mathbf{e}}_z \right) \delta x \delta y \delta z, \tag{2.20}$$

and we conclude that the net force arising from all six faces is

$$\delta F_i = \frac{\partial \tau_{ij}}{\partial x_j} \delta V. \tag{2.21}$$

We are now in a position to write down the viscous equation of motion for a fluid. Newton's second law applied to a moving blob of fluid of volume δV gives us

$$(\rho \delta V) \frac{D\mathbf{u}}{Dt} = (\rho \delta V) \mathbf{g} + \frac{\partial \tau_{ij}}{\partial x_j} \delta V, \tag{2.22}$$

where we have included the self-weight of the blob. We conclude that

$$\rho \frac{D\mathbf{u}}{Dt} = \frac{\partial \tau_{ij}}{\partial x_j} + \rho \mathbf{g}, \tag{2.23}$$

where the first term on the right represents both the pressure and viscous forces. This is called *Cauchy's equation*, and it is as far as Newton's second law will take us. If we are to make further progress, we need to find a *constitutive law* that relates τ_{ij} to S_{ij}.

2.4.2 Newton's Law of Viscosity and the No-Slip Condition

Navier and then Poisson were the first to derive the viscous equations of motion for a fluid. However, their derivations rested on various assumptions about the molecular origins of the viscous stresses, and these assumptions turn out not to be generally valid. Some 20 years later, in 1845, Stokes provided the first modern derivation of the three-dimensional constitutive law which relates viscous stresses to velocity gradients. The significance of Stokes' derivation is that it is based purely on a macroscopic argument, and makes no assumptions of a molecular nature. We now provide an outline of that derivation.

The first problem we face is that of defining pressure. This is not an issue in hydrostatics because *Pascal's law* tells us that the magnitude of the normal stress acting on any surface within the fluid is independent of the orientation of that surface. Pressure is then simply defined as (minus) that normal stress, so that $\tau_{ij} = -p\delta_{ij}$. However, because of the presence of shear stress in a moving fluid, Pascal's law does not apply and the three normal stresses at a given point, τ_{xx}, τ_{yy}, and τ_{zz}, need not be the same. Evidently, we can no longer make the naïve assumption that the normal stresses are all equal to (minus) the pressure, and indeed we might ask what is meant by pressure in a moving fluid. Nevertheless, it is convenient to construct a scalar quantity for a moving fluid which is analogous to hydrostatic pressure. Since the trace of τ_{ij}, *i.e.* τ_{ii}, is independent of the orientation of the coordinate system, we take the judicious step of defining the *mechanical pressure* to be (minus) the average of the three normal stresses, $p = -\tau_{ii}/3$. Of course, this reduces to our conventional notion of pressure when there is no motion.

We now assume that the stress tensor at any one location depends only on the *local* velocity gradients and, following Stokes, we conjecture that:

  (i) when $S_{ij} = 0$, so that there is no local deformation of the fluid, the stress tensor, τ_{ij}, reverts to the hydrostatic form demanded by Pascal's law, $\tau_{ij} = -p\delta_{ij}$;
 (ii) there is no preferred direction in the relationship between τ_{ij} and S_{ij};
(iii) the components of τ_{ij} are at most linear functions of the component of S_{ij}.

Note that point (i) is a consequence of the fact that, if we have a region of fluid in which the velocity gradients take the form of rigid-body rotation only (no distortion), then that region can always be viewed as locally stationary through a change in the frame of reference. Moreover, (ii) simply says that we shall restrict ourselves to fluids whose macroscopic properties are isotropic. So the key assumption is (iii), which is a natural generalization of (2.10) and is clearly analogous to Hook's law in elasticity.

Given assumptions (i) and (iii) we may write

$$\tau_{ij} = -p\delta_{ij} + \tau_{ij}^{\text{dev}}, \quad \tau_{ij}^{\text{dev}} = C_{ijmn}S_{mn}, \tag{2.24}$$

where τ_{ij}^{dev} is known as the *deviatoric stress tensor* and the components of the tensor C_{ijmn} are constants, symmetric in i and j as well as m and n. Note that, in the light of our definition of p, the deviatoric stress satisfies $\tau_{ii}^{\text{dev}} = 0$. Now, if the fluid is isotropic, then the principal axes of τ_{ij}^{dev} and S_{ij} must be coincident. That is to say, if we are in the principal axes of stress, then a spherical blob of fluid will experience pure tension or compression along those axes and so the blob will be stretched or compressed along each axis, but not sheared. It turns out that the only form of the linear relationship (2.24) which satisfies this constraint, as well as the symmetries required of C_{ijmn}, is

$$\tau_{ij}^{\text{dev}} = c_1 S_{kk}\delta_{ij} + c_2 S_{ij}, \tag{2.25}$$

where c_1 and c_2 are constants. Now, the first term on the right of (2.25) is zero in an incompressible fluid, since $S_{ii} = \nabla \cdot \mathbf{u} = 0$. Also, we must take $c_2 = 2\rho\nu$ for consistency with (2.12). We conclude that

$$\tau_{ij} = -p\delta_{ij} + 2\rho\nu S_{ij}, \tag{2.26}$$

which is Newton's law of viscosity as generalized to three dimensions by Stokes. Biology and chemical engineering apart, most common fluids are well approximated by (2.26).

There is an important boundary condition that accompanies (2.26). It is an empirical observation that the fluid velocity adjacent to a stationary solid surface is zero. If the surface moves, the fluid velocity next to the boundary acquires the same velocity as the surface. In short, viscous fluids stick to surfaces. This is known as the *no-slip boundary condition*. At a microscopic level, the no-slip condition arises because the atoms in the fluid ricochet off the stationary atoms in the solid, and so lose their momentum.

Now *all* fluids have a finite viscosity (except superfluid helium), although clearly some are more viscous than others. Indeed, many fluids, such as water and most gases, have very small viscosities, of the order of 10^{-5} or 10^{-6} m^{2}s^{-1}. It is sometimes useful, therefore, to consider an imaginary fluid that has zero viscosity, which is known as an *inviscid* (or *ideal*) fluid. Clearly, there are no shear stresses in such a fluid as the stress tensor reverts to $\tau_{ij} = -p\delta_{ij}$. More importantly, it turns out that the boundary condition for an inviscid fluid is different to that for a viscous fluid, being one of zero normal velocity at a solid boundary, $\mathbf{u} \cdot d\mathbf{S} = 0$. In short, inviscid fluids (which do not really exist) can *slip* over solid surfaces. It is this change in boundary condition, rather than the presence of weak, distributed viscous stresses, that often leads to a dramatic difference in the behaviour of a real fluid with a small but finite viscosity and a hypothetical inviscid fluid.

2.4.3 The Navier–Stokes Equation and the Reynolds Number

We now substitute Newton's law of viscosity into Cauchy's equation of motion. Noting that $\nabla \cdot \mathbf{u} = 0$ ensures

$$\frac{\partial \tau_{ij}}{\partial x_j} = \frac{\partial}{\partial x_j}\left(-p\delta_{ij} + 2\rho\nu S_{ij}\right) = -\nabla p + \rho\nu \frac{\partial^2 u_i}{\partial x_j^2},$$

we obtain

$$\frac{\partial \mathbf{u}}{\partial t} + (\mathbf{u} \cdot \nabla)\mathbf{u} = -\nabla\left(p/\rho\right) + \mathbf{g} + \nu\nabla^2\mathbf{u}, \qquad (2.27)$$

which is known as the *Navier–Stokes equation*. This is the key equation of motion for an incompressible fluid. For a hypothetical inviscid fluid this simplifies to

$$\frac{D\mathbf{u}}{Dt} = -\nabla\left(p/\rho\right) + \mathbf{g}, \qquad (2.28)$$

which is known as *Euler's equation*. As already noted, the boundary condition appropriate for (2.27) is $\mathbf{u} = 0$ at a stationary solid surface, while that for (2.28) is $\mathbf{u} \cdot d\mathbf{S} = 0$. (Mathematically, (2.28) cannot accommodate the stricter boundary condition of $\mathbf{u} = 0$ at a stationary solid surface because, when we dispense with the viscous term, we lose the highest derivative in $\mathbf{u}$.)

We note in passing that, for steady, inviscid flows in which gravity is neglected, (2.28) combines with (2.3) to give

$$\frac{\partial p}{\partial s} = -\rho V \frac{\partial V}{\partial s}, \qquad \frac{\partial p}{\partial n} = \rho \frac{V^2}{R},$$

from which we deduce that $p + \frac{1}{2}\rho V^2$ is constant along a streamline in an inviscid, steady flow. This is *Bernoulli's theorem*. Note also that the divergence of (2.27) or (2.28) yields

$$\nabla^2 (p/\rho) = -\nabla \cdot (\mathbf{u} \cdot \nabla \mathbf{u}). \tag{2.29}$$

In an infinite domain, this Poisson equation for pressure may be inverted to give p as an integral over all space of the instantaneous velocity field (see Appendix 1), so in this sense we may regard p as being slave to $\mathbf{u}$. On the other hand, we may regard (2.27) as an evolution equation for the velocity field $\mathbf{u}$, in which $-\nabla p$ appears as a source of momentum. In any event, we can advance (2.27) and (2.29) forward in time from given initial conditions.

It is of interest to estimate the relative sizes of the inertial and viscous forces in (2.27). The viscous forces per unit mass are of the order of $\nu \nabla^2 \mathbf{u} \sim \nu |\mathbf{u}|/\ell_\perp^2$, where $\ell_\perp$ is a characteristic length normal to the streamlines. The inertial forces per unit mass, on the other hand, are of the order of $\mathbf{u} \cdot \nabla \mathbf{u} \sim |\mathbf{u}|^2/\ell_\parallel$, where $\ell_\parallel$ is a characteristic length scale parallel to the streamlines. The ratio of the inertial to viscous forces is then of the order,

$$\frac{\text{inertial forces}}{\text{viscous forces}} \sim \frac{u\ell_\perp^2}{\nu\ell_\parallel}. \tag{2.30}$$

If we do not distinguish between length scales, and approximate both $\ell_\perp$ and $\ell_\parallel$ by some geometric length scale, say ℓ, then this ratio becomes the all-important *Reynolds number*,

$$\text{Re} = \frac{u\ell}{\nu}.$$

We now make an important observation: for nearly all flows of interest to applied physicists and engineers, the Reynolds number based on a characteristic geometric length scale is large. This reflects the fact that the kinematic viscosity of most common fluids is small. The large value of Re in most flows tentatively suggests that viscous stresses might be ignored, and that we adopt the hypothetical notion of an inviscid fluid. However, as suggested above, this frequently leads to trouble. The key point is that the no-slip condition for a viscous fluid requires that the tangential velocity falls to zero adjacent to any stationary solid surface. Typically, this occurs within a thin layer near the surface, called the *boundary layer*. Since it is the viscous stresses that cause the rapid reduction in tangential velocity, they must be of the same order of magnitude as the inertial forces within the boundary layer, and this is achieved by establishing a very high cross-stream gradient in velocity, so that the viscous forces remain order one, despite the smallness of ν. Consider, for example, the aerofoil shown in Figure 2.8. If we let $\ell_\parallel \sim \ell$ be the length of the wing, and $\ell_\perp \sim \delta$ be the boundary-layer thickness, then (2.30) tells us that the ratio of δ to ℓ is $\delta/\ell \sim \sqrt{\nu/u\ell}$, which is typically of the order of 10^{-3}. Crucially, the viscous stresses in a boundary layer lead to a drag force, which is absent in an inviscid flow.

There is, however, a second reason why viscous stresses remain important at large Reynolds numbers: flows at large Re cease to be *laminar* (non-chaotic) and develop a chaotic component of motion, that is to say, they become *turbulent*. For confined flows, such as flow in a pipe, the entire flow tends to become turbulent when Re is large enough, while in external flows the turbulence tends to be confined to restricted regions of space. For example, for the aerofoil shown in Figure 2.8, the boundary layer typically become turbulent towards the rear of the foil, or if the aerofoil is flying through patches of atmospheric turbulence, this can trigger turbulence across the entire boundary layer.

Virtually all naturally occurring flows are turbulent to varying degrees, and in all turbulent flows the viscous stresses are important. In particular, they are responsible for the intense dissipation of energy which inevitably accompanies turbulence. The way this happens, despite the large nominal value of Re, is that very thin vortices develop within the turbulence, usually in the form of a tangle of intense vortex tubes, whose typical diameter is only a tiny fraction

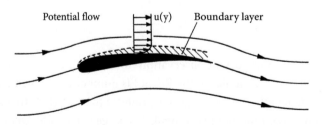

Figure 2.8 The boundary layer on an aerofoil.

of a millimetre. Such vortex tubes, which are embedded within larger eddies, are often referred to as 'worms'. In any event, the key point is that, as with a boundary layer, the small transverse scale of the worms gives rise to large velocity gradients, and hence the viscous stresses are large despite the smallness of ν. This then leads to the intense dissipation of mechanical energy within the turbulence.

2.5 Boundary Layers, Boundary-Layer Separation, and Turbulence

We should say a little more about boundary layers and turbulence. Inviscid fluid dynamics had some early successes, culminating in the seminal work on inviscid vortex dynamics by Helmholtz. However, by the time Stokes initiated the mathematical study of viscous flow, there was already a growing chasm between the applied mathematicians studying inviscid (ideal) hydrodynamics and engineers who were trying to apply the ideas of hydraulics to real flows. Indeed, the two subjects often seemed quite disconnected.

It was a young German engineer, Ludwig Prandtl, who eventually showed where the problems lay, how to circumvent the difficulties using the concept of the boundary layer, and how to analyse such layers. Prandtl argued as follows. Consider the flow over a streamlined body at large Re, such as that in Figure 2.8. In order to satisfy the no-slip condition, the viscous forces must compete with inertia to reduce the velocity of the external flow down to zero. This requires that the velocity gradients are large and the boundary-layer thickness, δ, is thin, so that the viscous stresses are significant, despite the smallness of ν. Since the ratio of the inertial to viscous forces is given by (2.30), and this ratio must be of order unity, the boundary-layer thickness scales as $\delta \sim \sqrt{\nu \ell_\parallel / u}$. Continuity now requires that the normal component of velocity within the boundary layer is of the order of

$$u_n \sim (\delta/\ell_\parallel) u \sim \sqrt{\nu/u\ell_\parallel}\, u \ll u,$$

and so the acceleration of the fluid normal the surface is negligible. This, in turn, demands that the normal gradient in pressure is very small. In short, because there is no significant transverse component of acceleration, the pressure of the external flow at the outer edge of the boundary layer is imposed on the fluid within the boundary layer. Moreover, the thinness of the boundary layer means that we may treat the surface as locally flat. Hence, for a steady, laminar, two-dimensional boundary layer, Prandtl gave the governing equation as

$$u_x \frac{\partial u_x}{\partial x} + u_y \frac{\partial u_x}{\partial y} = -\frac{1}{\rho}\frac{dp}{dx} + v\frac{\partial^2 u_x}{\partial y^2}, \tag{2.31}$$

where $p(x)$ is the external pressure and x and y are parallel and normal to the surface.

Prandtl now divided the flow into an external region, which may be treated as inviscid, and the boundary layer, where the shear stresses are large. One then solves the external problem subject to the free-slip boundary condition $\mathbf{u} \cdot d\mathbf{S} = 0$, which in turn provides outer boundary conditions for the boundary layer, both in terms of the pressure gradient, dp/dx, and the tangential velocity at the top of the boundary layer. Given these boundary conditions, we may solve (2.31) to find the flow within the boundary layer.

Prandtl went on to point out that the situation for a bluff body is more complicated because of the phenomenon of *boundary-layer separation*. Suppose that, instead of an aerofoil, we consider flow over a cylinder or sphere at large Reynolds number. If the fluid were inviscid we would get a symmetric flow pattern like that shown in Figure 2.9(a). The pressures at the *stagnation points* A and C are then equal and given by $p_\infty + \frac{1}{2}\rho V_\infty^2$, where p_∞ and V_∞ are the upstream pressure and velocity. The real flow at large Re, by contrast, looks something like that shown in Figure 2.9(b). A boundary layer forms at the leading stagnation point and this remains thin as the fluid moves to the edges of the cylinder or sphere. However, near the edges of the body the fluid in the boundary layer is ejected into the external flow to form a turbulent wake.

This boundary-layer separation is caused by pressure forces. Outside the boundary layer the fluid, which tries to follow the inviscid flow pattern, starts to slow down as it passes over the outer edges of the cylinder (points B and D).

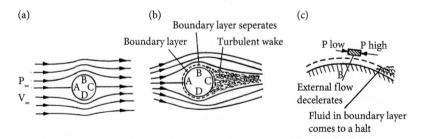

Figure 2.9 Flow over a cylinder. (a) Inviscid flow. (b) Real flow, Re $\gg$ 1. (c) Separation.

This deceleration is caused by a positive pressure gradient which opposes the external flow. The same pressure gradient acts on the fluid within the boundary layer in accordance with (2.31), and so the boundary-layer fluid also begins to decelerate (Figure 2.9c). However, the fluid in the boundary layer has less kinetic energy than that in the external flow and rapidly comes to a halt, moving off and into the external flow to form a wake.

The formation of a wake, which is usually turbulent, is a common feature of flow over a bluff body at large Re. The case of flow over a flat plate is shown schematically in Figure 2.10, with the inviscid flow on the left and the real flow on the right. Notice that in both Figures 2.9 and 2.10 the inviscid flow has upstream-downstream symmetry, and so the pressure field is also symmetric. It would seem, therefore, that there is no net pressure drag exerted on a body sitting in a uniform, inviscid flow. This is d'Alembert's paradox, which did so much to discredit inviscid theory.

So far, we have focussed on flows at large Re. It is instructive to consider how the flow in a particular geometry varies as Re is increased. Consider a uniform flow V approaching a cylinder of diameter d, as shown in Figure 2.11. At low values of $Re = Vd/\nu$ we get a symmetric flow pattern. However, once Re reaches ~10, steady vortices appear at the rear of the cylinder, and by the time Re ~ 100 these vortices peel off from the cylinder in a regular, periodic manner. This unsteady, but laminar, flow is called the *Kármán vortex street*.

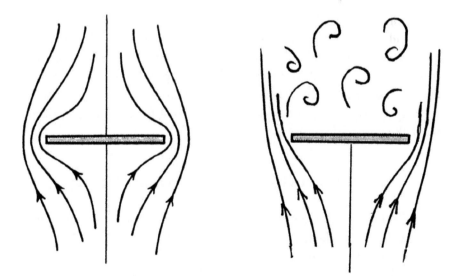

Figure 2.10 Schematic of flow over a flat plate at large Re, with the inviscid flow on the left and the real flow on the right.

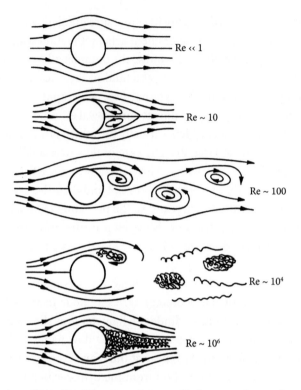

Figure 2.11 A uniform flow V approaches a cylinder at various values of Re.

At yet higher values of Re, low levels of turbulence appear in the boundary layer, and this turbulence is carried off in the shed vortices. Finally, for values of Re in excess of 10^5, the flow at the rear of the cylinder loses its periodic structure, becoming fully turbulent. Notice that upstream of the cylinder the fluid possesses no vorticity, whereas in the Kármán street there is clearly some vorticity. Moreover, this vorticity seems to have come from the boundary layer. We shall return to this point in §2.8.3.

2.6 The Viscous Dissipation of Mechanical Energy

It is useful to quantify the viscous dissipation of mechanical energy, which we now do. Taking the dot product of **u** with Cauchy's equation, (2.23), and ignoring gravity for simplicity, we find

$$\frac{\partial}{\partial t}\left(\frac{1}{2}\rho\mathbf{u}^2\right) + \nabla\cdot\left(\left(\frac{1}{2}\rho\mathbf{u}^2\right)\mathbf{u}\right) = u_i\frac{\partial\tau_{ij}}{\partial x_j}. \tag{2.32}$$

However, τ_{ij} is symmetric and so

$$u_i \frac{\partial \tau_{ij}}{\partial x_j} = \frac{\partial}{\partial x_j}(\tau_{ij}u_i) - \tau_{ij}\frac{\partial u_i}{\partial x_j} = \frac{\partial}{\partial x_j}(\tau_{ij}u_i) - \tau_{ij}S_{ij}, \qquad (2.33)$$

which allows us to rewrite (2.32) as

$$\frac{\partial}{\partial t}\left(\frac{1}{2}\rho\mathbf{u}^2\right) = -\nabla \cdot \left((\frac{1}{2}\rho\mathbf{u}^2)\mathbf{u}\right) + \frac{\partial}{\partial x_j}(\tau_{ij}u_i) - \tau_{ij}S_{ij}. \qquad (2.34)$$

Next, we substitute for the τ_{ij} using (2.26). Noting that the pressure contribution is

$$\frac{\partial}{\partial x_j}(-p\delta_{ij}u_i) + p\delta_{ij}S_{ij} = -\nabla \cdot (p\mathbf{u}) + pS_{ii} = -\nabla \cdot (p\mathbf{u}),$$

we find

$$\frac{\partial}{\partial t}\left(\frac{1}{2}\rho\mathbf{u}^2\right) = -\nabla \cdot \left((\frac{1}{2}\rho\mathbf{u}^2)\mathbf{u}\right) - \nabla \cdot (p\mathbf{u}) + \frac{\partial}{\partial x_j}(2\rho\nu S_{ij}u_i) - 2\rho\nu S_{ij}S_{ij}. \quad (2.35)$$

Finally, we integrate (2.35) over a control volume V fixed in space and with bounding surface S. This yields the energy equation

$$\frac{d}{dt}\int_V \frac{1}{2}\rho\mathbf{u}^2 dV = -\oint_S (\frac{1}{2}\rho\mathbf{u}^2)\mathbf{u} \cdot d\mathbf{S} - \oint_S p\mathbf{u} \cdot d\mathbf{S} + \oint_S (2\rho\nu S_{ij})u_i dA_j - \int_V \rho\varepsilon dV, \qquad (2.36)$$

where $\varepsilon = 2\nu S_{ij}S_{ij}$. The surface integrals on the right represent:

   (i) the material transport of kinetic energy across the bounding surface S;
   (ii) the rate of working of the pressure forces on S;
  (iii) the rate of working of the viscous stresses on S.

Conservation of energy now tells us that the volume integral of $\rho\varepsilon$ must represent the net rate of loss of mechanical energy to heat within V. It follows that the rate of increase of internal energy per unit mass due to viscous dissipation is $\varepsilon = 2\nu S_{ij}S_{ij}$. This is often simply referred to as the *viscous dissipation rate*. Finally, we note that

$$\varepsilon = 2\nu S_{ij}S_{ij} = \nu\boldsymbol{\omega}^2 + \nabla \cdot (2\,\nu\mathbf{u} \cdot \nabla\mathbf{u}). \qquad (2.37)$$

Since the divergence on the right often integrates to zero, $\nu\boldsymbol{\omega}^2$ is often used as a proxy for ε.

2.7 The Navier–Stokes Equation in Cylindrical Polar Coordinates

There are many flows which have axial symmetry and for which cylindrical polar coordinates, (r, θ, z), provide the most convenient framework. It is useful, therefore, to establish the form of the Navier–Stokes equation when written in terms of cylindrical polar coordinates. The polar form of the Navier–Stokes equation can be formally deduced through a routine (if somewhat tedious) coordinate transformation. However, our real interest is to understand why certain new and important terms arise during this transformation. We shall also introduce the azimuthal-poloidal decomposition for axisymmetric fields, which can be particularly useful in vortex dynamics.

The key to moving from Cartesian to polar coordinates is to recall that the unit vectors $(\hat{\mathbf{e}}_r, \hat{\mathbf{e}}_\theta, \hat{\mathbf{e}}_z)$ are not all constant. In particular,

$$\frac{\partial \hat{\mathbf{e}}_r}{\partial \theta} = \hat{\mathbf{e}}_\theta, \qquad \frac{\partial \hat{\mathbf{e}}_\theta}{\partial \theta} = -\hat{\mathbf{e}}_r, \tag{2.38}$$

which gives us

$$\frac{D\hat{\mathbf{e}}_r}{Dt} = \frac{u_\theta}{r}\hat{\mathbf{e}}_\theta, \qquad \frac{D\hat{\mathbf{e}}_\theta}{Dt} = -\frac{u_\theta}{r}\hat{\mathbf{e}}_r. \tag{2.39}$$

Now, the convective derivative operates on products in exactly the same way as a conventional derivative. It follows that

$$\frac{D\mathbf{u}}{Dt} = \frac{Du_r}{Dt}\hat{\mathbf{e}}_r + u_r\frac{D\hat{\mathbf{e}}_r}{Dt} + \frac{Du_\theta}{Dt}\hat{\mathbf{e}}_\theta + u_\theta\frac{D\hat{\mathbf{e}}_\theta}{Dt} + \frac{Du_z}{Dt}\hat{\mathbf{e}}_z,$$

which combines with (2.39) to give the acceleration on the left of the Navier–Stokes equation:

$$\frac{D\mathbf{u}}{Dt} = \frac{Du_r}{Dt}\hat{\mathbf{e}}_r + \frac{u_r u_\theta}{r}\hat{\mathbf{e}}_\theta + \frac{Du_\theta}{Dt}\hat{\mathbf{e}}_\theta - \frac{u_\theta^2}{r}\hat{\mathbf{e}}_r + \frac{Du_z}{Dt}\hat{\mathbf{e}}_z. \tag{2.40}$$

We now turn to the viscous term in the Navier–Stokes equation, which involves the Laplacian of $\mathbf{u}$. This is most simply evaluated in polar coordinates by noting that, since $\mathbf{u}$ is solenoidal, $\nabla^2 \mathbf{u} = -\nabla \times \nabla \times \mathbf{u}$. Applying the curl operator once yields

$$\boldsymbol{\omega} = \nabla \times \mathbf{u} = \left(\frac{1}{r}\frac{\partial u_z}{\partial \theta} - \frac{\partial u_\theta}{\partial z}\right)\hat{\mathbf{e}}_r + \left(\frac{\partial u_r}{\partial z} - \frac{\partial u_z}{\partial r}\right)\hat{\mathbf{e}}_\theta + \left(\frac{1}{r}\frac{\partial}{\partial r}(ru_\theta) - \frac{1}{r}\frac{\partial u_r}{\partial \theta}\right)\hat{\mathbf{e}}_z,$$

and taking the curl a second time gives (see Appendix 2)

$$\nabla^2 \mathbf{u} = \left(\nabla^2 u_r - \frac{u_r}{r^2} - \frac{2}{r^2}\frac{\partial u_\theta}{\partial \theta}\right)\hat{\mathbf{e}}_r + \left(\nabla^2 u_\theta - \frac{u_\theta}{r^2} + \frac{2}{r^2}\frac{\partial u_r}{\partial \theta}\right)\hat{\mathbf{e}}_\theta + (\nabla^2 u_z)\hat{\mathbf{e}}_z.$$

(2.41)

Combining (2.40) with (2.41) now yields the Navier–Stokes equation in cylindrical polar coordinates:

$$\frac{Du_r}{Dt} - \frac{u_\theta^2}{r} = -\frac{1}{\rho}\frac{\partial p}{\partial r} + \nu\left[\nabla^2 u_r - \frac{u_r}{r^2} - \frac{2}{r^2}\frac{\partial u_\theta}{\partial \theta}\right],$$

(2.42)

$$\frac{Du_\theta}{Dt} + \frac{u_r u_\theta}{r} = -\frac{1}{\rho r}\frac{\partial p}{\partial \theta} + \nu\left[\nabla^2 u_\theta - \frac{u_\theta}{r^2} + \frac{2}{r^2}\frac{\partial u_r}{\partial \theta}\right],$$

(2.43)

$$\frac{Du_z}{Dt} = -\frac{1}{\rho}\frac{\partial p}{\partial z} + \nu\nabla^2 u_z,$$

(2.44)

(see, also, Appendix 2). When the motion is axisymmetric, these simplify to

$$\frac{Du_r}{Dt} - \frac{u_\theta^2}{r} = -\frac{1}{\rho}\frac{\partial p}{\partial r} + \nu\frac{1}{r}\nabla_*^2(ru_r),$$

(2.45)

$$\frac{D\Gamma}{Dt} = \nu\nabla_*^2\Gamma,$$

(2.46)

$$\frac{Du_z}{Dt} = -\frac{1}{\rho}\frac{\partial p}{\partial z} + \nu\nabla^2 u_z,$$

(2.47)

where $\Gamma = ru_\theta$ is the angular momentum density, and the axisymmetric *Stokes operator*, ∇_*^2, is defined by

$$\nabla_*^2(\sim) = \left[r\frac{\partial}{\partial r}\frac{1}{r}\frac{\partial}{\partial r} + \frac{\partial^2}{\partial z^2}\right](\sim) = \nabla\cdot\left[r^2\nabla\left(\frac{(\sim)}{r^2}\right)\right] = r^2\nabla\cdot\left[\frac{1}{r^2}\nabla(\sim)\right].$$

(2.48)

Note the appearance of the centripetal acceleration in (2.45), as well as the inviscid conservation of Γ arising from the integration of (2.46), $\int \Gamma dV = $ constant.

Next we note that, in an axisymmetric flow, $\mathbf{u}$ may be decomposed into its *azimuthal*, $\mathbf{u}_\theta = (0, u_\theta, 0)$, and *poloidal*, $\mathbf{u}_p = (u_r, 0, u_z)$, components, each of which is individually solenoidal. Moreover, because $\nabla\cdot\mathbf{u}_p = 0$, the poloidal velocity may be expressed in terms of the Stokes streamfunction, $\Psi(r, z)$, first introduced in §2.2:

$$\mathbf{u}_p = \nabla \times \left[\frac{\Psi}{r}\hat{\mathbf{e}}_\theta\right] = \left(-\frac{1}{r}\frac{\partial \Psi}{\partial z}, \ 0, \ \frac{1}{r}\frac{\partial \Psi}{\partial r}\right). \tag{2.49}$$

Of course, the vorticity may be similarly decomposed into two fields, $\boldsymbol{\omega}_\theta$ and $\boldsymbol{\omega}_p$, which are also individually solenoidal when the flow is axisymmetric. Now, it is readily confirmed that, for an axisymmetric flow, the curl of a poloidal field is azimuthal, while the curl of an azimuthal field is poloidal. It follows that $\boldsymbol{\omega}_\theta = \nabla \times \mathbf{u}_p$, while $\boldsymbol{\omega}_p = \nabla \times \mathbf{u}_\theta$. The poloidal vorticity is therefore dictated by the instantaneous distribution of $\Gamma = ru_\theta$. In particular, a comparison of (2.49) with $\boldsymbol{\omega}_p = \nabla \times [(\Gamma/r)\,\hat{\mathbf{e}}_\theta]$ tells us that

$$\boldsymbol{\omega}_p = \left(-\frac{1}{r}\frac{\partial \Gamma}{\partial z}, \ 0, \ \frac{1}{r}\frac{\partial \Gamma}{\partial r}\right). \tag{2.50}$$

Moreover, combining (2.49) with $\boldsymbol{\omega}_\theta = \nabla \times \mathbf{u}_p$ yields

$$\omega_\theta = \frac{\partial u_r}{\partial z} - \frac{\partial u_z}{\partial r} = -\frac{1}{r}\nabla_*^2\Psi. \tag{2.51}$$

Thus, the Stokes streamfunction determines both the poloidal velocity field, through (2.49), and the azimuthal vorticity distribution, through (2.51).

Now, for a given distribution of ω_θ, the Poisson-like equation $\nabla_*^2\Psi = -r\omega_\theta$ can always be inverted to give the instantaneous distribution of Ψ (see Appendix 1), and hence of $\mathbf{u}_p$. It follows that the instantaneous velocity distribution is uniquely determined by the two scalar functions $\Gamma = ru_\theta$ and ω_θ/r. So, when considering axisymmetric flows, it is natural to focus on these two scalar fields. We have already shown that the evolution equation for Γ is the advection-diffusion equation (2.46), and we shall establish the equivalent evolution equation for ω_θ/r in §2.10.1.

2.8 Viscous Vortex Dynamics

We now turn to the intriguing and important topic of vortex dynamics, starting with the evolution equation for the vorticity field.

2.8.1 A Transport Equation for Vorticity

As noted in §2.4.3, we may regard the Navier–Stokes equation as an evolution equation for $\mathbf{u}(\mathbf{x}, t)$. However, each time we calculate a new velocity field from the old, we must update the pressure field by solving (2.29). A more powerful strategy, largely developed by Helmholtz, is to take the curl of the Navier–Stokes equation to give an evolution equation for $\boldsymbol{\omega}(\mathbf{x}, t)$. This eliminates pressure from the problem. In fact, as we shall see, there are several advantages to working with the vorticity field.

Our next task, then, is to determine the evolution equation for $\boldsymbol{\omega}(\mathbf{x}, t)$, which we shall do shortly. However, let us first see if we can predict the general form of this equation using only simple physical arguments. Recall that $\boldsymbol{\omega}(\mathbf{x}, t)$ is twice the intrinsic angular velocity, $\boldsymbol{\Omega}$, of a fluid blob passing through point $\mathbf{x}$ at time t. It seems appropriate, therefore, to focus on the angular momentum of a fluid element as it slides down a streamline. In particular, let us consider a blob of fluid which is *instantaneously* spherical, as shown in Figure 2.5. Since $\boldsymbol{\omega} = 2\boldsymbol{\Omega}$, the angular momentum of the sphere is $\mathbf{H} = I\boldsymbol{\omega}/2$, where I is its moment of inertia. Moreover, $\mathbf{H}$ can change only as a result of tangential surface stresses acting on the blob, the pressure playing no role while the blob is spherical. We conclude that, for as long as the fluid blob remains spherical, $D\mathbf{H}/Dt$ is determined by the viscous surface stresses only:

$$\frac{D}{Dt}\left(\frac{1}{2}I\boldsymbol{\omega}\right) = \text{viscous torque on sphere.} \tag{2.52}$$

Differentiating out the product $I\boldsymbol{\omega}$, and dividing through by $I/2$, now yields

$$\frac{D\boldsymbol{\omega}}{Dt} = -\boldsymbol{\omega}\frac{D}{Dt}(\ln I) + \text{ (viscous term).} \tag{2.53}$$

This suggests that the vorticity of a fluid particle can change either because its moment of inertia changes, or because the viscous stresses spin-up or spin-down the particle. The former mechanism is illustrated in Figure 2.12, where a blob is stretched out to reduce its moment of inertia, thus causing $\boldsymbol{\omega}$ to increase in line with conservation of angular momentum.

Actually, these arguments are not rigorous, merely suggestive, since the spherical blob will not stay spherical for long. The simplest way to get the exact evolution equation for $\boldsymbol{\omega}(\mathbf{x}, t)$ is to use the identity

$$\nabla\left(u^2/2\right) = \mathbf{u} \cdot \nabla\mathbf{u} + \mathbf{u} \times \boldsymbol{\omega} \tag{2.54}$$

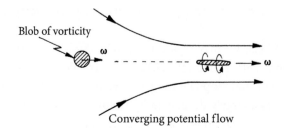

Figure 2.12 Stretching a fluid element can intensify its vorticity.

to rewrite (2.27) in the form

$$\frac{\partial \mathbf{u}}{\partial t} = \mathbf{u} \times \boldsymbol{\omega} - \nabla \left(p/\rho + \mathbf{u}^2/2 \right) + \mathbf{g} + \nu \nabla^2 \mathbf{u}. \tag{2.55}$$

Taking the curl of (2.55), and noticing that the Laplacian commutes with the curl, gives us our evolution equation

$$\frac{\partial \boldsymbol{\omega}}{\partial t} = \nabla \times (\mathbf{u} \times \boldsymbol{\omega}) + \nu \nabla^2 \boldsymbol{\omega}. \tag{2.56}$$

Moreover, since both $\mathbf{u}$ and $\boldsymbol{\omega}$ are solenoidal, we may use the expression

$$\nabla \times (\mathbf{u} \times \boldsymbol{\omega}) = (\boldsymbol{\omega} \cdot \nabla) \mathbf{u} - (\mathbf{u} \cdot \nabla) \boldsymbol{\omega}$$

to rewrite our evolution equation as

$$\frac{D\boldsymbol{\omega}}{Dt} = (\boldsymbol{\omega} \cdot \nabla) \mathbf{u} + \nu \nabla^2 \boldsymbol{\omega}. \tag{2.57}$$

Notice the similarity between (2.53) and (2.57). This suggests that $\boldsymbol{\omega} \cdot \nabla \mathbf{u}$ represents changes in $\boldsymbol{\omega}$ caused by changes in the moment of inertia of a fluid element, while $\nu \nabla^2 \boldsymbol{\omega}$ represents the action of viscous stresses spinning up or spinning down that element.

2.8.2 The Advection, Stretching, and Diffusion of Vorticity

To show that $\boldsymbol{\omega} \cdot \nabla \mathbf{u}$ is indeed related to changes in moment of inertia, consider the thin vortex tube shown in Figure 2.13. Let $u_{\parallel}$ be the component of velocity parallel to the tube and s a coordinate measured along the tube. Then we have

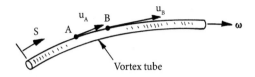

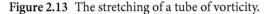

Vortex tube

Figure 2.13 The stretching of a tube of vorticity.

$$(\boldsymbol{\omega} \cdot \nabla)u_{\parallel} = |\boldsymbol{\omega}| \frac{du_{\parallel}}{ds}. \tag{2.58}$$

Now this vortex tube is stretched whenever $u_B > u_A$, *i.e.* when $du_{\parallel}/ds > 0$, and it is compressed when $du_{\parallel}/ds < 0$. It follows that, when $(\boldsymbol{\omega} \cdot \nabla)u_{\parallel} > 0$, the vortex tube is stretched and thinned, causing a reduction in its moment of inertia and a corresponding rise in angular velocity. This is consistent with $\boldsymbol{\omega} \cdot \nabla \mathbf{u}$ producing changes in $|\boldsymbol{\omega}|$ through changes in the moment of inertia of a fluid element, a process called *vortex-line stretching.*

There is, however, one special case where there is no vortex stretching: two-dimensional motion of the form $\mathbf{u}(x, y) = (u_x, u_y, 0)$. Here we have $\boldsymbol{\omega}(x, y) = (0, 0, \omega)$ and so (2.57) reduces to the simple scalar equation

$$\frac{D\omega}{Dt} = v\nabla^2\omega. \tag{2.59}$$

It is instructive to compare this with the governing equation for the temperature, T, in a thermally conducting fluid,

$$\frac{DT}{Dt} = \alpha\nabla^2 T. \tag{2.60}$$

Here α is the thermal diffusivity of the fluid, and the term on the right of (2.60) represents the diffusion of heat in or out of fluid elements as they move around. Equations of this type are referred to as advection-diffusion equations, with the left-hand side representing the ability of a flow to transport a quantity by virtue of material movement, and the right-hand side the tendency for T or ω to spread by diffusion. Evidently, in a two-dimensional flow, vorticity is materially transported (advected) by the motion, while simultaneously diffusing between adjacent fluid elements. For an inviscid fluid, the two-dimensional vorticity equation reduces to $D\omega/Dt = 0$, which says that each fluid element holds onto its vorticity as it moves around in an ideal, two-dimensional flow.

It is important to note that neither advection nor diffusion of an isolated patch of vorticity can change the net amount of vorticity within that patch, just

Figure 2.14 Vortex shedding behind a cylinder.

as the advection and diffusion of heat cannot change the net thermal energy contained in an isolated region of hot fluid. Rather, these two processes merely act to redistribute a given amount of vorticity, or heat, in space. This may be shown by rewriting (2.59) in the form

$$\frac{\partial \omega}{\partial t} = -\nabla \cdot (\omega \mathbf{u}) + \nabla \cdot (\nu \nabla \omega). \tag{2.61}$$

We now integrate (2.61) over a two-dimensional volume which encloses an isolated patch of vorticity, say one of the Kármán vortices in Figure 2.14. Applying Gauss' theorem, and noting that $\omega = 0$ on the bounding surface of that volume, gives the required result:

$$\frac{d}{dt} \int \omega \, dV = -\oint \omega \mathbf{u} \cdot d\mathbf{S} + \nu \oint \nabla \omega \cdot d\mathbf{S} = 0. \tag{2.62}$$

2.8.3 Where Does Vorticity Come From?

So far, we have established that vorticity may be redistributed throughout space by advection and diffusion, and it may be intensified or diminished through the stretching or compression of the vortex tubes. Crucially, though, none of these processes can create vorticity in a region of fluid which had none to start with. Yet most real flows are packed full of vorticity, and this begs the question: where does this vorticity come from?

Perhaps the analogy between the heat transport equation and the two-dimensional vorticity equation is useful here. Heat cannot be created or destroyed within the interior of a fluid by advection or diffusion. However, if the cylinder in Figure 2.14 is heated, then heat gets into the fluid by diffusing in from the surface of the cylinder. The same is true of vorticity. In the absence of a rotational body force, like buoyancy, vorticity can get into a fluid only by

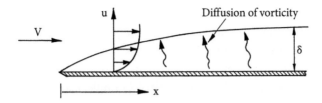

Figure 2.15 A boundary layer is a diffusion layer for vorticity.

diffusing in from the boundaries. So, just as there is a thermal boundary layer on a heated cylinder in a cross-flow, through which heat diffuses into the fluid, so there is a viscous boundary layer, through which vorticity diffuses into the flow. This explains the origin of the vorticity in the Kármán street shown in Figure 2.14. All of the vorticity downstream of the cylinder started off in thin viscous boundary layers on the upstream surface of the cylinder, and when those boundary layers separate, that vorticity is released into the main flow in the form of Kármán vortices.

Thus, viscous boundary layers are, in effect, diffusion layers for vorticity, just as thermal boundary layers are diffusion layers for heat. This is illustrated in Figure 2.15, which shows flow over a flat plate. The no-slip condition at the wall means that high velocity gradients form at the leading edge of the plate, and so that region acts as a source of vorticity. This vorticity then diffuses away from the plate and into the fluid, while simultaneously being swept downstream. When Re is large (*i.e.* the viscosity small), viscous diffusion is slow by comparison with advection, and this is why the boundary layer remains thin.

Returning to Figure 2.14, we might summarize events as follows. Vorticity is generated at the surface of the cylinder as the fluid passes over it. This then diffuses into the fluid, and so viscous boundary layers develop on the upper and lower surfaces of the cylinder, layers that progressively thicken as the fluid moves downstream. By the time we get to the edges of the cylinder, very high levels of vorticity have diffused out from the surface and into the boundary layers, and all of this vorticity is then released into the main flow when the boundary layers separate. Of course, similar processes occur for other bluff bodies.

2.8.4 The Biot–Savart Law

We now consider a useful kinematic analogy between vorticity and magnetostatics, first noticed by Helmholtz (1858). The point is the following. In general, the velocity field in a given region of space can be divided into two distinct parts. On the one hand, there can be an irrotational flow which is devoid of vorticity and is established in the far field. Such flows are governed by $\nabla \times \mathbf{u} = 0$ and $\nabla \cdot \mathbf{u} = 0$ and are called *potential flows*. On the other hand, a given distribution of vorticity establishes its own velocity field, and this owes its existence entirely to the presence of that vorticity. For example, a tornado, which is just a tube of vorticity passing from the ground to the cloud above, induces a horizontal swirling motion by virtue of the presence of the vertical vorticity within the tornado (Figure 2.16a). Likewise, a vortex ring induces its own velocity field, as shown in Figure 2.16(b). Of course, the tornado or vortex ring might sit in an irrotational cross-flow, governed by $\nabla \times \mathbf{u} = 0$ and $\nabla \cdot \mathbf{u} = 0$.

Now, in vortex dynamics, we often know more or less what the vorticity distribution looks like at some given instant, $\boldsymbol{\omega}(\mathbf{x})$, and we would like to find the

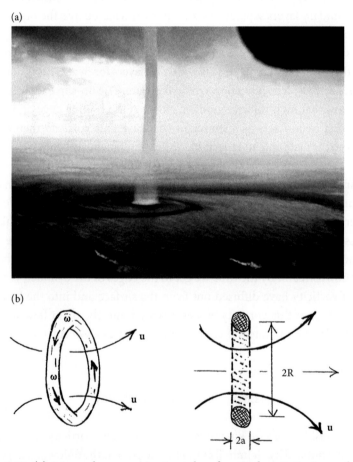

Figure 2.16 (a) A tornado over water is a tube of vertical vorticity. Image by J. Golden, NOAA. (b) A vortex ring is a hoop of azimuthal vorticity.

corresponding velocity field which owes its existence to the presence of that vorticity. However, the definition $\boldsymbol{\omega} = \nabla \times \mathbf{u}$ only allows us to calculate $\boldsymbol{\omega}$ from $\mathbf{u}$, and not $\mathbf{u}$ from $\boldsymbol{\omega}$. So it is natural to ask if we can invert $\nabla \times \mathbf{u} = \boldsymbol{\omega}$, subject to $\nabla \cdot \mathbf{u} = 0$, to find $\mathbf{u}$. The question is well posed, since a vector field is uniquely determined if its divergence and curl are specified, along with suitable boundary conditions (see Appendix 1). Fortunately, we can perform this inversion by borrowing a famous result from magnetostatics.

In magnetostatics the magnetic field, $\mathbf{B}$, is related to the current density, $\mathbf{J}$, through Ampère's law, $\nabla \times \mathbf{B} = \mu \mathbf{J}$, where μ is called the *permeability*. The magnetic field is also solenoidal, and so $\mathbf{B}$ is governed by $\nabla \times \mathbf{B} = \mu \mathbf{J}$ and $\nabla \cdot \mathbf{B} = 0$. Evidently, there is an analogy in which $\mathbf{B} \leftrightarrow \mathbf{u}$ and $\mu \mathbf{J} \leftrightarrow \boldsymbol{\omega}$. Moreover, often we know the distribution of $\mathbf{J}$ and would like to find the associated magnetic field, *i.e.* we would like to invert $\nabla \times \mathbf{B} = \mu \mathbf{J}$, subject to $\nabla \cdot \mathbf{B} = 0$. This is what the *Biot–Savart law* achieves, which states that, in an infinite domain, the magnetic field corresponding to a given distribution of current is

$$\mathbf{B}(\mathbf{x}) = \frac{\mu}{4\pi} \int \frac{\mathbf{J}(\mathbf{x}') \times \mathbf{r}}{r^3} d\mathbf{x}' \, , \qquad \mathbf{r} = \mathbf{x} - \mathbf{x}', \qquad (2.63)$$

where $\mathbf{x}'$ is a dummy variable which samples $\mathbf{J}$ at different points in space, $\mathbf{r} = \mathbf{x} - \mathbf{x}'$ takes us from $\mathbf{x}'$ to the point where $\mathbf{B}$ is measured, and $r = |\mathbf{r}|$, as shown in Figure 2.17(a).

It follows from our analogy that, in an infinite domain, the velocity field associated with a given distribution of $\boldsymbol{\omega}$ is

$$\mathbf{u}(\mathbf{x}, t) = \frac{1}{4\pi} \int \frac{\boldsymbol{\omega}(\mathbf{x}', t) \times \mathbf{r}}{r^3} d\mathbf{x}' \, , \qquad \mathbf{r} = \mathbf{x} - \mathbf{x}'. \qquad (2.64)$$

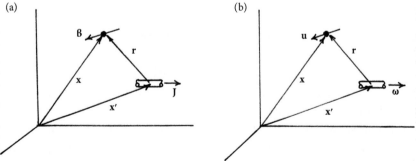

Figure 2.17 Coordinate system used for the Biot–Savart law in: (a) magnetism; (b) fluids.

Actually, it is readily confirmed that (2.64) not only ensures that $\nabla \times \mathbf{u} = \boldsymbol{\omega}$, but it also enforces $\nabla \cdot \mathbf{u} = 0$. Note that, if $\boldsymbol{\omega}$ is localized in space, as is often the case in practice, then $\mathbf{u}$ in (2.64) falls off with distance from the vorticity field as a power law, typically as r^{-3}, with $\mathbf{u} = 0$ at infinity. Note also that we can add to (2.64) a potential flow, which is governed by $\nabla \times \mathbf{u} = 0$ and $\nabla \cdot \mathbf{u} = 0$, and which is established by a non-zero distribution of $\mathbf{u}$ at infinity.

A direct proof of (2.64) proceeds as follows. Since $\nabla \cdot \mathbf{u} = 0$, we can introduce a vector potential, $\mathbf{A}$, for $\mathbf{u}$. This is governed by $\nabla \times \mathbf{A} = \mathbf{u}$, $\nabla \cdot \mathbf{A} = 0$, and $\mathbf{A} = 0$ at infinity, which uniquely determines $\mathbf{A}$. The definition $\nabla \times \mathbf{u} = \boldsymbol{\omega}$ now yields the Poisson equation $\nabla^2 \mathbf{A} = -\boldsymbol{\omega}$, which may be solved in an infinite domain using Green's integral,

$$\mathbf{A}(\mathbf{x}, t) = \frac{1}{4\pi} \int \frac{\boldsymbol{\omega}(\mathbf{x}', t)}{r} d\mathbf{x}', \qquad \mathbf{r} = \mathbf{x} - \mathbf{x}', \tag{2.65}$$

(see Appendix 1). To deduce (2.64) we now use the identity

$$\nabla \times \left(\boldsymbol{\omega}'/r \right) = -\boldsymbol{\omega}' \times \nabla (1/r) = \boldsymbol{\omega}' \times \mathbf{r}/r^3, \quad \boldsymbol{\omega}' = \boldsymbol{\omega}(\mathbf{x}', t), \tag{2.66}$$

where ∇ operates on $\mathbf{x}$ while treating $\mathbf{x}'$ as constant. This then yields the required result:

$$\mathbf{u}(\mathbf{x}, t) = \nabla \times \mathbf{A}(\mathbf{x}, t) = \frac{1}{4\pi} \int \nabla \times \left(\frac{\boldsymbol{\omega}'}{r} \right) d\mathbf{x}' = \frac{1}{4\pi} \int \frac{\boldsymbol{\omega}(\mathbf{x}', t) \times \mathbf{r}}{r^3} d\mathbf{x}'. \tag{2.67}$$

2.8.5 Flows Without Vorticity: The Dangers of Potential Flow Theory

On a windy day, any street is full of vorticity, vorticity that has diffused into boundary layers on the sides and roofs of the buildings, and then spilled out into the street. Indeed, nearly all naturally occurring flows are full of vorticity. One might ask, therefore, why some older textbooks devote considerable space to the theory of inviscid potential flow.

One motivation for potential flow theory is shown in Figure 2.8. Here there is a boundary layer full of vorticity, and an external flow. In aerodynamics, the flow upstream of the wing is usually *assumed* to be irrotational, and since the vorticity generated on the surface of the wing is confined to a boundary layer and wake, the entire external flow is then a potential flow. However, outside aerodynamics, such potential flows are rare.

One of the dangers of potential flow theory is illustrated in Figure 2.18. The upper image appears in many fluids texts as the potential flow at the entrance to a duct. The same figure also appears in books on magnetism, representing a static magnetic field, **B**, entering or leaving a pole, governed by $\nabla \cdot \mathbf{B} = 0$ and $\nabla \times \mathbf{B} = 0$. Of course, there is no inertia associated with **B**, and this reminds us that inertia plays no role in the calculation of an irrotational velocity field (other than to ensure that an upstream state of zero initial rotation is maintained by the fluid particles). That is, potential flows are governed by the two *kinematic* equations, $\nabla \cdot \mathbf{u} = 0$ and $\nabla \times \mathbf{u} = 0$. Newton's second law enters the problem only *retrospectively* when we use Bernoulli's theorem to infer a pressure distribution from $\mathbf{u}(\mathbf{x})$. Moreover, if we reverse the sign of **u** in Figure 2.18(a), the flow still satisfies $\nabla \cdot \mathbf{u} = 0$ and $\nabla \times \mathbf{u} = 0$, only now it represents fluid *leaving* a duct. Yet it is never presented as such, because the real flow looks very different. In fact, the real flow takes the form of a jet which is full of vorticity, vorticity which has been stripped off the walls of the duct (Figure 2.18b).

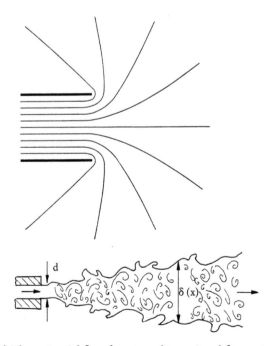

Figure 2.18 (a) The potential flow for a two-dimensional flow entering or leaving a duct. (b) Schematic of the actual exit flow at large Re. From Davidson (2017).

So, in some sense, potential flow theory is rather special, often owing more to kinematics than dynamics, at least when it comes to calculating $\mathbf{u}(\mathbf{x})$. In any event, aerodynamics and surface gravity waves apart, potential flows are rare in nature.

2.9 The Classical Theory of Inviscid Vortex Dynamics

The seminal work on vortex dynamics is Helmholtz (1858), which more or less defined the entire subject and is still well worth a read. A decade later, Kelvin published an important footnote to Helmholtz's laws, known as *Kelvin's theorem*. When Helmholtz wrote his 1858 paper, he was unaware that Stokes (1845) had successfully incorporated viscous stresses into the equation of motion for a fluid. Consequently, although well aware of the importance of friction, Helmholtz restricted himself to an inviscid fluid. In fact, both Helmholtz's laws and Kelvin's theorem are restricted to an inviscid fluid, free from rotational body forces.

2.9.1 Kelvin's Theorem

Let us start with Kelvin. Kelvin's theorem says that, in an inviscid flow, the *circulation*, $\Gamma = \oint \mathbf{u} \cdot d\mathbf{l}$, around any closed *material* curve is an invariant of the motion. *One* proof of the theorem proceeds as follows. (We shall provide a second proof shortly.) Let $d\mathbf{l}$ be a short material line that moves with the fluid, like a dyeline, with $\mathbf{x}$ and $\mathbf{x} + d\mathbf{l}$ locating the ends of $d\mathbf{l}$. Then the change in $d\mathbf{l}$ in a time δt is $[\mathbf{u}(\mathbf{x} + d\mathbf{l}) - \mathbf{u}(\mathbf{x})]\,\delta t$, from which we obtain

$$\frac{D}{Dt}(d\mathbf{l}) = \mathbf{u}(\mathbf{x} + d\mathbf{l}) - \mathbf{u}(\mathbf{x}) = (d\mathbf{l} \cdot \nabla)\mathbf{u}. \tag{2.68}$$

It follows that

$$\frac{D}{Dt}(\mathbf{u} \cdot d\mathbf{l}) = \frac{D\mathbf{u}}{Dt} \cdot d\mathbf{l} + \mathbf{u} \cdot (d\mathbf{l} \cdot \nabla \mathbf{u}) = \left(\mathbf{g} - \nabla\frac{p}{\rho}\right) \cdot d\mathbf{l} + d\mathbf{l} \cdot \nabla\left(u^2/2\right), \tag{2.69}$$

where we have used Euler's equation, (2.28), to substitute for $D\mathbf{u}/Dt$. Now let C_m be a closed material curve, always composed of the same fluid particles. On integrating (2.69) around C_m, and noting that D/Dt commutes with the integral sign, we obtain

$$\frac{d}{dt} \oint_{C_m} \mathbf{u} \cdot d\mathbf{l} = \oint_{C_m} \frac{D}{Dt} (\mathbf{u} \cdot d\mathbf{l}) = \oint_{C_m} \mathbf{g} \cdot d\mathbf{l} + \oint_{C_m} \nabla \left(u^2/2 - p/\rho \right) \cdot d\mathbf{l}. \quad (2.70)$$

However, the first line integral on the right is zero because $\mathbf{g}$ is conservative, while the second is zero since $u^2/2 - p/\rho$ is single valued. We conclude that, as claimed by Kelvin:

$$\frac{d\Gamma}{dt} = \frac{d}{dt} \oint_{C_m} \mathbf{u} \cdot d\mathbf{r} = 0. \quad (2.71)$$

Note that we require Euler's equation to hold only on the curve C_m, and so we may use (2.71) even when viscous effects are important, provided that those viscous effects do not occur on the curve C_m.

2.9.2 Helmholtz's Laws

Helmholtz's laws come in two parts, which we shall label H1 and H2, with H2 composed of two statements. The first law is couched in terms of vortex lines, and the second in terms of the flux of vorticity, Φ, along a vortex tube (see Figure 2.19). These laws assert that:

- H1—vortex lines are frozen into the fluid like dyelines, so fluid particles that lie on a vortex line at some initial instant continue to lie on that vortex line for all time;
- H2a—the flux of vorticity along a vortex tube is the same at all cross-sections of the tube;
- H2b—the flux of vorticity along a vortex tube is independent of time.

The origin of the first law, as well as part (a) of the second law, is straightforward. However, part (b) of the second law, which turns out to be closely related to Kelvin's circulation theorem, is a little less obvious. To prove Helmholtz's first law, we return to (2.68). Comparing this with the inviscid version of (2.57),

$$\frac{D\boldsymbol{\omega}}{Dt} = (\boldsymbol{\omega} \cdot \nabla) \mathbf{u}, \quad (2.72)$$

we see that $\boldsymbol{\omega}$ and $d\mathbf{l}$ obey the same evolution equation. It follows that $\boldsymbol{\omega}$ and $d\mathbf{l}$ evolve in identical ways under the influence of $\mathbf{u}$, and so, if they are initially coincident, then they must remain coincident. We conclude that vortex lines

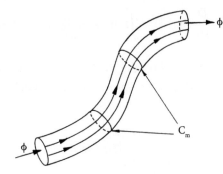

Figure 2.19 A vortex tube carries a vorticity flux Φ. According to Helmholtz's second law, Φ is the same at all cross-sections of the vortex tube and also independent of time.

move with the fluid, like dyelines. Moreover, part (a) of Helmholtz's second law follows directly from $\nabla \cdot \boldsymbol{\omega} = 0$ applied to a vortex tube, which in turn is a direct consequence of the definition $\boldsymbol{\omega} = \nabla \times \mathbf{u}$.

That leaves us with part (b) of Helmholtz's second law: the temporal invariance of the vorticity flux, Φ. This can be deduced in a number of different ways, one of which is to invoke the kinematic equation

$$\frac{d}{dt} \int_S \mathbf{G} \cdot d\mathbf{S} = \int_S \frac{\partial \mathbf{G}}{\partial t} \cdot d\mathbf{S} - \oint_{C_m} (\mathbf{u} \times \mathbf{G}) \cdot d\mathbf{r}. \qquad (2.73)$$

Here $\mathbf{G}(\mathbf{x}, t)$ is any solenoidal vector field that links an open surface S, which in turn spans a closed *material* curve C_m, as shown in Figure 2.20.

The reason for the surface integral on the right of (2.73) is clear cut, but the line integral requires some explanation. When the material curve C_m moves it may expand at points to include additional flux, or contract at other points to exclude flux. Now consider the surface element $d\mathbf{S}$ which is swept out by a line element $d\mathbf{r}$ of the curve C_m in a time δt. It has sides $d\mathbf{r}$ and $\mathbf{u}_\perp \delta t$, where $\mathbf{u}_\perp$ is the component of $\mathbf{u}$ perpendicular to $d\mathbf{r}$. This vector area, which is normal to both $d\mathbf{r}$ and $\mathbf{u}$, can be written as $d\mathbf{S} = (\mathbf{u}\delta t) \times d\mathbf{r}$. Thus, the change in the flux of $\mathbf{G}$ through S by virtue of movement of the boundary element $d\mathbf{r}$ is

$$\mathbf{G} \cdot d\mathbf{S} = \mathbf{G} \cdot (\mathbf{u} \times d\mathbf{r})\delta t \ = -[(\mathbf{u} \times \mathbf{G}) \cdot d\mathbf{r}]\,\delta t,$$

and summing all such contributions around C_m yields the line integral in (2.73).

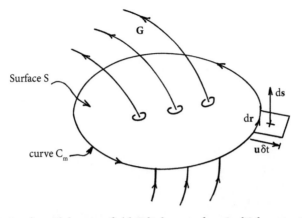

Figure 2.20 A solenoidal vector field **G** links a surface S which spans the material curve C_m. The surface element $d\mathbf{S}$, swept out by a line element $d\mathbf{r}$ in a time δt, is $d\mathbf{S} = (\mathbf{u}\delta t) \times d\mathbf{r}$.

We now apply the kinematic equation (2.73) to the vorticity field, using Stokes' theorem to rewrite the line integral as a surface integral,

$$\frac{d}{dt} \int_S \boldsymbol{\omega} \cdot d\mathbf{S} = \int_S \left(\frac{\partial \boldsymbol{\omega}}{\partial t} - \nabla \times (\mathbf{u} \times \boldsymbol{\omega}) \right) \cdot d\mathbf{S}.$$

However, the inviscid vorticity equation (2.72) can be rewritten in the form

$$\frac{\partial \boldsymbol{\omega}}{\partial t} = \nabla \times (\mathbf{u} \times \boldsymbol{\omega}), \tag{2.74}$$

and so we conclude that

$$\frac{d}{dt} \int_S \boldsymbol{\omega} \cdot d\mathbf{S} = \frac{d\Phi}{dt} = 0. \tag{2.75}$$

Evidently, the flux of vorticity, Φ, through any closed material curve, C_m, is an invariant. Of course, Stokes' theorem tells us that this is nothing more than Kelvin's circulation theorem applied to C_m. (We have, in effect, formulated a second proof of Kelvin's theorem.)

Now consider the vortex tube shown in Figure 2.19. From Helmholtz's first law the vortex tube moves with the fluid, and so we may draw a closed curve on the surface of the tube, encircling it, which is a material curve, C_m. Part (b) of Helmholtz's second law, the temporal invariance of the flux Φ, then follows directly from (2.75).

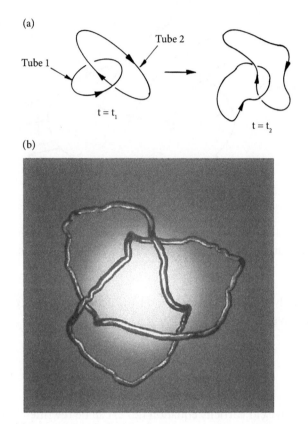

Figure 2.21 (a) Two inviscid vortex tubes preserve their topology and fluxes. (b) Two linked vortex tubes in a real fluid. Image courtesy of Dustin Kleckner & William Irvine.

2.9.3 Helicity and Helicity Conservation

Helmholtz's laws greatly constrain how an inviscid fluid can evolve. Consider, for example, the two thin interlinked vortex tubes shown in Figure 2.21(a). Each tube creates a velocity field via the Biot–Savart law and so they advect each other in some complicated manner. However, no matter how complex the motion, Helmholtz's first law tells us that the two tubes must remain linked in the same manner for all time, because to break the linkage would require different fluid particles to occupy the same region of space, which is not possible. Moreover, the second law tells us that the strength of each vortex tube, as measured by its flux Φ, is conserved.

This conservation of vortex-line topology in an inviscid fluid is captured by an integral invariant called *helicity*. This is defined as

$$h = \int_{V_m} \mathbf{u} \cdot \boldsymbol{\omega} dV, \tag{2.76}$$

where V_m is any material volume for which $\boldsymbol{\omega} \cdot d\mathbf{S} = 0$ on its surface S_m. We may confirm that h is an invariant by combining Euler's equation with the inviscid vorticity equation to give

$$\frac{D}{Dt}(\mathbf{u} \cdot \boldsymbol{\omega}) = \frac{D\mathbf{u}}{Dt} \cdot \boldsymbol{\omega} + \frac{D\boldsymbol{\omega}}{Dt} \cdot \mathbf{u} = -\nabla\left(\frac{p}{\rho}\right) \cdot \boldsymbol{\omega} + (\boldsymbol{\omega} \cdot \nabla \mathbf{u}) \cdot \mathbf{u}. \tag{2.77}$$

Since $\boldsymbol{\omega}$ is solenoidal, this yields

$$\frac{D}{Dt}(\mathbf{u} \cdot \boldsymbol{\omega}) = \nabla \cdot \left[\left(\frac{1}{2}u^2 - \frac{p}{\rho}\right)\boldsymbol{\omega}\right], \tag{2.78}$$

which may be integrated over our material volume, V_m, to give

$$\frac{dh}{dt} = \frac{d}{dt}\int_{V_m}(\mathbf{u} \cdot \boldsymbol{\omega})dV_m = \int_{V_m}\frac{D(\mathbf{u} \cdot \boldsymbol{\omega})}{Dt}dV_m = \oint_{S_m}\left(\frac{1}{2}u^2 - \frac{p}{\rho}\right)\boldsymbol{\omega} \cdot d\mathbf{S} = 0,$$

$$\tag{2.79}$$

where the surface integral vanishes by virtue of the boundary condition on S_m.

The connection between the invariance of h and Helmholtz's two laws may be illustrated by returning to Figure 2.21(a). Suppose that vortex tube 1 has volume V_1, flux Φ_1, and centreline C_1, and similarly for vortex tube 2. Integrating over all space, we find that the net helicity for this configuration is

$$\int_{V_1}\mathbf{u} \cdot \boldsymbol{\omega} dV + \int_{V_2}\mathbf{u} \cdot \boldsymbol{\omega} dV = \oint_{C_1}\mathbf{u} \cdot (\Phi_1 d\mathbf{l}) + \oint_{C_2}\mathbf{u} \cdot (\Phi_2 d\mathbf{l})$$

$$= \Phi_1\oint_{C_1}\mathbf{u} \cdot d\mathbf{l} + \Phi_2\oint_{C_2}\mathbf{u} \cdot d\mathbf{l}, \tag{2.80}$$

where we have used the fact that $\boldsymbol{\omega} dV = \Phi d\mathbf{l}$. Moreover, provided that the loops are interlinked once, Stokes' theorem tells us that the circulation around loop 1 is $\pm\Phi_2$, where the positive (negative) sign corresponds to a right-handed (left-handed) linkage. Similarly, the circulation around loop 2 is $\pm\Phi_1$. It follows that $h = \pm2\Phi_1\Phi_2$, and we conclude that the invariance of h in this simple example is a consequence of the conservation of the two fluxes, Φ_1 and Φ_2, and of the linkage of the vortex tubes.

Figure 2.22 Kelvin's sketches of linked and knotted vortex tubes.

Of course, one can construct more complex combinations of knotted and linked vortex tubes, as shown in Figure 2.22, and the conservation of helicity in each case reflects the conservation of the vortex-line topology. Kelvin was so impressed by the immutable nature of inviscid vortex-line topology that he proposed a 'vortex atom' theory of matter, in which the atoms of the various elements are composed of linked and knotted vortex tubes.

2.10 Axisymmetric Flow with Swirl

We now turn to the vortex dynamics of axisymmetric flows, extending the discussion of §2.7.

2.10.1 The Poloidal-Azimuthal Decomposition of the Viscous Vorticity Equation

In §2.7 we showed that the vorticity field of an axisymmetric flow can be decomposed into its azimuthal and poloidal components, $\boldsymbol{\omega}_\theta = \nabla \times \mathbf{u}_p$ and $\boldsymbol{\omega}_p = \nabla \times \mathbf{u}_\theta$, both of which are individually solenoidal. Moreover, ω_θ is related to the Stokes streamfunction, Ψ, by $\nabla_*^2 \Psi = -r\omega_\theta$, where ∇_*^2 is the Stokes operator (2.48), while $\Gamma = ru_\theta$ acts as an effective Stokes streamfunction for $\boldsymbol{\omega}_p$, *i.e.* $\boldsymbol{\omega}_p = \nabla \times [(\Gamma/r)\,\hat{\mathbf{e}}_\theta]$. It follows that the instantaneous vorticity distribution is uniquely determined by the two scalar fields Γ and ω_θ/r.

Of course, $\boldsymbol{\omega} = \nabla \times \mathbf{u}$ can always we inverted using the Biot–Savart law to give $\mathbf{u}$, and so we conclude that the instantaneous velocity field is also uniquely determined by Γ and ω_θ/r. Evolution equations for these two scalar fields can be found from the azimuthal components of the Navier–Stokes and vorticity equations. For example, the azimuthal component of the momentum equation shows that Γ is governed by the advection–diffusion equation (2.46). We shall now find the companion equation for ω_θ/r.

Since the curl of a poloidal field is azimuthal, the azimuthal component of the vorticity equation, (2.56), is evidently

$$\frac{\partial \boldsymbol{\omega}_\theta}{\partial t} = \nabla \times \left[\mathbf{u}_p \times \boldsymbol{\omega}_\theta\right] + \nabla \times \left[\mathbf{u}_\theta \times \boldsymbol{\omega}_p\right] + \nu\nabla^2\boldsymbol{\omega}_\theta. \tag{2.81}$$

Moreover, since $\mathbf{u}_p$, $\mathbf{u}_\theta$, $\boldsymbol{\omega}_p$, and $\boldsymbol{\omega}_\theta$ are all solenoidal, it is readily confirmed that

$$\nabla \times \left[\mathbf{u}_p \times \boldsymbol{\omega}_\theta\right] = \boldsymbol{\omega}_\theta \cdot \nabla\mathbf{u}_p - \mathbf{u}_p \cdot \nabla\boldsymbol{\omega}_\theta = -r\mathbf{u}_p \cdot \nabla\left(\omega_\theta/r\right)\hat{\mathbf{e}}_\theta, \tag{2.82}$$

and

$$\nabla \times \left[\mathbf{u}_\theta \times \boldsymbol{\omega}_p\right] = \boldsymbol{\omega}_p \cdot \nabla\mathbf{u}_\theta - \mathbf{u}_\theta \cdot \nabla\boldsymbol{\omega}_p = r\boldsymbol{\omega}_p \cdot \nabla\left(u_\theta/r\right)\hat{\mathbf{e}}_\theta, \tag{2.83}$$

where (2.38) has been used to evaluate $\partial\hat{\mathbf{e}}_r/\partial\theta$. Finally, we note that the Laplacian in (2.81) transforms in the same way as the Laplacian in the Navier–Stokes equation, and so (2.41) tells us that

$$\nabla^2\boldsymbol{\omega}_\theta = r^{-1}\nabla_*^2(r\omega_\theta)\hat{\mathbf{e}}_\theta. \tag{2.84}$$

Combining $(2.81)\rightarrow(2.84)$ now yields a scalar evolution equation for ω_θ/r,

$$\frac{D}{Dt}\left(\frac{\omega_\theta}{r}\right) = \boldsymbol{\omega}_p \cdot \nabla\left(\frac{u_\theta}{r}\right) + \frac{\nu}{r^2}\nabla_*^2\left(r\omega_\theta\right). \tag{2.85}$$

The most convenient form of this evolution equation is found by substituting for $\boldsymbol{\omega}_p$ using $\boldsymbol{\omega}_p = \nabla \times [(\Gamma/r)\,\hat{\mathbf{e}}_\theta]$. After a little algebra, this yields

$$\frac{D}{Dt}\left(\frac{\omega_\theta}{r}\right) = \frac{\partial}{\partial z}\left(\frac{\Gamma^2}{r^4}\right) + \frac{\nu}{r^2}\nabla_*^2(r\omega_\theta). \tag{2.86}$$

When combined with (2.46),

$$\frac{D\Gamma}{Dt} = \nu \nabla_*^2 \Gamma, \tag{2.87}$$

we have a complete description on the motion in terms of Γ and ω_θ/r.

Equations (2.86) and (2.87) are central to many axisymmetric flows, and we shall have reason to revisit them on many occasions. Note that (2.86) contains a source term in the form of axial gradients in angular momentum. This has its roots in the term $\nabla \times [\mathbf{u}_\theta \times \boldsymbol{\omega}_p]$ in (2.81). Physically, this represents the spiralling-up of the poloidal vortex lines by an axial gradient in swirl, which sweeps out a component of ω_θ from $\boldsymbol{\omega}_p$.

2.10.2 Inviscid Flow: The Squire–Long Equation and a Glimpse at Inertial Waves

For the special case of inviscid flow, our two evolution equations become

$$\frac{D\Gamma}{Dt} = 0, \qquad \frac{D}{Dt}\left(\frac{\omega_\theta}{r}\right) = \frac{\partial}{\partial z}\left(\frac{\Gamma^2}{r^4}\right), \tag{2.88}$$

and if the flow is also steady, these reduce to

$$\mathbf{u} \cdot \nabla \Gamma = 0, \qquad \mathbf{u} \cdot \left(\frac{\omega_\theta}{r}\right) = \frac{\partial}{\partial z}\left(\frac{\Gamma^2}{r^4}\right). \tag{2.89}$$

We shall now show how to use (2.89) to find a wide range of steady, axisymmetric, inviscid flows.

First, we note that $\mathbf{u} \cdot \nabla \Gamma = 0$ requires Γ to be constant on a streamline. In such cases we may write $\Gamma = \Gamma(\Psi)$, since Ψ is also constant on a streamline. The azimuthal vorticity equation then yields

$$\mathbf{u}_p \cdot \nabla \left(\frac{\omega_\theta}{r}\right) = \frac{2\Gamma\Gamma'(\Psi)}{r^4}\frac{\partial\Psi}{\partial z} = -\frac{2\Gamma\Gamma'(\Psi)}{r^3}u_r = \Gamma\Gamma'(\Psi)\mathbf{u}_p \cdot \nabla\left(\frac{1}{r^2}\right),$$

where the prime indicates a derivative with respect to Ψ. Given that $\mathbf{u}_p \cdot \nabla \Psi = 0$, this may be rewritten as

$$\mathbf{u}_p \cdot \nabla \left(\frac{\omega_\theta}{r} - \frac{\Gamma\Gamma'(\Psi)}{r^2}\right) = 0. \tag{2.90}$$

We conclude that the azimuthal vorticity in a steady, axisymmetric, inviscid flow satisfies

$$\frac{\omega_\theta}{r} = \frac{1}{r^2}\Gamma\Gamma'(\Psi) - H'(\Psi) = -\frac{1}{r^2}\nabla_*^2\Psi, \qquad (2.91)$$

where we have used $\nabla_*^2\Psi = -r\omega_\theta$ and $H(\Psi)$ is some function which has yet to be determined. Finally, this yields the *Squire–Long equation*,

$$\nabla_*^2\Psi = r^2 H'(\Psi) - \Gamma\Gamma'(\Psi), \qquad (2.92)$$

which may (in principle) be solved for Ψ, provided that $\Gamma(\Psi)$ and $H(\Psi)$ are specified at some upstream location.

It is not difficult to show that $H(\Psi)$ is Bernoulli's function, $H = \frac{1}{2}\mathbf{u}^2 + p/\rho$, which is constant along a streamline in a steady, inviscid flow. This is most readily seen from the steady, inviscid version of (2.55), with $\mathbf{g}$ neglected,

$$\mathbf{u} \times \boldsymbol{\omega} = \nabla\left(p/\rho + \mathbf{u}^2/2\right) = H'(\Psi)\nabla\Psi,$$

whose axial component can be rewritten as

$$H'(\Psi)\frac{\partial\Psi}{\partial z} = (\mathbf{u} \times \boldsymbol{\omega})_z = \omega_\theta u_r - u_\theta\omega_r = -\frac{\omega_\theta}{r}\frac{\partial\Psi}{\partial z} + \frac{\Gamma}{r^2}\frac{\partial\Gamma}{\partial z}.$$

Dividing through by $\partial\Psi/\partial z$, we find

$$H'(\Psi) = -\frac{\omega_\theta}{r} + \frac{\Gamma\Gamma'(\Psi)}{r^2}, \qquad (2.93)$$

which is (2.91).

The most important special case is where the upstream conditions take the form $u_\theta = \Omega r$ and $u_z = V$, i.e. rigid-body rotation plus uniform translation. In such a case the upstream distributions of Ψ, Γ, and p/ρ are, to within an unimportant constant in p,

$$\Psi = \frac{1}{2}Vr^2, \quad \Gamma = \Omega r^2, \quad p/\rho = \frac{1}{2}\Omega^2 r^2.$$

This, in turn, gives us

$$\Gamma = (2\Omega/V)\,\Psi, \quad H'(\Psi) = 2\Omega^2/V,$$

and so, for this particular upstream condition, the Squire–Long equation becomes

$$\nabla_*^2 \Psi + \frac{4\Omega^2}{V^2}\left(\Psi - \frac{Vr^2}{2}\right) = 0. \tag{2.94}$$

We now let F represent the departure of Ψ from its upstream value, defined through the expression $\Psi = \frac{1}{2}Vr^2 + rF(r, z)$. Substituting for Ψ in (2.94) then yields

$$\frac{\partial^2 F}{\partial z^2} + \frac{1}{r}\frac{\partial}{\partial r}r\frac{\partial F}{\partial r} + \left[\frac{4\Omega^2}{V^2} - \frac{1}{r^2}\right]F = 0. \tag{2.95}$$

It is extraordinary that, for these particular upstream conditions, our flow is governed by a simple, linear equation. Many inviscid flows are governed by (2.95), as discussed in Batchelor (1967).

Consider, for example, swirling flow in a pipe of radius b. A sphere of radius R is placed on the axis of the pipe, as shown in Figure 2.23, and we look for steady, axisymmetric solutions downstream of the sphere which are periodic in z. These periodic solutions are governed by (2.95) and it is readily confirmed that they take the form

$$F = AJ_1(k_r r)\cos(k_z z),$$

where A is an amplitude, J_1 is the usual Bessel function, and k_r and k_z are related by

$$k_r^2 + k_z^2 = (2\Omega/V)^2.$$

In a frame of reference moving at speed V, such periodic solutions may be regarded as progressive waves of *arbitrary magnitude* propagating *upstream* with a phase speed of V.

The admissible values of k_r are set by the requirement that the radial velocity disappears at $r = b$. This, in turn, demands that $k_r b = \gamma_n$, where γ_n are the zeros of J_1. It follows that the axial wavenumber is constrained to satisfy

$$k_z^2 b^2 = (2\Omega b/V)^2 - \gamma_n^2. \tag{2.96}$$

Evidently, such waves exist only when $2\Omega b/V > \gamma_1 = 3.83$, or equivalently $V/\Omega b < 0.522$. In experiments, such waves are indeed observed downstream of an axisymmetric blockage whenever the *Rossby number*, Ro $= V/\Omega b$, is less than ~0.4. These finite-amplitude waves are examples of so-called *inertial waves*, which are maintained by the background rotation. We shall have much more to say about inertial waves in due course.

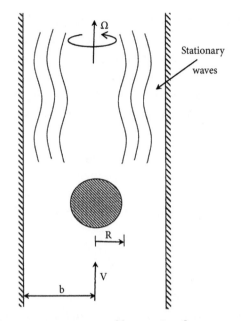

Figure 2.23 Stationary waves generated by rotating flow past a sphere. Reproduced from Davidson (2013).

2.11 Buoyancy-Driven Flow and the Boussinesq Approximation

In certain incompressible, rotating flows it is necessary to allow for spatial variations in density. Here it is convenient to adopt the so-called *Boussinesq approximation*, in which the variations in density are taken to be sufficiently small that they can be ignored in the momentum equation, except to the extent that they introduce a buoyancy force.

In order to formalize the Boussinesq approximation we write $\rho = \bar{\rho} + \rho'$, where $\bar{\rho}$ is a mean density and $\rho' \ll \bar{\rho}$. Incompressibility, $D\rho/Dt = 0$, plus mass conservation, (2.7), then give

$$D\rho'/Dt = 0, \quad \nabla \cdot \mathbf{u} = 0, \tag{2.97}$$

while (2.27) becomes

$$\bar{\rho}\frac{D\mathbf{u}}{Dt} = -\nabla p + \rho'\mathbf{g} + \bar{\rho}\nu\nabla^2\mathbf{u}, \tag{2.98}$$

where p must be interpreted as the departure from the hydrostatic pressure distribution. Equations (2.97) and (2.98) constitute, between them, the Boussinesq approximation.

A complication arises, however, when the buoyancy force, $\rho'\mathbf{g}$, is rewritten in terms of the temperature field, which is commonly done when dealing with natural convection, for example. The perturbation in density is then written as $\rho' = -\beta(T-T_0)\bar{\rho} = -\beta\Theta\bar{\rho}$, where T_0 is a reference temperature, $\Theta = T-T_0$, and β is the expansion coefficient for the fluid, $\beta = -(\partial\rho/\partial T)/\rho$, treated as a constant. The buoyancy force now becomes $-\beta\Theta\bar{\rho}\mathbf{g}$, and the momentum equation is

$$\frac{D\mathbf{u}}{Dt} = -\nabla\left(p/\bar{\rho}\right) - \beta\Theta\mathbf{g} + \nu\nabla^2\mathbf{u}. \tag{2.99}$$

The problem arises when we consider the advection-diffusion equation for temperature, (2.60), rewritten in terms of $\rho' = -\beta\bar{\rho}\Theta$,

$$\frac{D\rho'}{Dt} = \alpha\nabla^2\rho', \tag{2.100}$$

where α is the thermal diffusivity. Evidently, our fluid is no longer incompressible, since $D\rho'/Dt \neq 0$. This is because heat diffuses in and out of the fluid particles, changing their density. Moreover, mass conservation,

$$\frac{D\rho'}{Dt} + \rho\nabla\cdot\mathbf{u} = 0, \tag{2.101}$$

combines with (2.100) to give

$$\nabla\cdot\mathbf{u} = -\frac{\alpha}{\rho}\nabla^2\rho' \sim \frac{u}{\ell}\left(\frac{u\ell}{\alpha}\right)^{-1}\frac{\rho'}{\bar{\rho}}, \tag{2.102}$$

and so we lose the solenoidal constraint on $\mathbf{u}$. All is not lost, however, since we have $\rho' \ll \bar{\rho}$, and so we retain the approximation $\nabla\cdot\mathbf{u} = 0$, provided that $u\ell/\alpha$ is not too small.

A second complication is that often one has a stable background stratification, *i.e.* a density distribution which increases of depth. This is typical of the oceans, or the atmosphere at night. Suppose the density of the unperturbed fluid is $\rho_0(z)$, where z points upward and $d\rho_0/dz$ is uniform and negative. The corresponding hydrostatic pressure gradient is then $\nabla p_0 = \rho_0(z)\mathbf{g}$, and the density of the perturbed fluid $\rho(\mathbf{x}, t) = \rho_0(z) + \rho'$. Incompressibility now gives us

$$\frac{D\rho'}{Dt} + u_z\frac{d\rho_0}{dz} = 0, \tag{2.103}$$

while mass conservation, (2.7), still requires $\nabla \cdot \mathbf{u} = 0$. The Navier–Stokes equation is also unchanged, taking the form of (2.98). The new feature, however, is that ρ' is now controlled by (2.103), rather than (2.97). It is common in such cases to introduce the *Väisälä–Brunt frequency*, N, defined by

$$N^2 = \frac{g}{\bar{\rho}} \left| \frac{d\rho_0}{dz} \right|, \qquad (2.104)$$

so that (2.103) becomes

$$\frac{D\rho'}{Dt} = \frac{\bar{\rho} N^2}{g} u_z \qquad (2.105)$$

Note that N, which is a measure of the strength of the stratification, has the dimensions of a frequency. Indeed, as we shall see, N is indicative of the frequency of the internal gravity waves that can propagate through a stably stratified fluid.

2.12 The Analogy Between Buoyancy and Swirl

2.12.1 A Formal Analogy

Rayleigh (1916) was one of the first to notice the analogy between buoyancy and swirling flow. Moreover, he used the analogy to deduce his celebrated stability criterion for axisymmetric perturbations to the inviscid, rotating flow $\mathbf{u}_0 = u_\theta(r)\hat{\mathbf{e}}_\theta$.

Let us start by recalling the governing equations for inviscid, axisymmetric flow of a fluid of uniform density. These are

$$\frac{D\Gamma}{Dt} = 0, \; \nabla \cdot \mathbf{u}_p = 0, \qquad (2.106)$$

$$\frac{D\mathbf{u}_p}{Dt} = -\nabla \left(p/\rho \right) + \frac{\Gamma^2}{r^3}\hat{\mathbf{e}}_r, \qquad (2.107)$$

where we use cylindrical polar coordinates, $\mathbf{u}_p(r, z) = (u_r, 0, u_z)$ is the poloidal component of motion, and $\Gamma = ru_\theta$. Rayleigh noticed that there is an exact analogy between these equations and the axisymmetric (non-rotating) motion of an incompressible Boussinesq fluid driven by density gradients in a *radial* gravitational field. To see why, we write $\rho = \bar{\rho} + \rho'$ for the density, $\bar{\rho}$ being the mean density. The axisymmetric motion of our Boussinesq fluid is then governed by

$$\frac{D\rho'}{Dt} = 0, \quad \nabla \cdot \mathbf{u}_p = 0, \tag{2.108}$$

$$\frac{D\mathbf{u}_p}{Dt} = -\nabla \left(p/\bar{\rho} \right) + \left(\rho'/\bar{\rho} \right) \mathbf{g}, \tag{2.109}$$

where $\mathbf{g}$ is the gravitational acceleration. To establish the analogy we choose $\mathbf{g}$ to be the irrotational vector $\left(g^*/r^3 \right) \hat{\mathbf{e}}_r$, where g^* is a constant, and substitute Γ^2 for $g^* \rho'/\bar{\rho}$ in (2.108) and (2.109). This substitution brings us back to (2.106) and (2.107). Note that $\mathbf{g} = -\nabla \left(g^*/2r^2 \right)$, and so the potential energy density associated with (2.109) is $\rho' g^*/2r^2$. Under the analogy $g^* \rho'/\bar{\rho} \to \Gamma^2$, this becomes the kinetic energy $\bar{\rho} u_\theta^2/2$.

Two important observations follow from this analogy. First, since $\mathbf{g}$ is radial, our Boussinesq fluid has a static equilibrium of the form $\bar{\rho} + \rho' = \rho_0(r)$, and provided $\rho_0(r)$ increases monotonically with r, that equilibrium is stable. Conversely, if there exists a point at which $\rho_0(r)$ decreases with r, the equilibrium is unstable because potential energy can then be released by exchanging the radial position of two adjacent rings of fluid. Crucially, it follows from our analogy that a necessary and sufficient condition for the stability of the flow $\mathbf{u}_0 = u_\theta(r) \hat{\mathbf{e}}_\theta$ to inviscid, axisymmetric disturbances is that Γ^2 increases with radius. This is Rayleigh's centrifugal stability criterion, which is usually stated as follows: the inviscid flow $\mathbf{u}_0 = u_\theta(r) \hat{\mathbf{e}}_\theta$ is stable to axisymmetric disturbances if and only if, at all points in the flow,

$$\Phi(r) = \frac{1}{r^3} \frac{d\Gamma^2}{dr} \geq 0, \tag{2.110}$$

where Φ is known as *Rayleigh's discriminant*.

The second observation is that the stable equilibrium $\rho_0(r)$ supports internal gravity waves. This tells us that a rotating flow $\Gamma(r)$ can also support internal wave motion, and those waves are, in fact, a generalization of the inertial waves shown in Figure 2.23.

Rayleigh also used a simple energy argument to establish his stability criterion. Consider two, thin, circular rings of fluid of volume δV in the steady flow $\Gamma(r)$. One has radius r_1, angular momentum Γ_1, and kinetic energy $\frac{1}{2}\rho \delta V \left(\Gamma_1^2/r_1^2 \right)$, while the other has radius $r_2 = r_1 + \delta r$ and angular momentum $\Gamma_2 = \Gamma_1 + \delta\Gamma$. The rings now exchange position while conserving their angular momenta. The change in kinetic energy is then

$$\frac{1}{2}\rho\delta V\left(\Gamma_2^2 - \Gamma_1^2\right)\left(\frac{1}{r_1^2} - \frac{1}{r_2^2}\right) = \rho\delta V\left(\Gamma_2^2 - \Gamma_1^2\right)\frac{\delta r}{r_1^3} = \Phi(r)\rho\delta V(\delta r)^2. \qquad (2.111)$$

If $\Phi(r) < 0$, the kinetic energy associated with u_θ falls as a result of the perturbation and this releases energy to the disturbance, heralding an instability. Conversely, if $\Phi(r) > 0$, some external source of energy is required to establish the perturbation, and so the flow is stable.

2.12.2 An Illustrative Example: The Bursting Vortex and the Buoyant Thermal

As another illustration of Rayleigh's analogy, consider the rising *thermal* shown in Figure 2.24. This started out as a spherical blob of warm fluid, but as the blob rises it quickly adopts the characteristic mushroom-like shape of a thermal. In this case, $\mathbf{g}$ is vertical, $\mathbf{g} = -g\hat{\mathbf{e}}_z$, rather than radial, but nevertheless the analogy to rotating flows suggests that there should be an analogous swirling flow with a similar mushroom-like structure. We shall see that there is.

Figure 2.24 A buoyant blob evolves into a thermal.

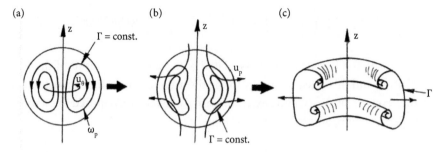

Figure 2.25 The bursting of an axisymmetric vortex. (a) The initial condition has $u_p = 0$. (b) The poloidal flow sweeps the angular momentum contours radially outward. (c) The asymptotic state takes the form of a poloidal vortex sheet which thins exponentially fast.

Consider an inviscid, axisymmetric flow of uniform density whose initial condition is a blob of swirling fluid, rather than a blob of hot fluid. For example, suppose that, in polar coordinates, we have $\mathbf{u}(t=0) = \Omega r \exp\left[-(r^2 + z^2)/\delta^2\right] \hat{\mathbf{e}}_\theta$, as shown schematically in Figure 2.25(a). From (2.88) in the form

$$\frac{D\Gamma}{Dt} = 0, \quad \frac{D}{Dt}\left(\frac{\omega_\theta}{r}\right) = \frac{\partial}{\partial z}\left(\frac{\Gamma^2}{r^4}\right) = \nabla \cdot \left(\frac{\Gamma^2}{r^4}\hat{\mathbf{e}}_z\right), \quad (2.112)$$

it is clear that ω_θ will start to grow, with $\omega_\theta < 0$ for $z > 0$ and $\omega_\theta > 0$ for $z < 0$. Hence, the poloidal velocity field, $\mathbf{u}_p$, has the structure shown in Figure 2.25(b). This flow sweeps the angular momentum, Γ, radially outward in accordance with $D\Gamma/Dt = 0$. Moreover, (2.112) integrates to give

$$\frac{d}{dt}\int_{z<0} (\omega_\theta/r)\, dV = \frac{d}{dt}\int_{z>0} (|\omega_\theta|/r)\, dV = 2\pi \int_0^\infty (\Gamma_0^2/r^3)\, dr > 0, \quad (2.113)$$

where $\Gamma_0 (r, t)$ is the angular momentum density on the plane $z = 0$. Evidently, the integral of $|\omega_\theta|/r$ rises monotonically. This causes the kinetic energy of the poloidal flow to increase steadily at the expense of the kinetic energy of the azimuthal motion, $\int (\Gamma^2/2r^2)dV$, which in turn must decline as Γ is swept radially outward.

Interestingly, as the vortex is centrifuged radially outward, the angular momentum density, Γ, is swept into a thin, mushroom-like vortex sheet, as shown in Figure 2.25(c). This vortex sheet thins exponentially fast and propagates radially outward at a constant speed. (See, for example, Davidson,

2021, and references therein.) However, what is of particular significance here is the similarity in the shape of the poloidal vortex sheet in Figure 2.25 and the rising thermal of Figure 2.24. We shall return to this topic in §11.1.5.

* * *

That completes our introduction to fluid dynamics. Readers seeking additional material will find excellent introductions in Acheson (1990) and Prandtl (1952), while more advanced texts include Batchelor (1967), Davidson (2021), and Kundu & Cohen (2004).

References

Acheson, D.J., 1990, *Elementary Fluid Dynamics*, Oxford University Press.

Batchelor, G.K., 1967, *An Introduction to Fluid Dynamics*, Cambridge University Press.

Davidson, P.A., 2013, *Turbulence in Rotating, Stratified and Electrically Conducting Fluids*, Cambridge University Press.

Davidson, P.A., 2017, *An Introduction to Magnetohydrodynamics*, 2nd Ed., Cambridge University Press.

Davidson, P.A., 2021, *Incompressible Fluid Dynamics*, Oxford University Press.

Helmholtz, H., 1858, On Integrals of the hydrodynamical equations which express vortex motion. Translated into English in: *Phil Mag.* (series 4), 1867, **33**, 485–512.

Kundu, P.K. & Cohen, I.M., 2004, *Fluid Mechanics*, 3rd Ed., Elsevier.

Prandtl, L., 1952, *Essentials of Fluid Dynamics*, Blackie & Son.

Rayleigh, Lord, 1916, On the dynamics of revolving fluids, *Proc. Royal Soc., A*, **93**, 148.

Stokes, G.G., 1845, On the theories of the internal friction of fluids in motion, and of the equilibrium and motion of elastic solids, *Trans. Camb. Phil. Soc.*, **8**, 287–305.

Chapter 3
Waves and Waves in Fluids

Waves play a central role in rapidly rotating fluids, dominating many of the striking phenomena observed in laboratory experiments and geophysical flows. So, in the chapters that follow, we shall make extensive use of the general theory of wave motion, particularly the concepts of *energy dispersion* and *group velocity* (*i.e.* the velocity at which wave energy disperses in the form of wave packets). By way of preparation, this chapter provides a brief introduction to the linear theory of waves, with particular emphasis on the idea of group velocity.

3.1 The Wave Equation and d'Alembert's Solution

There is a fundamental distinction in the theory of waves between those wave-bearing systems that are governed by *the* wave equation and those that are governed by a wave-like partial differential equation (PDE) which is similar to, but distinct from, the wave equation. The former host waves whose characteristics are relatively simple and intuitive, while the latter host waves whose energy disperses in complex and surprising ways.

Let us start with the wave equation in one dimension, one familiar example being acoustic waves in an organ pipe. These are governed by

$$\frac{\partial^2 p}{\partial t^2} = a^2 \frac{\partial^2 p}{\partial x^2}, \tag{3.1}$$

where x measures position along the axis of the organ pipe, p is the gas pressure, and a the speed of sound. Note that all disturbances travel at the same speed, *i.e.* at the speed a. More generally, the wave equation in one dimension takes the form

$$\frac{\partial^2 \eta}{\partial t^2} = c^2 \frac{\partial^2 \eta}{\partial x^2}, \tag{3.2}$$

The Dynamics of Rotating Fluids. P. A. Davidson, Oxford University Press. © Peter A Davidson (2024).
DOI: 10.1093/9780191994272.003.0003

where c is *the* wave speed for the system in question (the speed of sound, the speed of light, *etc.*). Waves which are governed by this equation have the special property that pulses of arbitrary profile can propagate without change of shape. This is known as *d'Alembert's solution* of the wave equation.

Consider the particular case of waves on a stretched string, as shown in Figure 3.1. These are governed by

$$\frac{\partial^2 \eta}{\partial t^2} = c^2 \frac{\partial^2 \eta}{\partial x^2}, \qquad c = \sqrt{T_s/\rho_s}, \tag{3.3}$$

where $\eta(x, t)$ is the lateral displacement of the string, T_s the tension in the string, and ρ_s its mass per unit length. A general solution of (3.3) is

$$\eta(x, t) = F_f(x - ct) + F_b(x + ct), \tag{3.4}$$

where F_f and F_b are arbitrary functions of their arguments. To confirm that this is so, we introduce the compound variable $\chi^{\pm} = x \pm ct$. Equation (3.4) then yields

$$\frac{\partial^2 \eta}{\partial t^2} = c^2 \left[F_f''(\chi^-) + F_b''(\chi^+) \right] = c^2 \frac{\partial^2 \eta}{\partial x^2}, \tag{3.5}$$

which evidently satisfies (3.3).

It is clear from Figure 3.2 that $F_f(x - ct)$ represents a *forward travelling* pulse that propagates at the speed c without any change of shape, while $F_b(x + ct)$ represents a *backward travelling* pulse that also propagates at speed c, again without change of shape. The generality of solution (3.4) is rather remarkable,

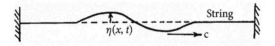

Figure 3.1 Waves on a stretched string.

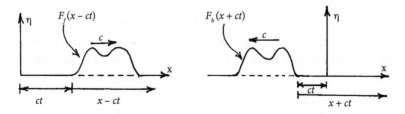

Figure 3.2 Pulses which travel along a string without any change of shape. (a) A forward travelling pulse, and (b) a backward travelling pulse.

and any one-dimensional wave governed by *the* wave equation supports just such a solution.

3.2 Some Wave-Bearing Systems Not Governed by the Wave Equation

Many types of waves are governed by a wave-like PDE which is *not* the wave equation. As we shall see, examples include flexural vibrations of a beam, waves on the surface of a deep pond, internal gravity waves in a stratified fluid, and inertial waves in a rapidly rotating fluid. Crucially, such waves do not, in general, admit d'Alembert's solution. This leads to one of the most important distinctions in the theory of wave propagation, which is between those waves that are governed by the wave equation and those which are governed by a distinct, wave-like PDE. The former are called *non-dispersive waves*, because wave packets travel without change of shape in accordance with d'Alembert's solution, while the latter are called *dispersive waves*, because wave packets change shape, typically spreading out as they propagate. We shall explain why this distinction is so important in §3.4, but first we introduce some examples of dispersive waves governed by a wave-like PDE. Let us start with flexural vibrations of a beam.

3.2.1 Flexural Vibrations of a Beam

Consider a thin, straight beam, as shown in Figure 3.3. It has a mass per unit length of ρ_b, a Young's modulus of E, and a cross-section whose second moment of inertia about its centreline is I. Transverse flexing of the beam is governed by

$$\rho_b \frac{\partial^2 \eta}{\partial t^2} + IE \frac{\partial^4 \eta}{\partial x^4} = 0, \tag{3.6}$$

where $\eta(x, t)$ is the lateral displacement of the beam and the coordinate x runs along the centreline. The double time derivative in (3.6) is a hint that this is a wave-bearing system, but the fourth-order spatial derivative tells us that this is not *the* wave equation.

Now consider a trial solution in the form of a travelling wave of wavelength λ, say

$$\eta(x, t) = \eta_0 \cos(kx - \varpi t) = \eta_0 \cos\left(k(x - c_p t)\right), \tag{3.7}$$

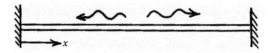

Figure 3.3 Transverse oscillations of a beam.

where $k = 2\pi/\lambda$ is the wavenumber of the disturbance and $c_p = \hat{\omega}/k$ is the speed of the wave crests. We shall adopt the sign convention that $\hat{\omega}$ is always positive (no negative frequencies), but k can be negative or positive, with positive k corresponding to forward travelling waves ($c_p > 0$) and negative k to backward travelling waves ($c_p < 0$). Substituting (3.7) into (3.6) we obtain the *dispersion relationship* for waves on a beam

$$\hat{\omega} = \sqrt{IE/\rho_b}k^2. \tag{3.8}$$

The velocity of the wave crests, $c_p = \hat{\omega}/k$, is called the *phase velocity*. In the case of flexural vibrations of a beam, c_p is a function of k, and so sinusoidal disturbances of different wavelengths travel at different speeds. This is in marked contrast to acoustic waves in an organ pipe, or waves on a string, where all disturbances travel at the same speed, *i.e.* at *the* wave speed for the system. This will turn out to be an important point.

More generally, if we substitute the trial solution (3.7) into the governing equation for a one-dimensional wave-bearing system, then we obtain a relationship between the angular frequency, $\hat{\omega}$, and the wavenumber, k. This is the dispersion relationship for the system in question, usually written in the form $\hat{\omega} = \hat{\omega}(k)$. If $c_p = \hat{\omega}/k$ is a function of k, then the waves are classed as dispersive, but if c_p is independent of k, which is the case for the wave equation (3.2), the waves are called *non-dispersive*.

3.2.2 Surface Gravity Waves on Water of Arbitrary Depth

Next, consider surface gravity waves of small amplitude on the surface of a pond of undisturbed depth h (see Figure 3.4). For simplicity, we ignore surface tension, take the motion to be in the x–y plane, and assume the flow is inviscid and irrotational, $\nabla \times \mathbf{u} = 0$, so that (2.55) simplifies to

$$\frac{\partial \mathbf{u}}{\partial t} = -\nabla\left(p/\rho + \mathbf{u}^2/2\right) + \mathbf{g}. \tag{3.9}$$

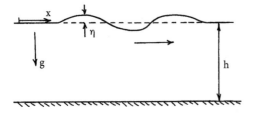

Figure 3.4 Wave on the surface of a pond.

Let us start by establishing the governing equations for such a wave, from which we can determine the dispersion relationship. Since $\nabla \cdot \mathbf{u} = 0$ and $\nabla \times \mathbf{u} = 0$, we may introduce a velocity potential for $\mathbf{u}$, say φ, defined by

$$\mathbf{u} = \nabla\varphi, \quad \nabla^2\varphi = 0. \tag{3.10}$$

The equation of motion, (3.9), now simplifies to

$$\nabla\left(\partial\varphi/\partial t + p/\rho + \mathbf{u}^2/2 + gy\right) = 0.$$

This, in turn, yields

$$\frac{\partial\varphi}{\partial t} + \frac{p}{\rho} + \frac{1}{2}\mathbf{u}^2 + gy = \text{constant}, \tag{3.11}$$

which is a generalized version of Bernoulli's equation. We shall need (3.11) to help establish the free-surface boundary conditions for such waves.

Now consider the situation shown in Figure 3.4, in which the origin of coordinates lies at the undisturbed surface and $\eta(x, t)$ is the upward displacement of the free surface from its equilibrium position. No restriction is placed on h/λ, and so we allow for both deep-water and shallow-water waves. A fluid particle that starts out on the surface stays on the surface, and so the free surface is characterized by $y_s = \eta(x_s, t)$, where $\mathbf{x}_s$ is the position of a given fluid particle on the surface. So, in a time δt,

$$\delta y_s = \delta\eta = \frac{\partial\eta}{\partial t}\delta t + \frac{\partial\eta}{\partial x}\delta x_s = \frac{\partial\eta}{\partial t}\delta t + \frac{\partial\eta}{\partial x}u_x\delta t.$$

This then yields the free-surface kinematic boundary condition

$$u_y = \frac{\partial\eta}{\partial t} + u_x\frac{\partial\eta}{\partial x}, \quad \text{on } y = \eta. \tag{3.12}$$

Since the waves are assumed to be of small amplitude, this may be linearized by dropping the quadratic term on the right and by applying (3.12) at $y = 0$, rather than at $y = \eta$. Our kinematic boundary condition is then

$$\frac{\partial \eta}{\partial t} = u_y = \frac{\partial \varphi}{\partial y}, \quad \text{on } y = 0. \tag{3.13}$$

Next, we apply Bernoulli's equation, (3.11), to the free surface. Setting the surface pressure to zero gives us

$$\frac{\partial \varphi}{\partial t} + g\eta + \frac{1}{2}\mathbf{u}^2 = 0, \quad \text{on } y = \eta,$$

which linearizes to

$$\frac{\partial \varphi}{\partial t} + g\eta = 0, \quad \text{on } y = 0. \tag{3.14}$$

Having established the free-surface boundary conditions, we now look for travelling wave solutions of the form

$$\varphi = \hat{\varphi}(y)\sin(kx - \omega t), \quad \eta(x, t) = \eta_0 \cos(kx - \omega t),$$

where η_0 is an amplitude. Since $\nabla^2 \varphi = 0$, $\hat{\varphi}(y)$ must satisfy $\hat{\varphi}''(y) = k^2 \hat{\varphi}$, whose general solution is

$$\hat{\varphi} = B \cosh k(y + h) + C \sinh k(y + h). \tag{3.15}$$

The two constants, B and C, can be determined from the boundary conditions. For example, the vanishing of u_y on the lower boundary demands that $\hat{\varphi}'(y = -h) = 0$, from which $C = 0$, while the free-surface boundary condition (3.13) requires

$$kB \sinh kh = \omega \eta_0. \tag{3.16}$$

It remains to apply the dynamic boundary condition (3.14), from which we obtain

$$\omega B \cosh kh = g\eta_0. \tag{3.17}$$

Eliminating B from these expressions now yields the dispersion relationship for surface gravity waves:

$$\omega^2 = gk \tanh kh. \tag{3.18}$$

For shallow-water waves, defined by $kh \ll 1$, this simplifies to $\omega = \sqrt{gh}\,|k|$, and the phase velocity is evidently $c_p = \pm\sqrt{gh}$. These are non-dispersive waves, in which all waves propagate at the same speed, irrespective of the value of k. By way of contrast, for deep-water waves, defined by $kh \gg 1$, (3.18) simplifies to $\hat{\omega} = \sqrt{g|k|}$, and the phase velocity is $c_p = \pm\sqrt{g/|k|}$. These are dispersive waves, in which the crests of the long waves travel faster than those of short wavelength.

3.3 Dispersive Versus Non-dispersive Waves: d'Alembert Revisited

Let us now explore, in a little more detail, the distinction between dispersive and non-dispersive waves. As already suggested, it is natural when investigating a one-dimensional wave-bearing system to look for travelling wave solutions of the form

$$\eta(x,\, t) = a \exp\left[j(kx - \hat{\omega}t)\right] = a \exp\left[jk(x - c_p t)\right], \qquad (3.19)$$

where a is an amplitude, the real part is understood, and c_p is the speed of the wave crests. In part, the idea behind using the trial solution (3.19) is that we can always Fourier decompose a more complicated initial condition into a series of such terms with varying wavenumber, and then take advantage of linearity to study each Fourier mode separately. Having determined how each Fourier mode in the initial condition propagates, we can then use superposition to reconstruct the full wave pattern at some later time. A common convention when writing expressions like (3.19) is to take $\hat{\omega}$ as positive, but allow k to take either sign. In such cases, $k > 0$ corresponds to forward travelling wave crests, $c_p > 0$, and $k < 0$ to backward travelling crests, $c_p < 0$. We shall conform to this convention.

Having adopted (3.19) as a trail solution, the governing PDE for the system in hand then yields the dispersion relationship, $\hat{\omega} = \hat{\omega}(k)$, and hence $c_p = \hat{\omega}/k$. Those systems for which c_p is independent of k, which are those waves governed by the wave equation, are labelled as non-dispersive, and those for which c_p is a function of k are classified as dispersive. The essential point about non-dispersive waves is that the phase speed and something we will meet in the next section called the *group velocity* are both independent of wavenumber and equal to *the* wave speed for the system (the speed of sound, the speed of waves on a string, or perhaps the speed of light). This, in turn, explains why the wave equation admits d'Alembert's solution in which wave pulses of arbitrary profile travel without change of shape. The idea is the following. Suppose our initial conditions consist of a localized pulse. To determine what happens to this pulse at later times the natural strategy is to Fourier decompose, or rather Fourier transform, the initial pulse into a set of Fourier modes, each of a given wavenumber, k, and amplitude, $a(k)$. As the system is linear, each mode then propagates independently of all the others. To determine the shape of the disturbance at some later time we simply reconstruct the disturbance by adding together all the Fourier modes, which mathematically amounts to performing the inverse Fourier transform. However, for non-dispersive waves governed by the wave equation, the phase speed and group velocity are the same for all Fourier modes, equal to c, say. So, after a given time t, all the Fourier modes will have moved the same distance, ct, and so when we reconstruct the pulse at time t it has exactly the same shape as it had at time $t = 0$, just shifted by the distance, ct.

The same argument explains why, in a dispersive system, a wave pulse changes shape as it propagates. After Fourier decomposing the initial disturbance, each Fourier mode will travel at a different speed and so will propagate a different distance in a time t. When we reconstruct the disturbance at a later time, by adding all the Fourier modes together, we get a shape which is different to that of the initial pulse because the various Fourier modes have travelled different distances, and so changed their relative positions.

3.4 The Concept of Group Velocity for Dispersive Systems

We now consider dispersive waves in more detail. Typically, in a dispersive system, a pulse starts out as localized. (Think of hitting a beam with a hammer.) The pulse then spreads out as it propagates. Eventually, the various Fourier modes become well separated in space, adopting the form of a *slowly modulated wave train*, sometimes called a *wave packet*. At any one location within such a wave train the disturbance looks like a single Fourier mode of given amplitude and wavenumber, but as we move along the wave train the amplitude and wavenumber slowly change. Typically, the amplitude of the modes of the various wavenumbers within a wave train, $a(k)$, is sharply peaked around the dominant wavenumber of the disturbance, say k_0, as indicated in Figure 3.5(b). For example, in the case of a stone thrown into a deep pond, we have $k_0 \sim 2\pi/$(stone size).

As already suggested, dispersive waves have many counterintuitive properties. For example, as we shall see, the speed at which a wave packet propagates

(a)

(b)

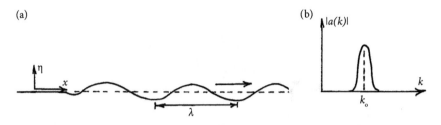

Figure 3.5 (a) A slowly modulated wave train of dominant wavelength λ. (b) The variation of the amplitude of the Fourier modes as a function of wavenumber, with $k_0 = 2\pi/\lambda$.

as a whole, which is called the *group velocity*, is *not* the speed at which individual wave crests move, *i.e.* not the phase speed, $c_p = \bar{\omega}/k$. Rather, it turns out that the group velocity is given by $c_g = d\bar{\omega}/dk$. Consider, for example, Figure 3.6. If you throw a stone into a deep pond, it turns out that the resulting wave crests travel at *twice* the speed of the wave packet within which they sit, or, if you strike a beam with a hammer, individual wave crests travel at *half* the speed of the overall wave packet.

So how can we rationalize such odd behaviour? Consider first the case of a stone striking the surface of a deep pond. The surface waves spread out in the form of an annulus as shown in Figure 3.7(a) and after some time the waves become well dispersed. A cross-section through the wave train then looks something like that shown in Figure 3.7(b), where the dominant wavelength is set by the size of the stone. Now consider a particular wave crest

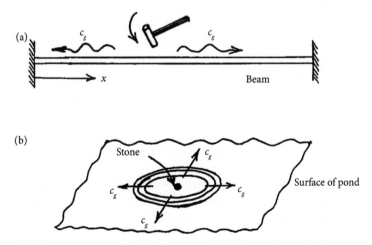

Figure 3.6 Waves generated in (a) a beam and (b) the surface of a deep pond.

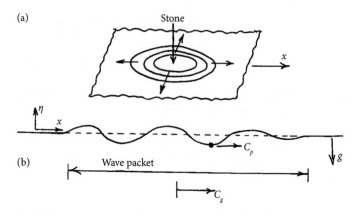

Figure 3.7 (a) Waves on the surface of a pond. (b) Cross-section through a wave packet.

within the wave packet. It will have an angular frequency of $\bar{\omega} = \sqrt{g|k|}$ and travel at a phase speed of $c_p = \sqrt{g/|k|}$. However, if you track the centre of the overall wave packet, then it turns out that it travels at the slower speed of $c_g = d\bar{\omega}/dk = \frac{1}{2}\sqrt{g/|k|}$. In order to accommodate this difference in speeds, wave crests appear out of nowhere at the rear of the wave packet, grow in amplitude as they ripple through the wave packet at a speed of $c_p = \sqrt{g/|k|}$, and then fade away as they approach the far end of the wave train.

Now consider flexural vibrations of a beam. Here the wave crests travel at $c_p = \sqrt{IE/\rho_b}k$, in accordance with (3.8), but an overall wave packet moves at twice the speed of the wave crests, with a group velocity of $c_g = d\bar{\omega}/dk = 2\sqrt{IE/\rho_b}k$. So, in a frame of reference moving with the wave packet, individual wave crests ripple *backwards* through the wave packet.

Of course, the key question is: why do wave packets travel at the speed $c_g = d\bar{\omega}/dk$, rather than the phase speed, $c_p = \bar{\omega}/k$? We now address this question. Suppose that we have a slowly modulated wave train in a dispersive system, as shown in Figure 3.7(b). Then waves of wavenumber k propagate as

$$\eta(x,\ t) = a(k) \exp\left[j\,(kx - \bar{\omega}t)\right], \quad \bar{\omega} = \bar{\omega}(k),$$

where $\bar{\omega} = \bar{\omega}(k)$ is the dispersion relationship for the system at hand, $a(k)$ is the amplitude of the waves of wavenumber k, and the real part is understood. Superposition now gives us

$$\eta(x,\ t) = \int_{-\infty}^{\infty} a(k) \exp\left[j\,(kx - \bar{\omega}t)\right] dk, \tag{3.20}$$

from which

$$\eta(x,\ 0) = \int_{-\infty}^{\infty} a(k)\exp(jkx)dk\,.\qquad(3.21)$$

Evidently, $a(k)$ is the Fourier transform of the initial condition, $\eta(x,\ 0)$.

Now, if $a(k)$ is centred on k_0, as shown in Figure 3.5(b), then we may write

$$\varpi(k) = \varpi(k_0) + (d\varpi/dk)_0\,(k - k_0) + O(k - k_0)^2,$$

or equivalently,

$$\varpi(k) = \varpi(k_0) + c_g\,(k_0)\,(k - k_0) + O(k - k_0)^2,\qquad(3.22)$$

where c_g is *defined* as $c_g = d\varpi/dk$. We now use expansion (3.22) to substitute for ϖ in (3.20), taking advantage of the fact that $a(k) \approx 0$ when k differs significantly from k_0 to neglect the higher order terms in (3.22). After a little work we find

$$\eta(x,\ t) \approx \exp\left[j\,(k_0 x - \varpi(k_0)t)\right] \int_{-\infty}^{\infty} a(k)\exp\left[j(k - k_0)\,(x - c_g(k_0)t)\right]d(k - k_0).$$

$$(3.23)$$

However, $k - k_0$ as a dummy variable, and so we can rewrite (3.23) as

$$\eta(x,\ t) = F\left(x - c_g(k_0)t\right) \cdot \exp\left[j\,(k_0 x - \varpi(k_0)t)\right],\qquad(3.24)$$

where the shape of the function F depends on the amplitude distribution $a(k)$.

This is the key result. It represents a travelling wave of wavenumber k_0 and frequency $\varpi(k_0)$ which is modulated by the envelope $F\left(x - c_g(k_0)t\right)$, as shown in Figure 3.8. It is clear that the overall wave packet, as represented by $F\left(x - c_g(k_0)t\right)$, propagates at the speed $c_g(k_0) = (d\varpi/dk)_0$, which is the group velocity corresponding to the dominant wavenumber in the wave packet.

A more careful analysis (which we shall not perform) shows that the group velocity, defined as $c_g = d\varpi/dk$, has three closely related properties:

(i) $c_g(k_0) = (d\varpi/dk)_0$ is the mean speed at which a wave packet as a whole travels, where k_0 is the dominant wavenumber in the wave packet;

(ii) $c_g(k) = d\varpi/dk$ is the speed at which you must travel to continually observe waves of wavenumber k;

(iii) $c_g(k)$ is the speed at which waves of wavenumber k transport energy.

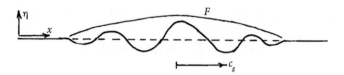

Figure 3.8 A traveling wave of wavenumber k_0 sits inside the envelope $F(x - c_g(k_0)t)$.

The strongest of these three statements is the second, and indeed we might consider properties (i) and (iii) as consequences of property (ii). In effect, property (i), which we have already established, tells us how to track the centre of a wave packet, while property (ii), which we have not proved, is a more refined statement. In particular, since there is always a range of wavenumbers in a given wave packet, statement (ii) tells us that wave packets inevitably spread out as they propagate, with the different Fourier modes travelling at slightly different speeds.

Although we have restricted the discussion to one-dimensional wave packets, the entire concept generalizes in an obvious way to three dimensions. For example, solutions of the three-dimensional wave equation,

$$\frac{\partial^2 \eta}{\partial t^2} = c^2 \nabla^2 \eta,$$

(3.25)

c being *the* wave speed for the system in question, are non-dispersive. This includes acoustic waves in three dimensions and electromagnetic waves in free space, where everything just travels at the speed of sound or the speed of light. When considering three-dimensional dispersive waves, on the other hand, one first looks for travelling waves of the form

$$\eta(x, t) \sim \exp\left[j\left(\mathbf{k} \cdot \mathbf{x} - \bar{\omega}t\right)\right],$$

(3.26)

where $\mathbf{k}$ is now a *wavevector*, which has a direction as well as a magnitude. Substituting this trial solution into the governing PDE for the particular system at hand yields the dispersion relationship, $\bar{\omega} = \bar{\omega}(\mathbf{k})$. The group velocity is then defined as the vector $\mathbf{c}_g$ whose components are

$$\left(\mathbf{c}_g\right)_i = \partial \bar{\omega} / \partial k_i.$$

(3.27)

The group velocity so defined has properties which are analogous to those in one dimension. Specifically:

(i) $c_g(\mathbf{k}_0)$ is the mean velocity at which a wave packet dominated by the wavevector $\mathbf{k}_0$ propagates away from the source of the disturbance;

(ii) $c_g(\mathbf{k})$ is the velocity at which you must travel in order to keep seeing waves of wavevector $\mathbf{k}$;

(iii) $c_g(\mathbf{k})$ is the velocity at which the energy of waves of wavevector $\mathbf{k}$ propagates through the medium.

Moreover, in three dimensions, the phase velocity (the velocity of the wave crests) points in the direction of $\mathbf{k}$ and has magnitude $|\mathbf{c}_p| = \bar{\omega}/k$. It is most conveniently written as

$$c_p(\mathbf{k}) = \left(\bar{\omega}/k^2\right)\mathbf{k}.$$

There is, however, one crucial difference between one-dimensional and three-dimensional dispersive waves. In three dimensions, the group velocity and phase velocity usually have different magnitudes *and* different directions. For example, internal gravity waves in a stably stratified fluid have c_g perpendicular to c_p, so the wave energy does *not* propagate in the same direction as the wave crests, but rather *parallel* to the crests!

3.5 An Example of Dispersion in Three Dimensions: Internal Gravity Waves

It is appropriate to consider internal gravity waves in a little more detail, as they have many similarities to inertial waves in a rapidly rotating fluid, and because there are geophysical flows where both rotation and stratification are important. As in §2.11, we adopt coordinates such that $\mathbf{g} = -g\hat{\mathbf{e}}_z$ and $\rho(\mathbf{x}, t) = \rho_0(z) + \rho'(\mathbf{x}, t)$, where $d\rho_0/dz < 0$ ensures a stable stratification. In order to keep the discussion brief, we shall assume that there is no background flow, the fluid is inviscid, $|d\rho_0/dz|$ is uniform, variations in density are sufficiently small for the Boussinesq approximation to apply, and the waves are of small amplitude, in the sense that

$$\rho' << |d\rho_0/dz|\, \lambda << \bar{\rho},$$

where $\bar{\rho}$ is a mean density and λ a typical wavelength.

Our governing equation is then the inviscid vorticity equation, (2.72), incorporating the curl of the buoyancy force,

$$\bar{\rho}\frac{D\boldsymbol{\omega}}{Dt} = \bar{\rho}(\boldsymbol{\omega}\cdot\nabla)\mathbf{u} + \nabla\times(\rho'\mathbf{g}). \tag{3.28}$$

To this we must add mass conservation for an incompressible fluid, *i.e.* (2.7), which demands $\nabla \cdot \mathbf{u} = 0$, and also incompressibility, $D\rho/Dt = 0$, in the form of (2.103),

$$\frac{D\rho'}{Dt} = \left|\frac{d\rho_0}{dz}\right| u_z = \frac{\bar{\rho} N^2}{g} u_z. \tag{3.29}$$

Here N is the Väisälä–Brunt frequency, as defined by (2.104).

The linearized version of these equations, in which both $\mathbf{u}$ and ρ' are assumed to be small, is evidently

$$\bar{\rho}\frac{\partial \boldsymbol{\omega}}{\partial t} = \nabla \left(\rho'\right) \times \mathbf{g}, \qquad \frac{\partial \rho'}{\partial t} = \frac{\bar{\rho} N^2}{g} u_z. \tag{3.30}$$

Eliminating ρ' now yields

$$\frac{\partial^2 \boldsymbol{\omega}}{\partial t^2} = \left(N^2 \hat{\mathbf{e}}_z\right) \times \nabla u_z,$$

whose curl is

$$\frac{\partial^2}{\partial t^2}\nabla^2\mathbf{u} + N^2\left[(\nabla^2 u_z)\hat{\mathbf{e}}_z - \nabla\left(\partial u_z/\partial z\right)\right] = 0. \tag{3.31}$$

Of particular significance is the z component of (3.31), which is the wave-like equation

$$\frac{\partial^2}{\partial t^2}\nabla^2 u_z + N^2\nabla_\perp^2 u_z = 0, \tag{3.32}$$

where $\nabla_\perp^2$ is the horizontal Laplacian, which contains no derivatives in z.

We now look for a trial solution of (3.32) in the form of (3.26), say

$$\mathbf{u}(\mathbf{x}, t) = \hat{\mathbf{u}} \exp\left[j\left(\mathbf{k}\cdot\mathbf{x} - \bar{\omega}t\right)\right], \tag{3.33}$$

where $\hat{\mathbf{u}}$ is an amplitude, $\nabla \cdot \mathbf{u} = 0$ demands that $\mathbf{k} \cdot \hat{\mathbf{u}} = 0$, and we adopt the convention that $\bar{\omega} > 0$. Substituting the z component of this trial solution into (3.32) yields the dispersion relationship

$$\bar{\omega}(\mathbf{k}) = \frac{Nk_\perp}{k} = N\left|\cos\theta\right|, \tag{3.34}$$

where $k = |\mathbf{k}|$, $k_\perp = \sqrt{k_x^2 + k_y^2}$, and θ is the angle $\mathbf{k}$ makes to the horizontal plane. Evidently, $\bar{\omega}$ is restricted to the range $0 \le \bar{\omega} \le N$. Moreover, the phase velocity is, by definition, in the direction of $\mathbf{k}$ and has magnitude $\hat{\omega}/k$, and hence

$$\mathbf{c}_p = \frac{Nk_\perp \mathbf{k}}{k^3}.$$

(3.35)

The group velocity, by contrast, has components $(\mathbf{c}_g)_i = \partial\hat{\omega}/\partial k_i$, and a little algebra yields

$$\mathbf{c}_g = \frac{N}{k^3 k_\perp}\mathbf{k} \times (\mathbf{k} \times \mathbf{k}_{//}) = \frac{N}{k^3 k_\perp}\left[k_{//}^2 \mathbf{k}_\perp - k_\perp^2 \mathbf{k}_{//}\right],$$

(3.36)

where $\mathbf{k}_{//} = k_z\hat{\mathbf{e}}_z$ and $\mathbf{k}_\perp = \mathbf{k} - \mathbf{k}_{//}$.

Evidently, the phase and group velocities are perpendicular, as noted above. That is to say, the direction of propagation of a wave packet is normal to the velocity of the wave crests which sit within that wave packet. It follows that wave energy travels *along* the wave crests, rather than normal to the crests, as shown in Figure 3.9. Note that $|\mathbf{c}_g| = (N/k)|\sin\theta|$, so the fastest wave packets travel horizontally, with $\mathbf{k}$ vertical. Note also that

$$c_p^2 + c_g^2 = \frac{N^2}{k^2}, \qquad \mathbf{c}_p + \mathbf{c}_g = \frac{N\mathbf{k}_\perp}{kk_\perp},$$

(3.37)

and so $\mathbf{c}_g$ and $\mathbf{c}_p$ have equal and opposite vertical components.

From (3.34) and (3.36) we see that low-frequency waves, in which $\hat{\omega} \ll N$, have $\mathbf{k}$ vertical and a horizontal group velocity of magnitude N/k. These constitute the fastest moving wave packets. Conversely, high-frequency waves,

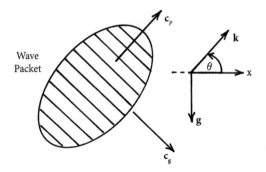

Figure 3.9 The relative orientation of the phase and group velocities for a two-dimensional wave packet. Notice that the wave energy travels *along* the wave crests.

defined by $\hat{\omega} \to N$, have a wavevector which is horizontal and a group velocity which is vertical and of vanishingly small magnitude. Waves of intermediate frequency propagate at an oblique angle.

It is instructive to consider the special case of two-dimensional waves, say confined to the x–z plane. Then (3.35) and (3.36) yield

$$c_p = \pm \frac{N \cos \theta}{k} (\cos \theta \hat{e}_x + \sin \theta \hat{e}_z), \qquad (3.38)$$

$$c_g = \pm \frac{N \sin \theta}{k} (\sin \theta \hat{e}_x - \cos \theta \hat{e}_z), \qquad (3.39)$$

where the upper (lower) signs correspond to positive (negative) k_x, and θ is the angle $\mathbf{k}$ makes to the x-axis, as shown in Figure 3.9. Notice that the horizontal components of $\mathbf{c}_p$ and $\mathbf{c}_g$ have the same sense, whereas the vertical components are equal and opposite.

These expressions for $\mathbf{c}_p$ and $\mathbf{c}_g$ are illustrated in Figure 3.10, where a circular cylinder of radius R, aligned with the y axis, oscillates horizontally at some intermediate frequency, say $\hat{\omega} \approx N/2$. Because there is only one frequency in the problem, and $|\cos \theta|$ is fixed by (3.34), there are only two allowable values

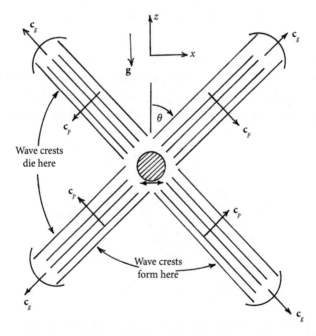

Figure 3.10 Dispersion pattern for gravity waves generated by an oscillating cylinder.

of θ. Given the $\pm$ signs in (3.39), we conclude that there are four directions in which wave energy can radiate away from the cylinder. It follows that the internal gravity waves are confined to four sheets, two directed to the right of the cylinder and two directed to the left. These sheets are bounded at large $|\mathbf{x}|$ by wave fronts which travel outward at the group velocity

$$|\mathbf{c}_g| = (N/k) |\sin \theta| \sim (NR/\pi) |\sin \theta| .\tag{3.40}$$

Since c_p and c_g are mutually perpendicular, the wave crests are aligned with the sheets, with crests appearing spontaneously on one side of each sheet, rippling across the sheet in the direction indicated, and then disappearing into thin air on the far side of the sheet.

We shall see that inertial waves in a rotating fluid produce a similar dispersion pattern when generated by a localized, oscillating source, as shown in Figure 1.7.

* * *

That concludes our brief introduction to dispersive waves and to the concept of group velocity. Readers seeking additional background material on waves in fluids could do worse than consult Acheson (1990) for a gentle introduction, or Lighthill (1978) for the definitive account.

References

Acheson, D.J., 1990, *Elementary Fluid Dynamics*, Oxford University Press.
Lighthill, J., 1978, *Waves in Fluids*, Cambridge University Press.

PART II
THE THEORY OF ROTATING FLUIDS

Chapter 4

Moving into a Rotating Frame of Reference, the Taylor–Proudman Theorem, and the Formation of Taylor Columns

Some flows are subject to a uniform background rotation, and in such cases it is often convenient to move into a rotating frame of reference. This is certainly the case when studying tropical cyclones or convective motion in the liquid core of the earth, where the earth's rotation plays an important role. On other occasions, however, there is no background rotation, or else the rigid-body rotation is too weak to influence the flow. Consequently, in the chapters which follow, we shall sometimes find ourselves in a rotating frame of reference, and sometimes in an inertial frame. In this chapter we discuss some of the more immediate consequences of moving into a rotating frame of reference.

4.1 Moving into a Rotating Frame of Reference and the Coriolis Force

Before discussing the behaviour of fluids viewed in a rotating frame, it is worth recalling some simple rigid-body mechanics. Consider an inertial frame of reference with unit vectors $(\mathbf{i}, \mathbf{j}, \mathbf{k})$, and a non-inertial frame with unit vectors $(\mathbf{i}^*, \mathbf{j}^*, \mathbf{k}^*)$. Let the starred system rotate relative to the inertial frame with angular velocity $\mathbf{\Omega}$, while the two systems share a common origin. A vector $\mathbf{A}$ can be represented in either system, as

$$\mathbf{A} = A_x\mathbf{i} + A_y\mathbf{j} + A_z\mathbf{k} = A_x^*\mathbf{i}^* + A_y^*\mathbf{j}^* + A_z^*\mathbf{k}^*. \qquad (4.1)$$

Moreover, the time derivatives of $\mathbf{A}$, as measured by distinct observers in the two reference frames, are

$$\frac{d\mathbf{A}}{dt} = \dot{A}_x\mathbf{i} + \dot{A}_y\mathbf{j} + \dot{A}_z\mathbf{k}, \qquad \frac{d^*\mathbf{A}}{dt} = \dot{A}_x^*\mathbf{i}^* + \dot{A}_y^*\mathbf{j}^* + \dot{A}_z^*\mathbf{k}^*, \qquad (4.2)$$

The Dynamics of Rotating Fluids. P. A. Davidson, Oxford University Press. © Peter A Davidson (2024).
DOI: 10.1093/9780191994272.003.0004

where a dot indicates the time derivative of a scalar. This is because the unit vectors $(\mathbf{i}, \mathbf{j}, \mathbf{k})$ are fixed in the inertial frame, while $(\mathbf{i}^*, \mathbf{j}^*, \mathbf{k}^*)$ are fixed in the rotating frame. Of course, dA/dt and d^*A/dt are not the same thing. For example, a vector $\mathbf{A}$ which is fixed in the starred system, $d^*\mathbf{A}/dt = 0$, rotates in the inertial frame, and indeed simple rigid-body dynamics tells us that

$$dA/dt = \mathbf{\Omega} \times \mathbf{A}.$$

More generally, the two time derivatives, dA/dt and d^*A/dt, are related by

$$\frac{d\mathbf{A}}{dt} = \frac{d^*\mathbf{A}}{dt} + \mathbf{\Omega} \times \mathbf{A}. \tag{4.3}$$

(See the discussion in Goldstein, 1980, for example.) Note that $d\mathbf{\Omega}/dt = d^*\mathbf{\Omega}/dt$.

Now any point in space is located by the same position vector, $\mathbf{x}$, in both coordinate systems. Applying (4.3) to $\mathbf{x}$ gives

$$\frac{d\mathbf{x}}{dt} = \frac{d^*\mathbf{x}}{dt} + \mathbf{\Omega} \times \mathbf{x} = \mathbf{u}^* + \mathbf{\Omega} \times \mathbf{x},$$

where $\mathbf{u}^*$ is the velocity of a particle at $\mathbf{x}$ in the starred system. A second application yields

$$\frac{d^2\mathbf{x}}{dt^2} = \left(\frac{d^*}{dt} + \mathbf{\Omega}\times\right)(\mathbf{u}^* + \mathbf{\Omega} \times \mathbf{x}) = \mathbf{a}^* + 2\mathbf{\Omega} \times \mathbf{u}^* + \frac{d\mathbf{\Omega}}{dt} \times \mathbf{x} + \mathbf{\Omega} \times (\mathbf{\Omega} \times \mathbf{x}),$$

$$\tag{4.4}$$

which is *Coriolis' theorem*. Here $\mathbf{a}^*$ is the acceleration in the starred system, while $2\mathbf{\Omega} \times \mathbf{u}^*$ is the *Coriolis acceleration*. The final term on the right can be rewritten as

$$\mathbf{\Omega} \times (\mathbf{\Omega} \times \mathbf{x}) = -\Omega^2 \mathbf{x}_\perp, \tag{4.5}$$

where $\mathbf{x}_\perp$ is the component of $\mathbf{x}$ perpendicular to the rotation axis. This is called the *centripetal acceleration*, where centripetal means 'towards the centre'.

Let us now consider a particle of mass m which is subject to a force $\mathbf{F}$ and has acceleration $\mathbf{a}$ in the inertial frame. Then Newton's second law applied in the inertial frame tells us that

$$\mathbf{F} = m\mathbf{a} = m\left[\mathbf{a}^* + 2\mathbf{\Omega} \times \mathbf{u}^* + \frac{d\mathbf{\Omega}}{dt} \times \mathbf{x} + \mathbf{\Omega} \times (\mathbf{\Omega} \times \mathbf{x})\right],$$

or equivalently,

$$m\mathbf{a}^* = \mathbf{F} + m\left[2\mathbf{u}^* \times \mathbf{\Omega} - \mathbf{\Omega} \times (\mathbf{\Omega} \times \mathbf{x}) + \mathbf{x} \times \frac{d\mathbf{\Omega}}{dt}\right]. \tag{4.6}$$

Now Newton's second law does not hold in the rotating frame, but we can pretend that it does provided that we add to $\mathbf{F}$ three fictitious forces, as indicated by the last three terms on the right of (4.6). The term $2\mathbf{u}^* \times \mathbf{\Omega}$ is the *Coriolis force* (per unit mass), while

$$-\mathbf{\Omega} \times (\mathbf{\Omega} \times \mathbf{x}) = \Omega^2 \mathbf{x}_\perp = \nabla\left(\frac{1}{2}\Omega^2 \mathbf{x}_\perp^2\right) \tag{4.7}$$

is called the *centrifugal force*, where centrifugal means 'away from the centre'. Finally, the term $\mathbf{x} \times (d\mathbf{\Omega}/dt)$ is often called the *Poincaré force*.

4.2 Governing Equations and the Rossby and Ekman Numbers

Let us now return to fluid mechanics and see how to adapt the Navier–Stokes equation to a non-inertial frame of reference which rotates with a *constant* angular velocity, $\mathbf{\Omega}$. Using (4.7) to combine the centrifugal and pressure forces, and dropping the superscript $*$ on the assumption that *all* quantities are measured in the rotating frame, we find that, for a fluid of uniform density,

$$\frac{D\mathbf{u}}{Dt} = 2\mathbf{u} \times \mathbf{\Omega} - \nabla\left(p/\rho - \frac{1}{2}\Omega^2 \mathbf{x}_\perp^2\right) + \mathbf{g} + \nu\nabla^2\mathbf{u}.$$

This is normally rewritten in the simpler form

$$\frac{D\mathbf{u}}{Dt} = 2\mathbf{u} \times \mathbf{\Omega} - \nabla\left(p/\rho\right) + \mathbf{g} + \nu\nabla^2\mathbf{u}, \tag{4.8}$$

on the assumption that p now represents the *reduced pressure*, $p_{\text{true}} - \frac{1}{2}\rho\Omega^2\mathbf{x}_\perp^2$.

At this point it is useful to introduce two dimensionless groups. The first, called the *Ekman number*, provides some indication as to the relative size of the viscous and Coriolis forces. It is defined as

$$\mathrm{Ek} = \frac{\nu}{\Omega \ell^2},$$

(4.9)

where ℓ is some representative length scale associated with the Laplacian in (4.8). The second group, called the *Rossby number*, provides an indication as to the relative strengths of the inertial force, as represented by $(\mathbf{u} \cdot \nabla)\mathbf{u}$, and the Coriolis force, $2\mathbf{u} \times \boldsymbol{\Omega}$. This is

$$\mathrm{Ro} = \frac{u}{\Omega \ell},$$

(4.10)

where u and ℓ are representative velocity and length scales associated with the inertial force.

There is, however, one important special case where it is inappropriate to think of (4.10) as indicative of the relative size of the inertial and Coriolis forces. This is when the flow is strongly anisotropic, with a large characteristic length scale parallel to the rotation axis, $\ell_{//}$, and a much shorter one normal to the axis, $\ell_\perp \ll \ell_{//}$. (As we shall discover, this is very common in a rapidly rotating fluid in a closed domain.) To see why (4.10) can be misleading in such cases, we introduce the vector potential for $\mathbf{u}$, defined by $\nabla \cdot \mathbf{A} = 0$ and $\nabla \times \mathbf{A} = \mathbf{u}$. Then, in line with Helmholtz's decomposition (see Appendix 1), the Coriolis force can be split into its solenoidal and irrotational parts using the identity

$$2\mathbf{u} \times \boldsymbol{\Omega} = 2(\boldsymbol{\Omega} \cdot \nabla)\mathbf{A} - \nabla(2\boldsymbol{\Omega} \cdot \mathbf{A}).$$

(4.11)

In a closed domain, the irrotational part of (4.11) is simply absorbed into the pressure gradient and consequently plays no active role. Moreover, the ratio of the solenoidal to the irrotational parts of $2\mathbf{u} \times \boldsymbol{\Omega}$ is $\ell_\perp / \ell_{//}$, and so the dynamically active part of the Coriolis force is actually much smaller than $u\Omega$, of the order of $(\ell_\perp / \ell_{//})\, u\Omega$. In such cases, the ratio of inertia to the dynamically active part of the Coriolis force is of order $(\ell_{//}/\ell_\perp)\,\mathrm{Ro}$.

4.3 Rapid Rotation, the Taylor–Proudman Theorem, and Taylor Columns

We now explore the case where the background rotation is strong, in the sense that the Coriolis force dominates over inertia, *i.e.* Ro << 1. We take $\boldsymbol{\Omega}$ to be aligned with the z-axis.

4.3.1 The Geostrophic Force Balance, Taylor's Experiment, and Taylor Columns

Consider an inviscid, rapidly rotating flow, with Ro << 1. Since Ro is small, we may neglect the inertial force, $\mathbf{u} \cdot \nabla \mathbf{u}$, in (4.8). So, in the rotating frame, we have

$$\frac{\partial \mathbf{u}}{\partial t} = 2\mathbf{u} \times \boldsymbol{\Omega} - \nabla \left(p/\rho \right). \tag{4.12}$$

The corresponding vorticity equation is

$$\frac{\partial \boldsymbol{\omega}}{\partial t} = 2(\boldsymbol{\Omega} \cdot \nabla)\mathbf{u}, \tag{4.13}$$

and so an inviscid, steady flow satisfies $2\mathbf{u} \times \boldsymbol{\Omega} = \nabla \left(p/\rho \right)$ and $(\boldsymbol{\Omega} \cdot \nabla)\mathbf{u} = 0$. In short, a steady velocity field in a rapidly rotating fluid must be two-dimensional, $\mathbf{u} = \mathbf{u}(x, y)$, while satisfying the so-called *geostrophic force balance*,

$$2\mathbf{u} \times \boldsymbol{\Omega} = \nabla \left(p/\rho \right). \tag{4.14}$$

In particular, the flow must satisfy $\partial u_z/\partial z = 0$, so that columns of fluid orientated with the z-axis cannot change their length. The constraint $(\boldsymbol{\Omega} \cdot \nabla)\mathbf{u} = 0$ is known as the *Taylor–Proudman theorem* and it is a very powerful constraint.

Perhaps the most famous illustration of the Taylor–Proudman theorem is an experiment first performed by G.I. Taylor (1923) and shown schematically in Figure 4.1. A tank of water is placed on a turntable and the experiment is observed in the rotating frame of reference. A small object is then *slowly* traversed across the base of the tank. One might have expected to see the fluid ahead of the object to move up and over the object to make way for it, as occurs in the non-rotating case. However, the flow is quasi-steady and must satisfy $(\boldsymbol{\Omega} \cdot \nabla)\mathbf{u} \approx 0$, which excludes such a three-dimensional motion. In particular, water rising up and over the object would require vertical columns of fluid to first contract and then expand, which is not allowed. Instead, what happens is that the water sitting above the object drifts across the tank keeping pace with the object below, as if the fluid sitting inside an imaginary cylinder which circumscribes the object is rigidly attached to that object. This cylinder of drifting fluid is known as a *Taylor column*, and the fluid outside the column flows around it as if the column were rigid, thus maintaining two-dimensional

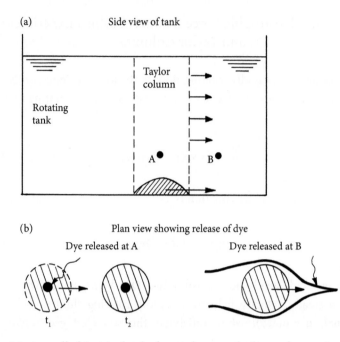

Figure 4.1 A small object is slowly dragged across the base of a rotating tank of water, producing a Taylor column. (a) Side view. (b) Plan view. From Davidson (2015).

motion. This counterintuitive behaviour can be confirmed by placing dye at the points A and B shown in Figure 4.1(a). The dye at A is seen to drift across the tank, always centred above the object, while that at B splits into two as the object passes below it, making way for the rigid Taylor column, as shown in Figure 4.1(b).

Actually, an equally elegant demonstration of the power of the constraint $\partial u_z/\partial z = 0$ is provided by Kelvin's less well-known 1868 experiment. This uses two well-separated corks placed on the central axis in a rotating tank, one above the other (see Figure 1.6). When the top cork is pushed down, the bottom cork also moves downward, maintaining a fixed distance between them, as demanded $\partial u_z/\partial z = 0$.

4.3.2 A Flaw in the Usual Rationalization of Taylor Columns: The Role of Waves

Let us return to Taylor's experiment. The interpretation of this experiment given in §4.3.1, which is based on the geostrophic force balance (4.14), is a plausible rationalization of events, but it leaves much unexplained. Consider, for example, the movement of the dye at point A. How does the fluid at A know the instantaneous location of the object below? Here there is a surprise: the information is transmitted upward by wave motion, and in particular by *inertial waves*. In fact, as we shall see in Chapter 6, Taylor columns are first established, and then subsequently maintained, by *low-frequency* inertial waves travelling along the rotation axis. In short, it is not correct to think of the flow in Figure 4.1 as steady, and the time derivative in (4.13) as unimportant. Rather, for as long as it moves, the object at the base of the tank acts like a radio antenna, continually emitting low-frequency inertial waves which travel upward, carrying the information that the object below is moving. When the object stops moving, the waves cease and the Taylor column vanishes. We shall provide some hint as to the importance of waves by considering a variant of Taylor's experiment. First, however, we should say a little about inertial waves.

We shall give a full account of inertial waves in Chapter 6. For the moment, we merely note that they are incompressible, internal waves in which the Coriolis force provides the restoring force (see Figure 1.7). Their governing equation is the curl of (4.13) in the form

$$\frac{\partial^2}{\partial t^2} \nabla^2 \mathbf{u} + (2\mathbf{\Omega} \cdot \nabla)^2 \mathbf{u} = 0. \tag{4.15}$$

Crucially, as noted in §1.2, low-frequency inertial waves (waves with a frequency much less than Ω) have a group velocity of $\mathbf{c}_g = \pm \Omega \lambda / \pi$ and a phase velocity of $\mathbf{c}_p \approx 0$. Here λ is the dominant wavelength, which, in the case of Taylor's experiment, is of the order of the diameter of the moving object, say d. So the claim is that the *slowly* moving object in Taylor's experiment triggers *low-frequency* inertial waves that travel vertically upward with a group velocity of $\mathbf{c}_g \sim \Omega d / \pi$. This is readily confirmed experimentally.

Consider, for example, the situation where the object is initially at rest and then suddenly moved to the right at the speed V, where $Ro = V/\Omega d \ll 1$. (When $V \ll \Omega d$, only low-frequency inertial waves are generated by the moving object.) If the Taylor column is indeed initiated by low-frequency inertial waves, then there should be a transition period during which the Taylor column is incomplete, as the waves travel up towards the free surface at the speed $c_g \sim \Omega d / \pi$. This is exactly what is observed (see Figure 4.2). If the water depth is H, then there is a transition period of $\sim \pi H / \Omega d$ during which a columnar structure is seen to emerge from the object and then elongate at the rate $\sim \Omega d / \pi$. Moreover, the flow within this growing column is helical, and we shall see in Chapter 6 that inertial waves have a helical structure. It is only

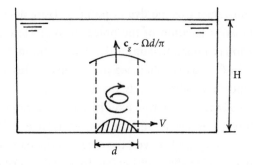

Figure 4.2 Transient growth of a Taylor column after the impulsive movement of an object.

after this transient concludes, and inertial waves have reflected back down from the water surface, that a fully developed Taylor column is observed.

We have also suggested that Taylor columns are maintained, as well as initiated, by low-frequency inertial waves. To show that this is so, we return to the configuration shown in Figure 4.1, in which the object is moved steadily to the right at the speed V, with Ro $= V/\Omega d \ll 1$. However, we now adopt a rotating frame of reference that moves slowly across the tank, also at the speed V. In such a frame the object is *stationary* while the fluid flows to the left. Now, if the Taylor column is maintained by low-frequency inertial waves streaming upward with a group velocity of $c_g \sim \Omega d/\pi$ (relative to stationary fluid), then the fact that the fluid moves to the left at the speed V means that the column should not be vertical, but rather tilted at the small angle $\theta = V/c_g \sim \pi V/\Omega d$ (Figure 4.3).

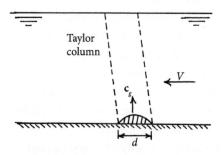

Figure 4.3 The slight tilting of a Taylor column due to the relative movement of the fluid and a slowly moving object. The frame of reference moves with the object.

This tilting of a Taylor column is indeed observed. However, the story is a little more complicated because a small but finite Ro leads to a small horizontal contribution to c_g, which causes additional tilting. In brief, a stationary wave pattern requires that the horizontal component of c_p is equal and opposite to V. However, as we shall see, inertial waves have the property $c_p + c_g = \Omega\lambda/\pi$, and so c_g has a horizontal component *equal* to V, which doubles the angle θ. (See Lighthill, 1970.)

4.4 Taylor Columns Associated with the Axial Movement of an Object

It would seem plausible, then, that Taylor columns are both initiated and maintained by low-frequency inertial waves, and indeed we shall confirm this in Chapter 6. In the meantime, we continue our qualitative discussion of Taylor column formation, turning now to the case where an object moves slowly *along* the rotation axis. We start by summarizing the results of Bretherton's (1967) analytical study.

Bretherton considered the case of a cylinder of radius R sitting in a rotating, inviscid fluid, with its axis perpendicular to the rotation axis. The cylinder is initially at rest, but at $t = 0$ it acquires a small velocity, V, directed parallel to the rotation axis, which is now taken to be *horizontal*. As usual, Ro = $V/\Omega R \ll 1$. The subsequent disturbance is found to propagate along the rotation axis as low-frequency inertial waves at the group velocity,

$$c_g = \pm\Omega\lambda/\pi \sim \pm 2R\Omega/\pi. \tag{4.16}$$

These waves fill a columnar region of space aligned with Ω, both ahead of and behind the cylinder, as shown in Figure 4.4. This region grows in length as $\ell_{//} \sim c_g t \sim 2R\Omega t/\pi$. Within this columnar region the fluid knows the cylinder

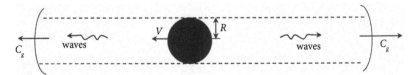

Figure 4.4 A cylinder moving slowly to the left generates low-frequency inertial waves which propagate along the rotation axis with a group velocity of $c_g \sim \pm 2R\Omega/\pi$.

is moving, because the waves carry that information, whereas outside the column the fluid is undisturbed. Moreover, it turns out that the fluid within this two-dimensional column moves with exactly the same velocity as the cylinder, thus satisfying the geostrophic constraint $\partial u_z/\partial z = 0$. This is, of course, a growing, 2D, Taylor column, and it has upstream–downstream symmetry.

Bretherton's (1967) inviscid solution is shown in Figure 4.5, in a frame of reference in which the cylinder is stationary. Note that, as in Figure 4.2, the propagation of inertial waves progressively establish a region of space that might be called quasi-geostrophic, in the sense that it satisfies $\partial u_z/\partial z = 0$. Thus, we see a theme emerging: an evolving quasi-geostrophic flow is reliant on the continual propagation of low-frequency inertial waves.

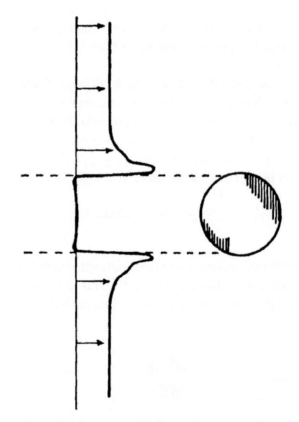

Figure 4.5 Bretherton's (1967) inviscid solution for rotating flow past a cylinder (with Ω horizontal) confirms that a Taylor column is established by low-frequency inertial waves.

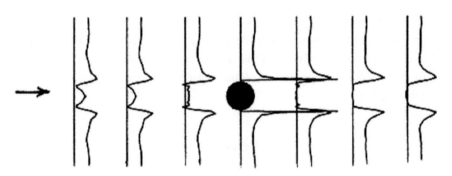

Figure 4.6 Numerical simulation of a Taylor column forming over a sphere at Ro ~ 0.04 and Re ~ 500. Note the upstream–downstream asymmetry. Adapted from Wang et al. (2004).

Bretherton's solution ignores viscosity and assumes Ro << 1. The question then arises as to what happens in a viscous fluid when Ro is not small. We shall consider the particular case of a sphere of radius R moving along a horizontal rotation axis, and adopt a frame of reference in which the sphere is stationary, as in Figure 4.6. Let Ro = $V/\Omega R$, where V is the speed of the oncoming fluid. Experiments show that, as Ro is increased, the upstream–downstream symmetry is quickly lost. For values of Ro up to around 0.3, the flow pattern *upstream* of the sphere is similar to that for Ro << 1, with the sphere blocking the column of fluid ahead of it. The length of the upstream column is limited by viscosity, which dissipates the inertial waves within the Taylor column, and for Ro < 0.3 the length of the column is around $0.06\Omega R^3/\nu$. For larger values of Ro, the upstream column becomes progressively shorter and disappears at Ro ~ 1.0 (see Maxworthy, 1970). This is to be expected, since the group velocity of the most energetic inertial waves coming off the sphere is $c_g \sim 2R\Omega/\pi$, so that when the speed of the oncoming fluid exceeds $V \sim 2R\Omega/\pi$, *i.e.* when Ro exceeds ~ $2/\pi$, no waves can propagate upstream.

The downstream behaviour is different, with a Taylor column forming only at very small values of Ro. This is seen in the experiments of Maxworthy (1970) and the numerical simulations of Wang et al. (2004), an example of which is shown in Figure 4.6.

* * *

That concludes our qualitative introduction to rapidly rotating fluids and to the formation of Taylor columns. Perhaps the key point to note is that there are two distinct ways of interpreting Taylor's experiment, and indeed many other quasi-steady flows. One approach is to treat the flow as steady and invoke the geostrophic equation $(\mathbf{\Omega} \cdot \nabla)\mathbf{u} = 0$. The other is to embrace the fact that the flow is weakly unsteady, retain the time derivative in (4.13), and interpret the results in terms of inertial waves. The first is simple and direct, but the second delivers more profound insights. We shall pick up this story again in Chapter 6.

References

Bretherton, F.P., 1967, The time-dependant motion due to a cylinder moving in an unbounded rotating or stratified fluid, *J. Fluid Mech.*, **28**, 545–70.

Davidson, P.A., 2015, *Turbulence: An Introduction for Scientists and Engineers*, 2nd Ed., Oxford University Press.

Goldstein, H., 1980, *Classical Mechanics*, 2nd Ed., Addison-Wesley.

Lighthill, M.J., 1970, The theory of trailing Taylor columns, *Proc. Cam. Phil. Soc.*, **68**, 485.

Maxworthy, T., 1970, The flow generated by a sphere moving along the axis of a rotating viscous fluid, *J. Fluid Mech.*, **40**(3), 453–79.

Taylor, G.I., 1923, Experiments on the motion of solid bodies in rotating fluids, *Proc. Roy. Soc., A*, **104**, 213–18.

Wang, Y.-X., Lu, X.-Y., & Zhuang, L.-X., 2004, Numerical analysis of the rotating viscous flow approaching a solid sphere, *Int. J. Num. Methods in Fluids*, **44**, 905–25.

Chapter 5
Ekman Boundary Layers

We now consider the various types of boundary layers that are associated with rotating flows in confined domains. As we shall discover, these boundary layers are quite unlike those encountered in, say, aerodynamics, because there is a continual exchange of mass between the boundary layer and the exterior flow. In particular, rotating boundary layers can either entrain exterior fluid, or else detrain boundary-layer fluid into the external flow. Either way, this exchange of mass sets up a weak secondary flow that couples the external flow to the boundary layer. Although this secondary flow is typically very weak, often difficult to see in an experiment, it nevertheless tends to dominate the dynamics of confined, rotating flows. Boundary layers that exchange mass with the exterior flow in this way are sometimes call *active* boundary layers, as opposed to the *passive* boundary layer shown in Figure 2.8.

5.1 Three Types of Ekman Layers

Roughly speaking, there are three distinct types of rotating boundary layers that can form on a flat surface oriented normal to the axis of rotation. A *Kármán layer* is set up when a disc rotates in an otherwise stationary fluid, whereas a *Bödewadt layer* is established when a fluid in rigid-body rotation encounters a stationary surface. Both of these layers are best viewed in an *inertial* frame of reference, where it is possible to find fully nonlinear solutions of the Navier–Stokes equation. A third common situation is when the fluid is in rigid-body rotation and the surface rotates at a slightly different rate, either faster or slower than the fluid. This is called an *Ekman layer* and it is best described in a frame of reference rotating with the fluid, as this allows the problem to be linearized. We shall, therefore, move back and forth between inertial and rotating frames in this chapter.

Of course, a more general case is where a fluid is in rigid-body rotation and the surface rotates at a very different rate to the fluid in the far field. However, as we shall see, this can usually be understood, at least qualitatively, in terms of the three canonical cases listed above.

The Dynamics of Rotating Fluids. P. A. Davidson, Oxford University Press. © Peter A Davidson (2024).
DOI: 10.1093/9780191994272.003.0005

In the geophysical literature, it has become common to refer to *any* boundary layer of the type described above as an Ekman layer, including the nonlinear solutions of Kármán and Bödewadt. The associated secondary flow is then referred to as *Ekman pumping*.

5.1.1 The Nonlinear Solutions of Kármán and Bödewadt

Let us start with Kármán's (1921) nonlinear solution for the flow generated by a disc of infinite radius rotating in an otherwise stationary fluid. We adopt an inertial frame of reference and make no assumption as to the size of the Rossby number, Ro.

Suppose the disc rotates with angular velocity Ω. Then a boundary layer forms on the surface of the disc within which the fluid is centrifuged radially outward. Thus, the streamlines within the boundary layer have a spiralled structure, and we shall see that the azimuthal and radial velocities having similar magnitudes, *i.e.* $u_r \sim u_\theta \sim \Omega r$ in (r, θ, z) coordinates centred on the surface of the disc.

Now, this flow is completely determined by just two physical parameters, v and Ω, from which we can construct the length scale $\hat{\delta} = \sqrt{v/\Omega}$. Since there is no geometric length scale associated with an infinite disc, dimensional analysis tells us that the boundary-layer thickness, δ, must be of the order of $\hat{\delta}$. The same estimate of δ can be obtained from a radial force balance. The driving force for the radial motion is the centrifugal force (per unit mass), u_θ^2/r. This is resisted by the viscous force $v\nabla^2 u_r$, and since the radial and azimuthal velocities scale as $u_r \sim u_\theta \sim \Omega r$, we have the force balance

$$\Omega^2 r \sim v\Omega r/\delta^2.$$

This then yields the estimate $\delta \sim \hat{\delta} = \sqrt{v/\Omega}$.

Kármán used the estimates of $\delta \sim \hat{\delta} = \sqrt{v/\Omega}$ and $u_r \sim u_\theta \sim \Omega r$ to propose a solution of the form

$$u_r = \Omega r F(\eta), \quad u_\theta = \Omega r G(\eta), \quad u_z = \Omega \hat{\delta} H(\eta), \quad \eta = z/\hat{\delta}, \tag{5.1}$$

in (r, θ, z) coordinates centred on the surface of the disc, with pressure assumed to be a function of z only. The functions F, G, and H are then determined by continuity, which demands $2F + H' = 0$, and by the radial and azimuthal components of the Navier–Stokes equation, which become

$$F^2 + F'H - G^2 = F'', \quad 2FG + HG' = G''. \tag{5.2}$$

(The axial component of the Navier–Stokes equation merely gives the axial pressure gradient.) The corresponding boundary conditions are no-slip at the surface of the disc and $u_r = u_\theta = 0$ in the far field, which require

$$F(0) = 0, \quad G(0) = 1, \quad H(0) = 0; \quad F(\infty) = G(\infty) = 0. \tag{5.3}$$

The integration of (5.2), subject to these boundary conditions, is straightforward and the solutions for F, G, and $-H$ are shown in Figure 5.1. We have, in effect, a centrifugal fan, with a flow that spirals radially outward within a boundary layer of thickness $\delta \approx 4\hat{\delta}$. The solution yields a finite axial velocity in the far field, with $H(\infty) = -0.885$, and so the radial mass flux within the boundary layer is fed by a *weak* axial flow towards the disc, of

$$u_z(z \to \infty) = H(\infty)\Omega\hat{\delta} = H_\infty\sqrt{\nu\Omega} = -0.885\sqrt{\nu\Omega}. \tag{5.4}$$

This secondary flow is shown in Figure 5.2(a). The solution also yields $G'(0) = -0.616$, and so the azimuthal stress acting on the disc is

$$|\tau_{\theta z}| = 0.616\rho\Omega^2 r\hat{\delta}, \tag{5.5}$$

from which the torque on the disc can be calculated. For a disc of radius R, that torque is

$$T = \int_0^R (r\,|\tau_{\theta z}|)\,2\pi r dr = 0.616\frac{\pi}{2}\rho\Omega^2 R^4\hat{\delta}. \tag{5.6}$$

Let us now consider a second case, in which the disc is stationary and the fluid remote from the disc rotates at Ω. The azimuthal velocity then adjusts from $u_\theta = \Omega r$ in the far field to $u_\theta = 0$ at $z = 0$. This problem was solved by

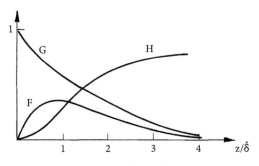

Figure 5.1 Kármán's solution of F, G, and $-H$ for the flow induced by a rotating disc.

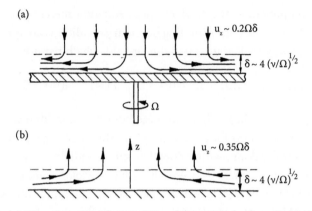

Figure 5.2 (a) The secondary flow in Kármán's problem of motion induced by a rotating disc. (b) The secondary flow induced by a rotating fluid above a stationary surface.

Bödewadt (1940) who noticed that, as in Kármán's problem, $\delta \sim \sqrt{\nu/\Omega}$ and $u_r \sim u_\theta \sim \Omega r$. It follows that the solution still takes the form of (5.1). However, there is now a radial pressure gradient in the far field, which adds a new term to the radial component of the Navier–Stokes equation. In particular, in order to satisfy $F(\infty) = 0$ and $G(\infty) = 1$, the term G^2 in (5.2) has to be replaced by $G^2 - 1$, and the new equations to be solved are $2F + H' = 0$ plus

$$F^2 + F'H - G^2 + 1 = F'', \quad 2FG + HG' = G''. \tag{5.7}$$

Integration of these equations, subject to $G(0) = 0$ and $G(\infty) = 1$, yields a boundary-layer thickness of $\delta \sim 4\hat{\delta}$, as before. However, the flow differs from Kármán's solution in that the secondary motion in the r–z plane is reversed, with the fluid spiralling radially *inward* within the boundary layer, eventually drifting up and out of the layer (Figure 5.2b). This radial inflow near the boundary is driven be a pressure gradient. That is, outside the boundary layer there is a radial force balance between $\partial p/\partial r$ and $\rho u_\theta^2/r$, resulting in a low pressure on the axis. This radial pressure gradient is imposed on the fluid within the boundary layer where the centrifugal force is smaller, and the resulting imbalance between $\partial p/\partial r$ and $\rho u_\theta^2/r$ drives the fluid radially inward.

Continuity now requires an axial flow away from the disc, and it turns out that $H(\infty) = 1.35$, so we have

$$u_z(z \to \infty) = H(\infty)\Omega\hat{\delta} = H_\infty\sqrt{\nu\,\Omega} = 1.35\sqrt{\nu\,\Omega}. \tag{5.8}$$

The solution also gives $G'(0) = 0.770$, and so the azimuthal stress on the disc surface is

$$\tau_{\theta z} = 0.770 \rho \Omega^2 r \hat{\delta}, \tag{5.9}$$

from which the torque on the disc can be calculated. For a disc of radius R, that torque is

$$T = \int_0^R (r\tau_{\theta z}) 2\pi r \, dr = 0.770 \frac{\pi}{2} \rho \Omega^2 R^4 \hat{\delta}. \tag{5.10}$$

5.1.2 Generalizing the Solutions of Kármán and Bödewadt

The more general case, where the fluid remote from the boundary rotates at Ω_f, while the boundary rotates at Ω_b, is readily solved in a similar manner. It is necessary only to change the boundary conditions. As one might have expected, in those cases where $\Omega_b > \Omega_f$, the flow is reminiscent of Kármán's solution, with the fluid spiralling radially outward within the boundary layer, accompanied by a weak axial flow towards the boundary. Conversely, when $\Omega_b < \Omega_f$, the flow is similar to Bödewadt's solution, with the fluid spiralling radially inward, eventually drifting up and out of the boundary layer.

Of particular interest here is the axial flow outside the boundary layer, for reasons that will become evident in the next section. To focus thoughts, consider the Kármán-like case where $\Omega_b > \Omega_f$, so the boundary layer entrains external fluid. Let $u_b = |u_z(z \to \infty)|$ be the remote axial flow towards the boundary. Then it is clear that u_b must take the form

$$u_b = \sqrt{\nu \Omega_b} \, h \left(\Omega_f / \Omega_b \right),$$

for some function h, where $h(0) = 0.885$ and $h(1) = 0$. The shape of h is given in Figure 5.3.

The function h does not have a simple analytical form, but there are various simple approximations to h in common use. One popular choice is

$$u_b = k\sqrt{\nu \Omega_b} \left(1 - \Omega_f / \Omega_b \right), \tag{5.11}$$

with some authors taking $k = 0.885$, to satisfy $h(0) = 0.885$, and yet others adopting $k = 1$, to be consistent with the linear analysis of §5.1.4 for $\Omega_f \to \Omega_b$. An alternative is

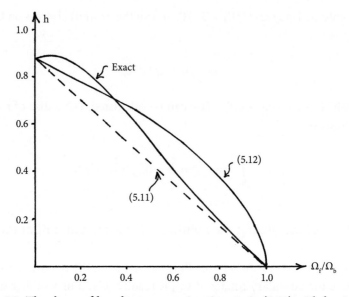

Figure 5.3 The shape of h and two approximations to it, (5.11) with $k = 0.885$ and (5.12).

$$u_b = 0.885\sqrt{\nu\,(\Omega_b - \Omega_f)},\tag{5.12}$$

which is clearly appropriate for $\Omega_f \ll \Omega_b$, but is poorly behaved in the limit of $\Omega_f \to \Omega_b$.

5.1.3   Combined Kármán and Bödewadt Layers: The Rotor–Stator Problem

In order to illustrate the utility of (5.11) or (5.12), consider the situation shown in Figure 5.4, consisting of two parallel discs. The top disc rotates at Ω, the bottom one is stationary, and the fluid between them rotates at the intermediate rate Ω_c, where the subscript c stands for 'core flow'. Of course, the question is, can we determine Ω_c? In order to answer this question, we note that, because $\Omega_c < \Omega$, we have a Bödewadt layer on the lower disc and a Kármán-like layer on the top plate. Moreover, the axial velocity leaving the lower boundary layer must match that entering the top layer. Thus, Ω_c must be chosen to satisfy

$$u_z = 1.35\sqrt{\nu\,\Omega_c} = \sqrt{\nu\,\Omega}\,h\,(\Omega_c/\Omega)\,.\tag{5.13}$$

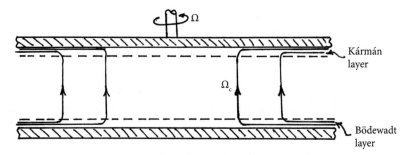

Figure 5.4 The Ekman pumping between two discs, where the upper disc rotates.

Approximation (5.11) then yields $\Omega_c = 0.283\Omega$ for the case of $k = 1$, or else $\Omega_c = 0.245\Omega$ when $k = 0.885$. On the other hand, approximation (5.12) predicts $\Omega_c = 0.301\Omega$, which turns out to be very close to the actual core rotation rate of $\Omega_c = 0.305\Omega$.

Of course, this is a rather trivial example. However, it provides the first hint that the weak secondary flow generated by these boundary layers plays an important role in controlling the core motion in such confined, rotating flows. We shall explore the importance of Ekman pumping for confined swirling flows in more detail in §5.2.

5.1.4 The Linear Solution of Ekman

In addition to the solutions of Bödewadt and Kármán, a third common situation is where the disc and the fluid rotate at slightly different rates. In such cases, it pays to adopt a rotating frame of reference in which the fluid at infinity is stationary. One then solves the low Rossby number version of the Navier–Stokes equation,

$$2\mathbf{u} \times \mathbf{\Omega} - \nabla\left(p/\rho\right) + \nu\nabla^2\mathbf{u} = 0, \qquad (5.14)$$

which is linear in $\mathbf{u}$. As mentioned earlier, boundary layers which are driven by small differences in rotation rate are called *Ekman layers*, although the same name is now commonly used to describe almost any rotating boundary layer.

The flow in an Ekman layer is similar to that in either a Bödewadt or Kármán layer. In particular, when the fluid rotates faster than the boundary, it resembles a Bödewadt layer, with the fluid spiralling radially inward, and when the boundary rotates faster than the fluid, it resembles a Kármán layer, with a radial outflow. The precise structure of the flow is readily found. Suppose that, in an inertial frame, the fluid remote from the disc rotates at Ω and the disc at $\Omega + \hat{\omega}$, with $|\hat{\omega}| \ll \Omega$. Then, in a frame rotating at Ω, we take

$$u_r = \hat{\omega} r F(\eta), \quad u_\theta = \hat{\omega} r G(\eta), \quad u_z = \hat{\omega}\hat{\delta} H(\eta), \quad \eta = z/\hat{\delta}, \qquad (5.15)$$

where $\hat{\delta} = \sqrt{\nu/\Omega}$ and $z = 0$ at the disc surface. This is subject to the boundary conditions

$$F(0) = 0, \; G(0) = 1; \quad F(\infty) = G(\infty) = 0.$$

Next, assuming that the reduced pressure is a function of z only, the radial and azimuthal components of (5.14) yield

$$F''(\eta) = -2G, \quad G''(\eta) = 2F, \qquad (5.16)$$

whose solution is

$$F = \exp(-\eta)\sin\eta, \quad G = \exp(-\eta)\cos\eta. \qquad (5.17)$$

It follows that u_r is (mostly) positive for $\hat{\omega} > 0$, and (mostly) negative for $\hat{\omega} < 0$, as suggested above. The velocity in a horizontal plane is evidently

$$(u_r, u_\theta) = \hat{\omega} r \exp(-\eta)\left[\sin\eta\,\hat{\mathbf{e}}_r + \cos\eta\,\hat{\mathbf{e}}_\theta\right], \qquad (5.18)$$

whose variation with z is known as the *Ekman spiral*. Finally, (5.17) plus continuity give

$$\int_0^\infty u_r dz = \tfrac{1}{2}\hat{\omega}\hat{\delta} r, \quad u_z(z \to \infty) = -\hat{\omega}\hat{\delta}, \qquad (5.19)$$

and the axial velocity can be found from u_r using continuity in the form $2F + H' = 0$:

$$H = \exp(-\eta)\left[\sin\eta + \cos\eta\right] - 1. \qquad (5.20)$$

A similar analysis can be performed when the normal to the surface is inclined to the rotation axis, although it is necessary to replace Ω by its component normal to the boundary, say Ω_n. The boundary-layer thickness then scales as $\hat{\delta} = \sqrt{\nu/\Omega_n}$. Clearly, the case where the boundary is parallel to Ω is a singular one, as $\delta \to \infty$ in this limit. This is the first hint that a completely different scaling is required for δ in the limit of $\Omega_n \to 0$.

5.2 Secondary Flows in Teacups, Rivers, and Ducts

When the Ekman number based on r, $Ek = \nu/\Omega r^2$, is small, the axial velocity induced by Ekman pumping is much weaker than the boundary-layer flow. Nevertheless, if the fluid is confined, this weak axial flow has important consequences for the motion as a whole, as noted in §5.1.3. This is often illustrated by the process of *spin-down* in a stirred cup of tea, a problem first discussed by James Thomson (Kelvin's brother) in 1857.

Suppose that tea in a flat-bottomed cup is set into a state of rotation and the spoon then removed. We might ask: how long does it take for the tea to stop spinning? To answer this, we first note that a Bödewadt-like layer is established on the bottom of the cup, inducing a radial inflow along the base of the teacup, with the fluid drifting up and out of the Ekman layer. Thomson (1857) describes the process thus:

> If a shallow circular vessel with flat bottom, be filled to a moderate depth with water, and if a few small objects, ..., suitable for indicating to the eye the motions of the water in the bottom, be put in, and if the water be set to revolve by being stirred round, then, on the process of stirring being terminated, and the water left to itself, the small particles in the bottom will be seen to collect at the centre. They are evidently carried there by a current determined towards the centre along the bottom in consequence of the centrifugal force of the lowest stratum of the water being diminished in reference to the strata above through a diminution of velocity of rotation in the lower stratum by friction on the bottom.
>
> (Thomson 1857)

This constitutes a surprisingly accurate description of a Bödewadt layer.

Thus, a secondary flow is established as shown in Figure 5.5, with the fluid spiralling radially inward through the Ekman layer, then up through the core, and finally back down through the sidewall boundary layer. Moreover, this

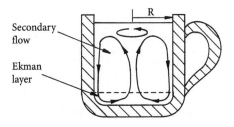

Figure 5.5 Spin-down of a stirred cup of tea.

secondary flow is not just some kind of inconsequential curiosity. As each fluid element passes through the Ekman layer, it gives up a significant fraction of its kinetic energy, and so the tea finally comes to rest when all of the contents of the cup have been flushed, once or twice, through the Ekman layer. It follows that the spin-down time is of the order of the turn-over time of the secondary flow, say $\tau \sim 2D/u_z \sim 2D/\left(1.4\sqrt{\nu\,\Omega}\right)$, where D is the depth of the water.

This might be compared with spin-down in a *long* cylinder where there is no Ekman pumping, and so the spin-down is controlled by the time taken for the centreline vorticity to diffuse a distance R to the sidewall. This happens to be $\tau \approx R^2/\gamma_1^2\nu$, where $\gamma_1 = 3.83$ is the first zero of the Bessel function J_1. Suppose, for example, that $\Omega = 4 \text{ s}^{-1}$, $R = D = 4$ cm, and $\nu = 10^{-6} \text{ m}^2/\text{s}$. Then $\tau \sim 30$ s in a teacup, as compared to $\tau \sim 2$ min in a long cylinder. It is clear from a comparison of these spin-down times that Ekman pumping provides a particularly efficient mechanism for destroying kinetic energy in a confined swirling flow.

Einstein (1926) was also interested in the teacup problem and noticed that it might help explain the meandering of rivers, although Thomson (1876, 1877) had already established the connection some 50 years earlier. The argument proceeds as follows. Consider the cross-section through a meandering river, as shown in Figure 5.6, where the inside of the curve is on the left and the outside on the right. Because the river is curved, the water rotates about an axis on the left, and that rotation causes a Bödewadt-like layer to form on the river bed. This then induces a secondary flow in the river cross-section, with the water flowing to the left along the river bed, up the river bank on the inside of the curve, back across the free surface, and finally down the river bank on the outside of the curve. Crucially, this secondary flow induces outer-bank scour and inner-bank deposition of sediment, causing the meandering of the river to increase with time. In short, there is a kind of eroding-boundary/fluid-flow instability. (See, for example, Seminara, 2006, for a review of the mechanics of meandering rivers.)

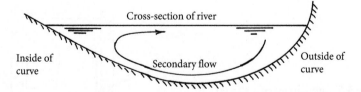

Inside of curve

Cross-section of river

Secondary flow

Outside of curve

Figure 5.6 The secondary flow in the cross-section of a meandering river.

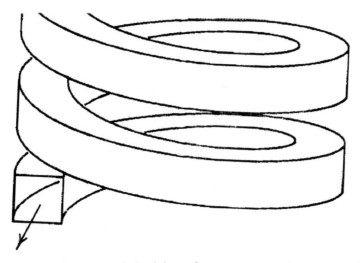

Figure 5.7 A helical duct of square cross-section.

As Thomson (1876, 1877) observed, a similar mechanism causes a secondary flow to develop in a curved duct. Consider Figure 5.7, which shows a helical duct of square cross-section. Suppose water is pumped down through the duct, exiting at the bottom. If the pitch of the helix is small, then the mean flow along the axis of the duct is more or less azimuthal, $u_{\mathrm{mean}} \approx u_\theta \approx \Omega r$ in cylindrical polar coordinates. Because of this rotation, Bödewadt-like layers develop on the top and bottom surfaces of the duct, creating a secondary flow, as shown in Figure 5.8.

As in the teacup problem, this secondary flow is not just a mere curiosity. Rather, it greatly enhances the rate of dissipation of mechanical energy by continually cycling the water through the thin, dissipative Ekman layers at the top and bottom of the duct. So the curvature of the duct causes a rise in the pumping power required to deliver a given flow rate, while simultaneously enhancing the heat transfer between the fluid and duct wall.

Essentially the same phenomenon occurs in curved pipes and tubes of circular or elliptical cross-section. This is important, not just in hydraulics, but also in biology, such as in the flow of blood in the human arterial system. Secondary flows are also common in rotating machinery, such as centrifuges and the housings of spinning discs, and they are particularly important in turbomachinery, where secondary flows occur within cooling passages and cascades of turbine blades. Because of its importance in engineering, there is a considerable literature on secondary flows in ducts, pipes, and turbomachinery, and useful reviews may be found in Berger & Talbot (1983) and Horlock & Lakshminarayana (1973).

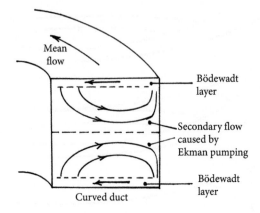

Figure 5.8 Water flows along a helical duct. Ekman pumping induces a secondary flow.

5.3 Spin-Down and Spin-Up in a Cylindrical Container

5.3.1 Spin-Down

Let us return to the problem of spin-down in a cylindrical container at low Ekman number, as shown in Figure 5.5. Our aim is to quantify the process and test the veracity of the assertion that the spin-down time is $\tau \sim 2D/u_b \sim 2D/\left(1.4\sqrt{\nu\,\Omega}\right)$, where D is the depth of the water and u_b the axial velocity exiting the Bödewadt-like layer at the base of the cylinder. We shall take the cylinder to have radius R and use cylindrical coordinates, (r, θ, z), centred on the base of the cylinder. We shall also assume that, outside the boundary layers, which are taken to be quasi-steady, the initial motion is one of rigid-body rotation. In practice, such an initial condition can be created by first co-rotating both the fluid and the cylinder, and then suddenly braking the motion of the cylinder.

One immediate complication is that inertial waves are typically generated when braking the cylinder, but let us assume those waves are weak and can be ignored. A more serious difficulty is that the sidewall boundary layer, across which the angular momentum density drops from $\Gamma = \Omega R^2$ to $\Gamma = 0$, is prone to Rayleigh's centrifugal instability, (2.110). Indeed, it is sometimes difficult to find a range of Ekman numbers, $\text{Ek} = \nu/\Omega D^2$, in which Ek is small enough to distinguish clearly between the core flow and the boundary layers, yet not so small that the sidewall boundary layer becomes unstable. However, let us set aside this complication and assume the flow is laminar and axisymmetric.

The structure of the flow in an *inertial* frame is, for small Ek, as shown in Figure 5.9. There is a Bödewadt-like boundary layer on the base of the cylinder and a sidewall boundary layer whose properties we shall discuss in §5.4. Since the secondary flow is very weak when Ek << 1, (2.86) tells us that the inviscid rotation in the core is independent of depth. Finally, following Weidman (1976a), and many other authors, we shall assume that the azimuthal velocity in the core takes the form of rigid-body rotation, $u_\theta = \Omega(t)r$.

Let us start with the *inviscid* core dynamics. Since we have $D\Gamma/Dt = 0$ and $\Gamma = \Omega(t)r^2$ throughout the core, Ω must satisfy

$$\frac{d\Omega}{dt} = -\frac{2u_r}{r}\Omega, \tag{5.21}$$

which tell us that u_r in the core is independent of z and linear in r. Continuity then demands

$$\frac{\partial u_z}{\partial z} = -\frac{u_b}{D} = -\frac{2u_r}{r} = f(t), \tag{5.22}$$

for some function f. Evidently, the axial velocity exiting from the Bödewadt-like layer, u_b, is independent of radius. Combining (5.21) and (5.22), we find

$$\frac{d\Omega}{dt} = \frac{\partial u_z}{\partial z}\Omega = -\frac{u_b}{D}\Omega, \tag{5.23}$$

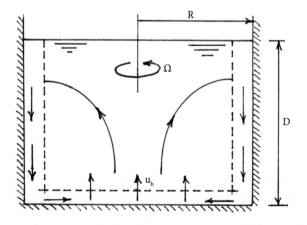

Figure 5.9 Spin-down in a cylindrical container. An inertial frame of reference is used.

which is the key result that allows the spin-down to be quantified. We recognize the first part of (5.23) as nothing more than the axial component of the inviscid vorticity equation

$$\frac{d\omega_z}{dt} = \omega_z \frac{\partial u_z}{\partial z}, \quad \omega_z = 2\Omega,$$

and so we may think of the decline in core rotation as a result of vortex-line compression.

It is also instructive to interpret (5.23) in terms of a global angular momentum balance. Let $\dot{\Gamma}_b$ be the net flux of angular momentum out of the Bödewadt-like layer and into the core, and $\dot{\Gamma}_s$ be the net flux of angular momentum out of the core and into the sidewall boundary layer. Then

$$\dot{\Gamma}_b = \rho \int_0^R (\Omega r^2) u_b 2\pi r dr = \frac{\pi}{2}\rho\Omega R^4 u_b, \tag{5.24}$$

and

$$\dot{\Gamma}_s = (\Omega R^2)\dot{m} = (\Omega R^2)\pi R^2 \rho u_b = \pi\rho\Omega R^4 u_b, \tag{5.25}$$

where $\dot{m}$ is the mass flux through the core. Moreover, the total angular momentum in the core is evidently

$$\int \rho\Gamma dV = \rho D \int (\Omega r^2)2\pi r dr = \frac{\pi}{2}\rho\Omega R^4 D. \tag{5.26}$$

Of course, the rate of change of angular momentum in the core must be the difference between $\dot{\Gamma}_b$ and $\dot{\Gamma}_s$, and hence

$$\frac{d}{dt} \int \rho\Gamma dV = \frac{\pi}{2}\frac{d\Omega}{dt}\rho R^4 D = \dot{\Gamma}_b - \dot{\Gamma}_s = -\frac{\pi}{2}\rho\Omega R^4 u_b. \tag{5.27}$$

Dividing through by $(\pi/2)\rho R^4 D$ brings us back to (5.23), as it must. However, (5.27) has an advantage over (5.23). We shall see in §5.4 that the sidewall boundary layer is too thick to exert any significant torque on the flow, and so (5.27) tells us that the total resistive torque exerted by the lower boundary on the flow is

$$T_b = \frac{\pi}{2}\rho\Omega R^4 u_b. \tag{5.28}$$

Returning now to (5.23), it is clear that we need to estimate u_b in order to determine the spin-down rate. For a Bödewadt layer on an infinite disc, (5.8) tells us that

$$u_b = H_\infty\sqrt{\nu\,\Omega} = 1.35\sqrt{\nu\,\Omega}. \tag{5.29}$$

However, we do not have an ideal Bödewadt layer here, since the inflow conditions at the edge of this layer are controlled by the sidewall boundary layer, and these are unlikely to match the velocity profile for a Bödewadt layer on an infinite disc. At best, we might hope to approach a classical Bödewadt layer towards the centre of the cylinder, once the inflow has had a chance to settle down. Nevertheless, u_b should scale as $u_b \sim \sqrt{\nu\,\Omega}$, like all such Ekman layers, and we have seen that it is independent of r. So let us write $u_b = H_b\sqrt{\nu\,\Omega}$, where H_b is a constant of order unity. Equation (5.23) then becomes

$$\frac{d\Omega}{dt} = -\frac{u_b}{D}\Omega = -\frac{H_b\sqrt{\nu\,\Omega}}{D}\Omega, \tag{5.30}$$

which integrates to give

$$\frac{\Omega}{\Omega_0} = \left[1 + \frac{H_b\sqrt{\nu\,\Omega_0}\,t}{2D}\right]^{-2}, \tag{5.31}$$

Ω_0 being the initial rotation rate.

Both Weidman (1976b) and Savas (1992) find (5.31) to be a good match to their experimental data for $r < R/2$, with $H_b \approx 1.37$. Interestingly, Savas finds that (5.31) is a good match even when the flow is turbulent! This curious result is attributed to the fact that laminar and turbulent Bödewadt layers have similar velocity profiles. In any event, it seems that the characteristic spin-down time is indeed $\tau \sim 2D/1.4\sqrt{\nu\,\Omega}$, as noted in §5.2.

There remains one issue, however. It is interesting that the measured value of $H_b \approx 1.37$ is close to the value $H_\infty \approx 1.35$ for a Bödewadt layer on an infinite disc. This tentatively suggests that the bottom boundary layer has a structure close to that of an ideal Bödewadt layer on an infinite disc, at least in the inner regions, say $r < R/2$. However, this *cannot* be the case for larger radii, since (5.28) demands that the resistive torque exerted on the flow by the base of the cylinder is,

$$T_b = \frac{\pi}{2}\rho\Omega R^4 u_b = H_b \frac{\pi}{2}\rho\Omega R^4 \sqrt{\nu\,\Omega}, \tag{5.32}$$

whereas (5.10) tells us that an ideal Bödewadt layer exerts a weaker torque, of magnitude

$$T_b = 0.770\frac{\pi}{2}\rho\Omega R^4 \sqrt{\nu\,\Omega}. \tag{5.33}$$

There is almost a factor of two difference in these two estimates. It seems likely, therefore, that the outer regions of the bottom boundary layer, which dominate the torque, have a structure that differs somewhat from that of the Bödewadt layer on an infinite disc. The existence of such a transition region should not come as a surprise, as the inflow conditions at the edge of the bottom layer are controlled by the sidewall boundary layer, and these will not match the velocity profile of an ideal Bödewadt layer.

5.3.2 Spin-Up

Let us now consider the reverse problem, that of the *spin-up* of a fluid in a cylindrical container. Here the fluid and container are initially at rest and then the container is suddenly set into rotation, say to Ω_b. As before, the water depth and radius are D and R, we use cylindrical polar coordinates centred on the base of the cylinder, and we assume that the flow is laminar and axisymmetric. However, this time we have a Kármán-like boundary layer near $z = 0$. This entrains core fluid, spins it up, and drives a flow radially outward and up into the sidewall boundary layer, as shown in Figure 5.10.

Physically, spin-up of the fluid proceeds as follows. Boundary layers develop during the first few rotations, after which they become quasi-steady. However, the core remains more or less stationary during the initial growth of the boundary layers, and so we shall ignore the initial transient and take $t = 0$ to correspond to the point where the boundary layers become quasi-steady. Stagnant fluid is then drawn into the Kármán-like layer, where it is spun up and carried into the sidewall boundary layer. The role of the sidewall layer is then to return spinning fluid to the interior. As in spin-down, the secondary flow is weak, and so (2.86) requires that the core rotation is independent of z, at least to leading order in Ek. Hence, an annular shell of rotating fluid slowly intrudes from the sidewall boundary layer into the interior. At any instant, then, the core flow may be divided into an inner cylindrical region, which is yet to be spun up, and an outer annulus, where the fluid rotates (see Figure 5.10). Spin-up is

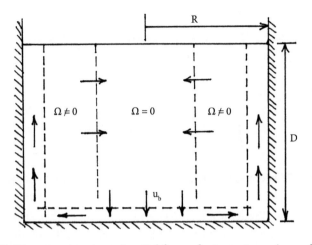

Figure 5.10 Flow structure in an inertial frame during spin-up in a cylindrical container.

then more or less complete when the entire contents of the cylinder have been flushed, once or twice, through the Kármán-like boundary layer.

From a mathematical point of view, the main difference between spin-up and spin-down is that the core angular velocity, $\Omega_c(r, t) = u_\theta/r$, is now a function of both r and t. This makes the analysis messier, and so we shall have to make a rather strong assumption in order to extract a closed-form solution from the governing equations.

Let us start with the inviscid core dynamics. We take $u_b(r, t) = |u_z|$ to be the downward speed of the fluid passing from the core into the Kármán-like layer. Since $D\Gamma/Dt = 0$ and $\Gamma = \Omega_c(r, t)r^2$ in the core, we have

$$\frac{\partial \Omega_c}{\partial t} + \frac{u_r}{r^2} \frac{\partial}{\partial r}\left(\Omega_c r^2\right) = 0. \tag{5.34}$$

Evidently, if we are to make progress, we need to say something about u_r, which is independent of z in the core.

Now, in §5.1.2 we saw that, when a fluid is in steady, rigid-body rotation, say at Ω_f, above an infinite plate which rotates at Ω_b, with $\Omega_b > \Omega_f$, a reasonable approximation to u_b is

$$u_b = \sqrt{\nu \Omega_b}\left(1 - \Omega_f/\Omega_b\right). \tag{5.35}$$

Moreover, u_r is linear in r for the case considered in §5.1.2, and so continuity applied to the boundary layer yields

$$u_b = -\int_0^\delta \frac{\partial u_z}{\partial z}\, dz = \int_0^\delta \frac{1}{r}\frac{\partial}{\partial r}(r u_r)\, dz = \frac{2}{r}\int_0^\delta u_r dz, \qquad (5.36)$$

from which

$$\int_0^\delta u_r dz = \tfrac{1}{2}\left(\Omega_b - \Omega_f\right)\sqrt{\nu/\Omega_b}\, r. \qquad (5.37)$$

Note that (5.37) is consistent with (5.19), which is exact whenever $(\Omega_b - \Omega_f) \ll \Omega_b$.

We now follow Wedemeyer (1964) and make the rather strong assumption that, despite the fact that the fluid is not in rigid-body rotation, the same relationship also holds in spin-up, in the sense that, at each radius,

$$\int_0^\delta u_r dz = \tfrac{1}{2}\left(\Omega_b - \Omega_c\right)\sqrt{\nu/\Omega_b}\, r. \qquad (5.38)$$

The logic behind this approximation is the following. The driving force for u_r in a Kármán-like boundary layer is the difference between the centrifugal force within the layer and the radial pressure gradient. However, the radial pressure gradient is just set by the centrifugal force outside the boundary layer, and so the driving force for u_r is $\rho\left(\Omega_b^2 - \Omega_c^2\right)r$. This is a *local* quantity that does *not* require $\Omega_c(r, t)$ to be rigid-body rotation. So perhaps (5.38) is a plausible assumption. In any event, if we accept (5.38), then continuity tells us that

$$(u_r)_c D = -\int_0^\delta u_r dz = -\frac{\sqrt{\nu\Omega_b}}{2}\left(1 - \Omega_c/\Omega_b\right)r, \qquad (5.39)$$

and substituting this into (5.34) yields our governing equation

$$\frac{\partial \Omega_c}{\partial t} = \frac{1}{r}\frac{\partial}{\partial r}\left(\Omega_c r^2\right)\frac{\sqrt{\nu\Omega_b}}{2D}\left(1 - \Omega_c/\Omega_b\right). \qquad (5.40)$$

This may be tidied up if we introduce a dimensionless time and core rotation rate, defined by

$$\hat{t} = \sqrt{\nu\Omega_b}\, t/2D, \qquad \hat{\Omega}_c(r, t) = \Omega_c(r, t)/\Omega_b. \qquad (5.41)$$

Equation (5.40) now becomes

$$\frac{\partial \hat{\Omega}_c}{\partial \hat{t}} = \frac{1}{r}\frac{\partial}{\partial r}\left(\hat{\Omega}_c r^2\right)\left(1 - \hat{\Omega}_c\right), \qquad (5.42)$$

which must be solved subject to $\Omega_c(r = R) = \Omega_b$ and $\Omega_c(t = 0) = 0$. The resulting solution is readily shown to be

$$\frac{\Omega_c}{\Omega_b} = \frac{\exp(2\hat{t}) - (R/r)^2}{\exp(2\hat{t}) - 1}, \quad \text{for } r > R\exp(-\hat{t}), \qquad (5.43)$$

and

$$\Omega_c = 0, \quad \text{for } r < R\exp(-\hat{t}), \qquad (5.44)$$

which represents an annulus of rotating fluid that spreads radially inward from the sidewall.

Because of the rather strong assumption inherent in (5.38), there remain questions as to the accuracy of (5.43). However, it does capture the qualitative aspects of spin-up with elegant simplicity, with a spin-up time scaling on $2D/\sqrt{\nu\Omega_b}$, and an outer annulus of spinning fluid, $r_{\text{inner}} < r < R$, whose inner boundary is convected inwards according to

$$r_{\text{inner}} = R\exp\left(-\sqrt{\nu\Omega_b}\,t/2D\right). \qquad (5.45)$$

Clearly, r_{inner} separates the rotating fluid, that has intruded into the core from the sidewall, from the non-rotating fluid, which is drained down into the Kármán-like layer below.

A more extensive discussion of spin-up may be found in Weidman (1976a), with theory compared against experiment in both Wedemeyer (1964) and Weidman (1976b).

5.4 Vertical Shear Layers and Spin-Down Revisited

So far, we have avoided saying anything about the sidewall boundary layers that arise during spin-up and spin-down. Perhaps the first point to note is that, when the normal to a surface is inclined to the rotation axis of the fluid, the Ekman layer thickness scales as $\hat{\delta} = \sqrt{\nu/\Omega_n}$, where Ω_n is the component of Ω normal to the surface (see §5.1.4). Clearly, this scaling is singular when the surface is parallel to $\mathbf{\Omega}$, and this is the first hint that sidewall boundary layers, called *Stewartson layers*, scale differently to Ekman layers.

To focus thoughts, let us consider the sidewall boundary layer that forms during spin-down at low Ek in a cylindrical container (Figure 5.9). Despite the fact that, at low Ek, such layers are notoriously unstable to Rayleigh's centrifugal instability, we shall take the flow to be laminar and axisymmetric. Moreover, to keep the algebra simple, we shall consider the linear problem in which the cylindrical container rotates at Ω_b and the fluid in the core at $\Omega_c(t) = \Omega_b + \hat{\omega}(t)$, where $\hat{\omega} \ll \Omega_b$. We adopt a *rotating* frame of reference that rotates with the cylindrical container and use cylindrical polar coordinates centred on the base of the cylinder. The water has a depth of D and the container a radius of R.

It turns out that there are two nested boundary layers embedded within the sidewall layer, and these perform quite different functions. One is associated with the primary flow, u_θ, while the other is associated with the secondary motion in the r–z plane. As we shall see, the outer layer has a thickness that scales as $(\delta_{//})_v = \text{Ek}^{1/4}D$, where $\text{Ek} = \nu/\Omega_b D^2$. Clearly, this is thicker than the Ekman layer at the base of the cylinder, which scales as $\hat{\delta} = \sqrt{\nu/\Omega_b}$. The function of this outer layer is the familiar one of allowing u_θ to drop down to zero at the sidewall in order to satisfy the no-slip condition.

The inner layer, by contrast, has a thickness that scales as $(\delta_{//})_m = \text{Ek}^{1/3}D$, which is also thicker than the Ekman layer. This inner layer is associated with the secondary flow and its function is to deflect the radial velocity downward, align the secondary flow with the sidewall, and ensure that u_z meets the no-slip boundary condition. Much of the mass flux associated with the secondary flow is then channelled down through the inner layer to the base of the cylinder, hence the subscript of m (for mass) on $(\delta_{//})_m$. The overall structure of the side layer, with its two nested boundary layers, is shown in Figure 5.11.

Let us now try to understand the origin of the two sidewall scaling laws,

$$(\delta_{//})_v = \text{Ek}^{1/4}D, \quad (\delta_{//})_m = \text{Ek}^{1/3}D. \tag{5.46}$$

As noted in §5.3.1, u_r in the core is independent of z and linear in r, and so continuity applied to the core flow requires

$$\frac{\partial u_z}{\partial z} = -\frac{u_b}{D} = -\frac{2u_r}{r}, \tag{5.47}$$

(see equation (5.22)), where u_b is the axial velocity exiting the Ekman layer. Moreover, according to (5.19), $u_b = \hat{\omega}(t)\hat{\delta}$. It follows that the radial velocity entering the side layer, u_R, is independent of z and equal to

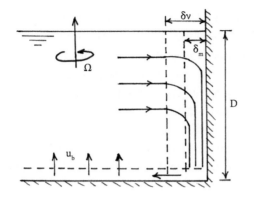

Figure 5.11 The sidewall layer that forms during spin-down has two parts to it.

$$u_R(t) = \frac{u_b R}{2D} = \frac{\hat{\omega}(t)\hat{\delta}R}{2D}, \qquad \hat{\delta} = \sqrt{\nu/\Omega_b}. \qquad (5.48)$$

We now turn to dynamics. In the *rotating frame of reference*, the low-Ro version of the azimuthal equation of motion is obtained from (2.46) by replacing the inertial forces with the Coriolis force. Similarly, the azimuthal component of the low-Ro vorticity equation is obtained from (2.85) by replacing the inertial terms by the curl of the Coriolis force. This yields

$$\frac{\partial \Gamma}{\partial t} + 2\Omega_b r u_r = \nu \nabla_*^2 \Gamma, \qquad r\frac{\partial \omega_\theta}{\partial t} = 2\Omega_b \frac{\partial \Gamma}{\partial z} + \nu \nabla_*^2 (r\omega_\theta), \qquad (5.49)$$

where $\Gamma = r u_\theta$. These are the equations of motion for the primary and secondary flows, respectively. Rewriting u_r and ω_θ in terms of the Stokes streamfunction, Ψ, using

$$u_r = -\frac{1}{r}\frac{\partial \Psi}{\partial z}, \qquad \omega_\theta = \frac{\partial u_r}{\partial z} - \frac{\partial u_z}{\partial r} = -\frac{1}{r}\nabla_*^2 \Psi,$$

and eliminating Γ (or Ψ) from (5.49), yields the governing equations for Ψ and Γ:

$$\left(\frac{\partial}{\partial t} - \nu\nabla_*^2\right)^2 \nabla_*^2 \Psi + (2\Omega_b)^2 \frac{\partial^2 \Psi}{\partial z^2} = 0, \qquad (5.50)$$

$$\left(\frac{\partial}{\partial t} - \nu\nabla_*^2\right)^2 \nabla_*^2 \Gamma + (2\Omega_b)^2 \frac{\partial^2 \Gamma}{\partial z^2} = 0. \qquad (5.51)$$

Note that, as the flow enters the outer layer, Ψ varies with height as $(\Delta\Psi)_{\text{outer}} = -Rzu_R$.

Now, we saw in §5.3.1 that, during spin-down, $\partial/\partial t \sim \mathrm{Ek}^{1/2}\Omega$. It follows from $(\delta_{//})_v = \mathrm{Ek}^{1/4}D$ that the viscous and time-dependant terms in (5.50) and (5.51) are of comparable magnitudes in the outer layer. Thus, all six terms in (5.49) are important in this layer. However, the scaling $(\delta_{//})_m = \mathrm{Ek}^{1/3}D$ tells us that the time derivatives in (5.50) and (5.51) are negligible in the inner layer, which is therefore *quasi-steady*.

We are now in a position to explain the origin of the scaling laws (5.46). The outer-layer scaling, which allows u_θ to adjust to the no-slip condition, is controlled by the azimuthal force balance in (5.49). When combined with (5.48), this requires

$$\nu\frac{\partial^2\Gamma}{\partial r^2} \sim 2\Omega_b R u_R \sim \Omega_b\frac{\hat{\omega}\hat{\delta}R^2}{D}, \tag{5.52}$$

from which,

$$\nu\frac{\hat{\omega}R^2}{(\delta_{//})_v^2} \sim \Omega_b\frac{\hat{\omega}R^2\hat{\delta}}{D}. \tag{5.53}$$

This then yields the outer-layer scaling law, $(\delta_{//})_v = \mathrm{Ek}^{1/4}D$. The inner-layer scaling, by contrast, is set by the quasi-steady version of (5.50),

$$\nu^2\frac{\partial^6\Psi}{\partial r^6} + (2\Omega_b)^2\frac{\partial^2\Psi}{\partial z^2} = 0. \tag{5.54}$$

Now axial gradients scale on D in the inner layer (but *not* in the outer layer, which is quasi-geostrophic—see below). So (5.54) yields

$$\frac{\nu^2}{(\delta_{//})_m^6} \sim \frac{\Omega_b^2}{D^2}, \tag{5.55}$$

from which we obtain the inner-layer scaling, $(\delta_{//})_m = \mathrm{Ek}^{1/3}D$.

We shall now show that the task of meeting the no-slip condition for u_θ falls almost exclusively to the outer layer, while much (but not all) of the mass flux associated with the secondary flow is channelled down through the thin inner layer. We start by noting that, if we combine (5.52) with the *inner-layer* scaling, $(\delta_{//})_m = \mathrm{Ek}^{1/3}D$, we find

$$\frac{\partial^2\Gamma}{\partial r^2} \sim \frac{\Omega_b}{\nu}\frac{\hat{\omega}\hat{\delta}R^2}{D} \sim \frac{(\Delta\Gamma)_{inner}}{\mathrm{Ek}^{2/3}D^2}, \tag{5.56}$$

where $(\Delta\Gamma)_{inner}$ is the change in Γ across the inner layer. This, in turn, yields

$$\frac{(\Delta\Gamma)_{inner}}{\hat\omega R^2} \sim \frac{(\Delta\Gamma)_{inner}}{(\Delta\Gamma)_{outer}} \sim \frac{\Omega_b D^2}{\nu}\frac{\hat\delta}{D}\mathrm{Ek}^{2/3} \sim \mathrm{Ek}^{1/6}. \qquad (5.57)$$

Since Ek << 1, we conclude that the drop in u_θ across the inner layer is negligible, with almost the entire adjustment to the no-slip condition occurring in the outer layer.

Moreover, in the *outer layer*, (5.50) and (5.51) yield the estimates

$$\frac{\partial^2}{\partial z^2}(\Psi,\Gamma) \sim \frac{\nu^2}{\Omega_b^2}\frac{\partial^6}{\partial r^6}(\Psi,\Gamma) \sim \frac{\nu^2}{\Omega_b^2}\frac{((\Delta\Psi)_{outer},\Gamma)}{(\delta_{//})_\nu^6} \sim \mathrm{Ek}^{1/2}\frac{((\Delta\Psi)_{outer},\Gamma)}{D^2}, \qquad (5.58)$$

where $(\Delta\Psi)_{outer} \sim RDu_R$ is now the change in Ψ *across* the outer layer. Evidently, unlike the inner layer, axial gradients are very weak in the outer layer, and in particular (5.58) gives us

$$\frac{\partial u_r}{\partial z} = -\frac{1}{r}\frac{\partial^2\Psi}{\partial z^2} \sim \mathrm{Ek}^{1/2}\frac{u_R}{D}, \qquad \frac{\partial^2 u_\theta}{\partial z^2} \sim \mathrm{Ek}^{1/2}\frac{u_\theta}{D^2}. \qquad (5.59)$$

The constraint $\partial u_r/\partial z \approx 0$ is a powerful one, requiring that u_z, and hence ω_θ, are both linear in z in the outer layer. In fact, it turns out that this constraint prevents the outer layer from deflecting all of the radial mass flux downward, as noted in Greenspan (1968, p. 100). Thus, the outer layer cannot satisfy the boundary conditions imposed on the secondary flow by the outer radial boundary. Evidently, that task falls to the inner layer.

In summary, we have two nested boundary layers adjacent to the sidewall. The outer layer scales as $(\delta_{//})_\nu = \mathrm{Ek}^{1/4}D$, and its function is to lower the azimuthal velocity to meet the no-slip condition at the wall. However, this layer cannot turn around all of the radial mass flux imposed on the sidewall layer by the core flow, and this necessitates a second layer. The inner layer scales as $(\delta_{//})_m = \mathrm{Ek}^{1/3}D$, and its role is to deflect the radial flow downward. The inner layer has negligible influence on u_θ, and so the task of meeting the no-slip condition for the azimuthal velocity falls almost exclusively to the outer layer.

These various characteristics are exhibited in other vertical shear layers. In particular, an $\mathrm{Ek}^{1/3}$ layer is often observed when mass is to be transferred vertically to or from an Ekman layer, while an $\mathrm{Ek}^{1/4}$ layer is required to smooth out any radial discontinuity in u_θ.

5.5 Boundary Layers Generated by Differentially Rotating Spheres

We now turn from cylinders to spheres and to a problem motivated by the fact that differential rotation may exist within certain planetary interiors. Suppose that fluid fills a spherical annulus, $R_i < |\mathbf{x}| < R_o$, formed by two, concentric, rotating spheres. The outer sphere rotates at the rate Ω (in an inertial frame) and the inner one at the slightly faster rate of $\Omega + \hat{\omega}_i$, with $\hat{\omega}_i \ll \Omega$. We assume the resulting motion is steady and axisymmetric and, as usual, we adopt cylindrical polar coordinates, (r, θ, z), with z aligned with Ω. It is convenient to let the coordinate system rotate with the outer sphere, so that the problem may be linearized on the assumption that Ro $\ll$ 1. The geometry is as shown in Figure 5.12.

We shall refer to the cylindrical surface $r = R_i$ as the *tangent cylinder*, since it is tangent to the inner sphere at the equator. It is clear that, inside the tangent cylinder, the fluid rotates at a rate intermediate between that of the two spheres, with the core rotating at $\hat{\omega}_c$, say. However, outside the tangent cylinder the fluid is stationary in the rotating frame of reference. Evidently, Ekman layers form on the two spherical boundaries, with inward radial motion within the outer boundary layer and outward radial flow within the inner Ekman layer. Moreover, following the discussion in §5.4, we might anticipate that the discontinuity in u_θ at the tangent cylinder will give rise to an $\mathrm{Ek}^{1/4}$ vertical

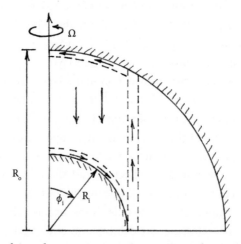

Figure 5.12 Flow driven by two, concentric, rotating spheres. In an inertial frame, the outer sphere rotates at the rate Ω and the inner one at the slightly faster rate of $\Omega + \hat{\omega}_i$.

shear layer, and if the inner Ekman layer possesses a finite mass flux at the tangent cylinder, we would expect an $Ek^{1/3}$ layer embedded within the $Ek^{1/4}$ shear layer, whose function is to exchange mass between the two Ekman layers. (Again, see Figure 5.12.)

If we are to determine the structure of the flow, we must first find $\hat{\omega}_c$. Now, the Taylor–Proudman theorem tells us that, since Ro << 1, both u_θ and u_z are independent of z within the core, and so $u_r = 0$ and $\hat{\omega}_c$ is a function of r only. We now choose $\hat{\omega}_c(r)$ to satisfy global continuity. That is to say, at each radius, r, the core rotation rate is chosen so that the opposing mass fluxes in the two Ekman layers are equal. In §5.1.4 we saw that Ekman layers on inclined surfaces behave like conventional Ekman layers provided that we use the component of rotation normal to the surface when evaluating the boundary-layer thickness. Thus, (5.19) tell us that

$$\dot{m}_o/2\pi r\rho = \int_{\text{outer}} u_t dn = \tfrac{1}{2}\hat{\omega}_c(r)\hat{\delta}_o r, \quad \hat{\delta}_o = \sqrt{\nu/(\Omega \cos \phi_o)}, \qquad (5.60)$$

and

$$\dot{m}_i/2\pi r\rho = \int_{\text{inner}} u_t dn = \tfrac{1}{2}(\hat{\omega}_i - \hat{\omega}_c)\hat{\delta}_i r, \quad \hat{\delta}_i = \sqrt{\nu/(\Omega \cos \phi_i)}, \qquad (5.61)$$

where u_t is the component of velocity tangential to the surface, n is a coordinate normal to the boundary, and ϕ is the polar angle measured from the rotation axis. Evidently,

$$\cos \phi_o = \sqrt{1 - (r/R_o)^2}, \quad \cos \phi_i = \sqrt{1 - (r/R_i)^2}. \qquad (5.62)$$

Equating the mass flow rates now yields,

$$\frac{\hat{\omega}_c(r)}{\sqrt{\cos \phi_o}} = \frac{\hat{\omega}_i - \hat{\omega}_c(r)}{\sqrt{\cos \phi_i}}, \quad r < R_i,$$

which requires $\hat{\omega}_c(r \to R_i) = \hat{\omega}_i$ and rearranges to give

$$\frac{\hat{\omega}_c(r)}{\hat{\omega}_i} = \frac{\sqrt{\cos \phi_o}}{\sqrt{\cos \phi_o} + \sqrt{\cos \phi_i}} = \frac{\left(1 - (r/R_o)^2\right)^{1/4}}{\left(1 - (r/R_o)^2\right)^{1/4} + \left(1 - (r/R_i)^2\right)^{1/4}}. \qquad (5.63)$$

It follows that

$$
\frac{\dot{m}_o}{2\pi r \rho} = \frac{\dot{m}_i}{2\pi r \rho} = \frac{\frac{1}{2}\hat{\omega}_i\, r\sqrt{\nu/\Omega}}{\left(1 - (r/R_o)^2\right)^{1/4} + \left(1 - (r/R_i)^2\right)^{1/4}}, \tag{5.64}
$$

and in particular, the mass flux approaching the tangent cylinder within the inner Ekman layer is

$$
\frac{\dot{m}_i(r \to R_i)}{2\pi R_i \rho} = \frac{\hat{\omega}_i R_i \sqrt{\nu/\Omega}}{2\left(1 - (R_i/R_o)^2\right)^{1/4}}. \tag{5.65}
$$

This flow is diverted upward through a $Ek^{1/3}$ Stewartson layer.

The axial velocity in the core can be found from (5.64) using continuity applied to an annulus:

$$
\rho u_z (2\pi r dr) = -\frac{d\dot{m}_i}{dr} dr.
$$

This can be rewritten in terms of the Stokes streamfunction, Ψ, as

$$
u_z = -\frac{1}{r}\frac{d}{dr}\left(\frac{\dot{m}_i}{2\pi\rho}\right) = \frac{1}{r}\frac{d\Psi}{dr}, \tag{5.66}
$$

from which we deduce

$$
\Psi = -\frac{\dot{m}_i}{2\pi\rho} = -\frac{\frac{1}{2}\hat{\omega}_i r^2 \sqrt{\nu/\Omega}}{\left(1 - (r/R_o)^2\right)^{1/4} + \left(1 - (r/R_i)^2\right)^{1/4}}, \tag{5.67}
$$

and

$$
\frac{4u_z}{\hat{\omega}_i\sqrt{\nu/\Omega}} = -\frac{\left[4 - 3\frac{r^2}{R_o^2}\right]\left[1 - \frac{r^2}{R_i^2}\right]^{3/4} + \left[4 - 3\frac{r^2}{R_i^2}\right]\left[1 - \frac{r^2}{R_o^2}\right]^{3/4}}{\left(\left[1 - \frac{r^2}{R_o^2}\right]^{1/4} + \left[1 - \frac{r^2}{R_i^2}\right]^{1/4}\right)^2 \left[1 - \frac{r^2}{R_o^2}\right]^{3/4}\left[1 - \frac{r^2}{R_i^2}\right]^{3/4}}, \tag{5.68}
$$

for $r < R_i$. Note that the axial velocity in the inviscid core is everywhere negative, as it should be. Note also that u_z is predicted to diverge as we approach the tangent cylinder, and this modifies somewhat the vertical shear layers, as discussed below.

The structure of the vertical shear layer near the tangent cylinder is notoriously complicated, but a summary of the various layers can be found in Marcotte et al. (2016). As might be expected from the discussion in §5.4, there

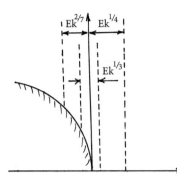

Figure 5.13 Schematic of the vertical Stewartson layers near the tangent cylinder.

is a relatively thick $Ek^{1/4}$ shear layer just outside the tangent cylinder, whose function is to smooth out the radial discontinuity in u_θ, of $\delta u_\theta = \hat{\omega}_i R_i$. However, it turns out that the contribution to the $Ek^{1/4}$ layer from immediately inside the tangent cylinder is modified and thinned by the intense Ekman pumping near the equator, and it scales as $Ek^{2/7}$ (see Figure 5.13). Embedded within these two layers there is the expected $Ek^{1/3}$ shear layer, which redirects the mass flux from the inner to the outer Ekman layer. Finally, there are transitional layers near the equator whose function is to merge the vertical $Ek^{1/3}$ shear layer with the inner sphere.

Of course, all of this assumes that Ro $<< 1$, so that inertia may be ignored. When allowance is made for the fact that Ro is finite, there is the possibility that the outward mass flux within the inner Ekman layer gives rise to an equatorial jet outside the tangent cylinder. Such jets are indeed seen in experiments with rotating spheres (Greenspan, 1968).

* * *

This concludes our discussion of Ekman layers and Ekman pumping. Those readers who are hungry for more details will find a comprehensive discussion of this rather classical subject in Greenspan (1968), while engineering applications are discussed in Berger & Talbot (1983) and Horlock & Lakshminarayana (1973). The role of Ekman pumping in the meandering of rivers is reviewed in Seminara (2006).

References

Berger, S.A., & Talbot, L., 1983, Flow in curved pipes, *Ann. Rev. Fluid Mech.*, **15**, 461.

Bödewadt, U.T., 1940, Die Drehstromung uber festem Grunde, *Z. angew Math Mech.*, **20**, 241–53.

Einstein, A., 1926, The cause of the formation of meanders in the courses of rivers and of the so-called Baer's law, *Die Naturwissenschaften*, **14,** 14–18.

Greenspan, H.P., 1968, *The Theory of Rotating Fluids*, Cambridge University Press.

Horlock, J.H., & Lakshminarayana, B., 1973, Secondary flows: theory, experiment, and application in turbomachinery aerodynamics, *Ann. Rev. Fluid Mech.*, **5**, 247–80.

Kármán, T., 1921, Uber laminare und turbulente Reibung, *Z. angew Math Mech.*, **1**, 233.

Marcotte, F., Dormy, E., & Soward, A., 2016, On the equatorial Ekman layer, *J. Fluid Mech.*, **803**, 395–435.

Savas, Ö., 1992, Spin-down to rest in a cylindrical cavity, *J. Fluid Mech.*, **234**, 529–52.

Seminara, G., 2006, Meanders, *J. Fluid Mech.*, **554**, 271–97.

Thomson, J., 1857, On the grand currents of atmospheric circulation. In: *Collected Papers in Physics and Engineering*, J. Larmor & J Thomson, eds. Cambridge University Press (1912), 144–7.

Thomson, J., 1876, On the origin of windings of rivers in alluvial plains, with remarks on the flow of water round bends in pipes, *Proc. R. Soc. London, Ser. A*, **25**, 5–8.

Thomson, J., 1877, Experimental demonstration in respect to the origin of windings of rivers in alluvial plains, and to the mode of flow of water round bends of pipes, *Proc. R. Soc. London, Ser. A*, **26**, 356–7.

Wedemeyer, E.H., 1964, The unsteady flow within a spinning cylinder, *J. Fluid Mech.*, **20**, part 3, 383–99.

Weidman, P.D., 1976a, On the spin-up and spin-down of a rotating fluid. Part 1: extending the Wedemeyer model, *J. Fluid Mech.*, **77**, part 4, 685–708.

Weidman, P.D., 1976b, On the spin-up and spin-down of a rotating fluid. Part 2: measurements and stability, *J. Fluid Mech.*, **77**, part 4, 709–35.

Chapter 6

Inertial Waves I

Inviscid Progressive Waves

We have already introduced the notion of inertial waves, albeit in a strictly qualitative fashion. For example, in §2.12.1, we saw that Rayleigh's analogy between stratified and swirling flows makes it clear that internal waves are inevitable in a rapidly rotating fluid undergoing axisymmetric motion. Moreover, in §4.3.2 we noted, but did not prove, that low-frequency inertial waves underpin the slow evolution of certain quasi-geostrophic flows, such as drifting Taylor columns. The purpose of this chapter is to establish the formal properties of inertial waves, with an emphasis on their dispersion characteristics and helical structure. We shall restrict ourselves to progressive waves, leaving standing waves for Chapter 7.

6.1 The Physical Origin of Inertial Waves

Before discussing the mathematical properties of inertial waves, perhaps it is worth asking how, in a physical sense, a rapidly rotating fluid can sustain internal wave motion. We offer two alternative answers to that question, one which adopts an inertial frame of reference and another which uses a rotating frame. In both cases, we restrict ourselves to inviscid, incompressible, axisymmetric motion. We start in an inertial frame of reference.

6.1.1 Rayleigh's Analogy to Buoyancy (Revisited)

Let us return to §2.12.1, and to Rayleigh's analogy between stratified and rotating flow. Recall that the governing equations for an inviscid, axisymmetric, constant density flow are

$$\frac{D\Gamma}{Dt} = 0, \ \nabla \cdot \mathbf{u}_p = 0 \tag{6.1}$$

$$\frac{D\mathbf{u}_p}{Dt} = -\nabla \left(p/\rho \right) + \frac{\Gamma^2}{r^3} \hat{\mathbf{e}}_r, \tag{6.2}$$

The Dynamics of Rotating Fluids. P. A. Davidson, Oxford University Press. © Peter A Davidson (2024).
DOI: 10.1093/9780191994272.003.0006

where $\Gamma = ru_\theta$, $\mathbf{u}_p(r, z) = (u_r, 0, u_z)$ is the poloidal component of motion, and we use cylindrical polar coordinates, (r, θ, z). Rayleigh (1916) noticed that there is an exact analogy between these equations and the axisymmetric motion of an incompressible, Boussinesq fluid driven by density gradients in the radial gravitational field $\mathbf{g} = (g^*/r^3)\,\hat{\mathbf{e}}_r$. (Here g^* is a constant.) That is to say, if $\bar{\rho}$ is the mean density, and $\rho = \bar{\rho} + \rho'$, where $\rho' << \bar{\rho}$, then the axisymmetric motion of our Boussinesq fluid is governed by

$$\frac{D\rho'}{Dt} = 0, \quad \nabla \cdot \mathbf{u}_p = 0, \tag{6.3}$$

$$\frac{D\mathbf{u}_p}{Dt} = -\nabla\left(p/\bar{\rho}\right) + \frac{\rho'}{\bar{\rho}} \cdot \frac{g^*\hat{\mathbf{e}}_r}{r^3}, \tag{6.4}$$

Evidently, there is an exact analogy in which $\Gamma^2 \leftrightarrow g^*\rho'/\bar{\rho}$, or equivalently,

$$\frac{u_\theta^2}{r}\hat{\mathbf{e}}_r \leftrightarrow \frac{\rho'}{\bar{\rho}}\mathbf{g}. \tag{6.5}$$

Now, our Boussinesq fluid admits static equilibria of the form $\bar{\rho} + \rho' = \rho_0(r)$, and those equilibria are stable if and only if $\rho_0(r)$ increases monotonically with r. It follows from Rayleigh's analogy that the swirling flow $\mathbf{u}_0 = u_0(r)\hat{\mathbf{e}}_\theta$ is stable if and only if Γ^2 increases with r. Of course, this is Rayleigh's centrifugal stability criterion. Moreover, the stable equilibria $\rho_0(r)$ support internal gravity waves in which

$$\rho = \bar{\rho} + \rho' = \rho_0(r) + \rho''(\mathbf{x}, t), \tag{6.6}$$

where $\rho''(\mathbf{x}, t)$ represents the wave motion, assumed to be infinitesimally small. Rayleigh's analogy now tells us that the rotating flow $\mathbf{u}_0 = u_0(r)\hat{\mathbf{e}}_\theta$ can support internal wave motion of the form $u_\theta = u_0(r) + u_\theta'(\mathbf{x}, t)$, with $u_\theta' << u_0$, provided that $\mathbf{u}_0$ is stable. Of course, these waves are inertial waves.

It is tempting to infer from (6.5) that the analogue of the gravitational restoring force is the centrifugal force. However, when viewed in a rotating frame of reference, that is not the case. Suppose that the background flow is rigid-body rotation, $u_0(r) = \Omega r$, so that $u_\theta = \Omega r + u_\theta'(\mathbf{x}, t)$. Then

$$\frac{u_\theta^2}{r}\hat{\mathbf{e}}_r = \Omega^2 r\hat{\mathbf{e}}_r + 2\Omega u_\theta'\hat{\mathbf{e}}_r, \tag{6.7}$$

where we have neglected the quadratic term in u'_θ. Evidently, $\Omega^2 r\hat{\mathbf{e}}_r$ is static and balanced by a radial pressure gradient, and so the dynamically active part of the restoring force in (6.5) is the second term on the right of (6.7), i.e. $2\Omega u'_\theta \hat{\mathbf{e}}_r$.

Now suppose we move into a rotating frame of reference which rotates with the background motion. Then the centrifugal and Coriolis forces are, according to §4.1,

$$-\boldsymbol{\Omega} \times (\boldsymbol{\Omega} \times \mathbf{x}) = \Omega^2 r\hat{\mathbf{e}}_r, \quad 2\mathbf{u} \times \boldsymbol{\Omega} = 2\Omega u'_\theta \hat{\mathbf{e}}_r - 2\Omega u'_r \hat{\mathbf{e}}_\theta, \qquad (6.8)$$

respectively. We conclude that the restoring force in (6.5) is, in fact, part of the Coriolis force. The remaining part, $-2\Omega u'_r \hat{\mathbf{e}}_\theta$, has its origins in $D\Gamma/Dt = 0$, whose linearized version is

$$\frac{\partial u'_\theta}{\partial t} = -2\Omega u'_r. \qquad (6.9)$$

6.1.2 The Coriolis Force as a Restoring Force

Rayleigh's analogy makes internal wave motion in a rotating fluid seem quite natural. However, it still does not explain how, in a rotating frame of reference, the Coriolis force can act as a restoring force. We shall address that point in §6.3.1, but in the meantime we offer here a simplified explanation, restricted to axisymmetric waves.

Consider small perturbations to a state of rigid-body rotation, viewed in a frame of reference that rotates with the background motion. For axisymmetric flow, the low-Ro equations of motion are (6.9) plus

$$\frac{\partial u'_r}{\partial t} = 2\Omega u'_\theta - \frac{1}{\rho}\frac{\partial p}{\partial r}, \qquad (6.10)$$

where p is the reduced pressure. Now consider a circular hoop of fluid, C, which sits in the $r-\theta$ plane and is initially stationary in the rotating frame. Suppose the loop is perturbed, causing it to spread radially outward, as shown in the top ring in Figure 6.1(a). Then the Coriolis force in (6.9), $-2u'_r\Omega\hat{\mathbf{e}}_\theta$, accelerates the material loop to produce $u'_\theta < 0$, and hence $\oint_C \mathbf{u}' \cdot d\mathbf{r} = \int \omega'_z dA < 0$. The resulting negative value of ω'_z is readily understood from Kelvin's theorem, provided we temporarily move back to an inertial frame of reference. In an inertial frame, the axial vorticity is $2\Omega + \omega'_z$ and Kelvin's theorem demands that the flux of vorticity through the material loop C is conserved. Thus, if $A(t)$

is the area enclosed by C, we have $2\Omega A(t) + \int \omega'_z dA = $ const. Evidently, as the loop expands, $\int \omega'_z dA$ must fall in order to conserve the vorticity flux in an inertial frame.

Returning to the rotating coordinate system, this negative value of u_θ gives rise to a second Coriolis force, $2u'_\theta \Omega \hat{e}_r$, which acts to reduce u'_r in accordance with (6.10), albeit moderated by the radial pressure gradient. This is shown in the upper hoop in Figure 6.1(b). So the radial expansion eventually comes to a halt and the ring starts to contract back to its initial radius. However, inertia ensures an overshoot and the ring passes through its initial equilibrium position and continues to contract. The sequence of events is then reversed, with u_r now negative, and we find that the Coriolis force once again acts to oppose the radial motion, eventually halting the contraction of C and forcing the loop back towards its equilibrium position. Evidently, the hoop oscillates back and forth, and this is the basic mechanism underpinning axisymmetric inertial waves.

Some hint as to the dynamics of this process may be obtained by combining (6.9) with (6.10), which yields

$$\frac{\partial^2 u'_\theta}{\partial t^2} + (2\Omega)^2 u'_\theta = \frac{2\Omega}{\rho} \frac{\partial p}{\partial r}. \tag{6.11}$$

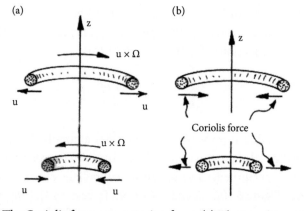

Figure 6.1 The Coriolis force as a restoring force. (a) The top ring moves radially outward (inward, for the lower ring). The Coriolis force induces a negative u_θ in the top ring (positive in the lower ring). (b) The negative (or positive) u_θ in the ring induces an inward (or outward) Coriolis force which halts the expansion (or contraction).

The left-hand side of (6.11) suggests oscillations with an angular frequency of 2Ω. However, it turns out that the radial pressure gradient in (6.11) acts to diminish the effects of the Coriolis term on the left, and it does so quite effectively when radial gradients in pressure are relatively strong, but less so when radial gradients are weak. As we shall discover, the end result is that inertial waves have an angular frequency limited to the range $0 < \varpi < 2\Omega$.

6.2 The Fundamental Properties of Inertial Waves

We now establish the formal properties of inertial waves, with particular emphasis on their dispersion characteristics and helical structure. We adopt a rotating frame, in which Ro << 1.

6.2.1 The Group Velocity and Spatial Structure of Inertial Waves

Consider an inviscid, rapidly rotating fluid. When Ro << 1 we have, in a rotating frame,

$$\frac{\partial \mathbf{u}}{\partial t} = 2\mathbf{u} \times \mathbf{\Omega} - \nabla\left(p/\rho\right), \quad \mathbf{\Omega} = \Omega \hat{\mathbf{e}}_z, \tag{6.12}$$

from which the vorticity equation is

$$\frac{\partial \boldsymbol{\omega}}{\partial t} = 2(\mathbf{\Omega} \cdot \nabla)\mathbf{u}. \tag{6.13}$$

Applying the operator $\nabla \times (\partial/\partial t)$ to (6.13) now yields the wave-like equation

$$\frac{\partial^2}{\partial t^2}\nabla^2 \mathbf{u} + (2\mathbf{\Omega} \cdot \nabla)^2 \mathbf{u} = 0, \tag{6.14}$$

which might be compared with (3.32) for internal gravity waves. Equation (6.14) supports plane waves of the form

$$\mathbf{u}(\mathbf{x},\, t) = \hat{\mathbf{u}} \exp\left[j\left(\mathbf{k} \cdot \mathbf{x} - \varpi t\right)\right], \tag{6.15}$$

where $\mathbf{k}$ is the wavevector, ϖ the angular frequency, and the real part is understood. Note that we adopt the convention $\varpi > 0$, as discussed in Chapter 3. Note also that $\nabla \cdot \mathbf{u} = 0$ demands $\mathbf{k} \cdot \hat{\mathbf{u}} = 0$. Substituting (6.15) into (6.14) yields the dispersion relationship

$$\varpi = \pm 2 \, \frac{\mathbf{\Omega} \cdot \mathbf{k}}{k} = 2\Omega \, |\sin\theta| \, , \qquad 0 < \varpi < 2\Omega, \tag{6.16}$$

where θ is the angle $\mathbf{k}$ makes to the horizontal, $-\pi/2 \le \theta \le \pi/2$. The velocity of the wave crests, *i.e.* the *phase velocity*, is in the direction of $\mathbf{k}$ and has magnitude ϖ/k. It follows that

$$\mathbf{c}_p = \pm 2 \frac{(\mathbf{\Omega} \cdot \mathbf{k})\,\mathbf{k}}{k^3}, \qquad |\mathbf{c}_p| = \frac{2\Omega \, |\sin\theta|}{k}. \tag{6.17}$$

On the other hand, the *group velocity*, which is the velocity at which energy disperses in the form of wave packets, is given by $\mathbf{c}_g = \partial\varpi/\partial k_i$. Given (6.16), this is readily shown to be

$$\mathbf{c}_g = \pm 2 \frac{\mathbf{k} \times (\mathbf{\Omega} \times \mathbf{k})}{k^3} = \pm 2 \frac{k^2 \mathbf{\Omega} - (\mathbf{k} \cdot \mathbf{\Omega})\mathbf{k}}{k^3}, \tag{6.18}$$

with $|\mathbf{c}_g| = (2\Omega/k) \cos\theta$. This might be compared with (3.36) for internal gravity waves.

Perhaps a few comments are in order. First, we note that $\mathbf{c}_g$ and $\mathbf{c}_p$ are related by

$$\mathbf{c}_p^2 + \mathbf{c}_g^2 = 4\Omega^2/k^2, \qquad \mathbf{c}_g + \mathbf{c}_p = \pm 2\Omega/k, \tag{6.19}$$

so that $\mathbf{c}_p$ and $\mathbf{c}_g$ have equal and opposite horizontal components. Second, inertial waves, like internal gravity waves, have their group velocity perpendicular to the phase velocity. Hence, the direction of propagation of a wave packet is perpendicular to the velocity of the wave crests which sit within that wave packet! In short, wave energy travels *along* the wave crests. Third, (6.18) yields

$$\mathbf{c_g} \cdot \mathbf{\Omega} = \pm 2 k^{-3} \big[k^2 \Omega^2 - (\mathbf{k} \cdot \mathbf{\Omega})^2 \big], \tag{6.20}$$

so the positive (negative) sign in (6.16) corresponds to wave energy travelling upwards (downwards). Finally, (6.18) tells us that wave packets of intermediate frequency travel obliquely and at an angle fixed by the dispersion relationship. For example, if $\varpi = \Omega$ then (6.16) fixes the orientation of $\mathbf{k}$ through $k_z/k = \pm \varpi/2\Omega = \pm 1/2$, and then $\mathbf{c}_g$ is chosen to be normal to $\mathbf{k}$. This is illustrated in Figure 6.2, where the waves generated by an oscillating disc spread to fill two

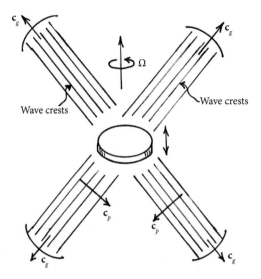

Figure 6.2 An oscillating disc radiates inertial waves at $\varpi \approx \Omega$. The radiation pattern consists of two conical annuli, one above the disc and one below.

conical annuli, one above the disc and one below. This might be compared with Figure 3.10, which shows the dispersion of internal gravity waves.

It is worth considering briefly the limiting cases of low and high frequency. Low-frequency waves have $\mathbf{k} \cdot \mathbf{\Omega} \approx 0$ and a group velocity of $\mathbf{c}_g = \pm 2\Omega/k$, so $\mathbf{k}$ is horizontal and the wave packets disperse up and down the rotation axis. These are the fastest wave packets. Conversely, high-frequency waves are those for which $\varpi \to 2\Omega$. These have $\mathbf{k}$ vertical and a group velocity which is horizontal and vanishes in the limit of $\varpi \to 2\Omega$.

It is important to note that inertial waves have a helicity, $h = \mathbf{u} \cdot \boldsymbol{\omega}$, which is maximal, *i.e.* $|h| = |\mathbf{u}|\,|\boldsymbol{\omega}|$. This follows from (6.13), which requires $2(\mathbf{\Omega} \cdot \mathbf{k})\hat{\mathbf{u}} = -\varpi \hat{\boldsymbol{\omega}}$, where $\hat{\boldsymbol{\omega}}$ is the amplitude of the vorticity, $\hat{\boldsymbol{\omega}} = j\mathbf{k} \times \hat{\mathbf{u}}$. When combined with (6.16), this yields

$$\hat{\boldsymbol{\omega}} = j\mathbf{k} \times \hat{\mathbf{u}} = \mp k\hat{\mathbf{u}}. \qquad (6.21)$$

It follows that the velocity and vorticity fields are parallel and in phase. Moreover, the positive sign in (6.20) corresponds to negative helicity, and the negative sign to positive helicity, so a wave packet with negative helicity travels upward ($\mathbf{c}_g \cdot \mathbf{\Omega} > 0$) while a wave packet with positive helicity travels downward ($\mathbf{c}_g \cdot \mathbf{\Omega} < 0$), as indicated in Figure 6.3.

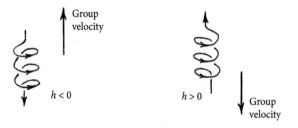

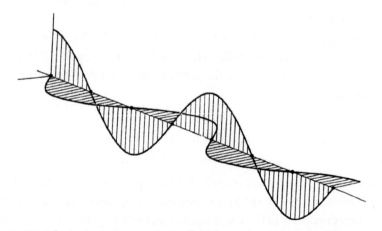

Figure 6.3 A wave packet with negative helicity travels upward, while one with positive helicity travels downward. The helical lines are the streamlines corresponding to (6.28).

Let us now determine the structure of a monochromatic inertial wave. To that end, we adopt coordinates (X, Y, Z) with Z parallel to **k**. Then $\mathbf{k} \cdot \hat{\mathbf{u}} = 0$ requires that $\hat{\mathbf{u}} = (\hat{u}_X, \hat{u}_Y, 0)$, while (6.21) demands $\hat{u}_Y = \mp j\hat{u}_X$. It follows that $\hat{\mathbf{u}} = \hat{u}_X(1, \mp j, 0)$, and so

$$\mathbf{u} = \hat{u}_X \left(\cos(kZ - \varpi t), \pm \sin(kZ - \varpi t), 0 \right), \tag{6.22}$$

where we have chosen $\hat{u}_X$ to be real. This is a circularly polarized wave whose velocity rotates about the z-axis as the wave propagates. Moreover, the sense of rotation is determined by the helicity. For example, if the wave has negative helicity, so that the upper sign applies in (6.22), then the velocity distribution at $t = 0$ is as shown in Figure 6.4.

Figure 6.4 The instantaneous velocity field for an inertial wave with negative helicity.

6.2.2 The Formation of Taylor Columns by Low-Frequency Waves

We now return to the topic of Taylor column formation, first introduced in §4.3.2. Suppose that, at $t = 0$, a disc of radius R is set into slow, steady motion. In particular, suppose it moves along the z-axis with a speed V. Provided that $V << \Omega R$, low-frequency waves will be excited at the surface of the disc, and these will propagate up and down the rotation axis with a group velocity of $\mathbf{c}_g = \pm 2\Omega/k$. Since the dominant wavenumber for a disc is $k \approx \pi/R$, the group velocity has a magnitude of $c_g \approx (2/\pi)\Omega R$. It follows that, after a time t, there will be a cylindrical region of space above and below the disc, of overall length $\ell \approx (4/\pi)\Omega R t$, which is filled with inertial waves (see Figure 6.5).

Now the fluid inside this cylindrical region knows that the disc is moving because the waves carry that information. Outside the cylindrical region, however, there are no waves, and so the fluid there does not know the disc has been set in motion. Thus, the fluid outside the cylinder region is quiescent in the rotating frame of reference. If we now ask what the fluid within the cylindrical region is doing, we get a surprise: it turns out that all of the fluid within the cylinder, both above and below the disc, moves upward with the speed V. (See, for example, Figure 4.2 in Greenspan, 1968).

The reason for this unusual behaviour is the following. The velocity $V\hat{\mathbf{e}}_z$ is imposed on fluid at the top and bottom faces of the disc. Moreover, the waves within the growing cylindrical region are of low frequency, so (6.13) reduces to $(\mathbf{\Omega} \cdot \nabla)\mathbf{u} \approx 0$, meaning this cylindrical region of space is quasi-geostrophic, with axial variations in $\mathbf{u}$ suppressed. (This is *not* true, however, close to the wave fronts that mark the top and bottom of the growing cylinder.) It follows that the upward velocity, $V\hat{\mathbf{e}}_z$, imposed on the fluid at the surfaces of the disc, is maintained throughout the quasi-geostrophic region.

Now suppose that we conduct a slightly different experiment and move a sphere *slowly to the right*, rather than upwards, at a speed V. Then we obtain essentially the same result, in the sense that, at time t, waves fill a cylindrical region of space aligned with $\mathbf{\Omega}$ and of axial length $\ell \sim \Omega R t$. Moreover, within that cylindrical region the fluid mimics the behaviour of the sphere, drifting to the right at the speed V. Of course, we are generating a Taylor column, as shown in Figure 4.1. We have returned to §4.3.2, and to our explanation of Taylor column formation in terms of the propagation of low-frequency inertial waves.

Note that the columnar structure in Figure 6.5 is a consequence of the fact that we chose to move the disc *slowly*, and so only low-frequency waves are generated. If we were to oscillate the disc at some intermediate frequency, say

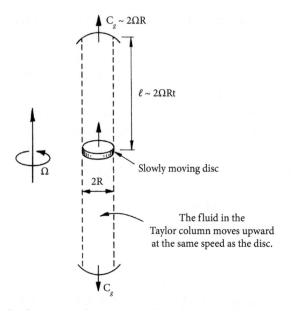

Figure 6.5 A slowly moving disc generates low-frequency inertial waves which fill a cylindrical region of space above and below the disc, of overall length $\ell \approx (4/\pi)\Omega Rt$.

$\tilde{\omega} \approx \Omega$, then waves would propagate at an oblique angle, filling two conical annuli, as shown in Figure 6.2.

6.2.3 The Spontaneous Focussing of Inertial Waves to Form Columnar Wave Packets

So far, we have focussed on waves generated by a moving boundary. In such boundary-value problems one can control the direction of wave propagation through the choice of frequency. We now turn to cases where the wave source *does not impose a timescale.*

Consider the case where the waves are excited by some initial perturbation in the velocity field; for example, by a localized vortex in the rotating frame of reference. One normally solves such an initial-value problem by taking a three-dimensional Fourier transform of the initial velocity field. This decomposes $\mathbf{u}_0(\mathbf{x})$ into a sum of Fourier modes of amplitude $\hat{\mathbf{u}}_0(\mathbf{k})$, with each mode looking like a plane wave whose orientation is fixed by its wavevector, $\mathbf{k}$. The transform also attributes a certain amount of kinetic energy to each mode. These modes then propagate as plane waves, carrying their energy with them in accordance

(6.18). The disturbance field at a later time is found by summing the modes, which amounts to performing an inverse Fourier transform on $\hat{\mathbf{u}}(\mathbf{k}, t)$.

Now we can vary $\hat{\mathbf{u}}_0(\mathbf{k})$, and hence the way that energy is distributed across the various Fourier modes, through the choice of $\mathbf{u}_0(\mathbf{x})$. So we might have anticipated that we can control the direction in which wave energy disperses from a localized source through our choice of the modes excited at $t = 0$. Surprisingly, however, it turns out that this is *not* the case. Rather, for almost *any* localized disturbance, the dispersion pattern is such that the wave energy density is largest on the rotation axis. This ability to spontaneously focus the wave energy flux onto the Ω-axis, irrespective of the initial condition, explains why inertial-wave dispersion patterns are often dominated by columnar vortices aligned with Ω.

There are two rather different ways of understanding why the radiation of inertial waves from an isolated source tends to be dominated by dispersion along the rotation axis. The simplest argument is geometrical in nature, and was given first in Davidson (2013), whereas the second is dynamic, and rests on angular momentum conservation. We start with the geometrical argument. Consider a wave packet which has dominant wavevector $\mathbf{k}$, leaves the wave source at $t = 0$, and propagates upward. Then, at the time t, the packet is centred on the point

$$\mathbf{x} = \mathbf{c}_g t = \frac{2\Omega t}{k^3} \left[k^2 \hat{\mathbf{e}}_z - k_z \mathbf{k} \right]. \tag{6.23}$$

Note that $\mathbf{x}$ is normal to $\mathbf{k}$ and coplanar with $\mathbf{k}$ and Ω. Now, for any $\mathbf{x}$ which lies off the axis, such as point B in Figure 6.6, there is only one orientation of $\mathbf{k}$ that takes wave energy from the source to that point (the inclined $\mathbf{k}$ in the figure). However, if $\mathbf{x}$ is located on the rotation axis, such as point A, then any $\mathbf{k}$ which lies in the horizontal plane transports energy to that location. It follows that all of the energy which lies within a thin horizontal disc in $\mathbf{k}$-space is folded up into a narrow cylinder in real space. It is this process of channelling energy from a two-dimensional object in $\mathbf{k}$-space into a one-dimensional object in real space that is responsible for the concentration of energy on the rotation axis.

This explanation for the focussing of wave energy onto the rotation axis has the advantage of simplicity. Another advantage is that it is independent of how the waves are excited, and requires only that the excitation is local and has no imposed timescale. However, unlike the explanation given below, it gives no quantitative information about the relative wave energy densities on and off the rotation axis.

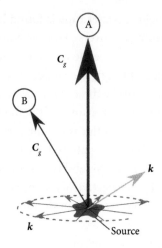

Figure 6.6 The group velocity, $\mathbf{c}_g$, and associated $\mathbf{k}$, for on-axis and off-axis wave packets.

An alternative explanation for the self-focussing of inertial waves, which is based on angular momentum conservation, is given in Davidson et al. (2006). This yields more information than the simple geometrical argument, but is limited to certain types of wave sources. Consider an initial condition in the rotating frame that consists of a vortex of scale δ which is located near the origin and is circumscribed by an imaginary cylinder of diameter δ, as shown in Figure 6.7(a). The cylinder, V_c, is orientated parallel to the rotation axis and is infinitely long. If we treat the fluid as inviscid, the angular momentum density of the fluid is governed by the cross product of $\mathbf{x}$ with (6.12),

$$\frac{\partial}{\partial t}(\mathbf{x} \times \mathbf{u}) = 2\mathbf{x} \times (\mathbf{u} \times \mathbf{\Omega}) + \nabla \times (p\mathbf{x}/\rho).$$

The z component of this may be rewritten using the identity

$$\nabla \cdot \left[\left(x^2 - z^2\right)\Omega\mathbf{u}\right] = 2\Omega(\mathbf{x} \cdot \mathbf{u} - zu_z) = -[2\mathbf{x} \times (\mathbf{u} \times \mathbf{\Omega})]_z,$$

which yields

$$\frac{\partial}{\partial t}(\mathbf{x} \times \mathbf{u})_z = -\nabla \cdot \left[r^2 \Omega\mathbf{u}\right] + \left[\nabla \times (p\mathbf{x}/\rho)\right]_z, \qquad (6.24)$$

in (r, θ, z) coordinates. We now integrate (6.24) over the cylinder V_c and note that the pressure term converts to a surface integral, which is readily shown to be zero. (Physically, there is no axial torque acting on V_c arising from pressure.) The end result is

$$\frac{dH_z}{dt} = \frac{d}{dt} \int_{V_c} \rho(\mathbf{x} \times \mathbf{u})_z dV = -(\delta/2)^2 \rho \Omega \oint \mathbf{u} \cdot d\mathbf{S} = 0, \qquad (6.25)$$

where $\mathbf{H}$ is the angular momentum in V_c and the term on the right is zero because $\nabla \cdot \mathbf{u} = 0$.

It would appear that, despite the radiation of energy in the form of inertial waves, the axial component of angular momentum within V_c is conserved. This greatly constrains the dispersion pattern. For example, after a time t, inertial waves will have travelled a distance of order $\Omega \delta t$, and so the initial kinetic energy fills a volume of order $V_{3D} \sim (\Omega \delta t)^3$. Conservation of energy now requires that a characteristic velocity at time t satisfies $u^2 (\Omega \delta t)^3 \sim u_0^2 \delta^3$, where u_0 is a typical velocity at $t = 0$. It follows that, as the wave energy disperses, the velocity falls off as $u \sim u_0 (\Omega t)^{-3/2}$. This is true everywhere except within V_c, where angular momentum conservation limits the rate of decline of u. Specifically, as the inertial waves disperse, the angular momentum is spread over a cylindrical volume of the order of $V_{1D} \sim (\Omega \delta t) \delta^2$, and so conservation of H_z requires that, within V_c, $(u\delta)(\Omega \delta t)\delta^2 \sim (u_0 \delta)\delta^3$. It follows that a typical velocity near the rotation axis declines no faster than $u \sim u_0 (\Omega t)^{-1}$, as noted in Figure 6.7(a). These scaling laws, $u \sim u_0 (\Omega t)^{-1}$ for on-axis radiation versus $u \sim u_0 (\Omega t)^{-3/2}$ for off-axis waves, are readily confirmed by an asymptotic study of the generalized dispersion integral for inertial waves.

A typical dispersion pattern is shown schematically in Figure 6.7(b). While there is some off-axis radiation, this tends to fall off quickly and the dominant pattern consists of two lobes, or wave packets, propagating along the rotation axis, one upwards and one downwards. The centres of the two lobes propagate at the group velocity corresponding to the dominant wavenumber, $c_g = 2\Omega/k_{dom} \sim \Omega \delta$, and so are located at $|z| = \ell_c \sim \Omega \delta t$. Moreover, there is a spread of transverse wavenumbers in the initial condition, say $\Delta k_\perp \sim 1/\delta$, and for each transverse wavenumber we have $c_{gz} = 2\Omega/k_\perp$. Consequently, the lobes elongate as they propagate, growing in length as $\ell_z \sim \Omega \delta t$. In summary, then, the spontaneous radiation of energy from a localized vortex generates a pair of columnar wave packets, as shown in Figure 6.7(b).

These arguments are rather general, and so it is instructive to consider a specific example. Consider the case of axisymmetric dispersion initiated by

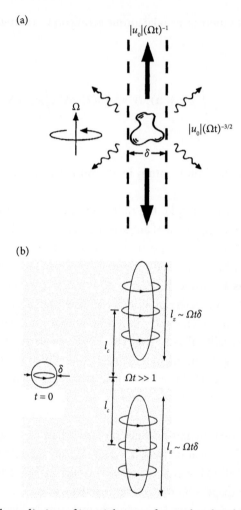

Figure 6.7 (a) The radiation of inertial waves from a localized vortex. The solid arrows represent the dispersion of angular momentum and the wiggly arrows the dispersion of energy. (b) The dispersion pattern after some time is dominated by two columnar vortices.

the initial condition

$$\Gamma = ru_\theta = \Lambda r^2 \exp\left[-(r^2 + z^2)/\delta^2\right], \quad u_r = u_z = 0, \tag{6.26}$$

in (r, θ, z) coordinates. Now (2.41) tells us that, for axisymmetric flows,

$$r(\nabla^2 \mathbf{u})_\theta = r\frac{\partial}{\partial r}\frac{1}{r}\frac{\partial \Gamma}{\partial r} + \frac{\partial^2 \Gamma}{\partial z^2} = \nabla_*^2 \Gamma,$$

and so the azimuthal component of the wave equation (6.14) takes the form

$$\frac{\partial^2}{\partial t^2} \nabla_*^2 \Gamma + (2\Omega)^2 \frac{\partial^2 \Gamma}{\partial z^2} = 0. \tag{6.27}$$

We shall see in §6.4 that this may be solved using a Hankel-cosine transform. This yields the dispersion integral

$$u_\theta \approx \Lambda\delta \int_0^\infty \kappa^2 e^{-\kappa^2} J_1\left(\frac{2\kappa r}{\delta}\right) \left\{ \exp\left[-\left(\frac{z}{\delta} - \frac{\Omega t}{\kappa}\right)^2\right] + \exp\left[-\left(\frac{z}{\delta} + \frac{\Omega t}{\kappa}\right)^2\right] \right\} d\kappa, \tag{6.28}$$

where J_1 is the usual Bessel function and $\kappa = k_r \delta/2$. (Again, see §6.4 for the details.)

The term $\kappa^2 e^{-\kappa^2}$ ensures that the integrand is dominated by $\kappa = O(1)$. Consequently, in order to keep the argument in one of the other two exponentials of order unity for $\Omega t \gg 1$, we require $z \sim \pm\Omega\delta t$. This tells us that the wave energy is centred at $|z| \sim \Omega\delta t$, exactly as predicted above. Figure 6.8 shows the kinetic energy distribution at various times in the upper half of the r–z plane. The dominance of the on-axis radiation is evident.

The exact distribution of u_θ for $\Omega t \gg 1$ may be found by demanding that the argument in one of the exponentials remain of order unity as $\Omega t \to \infty$. At the locations $z = \pm\delta\Omega t$, this requires $\kappa \to 1$ as $\Omega t \to \infty$, and it is readily confirmed that (6.28) then gives $u_\theta \sim \Lambda\delta\, J_1(2r/\delta)(\Omega t)^{-1}$. Given that $J_1(x \to 0) = x/2$ and $J_1(x \to \infty) \sim x^{-1/2}$, this in turn yields $u_\theta \sim \Lambda r(\Omega t)^{-1}$ near the axis and $u_\theta \sim \Lambda\delta(\Omega t)^{-3/2}(z/r)^{1/2}$ off the axis, exactly as predicted above.

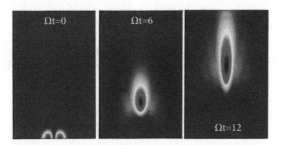

Figure 6.8 The radiation pattern for inertial waves dispersing from a Gaussian vortex.

6.2.4 The Generation and Segregation of Helicity

It is clear from (6.21) that a monochromatic inertial wave has maximal helicity, with $h < 0$ when the wave energy propagates upward and $h > 0$ for the downward propagation of energy. It does not immediately follow that a wave packet, which contains a range of wavenumbers, behaves in the same way. However, it turns out that wave packets do indeed exhibit the same behaviour (Davidson & Ranjan, 2015), with the helicity of the packet close to maximal and $h < 0$ for upward propagating packets ($c_g \cdot \Omega > 0$) and $h > 0$ for downward propagating wave packets ($c_g \cdot \Omega < 0$).

This is illustrated in Figure 6.9, which shows the evolution of a slab of turbulence in a rotating fluid. The initial condition, which is shown in Figure 6.9(a), consists of a random collection of eddies located near the central plane of the computational domain, with Ro≈0.1 based on the rms fluctuating velocity and the size of the eddies. The flow then evolves as the turbulent eddies radiate waves, with Figure 6.9(b) showing the motion at $\Omega t = 6$. As expected, the wave packets take the form of columnar vortices aligned with the rotation axis. However, the key point about Figure 6.9 is that the vortices are coloured according to their helicity, with red for $h < 0$ and green for $h > 0$. Clearly, the upward propagating wave packets carry negative helicity and the downward propagating packets positive helicity, so inertial waves *spatially segregate*, as well as generate, the helicity field.

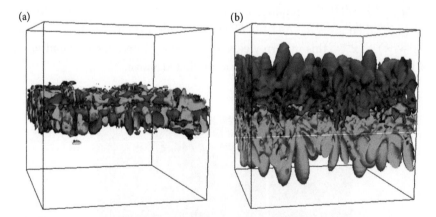

Figure 6.9 (a) The initial condition consists of a random collection of eddies confined to the central plane of the computational domain, with Ro ≈ 0.1. (b) The flow at $\Omega t = 6$ coloured by helicity, with red for $h < 0$ and green for $h > 0$. Reproduced from Davidson (2013).

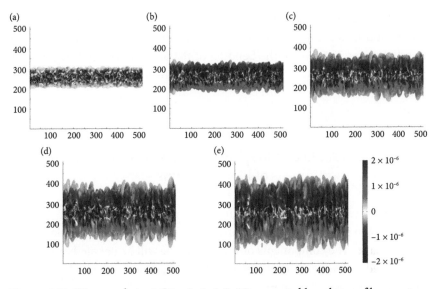

Figure 6.10 Wave packets at $\Omega t = 2, 4, 6, 8, 10$ generated by a layer of buoyant blobs slowly drifting in a Boussinesq fluid. Colour represents helicity. Reproduced from Davidson & Ranjan (2015).

The same behaviour is seen if the wave source consists of a layer of buoyant blobs slowly drifting through a rotating, Boussinesq fluid at low Ro under the influence of gravity. Such a situation is shown in Figure 6.10, where the gravitational field is orthogonal to Ω, and so the buoyant blobs drift horizontally across the central plane of the domain. As expected, the flow evolves by radiating wave packets up and down the rotation axis. The vortices are again coloured by helicity, this time with blue for $h < 0$ and red for $h > 0$, and the upward propagating wave packets clearly carry negative helicity, while the downward propagating packets have positive helicity. Evidently, inertial waves triggered by a localized source not only generate and disperse helicity, but also segregate that helicity, with left-handed spirals above the source and right-handed spirals below.

6.3 The Physical Nature of Inertial Waves Revisited

Having established the formal properties of inertial waves, we now return to the question of their physical origin. In particular, the discussion in §6.1 is limited to axisymmetric wave motion and our task now is to remove that artificial restriction. We start, in §6.3.1, by exploring the mechanism by which the

Coriolis force acts as a restoring force. This is a direct generalization of the discussion in §6.1.2. Next, in §6.3.2, we seek an alternative interpretation of inertial waves in terms of the bending of the absolute vortex lines, $2\Omega + \omega$. It turns out that this has the advantage of exposing the helical nature of the waves.

6.3.1 An Elastic Response to the Axial Compression of Fluid Columns

Consider an inviscid, rapidly rotating fluid at low Rossby number. As usual, we take Ω to point in the z direction and we shall refer to the x-y plane as the lateral plane. The components of $\mathbf{u}$ in this plane will be denoted by $\mathbf{u}_\perp$, while $\nabla_\perp^2$ is used to denote the horizontal Laplacian. In a rotating frame of reference our governing equation of motion is

$$\frac{\partial \mathbf{u}}{\partial t} = 2\mathbf{u} \times \Omega - \nabla\left(p/\rho\right), \tag{6.29}$$

while the corresponding vorticity equation is (6.13). From these we deduce

$$\frac{\partial}{\partial t}\left(\nabla \cdot \mathbf{u}_\perp\right) = 2\Omega\omega_z - \nabla_\perp^2\left(p/\rho\right), \tag{6.30}$$

$$\frac{\partial \omega_z}{\partial t} = 2\Omega\frac{\partial u_z}{\partial z} = -2\Omega\left(\nabla \cdot \mathbf{u}_\perp\right), \tag{6.31}$$

and

$$\nabla^2\left(p/\rho\right) = 2\Omega\omega_z. \tag{6.32}$$

Note that, if we introduce the vector potential for $\mathbf{u}$, defined by $\nabla \cdot \mathbf{A} = 0$ and $\nabla \times \mathbf{A} = \mathbf{u}$, then (6.32) tells us that $p/\rho + 2\Omega A_z$ is harmonic. Moreover, if the fluid is unbounded and stationary at infinity, this then yields $p/\rho = -2\Omega A_z$. These are the key equations that will help us to establish the mechanism by which the Coriolis force can act as a restoring force.

Let C be the projection of a closed material curve onto the lateral plane. Also, let C enclose an area A and $\mathbf{n}$ be the unit normal to C which lies in the lateral plane, pointing outward. If C is composed of the line elements $d\mathbf{r}$, then we may introduce the vector area $d\mathbf{S} = |d\mathbf{r}|\mathbf{n}$, and Gauss' theorem gives us

$$\int_A (\nabla \cdot \mathbf{u}_\perp)\, dxdy = \oint_C \mathbf{u} \cdot d\mathbf{S}.$$

Integrating (6.31) and (6.30) across the area A then yields, for Ro $\ll 1$,

$$\frac{d}{dt}\oint_C \mathbf{u} \cdot d\mathbf{r} = -2\Omega \oint_C \mathbf{u} \cdot d\mathbf{S}, \tag{6.33}$$

$$\frac{d}{dt}\oint_C \mathbf{u} \cdot d\mathbf{S} = 2\Omega \oint_C \mathbf{u} \cdot d\mathbf{r} - \oint_C \nabla_\perp (p/\rho) \cdot d\mathbf{S}. \tag{6.34}$$

The first of these tells us that an expansion (or contraction) of the loop C in the lateral plane results in a decline (or growth) of circulation around the loop. This can be thought of as a form of Kelvin's theorem applied in an inertial frame, where the total axial vorticity is $2\Omega + \omega_z$. That is to say, (6.33) demands $2\Omega A(t) + \oint \mathbf{u} \cdot d\mathbf{r} = $ const., which tells us that the net flux of vorticity through the loop C is conserved in an inertial frame. Thus, as the loop expands or contracts, so $\oint \mathbf{u} \cdot d\mathbf{r}$ must fall or rise to conserve the net circulation.

The second integral equation tells us that, pressure forces apart, a positive (or negative) circulation tends to cause an expansion (or contraction) of the loop. Taken together with (6.33), we have a restoring mechanism that tends to suppress any expansion or contraction in the lateral plane. For example, if C expands, then (6.33) tells us that the circulation falls, and if the circulation goes negative, (6.34) tends to drive a contraction of the loop, thus countering the initial expansion (see Figure 6.11). Indeed, (6.33) and (6.34) combine to give

$$\frac{d^2}{dt^2}\oint_C \mathbf{u} \cdot d\mathbf{r} + (2\Omega)^2 \oint_C \mathbf{u} \cdot d\mathbf{r} = 2\Omega \oint_C \nabla_\perp (p/\rho) \cdot d\mathbf{S}, \tag{6.35}$$

which, pressure forces apart, is suggestive of oscillations at a frequency of 2Ω.

We can reach the same conclusion directly from (6.30) and (6.31), which combine to give

$$\frac{\partial^2 \omega_z}{\partial t^2} + (2\Omega)^2 \omega_z = 2\Omega \nabla_\perp^2 (p/\rho). \tag{6.36}$$

Of course, the problem with (6.35) and (6.36) is the unknown role of the pressure terms. However, recalling that both the pressure and the vorticity can

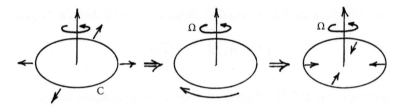

Figure 6.11 The restoring mechanism of the Coriolis force. An expansion of the loop C causes the circulation around the loop to fall. When the circulation goes negative, this drives a contraction of the loop, thus countering the initial expansion.

be written in terms of the vector potential, according to $p/\rho = -2\Omega A_z$ and $\omega_z = -\nabla^2 A_z$, we can rewrite (6.36) as

$$\frac{\partial^2}{\partial t^2}\nabla^2 A_z + (2\Omega)^2\nabla^2 A_z = (2\Omega)^2\nabla_\perp^2 A_z, \tag{6.37}$$

which exposes the relative roles played by the Coriolis term on the left and the pressure term on the right. On combining those terms, (6.37) reverts to the governing equation for inertial waves,

$$\frac{\partial^2}{\partial t^2}\nabla^2 A_z + (2\Omega)^2\frac{\partial^2 A_z}{\partial z^2} = 0. \tag{6.38}$$

The tendency for the Coriolis force to suppress any divergence in the lateral plane can equally be thought of as a tendency to suppress any change in the length of a fluid column aligned with $\mathbf{\Omega}$. This follows directly from continuity, i.e. $\nabla \cdot \mathbf{u}_\perp + \partial u_z/\partial z = 0$. Thus, rapid rotation resists the axial extension or contraction of fluid columns, resulting in an elastic response to axial compression. When combined with the idea of inertial overshoot, we have all the ingredients needed for inertial waves. Of course, this resistance to the axial extension or contraction of fluid columns also lies at the heart of the Taylor–Proudman theorem.

6.3.2 The Bending of Absolute Vortex Lines and the Generation of Helical Flow

While the discussion above has the merit of endowing a rotating fluid with a kind of elasticity, it does not explain the helical nature of inertial waves.

To do that, it is helpful to consider the bending of the absolute vortex lines, $\boldsymbol{\omega}_{net} = 2\boldsymbol{\Omega} + \boldsymbol{\omega}$. Let us start by deriving the governing equation for $\boldsymbol{\omega}_{net}$ in a rotating frame. Ignoring viscosity, we can rewrite our fully nonlinear equation of motion in the rotating frame, (4.8), as

$$\frac{\partial \mathbf{u}}{\partial t} = 2\mathbf{u} \times \boldsymbol{\Omega} + \mathbf{u} \times \boldsymbol{\omega} - \nabla H = \mathbf{u} \times \boldsymbol{\omega}_{net} - \nabla H, \tag{6.39}$$

whose curl yields

$$\frac{D\boldsymbol{\omega}_{net}}{Dt} = (\boldsymbol{\omega}_{net} \cdot \nabla)\mathbf{u}, \tag{6.40}$$

where D/Dt and $\mathbf{u}$ are both measured in the rotating frame. Evidently, the evolution equation for the absolute vorticity in the rotating frame has the same form as that in an inertial frame. Now (2.68) tells us that a dyeline in either an inertial *or* a rotating frame of reference is governed by

$$\frac{D}{Dt}(d\mathbf{l}) = (d\mathbf{l} \cdot \nabla)\mathbf{u}, \tag{6.41}$$

where D/Dt and $\mathbf{u}$ are measured in the appropriate frame. A comparison of (6.41) with (6.40) tells us that $\boldsymbol{\omega}_{net}$ is frozen into the fluid in the rotating frame, as it must be.

Let us now consider the case where Ro << 1, so that $\mathbf{u} \times \boldsymbol{\omega}$ may be dropped from (6.39) and our equation of motion simplifies to (6.12). Suppose that a horizontal gust bends the absolute vortex lines, as shown on the left of Figure 6.12. Then, as the vortex lines bow out, the Coriolis force, $2\mathbf{u} \times \boldsymbol{\Omega}$, induces a velocity directed out of the page. The net effect is that the fluid, as well as the absolute vortex lines, rotate as shown. After a time of order Ω^{-1} the velocity, as well as the bulge in $\boldsymbol{\omega}_{net}$, is directed out of the page. Indeed, it is clear that the Coriolis force acts to continually rotate both $\mathbf{u}$ and $\boldsymbol{\omega}$ in a *retrograde* direction, that is, in a direction opposite to $\boldsymbol{\Omega}$. Note that the directions of $\mathbf{u}$ and $\boldsymbol{\omega}$ are such that the helicity, $h = \mathbf{u} \cdot \boldsymbol{\omega}$, is negative in the upper half of the disturbance and positive in the lower half. Note also that the induced perturbation in $\boldsymbol{\omega}_{net}$ creates motion above and below the initial disturbance through the Biot–Savart law, causing the disturbance to spread.

In summary, the Coriolis force causes $\mathbf{u}$ to rotate, and the freezing-in of the absolute vortex lines means that $\boldsymbol{\omega}_{net}$ co-rotates with the fluid. The net effect is that both $\mathbf{u}$ and $\boldsymbol{\omega}$ rotate in a retrograde direction, with h negative in the upper half of the disturbance and positive in the lower half. It is a combination of the

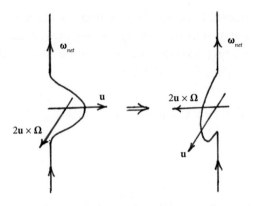

Figure 6.12 A horizontal gust bends the absolute vortex lines while inducing a velocity out of the page. The velocity, as well as the bowed vortex lines, rotates in a retrograde sense.

Coriolis force acting on the fluid, and the freezing-in of the absolute vortex lines, that induces helicity.

6.4 A More Detailed Look at the Dispersion Pattern Generated by a Gaussian Vortex

In §6.2.3 we discussed briefly the dispersion of axisymmetric inertial waves initiated by the Gaussian vortex

$$\Gamma = r u_\theta = \Lambda r^2 \exp\left[-(r^2 + z^2)/\delta^2\right], \quad u_r = u_z = 0, \tag{6.42}$$

that dispersion being governed by the azimuthal component of the axisymmetric wave equation

$$\frac{\partial^2}{\partial t^2}\left[\frac{\partial}{\partial r}\frac{1}{r}\frac{\partial \Gamma}{\partial r} + \frac{\partial^2 \Gamma}{\partial z^2}\right] + (2\Omega)^2 \frac{\partial^2 \Gamma}{\partial z^2} = 0. \tag{6.43}$$

(See equations (6.27) and (6.28), as well as Figure 6.8.) This turns out to be a particularly informative model problem; one that merits a more detailed examination.

Let us start by recalling the relevant governing equations for axisymmetric inertial waves. First, (2.49), (2.50), and (2.51) give us the general kinematic relationships,

$$\mathbf{u}_p = \nabla \times [(\Psi/r)\hat{\mathbf{e}}_\theta], \quad \boldsymbol{\omega}_p = \nabla \times [(\Gamma/r)\,\hat{\mathbf{e}}_\theta], \quad \nabla_*^2\Psi = -r\omega_\theta, \tag{6.44}$$

where Ψ is the Stokes streamfunction, the subscript p stands for poloidal, and ∇_*^2 is the Stokes operator, (2.48). Second, the azimuthal components of (6.12) and (6.13) yield

$$\frac{\partial\Gamma}{\partial t} = -2\Omega r u_r = 2\Omega\frac{\partial\Psi}{\partial z}, \tag{6.45}$$

and

$$\frac{\partial}{\partial t}r\omega_\theta = -\frac{\partial}{\partial t}\nabla_*^2\Psi = 2\Omega\frac{\partial\Gamma}{\partial z}, \tag{6.46}$$

which combine to give the wave-like equation (6.43). These are all the equations we need.

Now the natural way to solve this kind of initial value problem is to take a 3D Fourier transform of (6.43), solve the resulting second-order (in time) ODE, and then perform the inverse Fourier transform. However, for axisymmetric problems a 3D Fourier transform reduces to a Hankel-cosine transform in r and z, as discussed in Bracewell (1986). It is simpler, therefore, to start with the Hankel-cosine transform pair

$$\hat{u}_\theta(k_r, k_z) = \frac{1}{2\pi^2}\int_0^\infty\int_0^\infty u_\theta J_1(k_r r)\cos(k_z z)r\,dr\,dz, \tag{6.47}$$

$$u_\theta(r, z) = 4\pi\int_0^\infty\int_0^\infty \hat{u}_\theta J_1(k_r r)\cos(k_z z)k_r\,dk_r\,dk_z, \tag{6.48}$$

where J_1 is the usual Bessel function. Noting that

$$\nabla_*^2\left[rJ_1(k_r r)\right] = -k_r^2\left[rJ_1(k_r r)\right],$$

the transform of (6.43) is evidently

$$\frac{\partial^2\hat{u}_\theta}{\partial t^2} + \varpi^2\hat{u}_\theta = 0, \quad \varpi = \frac{2\Omega k_z}{k}, \tag{6.49}$$

where $k^2 = k_r^2 + k_z^2$. The initial conditions are $\hat{u}_\theta = \hat{u}_\theta^{(0)}$ and $\hat{u}_r = \hat{u}_z = 0$, the second of which requires $\partial\hat{u}_\theta/\partial t = 0$ at $t = 0$. It follows that $\hat{u}_\theta = \hat{u}_\theta^{(0)}\cos\varpi t$, and so the inverse transform yields

$$u_\theta(r, z, t) = 4\pi \int_0^\infty \int_0^\infty \hat{u}_\theta^{(0)} J_1(k_r r) \cos(k_z z) \cos(\varpi t) k_r dk_r dk_z,$$

or equivalently,

$$u_\theta = 2\pi \int_0^\infty \int_0^\infty \hat{u}_\theta^{(0)} k_r J_1(k_r r) \left[\cos\left(k_z \left(z - \frac{2\Omega t}{k}\right)\right) + \cos\left(k_z \left(z + \frac{2\Omega t}{k}\right)\right) \right] dk_r dk_z.$$

Moreover, it is readily confirmed that the Gaussian initial condition (6.42) transforms to

$$\hat{u}_\theta^{(0)} = \frac{\Lambda \delta^5}{16\pi^{3/2}} k_r \exp\left(-k^2 \delta^2/4\right), \tag{6.50}$$

and hence u_θ is given by,

$$u_\theta = \frac{\Lambda \delta^5}{8\sqrt{\pi}} \int_0^\infty k_r^2 \exp\left(-k_r^2 \delta^2/4\right) J_1(k_r r) I(k_r, z, t) dk_r, \tag{6.51}$$

where

$$I = \int_0^\infty \exp\left(-k_z^2 \delta^2/4\right) \left[\cos\left(k_z\left(z - 2\Omega t/k\right)\right) + \cos\left(k_z\left(z + 2\Omega t/k\right)\right)\right] dk_z.$$

Now $\hat{u}_\theta^{(0)}$ is dominated by wavevectors in the vicinity of $k_z \approx 0$ and $k_r \sim \delta^{-1}$, and so a reasonable approximation to I is obtained by replacing k_z/k by k_z/k_r in the argument of the cosines (see Davidson et al., 2006). The integral I can then be evaluated exactly to give

$$u_\theta \approx \Lambda \delta \int_0^\infty \kappa^2 e^{-\kappa^2} J_1\left(\frac{2\kappa r}{\delta}\right) \left\{ \exp\left[-\left(\frac{z}{\delta} - \frac{\Omega t}{\kappa}\right)^2\right] + \exp\left[-\left(\frac{z}{\delta} + \frac{\Omega t}{\kappa}\right)^2\right] \right\} d\kappa,$$

$$\tag{6.52}$$

where $\kappa = k_r \delta/2$. Equations (6.44), (6.45), and (6.46) now yield

$$u_z \approx \Lambda\delta \int_0^\infty \kappa^2 e^{-\kappa^2} J_0\left(\frac{2\kappa r}{\delta}\right)\left\{-\exp\left[-\left(\frac{z}{\delta} - \frac{\Omega t}{\kappa}\right)^2\right] + \exp\left[-\left(\frac{z}{\delta} + \frac{\Omega t}{\kappa}\right)^2\right]\right\} d\kappa,$$

(6.53)

$$\omega_\theta \approx 2\Lambda \int_0^\infty \kappa^3 e^{-\kappa^2} J_1\left(\frac{2\kappa r}{\delta}\right)\left\{-\exp\left[-\left(\frac{z}{\delta} - \frac{\Omega t}{\kappa}\right)^2\right] + \exp\left[-\left(\frac{z}{\delta} + \frac{\Omega t}{\kappa}\right)^2\right]\right\} d\kappa,$$

(6.54)

$$\omega_z \approx 2\Lambda \int_0^\infty \kappa^3 e^{-\kappa^2} J_0\left(\frac{2\kappa r}{\delta}\right)\left\{\exp\left[-\left(\frac{z}{\delta} - \frac{\Omega t}{\kappa}\right)^2\right] + \exp\left[-\left(\frac{z}{\delta} + \frac{\Omega t}{\kappa}\right)^2\right]\right\} d\kappa.$$

(6.55)

These expressions are rather informative. The presence of the term $\kappa^2 e^{-\kappa^2}$ tells us that the integrands are dominated by $\kappa = O(1)$. Consequently, in order to keep the argument in one of the exponentials of order unity for $\Omega t \gg 1$, we require $z \sim \pm\Omega\delta t$. Thus, the wave energy is centred at $z \sim \pm\Omega\delta t$. Moreover, because a range of wavenumbers contribute to the integrals, of the order of $\Delta k_r \sim 1/\delta$, each of which has a different group velocity, the energy is spread over an increasingly large area, which grows as $\Delta z \sim \Omega\delta t$. In short, we have two wave packets propagating along the rotation axis, one upwards and one downwards, whose centres are located at $|z| = \ell_c \sim \Omega\delta t$ and whose axial lengths grow as $\ell_z \sim \Omega\delta t$. This is shown in Figure 6.7(b). Moreover, as noted in §6.2.3, at $z = \pm\delta\Omega t$ we have $u_\theta \sim \Lambda\delta J_1(2r/\delta)(\Omega t)^{-1}$ for $\Omega t \gg 1$. This yields $u_\theta \sim \Lambda r(\Omega t)^{-1}$ near the axis and $u_\theta \sim \Lambda\delta(\Omega t)^{-3/2}(z/r)^{1/2}$ off the axis, confirming the dominance of the on-axis radiation.

Note that Γ and ω_z are symmetric about the x–y plane and positive near the z-axis, while u_z and ω_θ are antisymmetric and negative for $z > 0$. These signs are consistent with $\int \Gamma dV$ being conserved and positive at $t = 0$, $\int \omega_\theta dV$ being conserved and zero at $t = 0$, and $h \approx u_\theta\omega_\theta + u_z\omega_z$ being negative (positive) in the upward (downward) propagating wave packet. Note also that u_z near the axis takes the opposite sign to $c_{g,z}$. So the streamlines in the upward propagating wave packet are left-handed spirals pointing downward, while those in the downward propagating packet are right-handed spirals pointing upward, as shown in Figure 6.3. It follows that, near the axis, any material movement caused by the passage of a wave packet is in the *opposite* direction to that of the energy propagation.

Now equations (6.44), (6.45), and (6.46) have the useful property that if one solution is known, then another can be obtained through the transformation $\mathbf{u} \to \boldsymbol{\omega}$, $\Psi \to \Gamma$. It follows that we can convert (6.42) into an initial condition consisting of the vortex ring

$$\omega_\theta = (2\Lambda/\delta)\, r \exp\left[-(r^2 + z^2)/\delta^2\right], \quad u_\theta = 0, \tag{6.56}$$

as shown in Figure 6.13. The associated wave pattern follows from (6.52) and (6.53), and is

$$u_\theta \approx \Lambda\delta \int_0^\infty \kappa e^{-\kappa^2} J_1\left(\frac{2\kappa r}{\delta}\right)\left\{-\exp\left[-\left(\frac{z}{\delta} - \frac{\Omega t}{\kappa}\right)^2\right] + \exp\left[-\left(\frac{z}{\delta} + \frac{\Omega t}{\kappa}\right)^2\right]\right\} d\kappa, \tag{6.57}$$

$$u_z \approx \Lambda\delta \int_0^\infty \kappa e^{-\kappa^2} J_0\left(\frac{2\kappa r}{\delta}\right)\left\{\exp\left[-\left(\frac{z}{\delta} - \frac{\Omega t}{\kappa}\right)^2\right] + \exp\left[-\left(\frac{z}{\delta} + \frac{\Omega t}{\kappa}\right)^2\right]\right\} d\kappa, \tag{6.58}$$

$$\omega_\theta \approx 2\Lambda \int_0^\infty \kappa^2 e^{-\kappa^2} J_1\left(\frac{2\kappa r}{\delta}\right)\left\{\exp\left[-\left(\frac{z}{\delta} - \frac{\Omega t}{\kappa}\right)^2\right] + \exp\left[-\left(\frac{z}{\delta} + \frac{\Omega t}{\kappa}\right)^2\right]\right\} d\kappa, \tag{6.59}$$

$$\omega_z \approx 2\Lambda \int_0^\infty \kappa^2 e^{-\kappa^2} J_0\left(\frac{2\kappa r}{\delta}\right)\left\{-\exp\left[-\left(\frac{z}{\delta} - \frac{\Omega t}{\kappa}\right)^2\right] + \exp\left[-\left(\frac{z}{\delta} + \frac{\Omega t}{\kappa}\right)^2\right]\right\} d\kappa. \tag{6.60}$$

As before, we have wave packets propagating up and down the rotation axis whose centres are located at $|z| = \ell_c \sim \Omega\delta t$ and whose axial lengths grow as $\ell_z \sim \Omega\delta t$. This time, however, u_z and ω_θ are symmetric about the x–y plane, while Γ and ω_z are antisymmetric. These signs are consistent with $\int \Gamma dV$ being conserved and zero at $t = 0$, $\int \omega_\theta dV$ being conserved and positive at $t = 0$, and $h \approx u_\theta\omega_\theta + u_z\omega_z$ being negative in the upward propagating wave packet and positive in the downward propagating packet.

Finally, we note that, in these two examples, the distributions of $|u_z|$ and $|\omega_z|$ are very similar but not identical, and the same is true of $|u_\theta|$ and $|\omega_\theta|$. It follows that, unlike a monochromatic inertial wave, the helicity is not maximal, merely close to maximal.

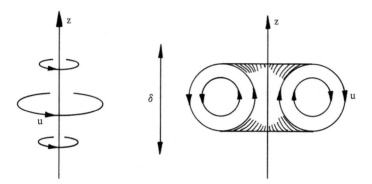

Figure 6.13 The initial condition (6.42) is on the left and (6.56) on the right.

6.5 The Dispersion Patterns Generated by Buoyant Blobs in a Boussinesq Fluid

We now consider the dispersion of inertial waves from buoyant blobs slowly drifting through a rotating, Boussinesq fluid. As we shall see in Chapter 17, this is a ubiquitous feature of motion in the core of the earth and it may play an important role in the geodynamo.

6.5.1 A Single Buoyant Blob Drifting Normal or Parallel to the Rotation Axis

Consider a single, isolated blob of buoyant material of scale δ which sits in a rotating, incompressible, Boussinesq fluid. We assume that Ro = $u/\Omega\delta \ll$ 1 and take $\boldsymbol{\Omega} = \Omega\hat{\mathbf{e}}_z$. There are two configurations of interest: one where $\mathbf{g}$ is normal to $\boldsymbol{\Omega}$, say $\mathbf{g} = -g\hat{\mathbf{e}}_x$, and one where $\mathbf{g}$ and $\boldsymbol{\Omega}$ are aligned, $\mathbf{g} = -g\hat{\mathbf{e}}_z$. Let us start with $\mathbf{g} = -g\hat{\mathbf{e}}_x$, which characterizes a buoyant blob drifting out along the equatorial plane in the liquid core of the earth. This is governed by

$$\frac{\partial \mathbf{u}}{\partial t} = 2\mathbf{u} \times \boldsymbol{\Omega} - \nabla\left(p/\bar{\rho}\right) + b\mathbf{g}, \quad \nabla \cdot \mathbf{u} = 0, \tag{6.61}$$

where $\rho = \bar{\rho} + \rho'$ and $b = \rho'/\bar{\rho}$. The corresponding vorticity equation is evidently

$$\frac{\partial \boldsymbol{\omega}}{\partial t} = 2\left(\boldsymbol{\Omega} \cdot \nabla\right)\mathbf{u} + \nabla b \times \mathbf{g}. \tag{6.62}$$

Since ρ' is governed by an advection-diffusion equation, it evolves on a slow timescale set by **u**, while the inertial wave packets evolve on the fast timescale of Ω^{-1}. We may therefore treat ρ' as quasi-steady as far as the initiation of inertial waves is concerned. Taking ρ' to be independent of time, and applying the operator $\nabla \times (\partial/\partial t)$ to (6.62), yields an inhomogeneous version of our inertial wave equation

$$\frac{\partial^2}{\partial t^2}\left(\nabla^2 \mathbf{u}\right) + (2\boldsymbol{\Omega} \cdot \nabla)^2 \mathbf{u} = (2\boldsymbol{\Omega} \cdot \nabla)\left(\mathbf{g} \times \nabla b\right). \tag{6.63}$$

Evidently, our buoyant blob acts as a source of inertial waves and our task is to determine the resulting dispersion pattern.

Let us return briefly to §6.2.3, and to the focussing of wave energy onto the rotation axis. The geometrical argument given in §6.2.3 is independent of how the waves are excited, and requires only that the excitation is local and has no imposed timescale. It follows that the dispersion pattern generated by our buoyant blob is necessarily dominated by low-frequency waves propagating up and down the rotation axis, in the sense that the radiation density will be highest on the rotation axis. We can determine the structure of these axially propagating wave packets by considering the vertical jump conditions across the buoyant blob. Since ρ' is quasi-steady, and the inertial waves of interest are of low frequency, the motion within the buoyant blob is governed by

$$2\left(\boldsymbol{\Omega} \cdot \nabla\right)\mathbf{u} \approx \mathbf{g} \times \nabla b = g\left(0, \partial b/\partial z, -\partial b/\partial y\right), \tag{6.64}$$

whose curl gives

$$2\left(\boldsymbol{\Omega} \cdot \nabla\right)\boldsymbol{\omega} \approx g\nabla^2 b - (\mathbf{g} \cdot \nabla)\nabla b, \tag{6.65}$$

and, in particular,

$$2\left(\boldsymbol{\Omega} \cdot \nabla\right)\omega_z \approx g\frac{\partial^2 b}{\partial x \partial z}.$$

It follows that the vertical jump conditions across the buoyant blob are $\Delta\omega_z \approx 0$ and

$$\Delta u_x \approx 0, \ \Delta u_y \approx 0, \ \Delta u_z \approx -\frac{g}{2\Omega}\int\limits_{-\infty}^{\infty}\left(\partial b/\partial y\right)dz. \tag{6.66}$$

For a Gaussian blob, say $b = -|b_0|\exp\left(-\mathbf{x}^2/\delta^2\right)$, (6.66) requires that Δu_z is positive for $y < 0$ and negative for $y > 0$. Hence, u_z, which is antisymmetric about $z = 0$, *diverges* from the x–y plane for $y < 0$, but *converges* for $y > 0$. Moreover, upward propagating wave packets have negative helicity, $u_z\omega_z < 0$, while downward propagating wave packets have positive helicity, $u_z\omega_z > 0$. It follows that, for $y < 0$, the diverging axial flow is anticyclonic (*i.e.* $\omega_z < 0$), both above and below the x–y plane. Conversely, for $y > 0$, the converging axial flow is cyclonic ($\omega_z > 0$) above and below $z = 0$. This is consistent with the jump condition for ω_z, which requires that a cyclonic (anticyclonic) flow above the buoyant blob must correspond to a cyclonic (anticyclonic) flow below.

Putting all of this together, we find that the inertial-wave dispersion pattern consists of a cyclonic–anticyclonic pair of columnar vortices above the blob, matched to a cyclone–anticyclone pair below, with the anticyclones located at negative y, and the cyclones at positive y. Moreover, the flow diverges from the x–y plane for $y < 0$, and converges for $y > 0$.

A fully nonlinear numerical simulation of this flow is shown in Figure 6.14(a, b). Here y points to the right, blue indicates $u_z < 0$ and red is $u_z > 0$, the buoyancy field is black, gravity points into the page, and the image corresponds to a time $\Omega t = 12$ after the release of the blob. The direction of the axial flow is as predicted, and the cyclones (anticyclones) are on the right (left) of the blob, also as predicted. Note that the divergent axial flow for $y < 0$ results in the buoyant blob being stretched along the rotation axis, while the convergent axial flow for $y > 0$ results in the blob being compressed.

As we shall see in Chapter 17, such a pattern is highly reminiscent of the kind of flow observed in numerical simulations of the geodynamo, as indicated in Figure 6.14(c).

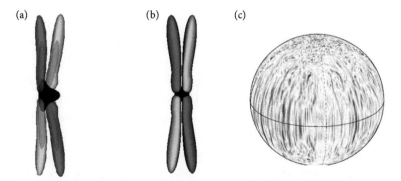

(a)   (b)   (c)

Figure 6.14 (a) A buoyant blob drifts out of the page at Ro = 0.1. It emits inertial waves, with $u_z < 0$ shown blue and $u_z > 0$ shown red. (b) As for (a) but with Ro = 0.01. (c) Simulation of a planetary core showing alternating cyclones–anticyclones. Reproduced from Davidson & Ranjan (2015).

The wave pattern is quite different when **g** and $\boldsymbol{\Omega}$ are aligned, say $\mathbf{g} = -g\hat{\mathbf{e}}_z$. Consider the case where the blob is buoyant and has a Gaussian profile. Equations (6.64) and (6.65) now yield

$$2\left(\boldsymbol{\Omega} \cdot \nabla\right) u_z \approx 0, \quad 2\left(\boldsymbol{\Omega} \cdot \nabla\right) \omega_z \approx -g\nabla_\perp^2 b,$$

where $\nabla_\perp^2$ is the Laplacian in the x–y plane. This gives the vertical jump conditions as

$$\Delta u_z \approx 0, \quad \Delta \omega_z \approx -\frac{g}{2\Omega} \int\limits_{-\infty}^{\infty} \nabla_\perp^2 b\, dz.$$

The dispersion pattern is now axisymmetric, with $\Delta\omega_z < 0$ and $\Delta u_z \approx 0$ close to the z-axis, while further away from the z-axis we have $\Delta\omega_z > 0$ and $\Delta u_z \approx 0$. Moreover, upward propagating wave packets carry negative helicity, $u_z\omega_z < 0$, while downward propagating wave packets carry positive helicity, $u_z\omega_z > 0$. These various constraints are satisfied as follows. Close to the z-axis, where $\Delta\omega_z < 0$, the dispersion pattern consists of an upward propagating wave packet above the blob, with anticyclonic rotation ($\omega_z < 0$) and positive u_z, combined with a downward propagating wave packet below the blob, with cyclonic rotation ($\omega_z > 0$) and positive u_z. Conversely, we have $\Delta\omega_z > 0$ in an annular region away from the axis. Here the upward propagating wave packet has cyclonic rotation and negative u_z, while the downward propagating wave packet has anticyclonic rotation and negative u_z. Interestingly, this particular arrangement of helical vortices is more or less that seen in rapidly rotating, Rayleigh–Bénard convection, as shown in Figure 6.15.

6.5.2 A Layer of Drifting Buoyant Blobs

Let us return to the case where $\boldsymbol{\Omega} = \Omega\hat{\mathbf{e}}_z$, $\mathbf{g} = -g\hat{\mathbf{e}}_x$, and Ro << 1, only now we consider a layer of randomly distributed buoyant blobs confined to the central plane of the computational domain. As in Figure 6.14, the blobs drift slowly in the x–y plane, while emitting wave packets that propagate up and down the rotation axis. A fully nonlinear numerical simulation of this flow is shown in Figure 6.16, where the panels show surfaces of u_z (blue for $u_z < 0$, red for $u_z > 0$) at various times after the release of the blobs. As expected, we find a forest of cyclone–anticyclone pairs propagating axially away from the layer of buoyant blobs. The same flow is shown in Figure 6.10, coloured by helicity, which reminds us that the upward propagating wave packets carry a sea of

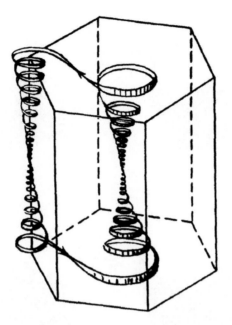

Figure 6.15 Convection pattern for rapidly rotating, Rayleigh–Bénard convection when the convection cells are hexagonal. Adapted from Veronis (1959).

negative helicity and the downward propagating wave packets a sea of positive helicity.

We can quantify how helicity disperses from such a localized source of buoyancy as follows. Let us rewrite (6.61) as

$$\frac{\partial \mathbf{u}}{\partial t} = 2\mathbf{\Omega} \cdot \nabla \mathbf{a} - \nabla \left(p^*/\bar{\rho} \right) + b\mathbf{g},$$

where **a** is the solenoidal vector potential for **u** and the modified pressure, p^*, satisfies

$$\nabla^2 \left(p^*/\bar{\rho} \right) = \nabla \cdot (b\mathbf{g}) .$$

Combining this with (6.62) yields, after a little algebra,

$$\frac{\partial}{\partial t} (\mathbf{u} \cdot \mathbf{\omega}) = -\nabla \cdot \mathbf{F} + S_h,$$

where the flux, **F**, and helicity source, S_h, are

$$\mathbf{F} = - \left(2\mathbf{u}^2 \right) \mathbf{\Omega} - 2\mathbf{u} \times (\mathbf{\Omega} \cdot \nabla \mathbf{a}) + p^*\mathbf{\omega}/\bar{\rho},$$

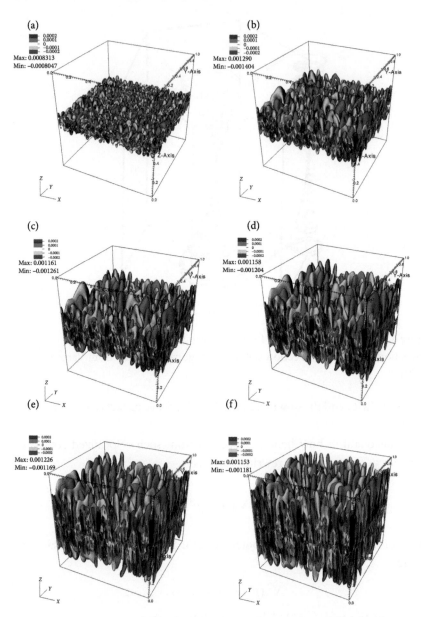

Figure 6.16 The radiation of inertial waves packets from a layer of buoyant blobs slowly drifting through a rotating, Boussinesq fluid under the influence of gravity, with Ro << 1 and $\mathbf{g} = -g\hat{\mathbf{e}}_x$. The panels show surfaces of u_z, with blue for $u_z < 0$ and red for $u_z > 0$, at the times $\Omega t = 2, 4, 6, 8, 10, 12$. Reproduced from Davidson & Ranjan (2015).

and

$$S_h = \nabla \cdot (\mathbf{u} \times (b\mathbf{g})) + 2\mathbf{g} \cdot (\mathbf{u} \times \nabla b).$$

Now suppose we are well removed from the localized source of buoyancy. Then the modified pressure is weak, $p^* \sim |\mathbf{x}|^{-2}$, and so these equations can be approximated by

$$\frac{\partial}{\partial t} (\mathbf{u} \cdot \boldsymbol{\omega}) = -\nabla \cdot \mathbf{F}, \quad \mathbf{F} \approx - \left(2\mathbf{u}^2\right) \boldsymbol{\Omega} - \mathbf{u} \times \frac{\partial \mathbf{u}}{\partial t}. \tag{6.67}$$

Moreover, as we have seen, dispersion from a localized buoyancy source is usually dominated by low-frequency wave packets, $\mathbf{k} \cdot \boldsymbol{\Omega} \approx 0$, and in such cases $\tilde{\omega} \ll \Omega$, so that the helicity flux is simply $\mathbf{F} \approx - \left(2\mathbf{u}^2\right) \boldsymbol{\Omega}$. This is consistent with upward (downward) propagating wave packets carrying with them negative (positive) helicity. To see why, suppose the buoyancy source is localized near the plane $z = 0$, confined to the region $-\delta < z < \delta$. If we integrate (6.67) over all space that lies above the horizontal plane $z = \ell \gg \delta$, or below the horizontal plane $z = -\ell$, then we find

$$\frac{d}{dt} \int_{V_\pm} \mathbf{u} \cdot \boldsymbol{\omega} \, dV = - \oint \mathbf{F} \cdot d\mathbf{S} \approx \mp 2\Omega \int_{z=\pm\ell} \mathbf{u}^2 dA,$$

where the upper sign corresponds to $z > \ell$ and the lower one to $z < -\ell$. So, we obtain negative helicity above the source and positive helicity below, as expected from an analysis of monochromatic waves.

6.6 Waves in a Rotating, Stratified Fluid and Near-Inertial Waves in the Oceans

We now combine the results of §3.5 and §6.2.1 and consider waves in a rotating Boussinesq fluid which is stably stratified, with $\boldsymbol{\Omega} = \Omega \hat{\mathbf{e}}_z$ and $\mathbf{g} = -g\hat{\mathbf{e}}_z$. Such waves are particularly important in the oceans. We continue to treat the flow as inviscid and, for simplicity, we take the stratification to be uniform, so the Väisälä–Brunt frequency, N, is a constant. (See §3.5 for the definition of N.) As usual, our first step is to find the governing equation for the waves and then determine the dispersion relationship and group velocity.

Let us return to §3.5. Adding the Coriolis force to (3.30), our governing equations are

$$\frac{\partial \boldsymbol{\omega}}{\partial t} = 2\boldsymbol{\Omega} \cdot \nabla \mathbf{u} + \nabla \left(\rho'/\bar{\rho} \right) \times \mathbf{g}, \qquad \frac{\partial \rho'}{\partial t} = \frac{\bar{\rho} N^2}{g} u_z, \qquad (6.68)$$

and $\nabla \cdot \mathbf{u} = 0$. Eliminating ρ' now yields

$$\frac{\partial^2 \boldsymbol{\omega}}{\partial t^2} = 2\boldsymbol{\Omega} \cdot \nabla \frac{\partial \mathbf{u}}{\partial t} + N^2 \hat{\mathbf{e}}_z \times \nabla u_z, \qquad (6.69)$$

whose curl gives us

$$\frac{\partial^2}{\partial t^2} \nabla^2 u_z + (2\boldsymbol{\Omega} \cdot \nabla)^2 u_z + N^2 \nabla_\perp^2 u_z = 0, \qquad (6.70)$$

where $\nabla_\perp^2$ is the horizontal Laplacian. This supports plane waves of the form

$$\mathbf{u}(\mathbf{x}, t) = \hat{\mathbf{u}} \exp \left[j \left(\mathbf{k} \cdot \mathbf{x} - \hat{\omega} t \right) \right], \quad \mathbf{k} \cdot \hat{\mathbf{u}} = 0,$$

and the resulting dispersion relationship is

$$\hat{\omega}^2 = (2\Omega)^2 \frac{k_z^2}{k^2} + N^2 \frac{k_\perp^2}{k^2} = (2\Omega)^2 + \left(N^2 - (2\Omega)^2 \right) \frac{k_\perp^2}{k^2}, \qquad (6.71)$$

where $k_\perp^2 = k^2 - k_z^2$. The phase and group velocities are readily found from (6.71). They are

$$\mathbf{c}_p = \frac{\hat{\omega} \mathbf{k}}{k^2} = \frac{\mathbf{k}}{\hat{\omega} k^4} \left[(2\Omega)^2 k_z^2 + N^2 k_\perp^2 \right], \qquad (6.72)$$

and

$$\mathbf{c}_g = \frac{N^2 - (2\Omega)^2}{\hat{\omega} k^4} \mathbf{k} \times (\mathbf{k} \times \mathbf{k}_{//}) = \frac{N^2 - (2\Omega)^2}{\hat{\omega} k^4} \left[k_z^2 \mathbf{k}_\perp - k_\perp^2 \mathbf{k}_{//} \right], \qquad (6.73)$$

where $\mathbf{k}_{//} = k_z \hat{\mathbf{e}}_z$ and $\mathbf{k}_\perp = \mathbf{k} - \mathbf{k}_{//}$. It is readily confirmed that (3.36) and (6.18) are special cases of (6.73). These two relationships in turn give us

$$\mathbf{c}_g \cdot \boldsymbol{\Omega} = -\frac{N^2 - (2\Omega)^2}{\hat{\omega} k^4} k_\perp^2 (\boldsymbol{\Omega} \cdot \mathbf{k}), \qquad (6.74)$$

and

$$\mathbf{c}_g + \mathbf{c}_p = \frac{(2\Omega)^2}{\hat{\omega} k^2}\mathbf{k}_{//} + \frac{N^2}{\hat{\omega} k^2}\mathbf{k}_{\perp}, \tag{6.75}$$

the second of which reduces to (3.37), or else (6.19), in the appropriate limit. Finally, we note that the amplitude of the vorticity, $\hat{\boldsymbol{\omega}}$, is related to $\hat{\mathbf{u}}$ through (6.69), which yields

$$\hat{\omega}^2\hat{\boldsymbol{\omega}} = -(2\boldsymbol{\Omega} \cdot \mathbf{k})\hat{\omega}\hat{\mathbf{u}} + jN^2\mathbf{k} \times \hat{\mathbf{u}}_{//}. \tag{6.76}$$

Perhaps some comments are in order. First, the angular frequency lies between N and 2Ω, so there are no low-frequency modes. Second, as with inertial waves and internal gravity waves, the phase and group velocities are orthogonal. Third, the group velocity falls to zero when $N = 2\Omega$. Fourth, for $k_z > 0$, the energy propagation is upward ($\mathbf{c}_g \cdot \boldsymbol{\Omega} > 0$) when $2\Omega > N$, but downward ($\mathbf{c}_g \cdot \boldsymbol{\Omega} < 0$) when $N > 2\Omega$. This reversal in direction will turn out to be significant. Fifth, the first term on the right of (6.76) tells us that at least part of the velocity is in phase with, and parallel to, the vorticity, and so these waves are in general helical, but they do not have maximal helicity (unless $N = 0$).

In the oceans, the most energetic internal waves are often observed to have frequencies just marginally greater than 2Ω. These waves, which are referred to as *near-inertial waves* in the oceanographic literature, are characterized by the duel constraints

$$N^2 >> (2\Omega)^2, \quad N^2 k_{\perp}^2 << (2\boldsymbol{\Omega} \cdot \mathbf{k})^2. \tag{6.77}$$

Of course, this requires $k_z^2 >> k_{\perp}^2$, and so these waves correspond to almost horizontal motion in a strongly stratified environment. When (6.77) is combined with the expressions above, we find that near-inertial waves are characterized by

$$\hat{\omega}^2 = (2\Omega)^2 + N^2\frac{k_{\perp}^2}{k^2}, \quad \hat{\omega}\hat{\boldsymbol{\omega}} \approx -(2\boldsymbol{\Omega} \cdot \mathbf{k})\hat{\mathbf{u}}, \tag{6.78}$$

and

$$\mathbf{c}_g = \frac{N^2}{\hat{\omega} k^4}\left[k_z^2\mathbf{k}_{\perp} - k_{\perp}^2\mathbf{k}_{//}\right], \quad \mathbf{c}_g \cdot \boldsymbol{\Omega} = -\frac{N^2 k_{\perp}^2}{\hat{\omega} k^4}(\boldsymbol{\Omega} \cdot \mathbf{k}). \tag{6.79}$$

The simplified dispersion relationship in (6.78) tells us that near-inertial waves have a frequency which is marginally larger than 2Ω, consistent with the observations.

Notice that $\mathbf{k}$ and $\mathbf{c}_g$ point in the same horizontal direction, but in opposite vertical directions (see Figure 6.17), and that the sign of h is fixed by the sign of k_z. Moreover, the expression for the group velocity confirms that the direction of energy propagation is very close to horizontal, which it has to be since $\mathbf{k}$ is almost vertical and $\mathbf{c}_g$ is perpendicular to $\mathbf{k}$. One surprising consequence of the expressions above comes when we combine (6.78) and (6.79) to give

$$\hat{\boldsymbol{\omega}} \approx \frac{2(\mathbf{c}_g \cdot \boldsymbol{\Omega})k^4}{N^2 k_\perp^2}\hat{\mathbf{u}}. \tag{6.80}$$

This tells us that the helicity is positive in wave packets propagating parallel to $\boldsymbol{\Omega}$, $\mathbf{c}_g \cdot \boldsymbol{\Omega} > 0$, and negative in wave packets propagating antiparallel to $\boldsymbol{\Omega}$, $\mathbf{c}_g \cdot \boldsymbol{\Omega} < 0$, which is precisely the opposite to the helicity in conventional inertial waves. This surprising result is related to (6.74), and to the observation that, for a given k_z (*i.e.* a given sign of h), the sign of $\mathbf{c}_g \cdot \boldsymbol{\Omega}$ depends on which is the larger of 2Ω and N.

Near-inertial waves are generated primarily by wind shear, particularly during tropical storms, and can travel many hundreds of kilometres. They have large horizontal wavelengths, between 10 km and 100 km, though their vertical wavelength is only a few hundred metres. Following hurricanes and typhoons, they can acquire amplitudes as large as $|\mathbf{u}| \sim 1$ m/s, although the vertical component of their group velocity,

$$c_{g,z} = -\frac{N^2 k_\perp^2 k_z}{\hat{\omega} k^4}, \tag{6.81}$$

is small, say ~ 1 mm/s. Surface-generated wave packets travelling downward in the northern hemisphere have $\mathbf{c}_g \cdot \boldsymbol{\Omega} < 0$, and hence a negative helicity. So, when viewed from above, $\mathbf{u}$ rotates in a clockwise sense (see Figures 6.4 and

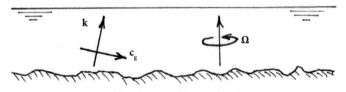

Figure 6.17 Near-inertial waves propagating downward in the northern hemisphere possess negative helicity and so $\mathbf{u}$ rotates in a clockwise sense when viewed from above.

6.17). However, the rotation is anti-clockwise in the south, where $\mathbf{c}_g \cdot \mathbf{\Omega} > 0$. Near-inertial waves are reviewed in Alford et al. (2016).

6.7 Evanescent Inertial Waves

Consider an inviscid, rotating fluid which is bounded internally by the cylindrical surface $r = R$. Axisymmetric motion of the fluid is driven by radial oscillations of that surface, say

$$\eta(z, t) = \hat{\eta} \cos(k_z z) \cos \varpi t, \quad \hat{\eta} << R, \tag{6.82}$$

which requires

$$u_r(z, t) = \partial \eta / \partial t = -\varpi \hat{\eta} \cos(k_z z) \sin \varpi t, \tag{6.83}$$

on $r = R$. Moreover, u_r is related to u_θ by the azimuthal component of Euler's equation,

$$\frac{\partial u_\theta}{\partial t} = -2\Omega u_r, \tag{6.84}$$

and so the boundary condition for u_θ is

$$u_\theta(z, t) = -2\Omega \hat{\eta} \cos(k_z z) \cos \varpi t. \tag{6.85}$$

Clearly these surface undulations will excite inertial waves when $0 < \varpi < 2\Omega$, and we now ask what happens if $\varpi > 2\Omega$. We might expect some kind of oscillatory disturbance close to the boundary, and such localized oscillations are generally referred to as *evanescent*.

The governing equations for the disturbance are, in cylindrical polar coordinates, (6.43)→(6.46). In particular, the angular momentum density, $\Gamma = ru_\theta$, is governed by the wave-like equation

$$\frac{\partial^2}{\partial t^2} \left[\frac{\partial}{\partial r} \frac{1}{r} \frac{\partial \Gamma}{\partial r} + \frac{\partial^2 \Gamma}{\partial z^2} \right] + (2\Omega)^2 \frac{\partial^2 \Gamma}{\partial z^2} = 0. \tag{6.86}$$

Given the boundary condition (6.85), we look for disturbances of the form

$$\Gamma = r\hat{u}_\theta(r) \cos(k_z z) \cos(\varpi t), \tag{6.87}$$

and (6.86) then yields

$$r^2 \hat{u}_\theta''(r) + r\hat{u}_\theta'(r) - \hat{u}_\theta - (\gamma r)^2 \hat{u}_\theta = 0,$$

where

$$\gamma = k_z \sqrt{1 - (2\Omega/\bar{\omega})^2}. \tag{6.88}$$

This has the solution

$$u_\theta = -2\Omega\hat{\eta} \frac{K_1(\gamma r)}{K_1(\gamma R)} \cos(k_z z) \cos(\bar{\omega} t), \tag{6.89}$$

where K_1 is the usual modified Bessel function. Finally, equation (6.84) gives us

$$u_r = -\bar{\omega}\hat{\eta} \frac{K_1(\gamma r)}{K_1(\gamma R)} \cos(k_z z) \sin(\bar{\omega} t). \tag{6.90}$$

Now K_1 decays like an exponential, and so our evanescent disturbance is indeed localized near the boundary. Crucially, however, $\gamma \to 0$ as $\bar{\omega}$ approaches 2Ω from above, and so in this limit (6.89) and (6.90) represent motion which is highly elongated in the radial direction. Indeed, as we approach $\bar{\omega} \to 2\Omega$ from above, (6.89) and (6.90) become

$$\Gamma = -2\Omega\hat{\eta}R \cos(k_z z) \cos(2\Omega t), \tag{6.91}$$

$$u_r = -\bar{\omega}\hat{\eta}\frac{R}{r} \cos(k_z z) \sin(2\Omega t). \tag{6.92}$$

Just such radially elongated disturbances have been observed by Nosan et al. (2021), as shown in Figure 6.18. Nosan et al. have also shown that there is a smooth analytical transition from inertial waves to evanescent motion as $\bar{\omega} \to 2\Omega$ from below and from above. Note that, at $\bar{\omega} = 2\Omega$, the radial variation in velocity is neither oscillatory nor exponential, but rather a power law. Note also that, in principle, the angular momentum extends out to infinity, although in practice there is a viscous decay, as described in Nosan et al. (2021).

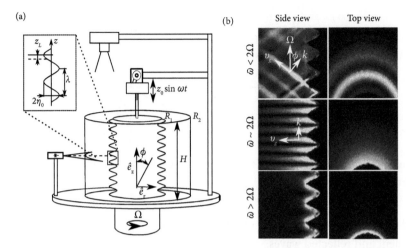

Figure 6.18 (a) Apparatus used by Nosan et al. (b) Kinetic energy in vertical and horizontal planes for $\bar{\omega}/2\Omega = 0.72$ (top), 1 and 1.5. Adapted from Nosan et al. (2021).

6.8 The Reflection of Inertial Waves from Plane Surfaces

We now consider the reflection of inviscid inertial waves from a plane surface. Because inertial waves are dispersive, their reflection properties are more complicated than those encountered in, say, geometrical optics. In the interests of simplicity, we shall restrict ourselves to cases where Ω, $\mathbf{k}$, and $\mathbf{k'}$ are coplanar, the prime indicating a reflected wave.

Let ϕ be the angle measured from the rotation axis to the wavevector $\mathbf{k}$ of a plane inertial wave, with $-\pi \leq \phi \leq \pi$. Then, as discussed in §6.2.1, (6.16) and (6.18) require

$$\bar{\omega} = \pm 2\Omega \cos \phi, \quad |\mathbf{c}_g| = (2\Omega/k)|\sin \phi|, \tag{6.93}$$

where the negative sign applies when $|\phi| > \pi/2$. Now consider an incident wave whose group velocity is directed upward, as shown in Figure 6.19(a). The incident wave reflects off a flat boundary and we wish to find the form of the reflected wave. Equation (6.18) tells us that the group velocity of the incident wave is co-planar with Ω and $\mathbf{k}$, and so we consider the plane containing $\mathbf{c}_g$, Ω, and $\mathbf{k}$. Moreover, (6.19) requires that

$$\mathbf{c}_g + \mathbf{c}_p = \pm 2\Omega/k, \tag{6.94}$$

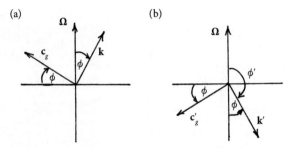

Figure 6.19 The orientations of the incident and reflected waves for a horizontal boundary.

and so the horizontal components of $\mathbf{k}$ and $\mathbf{c}_g$ are of opposite signs in both the incident and reflected waves.

If we use a prime to denote the reflected wave, then the velocity associated with the combined incident and reflected wave is

$$\mathbf{u} = \hat{\mathbf{u}} \exp j(\mathbf{k} \cdot \mathbf{x} - \bar{\omega}t) + \hat{\mathbf{u}}' \exp j(\mathbf{k}' \cdot \mathbf{x} - \bar{\omega}'t),$$

and we must choose the reflected wave so as to ensure $\mathbf{u} \cdot \mathbf{n} = 0$ at the boundary, where $\mathbf{n}$ is a unit normal to the surface. Since this must be satisfied at all times and at all points on the surface, we require that the arguments in the two exponentials are identical, which in turn demands that $\bar{\omega} = \bar{\omega}'$ and $\mathbf{k} \cdot \mathbf{t} = \mathbf{k}' \cdot \mathbf{t}$, where $\mathbf{t}$ is a unit tangential vector to the surface. When both of these prerequisites are satisfied, the boundary condition simplifies to $(\hat{\mathbf{u}} + \hat{\mathbf{u}}') \cdot \mathbf{n} = 0$. The first of these two restrictions, $\bar{\omega} = \bar{\omega}'$, tells us that $\cos \phi = \pm \cos \phi'$, and when $\phi' > 0$, this requires $\phi' = \pi - \phi$ and $\sin \phi = \sin \phi'$. The orientations of $\Omega, \mathbf{c}_g, \mathbf{k}, \mathbf{c}'_g$ and $\mathbf{k}'$, which are coplanar, are shown in Figure 6.19 for just such a case. Note that, for the important case of low-frequency waves (*i.e.* $\phi = \pi/2$), the incident and reflected waves are parallel and antiparallel to Ω, irrespective of the orientation of the boundary.

So far we have said nothing about the orientation of the boundary. Consider first the case where the boundary is horizontal. Then $\mathbf{k} \cdot \mathbf{t} = k \sin \phi$ and $\mathbf{k}' \cdot \mathbf{t} = k' \sin \phi'$, and so $\mathbf{k} \cdot \mathbf{t} = \mathbf{k}' \cdot \mathbf{t}$ requires that $k \sin \phi = k' \sin \phi'$. Since $\sin \phi = \sin \phi'$, we deduce that $k = k'$ and

$$\left| \mathbf{c}'_g \right| = \left(2\Omega/k' \right) \sin \phi' = (2\Omega/k) \sin \phi = \left| \mathbf{c}_g \right|. \tag{6.95}$$

It follows that $\mathbf{c}_g$ and $\mathbf{c}'_g$ have equal horizontal components but opposite vertical components, as shown in Figure 6.20.

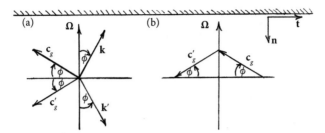

Figure 6.20 The reflection of a monochromatic inertial wave off a plane, horizontal boundary: (a) c_g and c'_g are tail to tail; (b) c_g and c'_g are nose to tail.

The case where the boundary is not horizontal is more complicated. Here the ratio of k' to k is no longer unity, and this in turn changes the ratio of c'_g to c_g. Consider the case where the boundary is rotated about an axis which is normal to the plane containing Ω and $\mathbf{k}$, which ensures Ω, $\mathbf{k}$, and $\mathbf{k}'$ remain coplanar. Also, let β be the clockwise angle the surface makes to the horizontal (see Figure 6.21), with β restricted to $-\phi < \beta < \phi$. (The case of $|\beta| > \phi$ is discussed below.) We now have $\phi' = \pi - \phi$ and $\mathbf{k} \cdot \mathbf{t} = k \sin(\phi - \beta)$, and so

$$\mathbf{k}' \cdot \mathbf{t} = k' \sin(\phi' - \beta) = k' \sin\left(\pi - (\phi + \beta)\right) = k' \sin\left(\phi + \beta\right).$$

The expression $\mathbf{k} \cdot \mathbf{t} = \mathbf{k}' \cdot \mathbf{t}$ now demands

$$\frac{k'}{k} = \frac{\sin(\phi - \beta)}{\sin(\phi + \beta)}. \tag{6.96}$$

Finally, since $\sin \phi = \sin \phi'$, we conclude that

$$|c'_g| = \frac{2\Omega}{k'} \sin \phi' = \frac{2\Omega}{k} \sin \phi \frac{k}{k'} = |c_g| \frac{\sin(\phi + \beta)}{\sin(\phi - \beta)}. \tag{6.97}$$

For $\beta > 0$, the reflected wave has a longer wavelength than the incident wave ($k' < k$), and hence a greater group velocity. This case is shown in Figure 6.21. Since the incident and reflected energy fluxes must be equal, this increase in group velocity is accompanied by a relative fall in the amplitude of the reflected wave. Note that c_g and c'_g are bisected by Ω, and *not* by the normal $\mathbf{n}$, as would be expected in geometrical optics.

Conversely, for $\beta < 0$, the wavelength of the reflected wave is shorter than that of the incident wave. The group velocity is now reduced and the wave amplitude increased relative to the incident wave. The limiting case of

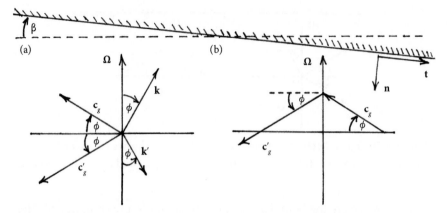

Figure 6.21 The reflection of an inertial wave off an inclined boundary for the case where β is restricted to $-\phi < \beta < \phi$. (a) $\mathbf{c}_g$ and $\mathbf{c}'_g$ are tail to tail; (b) $\mathbf{c}_g$ and $\mathbf{c}'_g$ are nose to tail.

$\beta \to -\phi$ (or $\bar{\omega} = -2\boldsymbol{\Omega} \cdot \mathbf{n}$) corresponds to the reflected wave grazing the surface ($\mathbf{c}'_g$ antiparallel to $\mathbf{t}$), and to $c'_g \to 0$ and $\hat{u}' \to \infty$. In such cases an intense boundary layer can develop that absorbs the reflected energy. Also, the large wave amplitude can lead to boundary-layer instabilities (Greenspan, 1968).

When $|\beta| > \phi$ the reflection pattern shown in Figure 6.21 does not apply, because either $\mathbf{c}_g$ points away from the boundary ($\beta > 0$), or else $\mathbf{c}'_g$ points towards the boundary ($\beta < 0$). Consider the case where $\beta < 0$ and we have a steep slope, $|\beta| > \phi$. To ensure that $\mathbf{c}'_g$ points away from the boundary, $\mathbf{k}'$ must be repositioned as shown in Figure 6.22. Note that $\cos \phi' = \cos \phi$, and so we still satisfy $\bar{\omega} = \bar{\omega}'$. For this new configuration we have

$$\mathbf{k} \cdot \mathbf{t} = k \sin \left(|\beta| + \phi\right), \quad \mathbf{k}' \cdot \mathbf{t} = k' \sin \left(|\beta| - \phi\right),$$

and so $\mathbf{k} \cdot \mathbf{t} = \mathbf{k}' \cdot \mathbf{t}$ now demands

$$\frac{k'}{k} = \frac{\sin \left(|\beta| + \phi\right)}{\sin \left(|\beta| - \phi\right)}, \tag{6.98}$$

and we conclude that

$$|\mathbf{c}'_g| = \frac{2\Omega}{k'} \left|\sin \phi'\right| = \frac{k}{k'} \frac{2\Omega}{k} \left|\sin \phi\right| = \frac{\sin \left(|\beta| - \phi\right)}{\sin \left(|\beta| + \phi\right)} |\mathbf{c}_g| . \tag{6.99}$$

Note that, in the limiting case of $|\beta| \to \phi$, the reflected wave grazes the surface and $c'_g \to 0$.

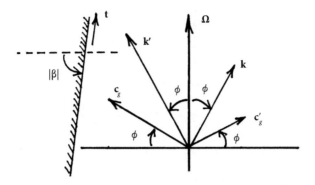

Figure 6.22 Reflection of an inertial wave off an inclined boundary for the case of $|\beta| > \phi$.

6.9 Finite Amplitude Inertial Waves

In §2.10.2 we used the Squire–Long equation to show that axisymmetric inertial waves of arbitrary amplitude can, in principle, exist in certain geometries. However, that discussion is restricted to axisymmetric waves. To explore the more general case of finite-amplitude inertial waves without axial symmetry, it is helpful to look at the experimental evidence.

Consider, for example, the experiment of Davidson et al. (2006), shown schematically in Figure 6.23. Here turbulence is created in the upper part of a tank of water by dragging a planar grid (or mesh) of bars part way through the tank and then removing the grid. This generates turbulence in which $Ro = u/\Omega\delta >> 1$, u being the rms velocity fluctuation and δ a measure of the size of the large eddies. Since $Ro >> 1$, no inertial waves are created during the generation of the turbulence. However, as the turbulent energy decays, Ro drifts downwards and eventually we enter the regime in which $Ro \sim O(1)$. When Ro reaches a value of ≈ 0.4, columnar vortices start to emerge from the turbulent cloud, propagating in the axial direction, as shown schematically in Figure 6.23. Of course, this experiment is similar to the numerical simulation shown in Figure 6.9, only now we have $Ro \sim O(1)$, rather than $Ro << 1$.

Given that the columnar structures in Figure 6.9 are nothing more than inertial wave packets of small amplitude, it seems likely that the columnar structures in this experiment are also inertial wave packets, though of finite amplitude. This conjecture is readily confirmed. Measurements show that the columnar vortices elongate at a constant rate. This can be seen from Figure 6.24(a), which shows the axial length, Δz, versus time of the dominant columnar vortex in six experiments which had different rotation rates and different

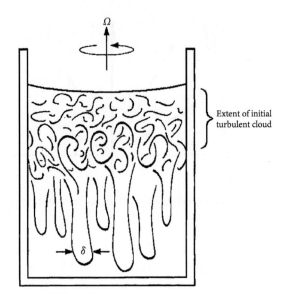

Figure 6.23 Schematic of the experiment of Davidson et al. (2006)

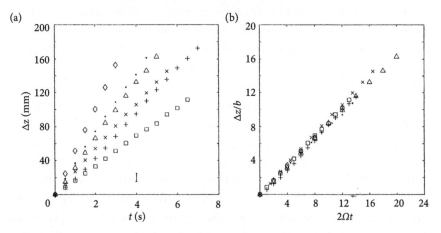

Figure 6.24 The variation of axial length, Δz, versus time of the dominant columnar vortex in six experiments which had different rotation rates and different bar sizes, b. (a) Δz versus t shows that each columnar vortex grows at a constant rate. (b) $\Delta z/b$ versus $2\Omega t$ confirms that the columnar vortices grow as $\Delta z \sim b\Omega t$.

bar sizes, b. Moreover, the growth rate is proportional to Ω and to the transverse scale of the vortices, δ, for which the bar size b acts as a proxy. This is confirmed in Figure 6.24(b), which shows the same data, but plotted as $\Delta z/b$ versus $2\Omega t$. Thus, we have $\Delta z \sim \delta\Omega t$, which confirms that the columnar vortices are indeed inertial wave packets.

One of the interesting features of this experiment is that it suggests that inertial wave packets can propagate with amplitudes of up to Ro = $\hat{u}/\Omega\lambda \approx 0.4$, where λ is the dominant horizontal wavelength and $\hat{u}$ is the wave amplitude. This has since been confirmed in numerical simulations. Indeed, there is some evidence that the transition from linear-like wave propagation to the complete suppression of inertial waves is surprisingly abrupt, with a threshold of around $u/\Omega\lambda \sim 0.4$.

$$* * *$$

This concludes our introduction to progressive inertial waves. The subject is a large one, and so we had to be selective in the choice of material. Readers looking for additional sources will find helpful overviews of inertial waves in Greenspan (1968) and Moffatt (1978), with Greenspan particularly good on their relationship to Taylor columns. Moreover, near-inertial waves in the oceans are reviewed by Alford et al. (2016), while evanescent inertial waves are discussed in Nosan et al. (2021).

References

Alford, M.H., MacKinnon, J.A., Simmons, H.L., & Nash, J.D., 2016, Near-inertial internal gravity waves in the ocean, *Ann. Rev. Mar. Sci.*, **8**, 95–123.

Bracewell, R.N., 1986, *The Fourier Transform and its Applications*, 2nd Ed., McGraw-Hill.

Davidson, P.A., 2013, *Turbulence in Rotating, Stratified and Electrically Conducting Fluids*, Cambridge University Press.

Davidson, P.A., & Ranjan, A., 2015, Planetary dynamos driven by helical waves: Part 2, *Geophys. J. Int.*, **202**, 1646–62.

Davidson, P.A., Staplehurst, P.J., & Dalziel, S.B., 2006, On the evolution of eddies in a rapidly-rotating system, *J. Fluid Mech.*, **557**, 135–45.

Greenspan, H.P., 1968, *The Theory of Rotating Fluids*, Cambridge University Press.

Moffatt, H.K., 1978, *Magnetic Field Generation in Electrically Conducting Fluids*, Cambridge University Press.

Nosan, Ž., Burmann, F., Davidson, P.A., & Noir, J., 2021, Evanescent inertial waves. *J. Fluid Mech.*, **918**, R2.

Rayleigh, Lord, 1916, On the dynamics of revolving fluids, *Proc. Royal Soc., A*, **93**, 148.

Veronis, G., 1959, Cellular convection with finite amplitude in a rotating fluid, *J. Fluid Mech.*, **5**(3), 401–35.

Chapter 7

Inertial Waves II

Inviscid, Linear Modes in Bounded, Axisymmetric Domains

We turn now to inviscid inertial modes in closed, axisymmetric domains where the symmetry axis is aligned with Ω. As we shall see, such modes include standing waves and azimuthally drifting waves. We start with some general observations.

7.1 General Properties of Inertial Modes in Bounded Domains

7.1.1 From Progressive Waves to Modes

Recall that a monochromatic, progressive inertial wave takes the form

$$\mathbf{u} = \hat{u}_X \left(\cos(kZ - \varpi t),\ \pm \sin(kZ - \varpi t), 0 \right), \tag{7.1}$$

in (X, Y, Z) coordinates with Z parallel to $\mathbf{k}$. (See equation (6.22).) This is a circularly polarized wave with a velocity that rotates about the Z-axis as the wave propagates, as shown in Figure 7.1. All of the energy is kinetic, which is the origin of the name 'inertial' wave, and that energy bounces back and forth between u_X and u_Y, there being an equipartition of energy between the two velocity components. Recall also that the wave frequency is preserved when such a wave reflects off a plane surface. It is natural, therefore, to look for inertial modes in a confined domain of the form $\mathbf{u} = \Re \left\{ \hat{\mathbf{u}}(\mathbf{x}) \exp(j\varpi t) \right\}$, where $\hat{\mathbf{u}}$ is complex and $\Re$ indicates the real part. Moreover, we might anticipate that there is some form of equipartition of kinetic energy between velocity components in such modes.

We shall restrict ourselves to axisymmetric domains and use cylindrical polar coordinates, (r, θ, z), with $\Omega = \Omega \hat{\mathbf{e}}_z$. Homogeneity in θ then allows us to write all inertial modes in the form $\mathbf{u} = \hat{\mathbf{u}}(r, z) \exp \left(j(\varpi t + m\theta) \right)$, where m is the azimuthal wavenumber and $m = 0, 1, 2, 3....$. There are clearly two distinct possibilities. Axisymmetric modes ($m = 0$) take the form $\mathbf{u} = \hat{\mathbf{u}}(r, z) \exp \left(j\varpi t \right)$,

The Dynamics of Rotating Fluids. P. A. Davidson, Oxford University Press. © Peter A Davidson (2024).
DOI: 10.1093/9780191994272.003.0007

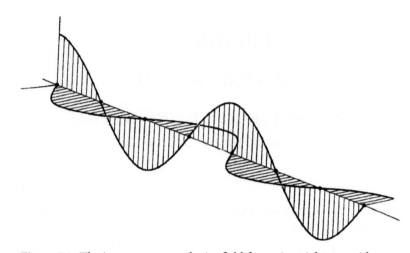

Figure 7.1 The instantaneous velocity field for an inertial wave with negative helicity.

which we shall refer to as *standing inertial waves*. However, non-axisymmetric modes propagate in the azimuthal direction, either as *prograde waves* (in the same sense as Ω), or else as *retrograde waves* (in the opposite sense to Ω). We shall refer to these as *drifting inertial waves*. It turns out to be convenient to restrict m to be positive. This, in turn, requires that we break with our convention of insisting that $\varpi > 0$. So, in this chapter, but *only* in this chapter, we shall relax our restriction on ϖ and allow it to take either sign, with $\varpi < 0$ representing a prograde wave and $\varpi > 0$ a retrograde wave. Of course, $0 < |\varpi| < 2\Omega$, as for progressive waves.

7.1.2 The Orthogonality of Distinct Modes and the Equipartition of Kinetic Energy

Consider two distinct solutions, $\mathbf{u}$ and $\mathbf{u}'$, of the governing equation of motion

$$\frac{\partial \mathbf{u}}{\partial t} = 2\mathbf{u} \times \Omega - \nabla \left(p/\rho \right), \quad \mathbf{u} \cdot d\mathbf{S} = 0, \tag{7.2}$$

where $d\mathbf{S}$ is part of the boundary. Then

$$\rho \frac{\partial}{\partial t} (\mathbf{u} \cdot \mathbf{u}') = 2\rho \left[\mathbf{u}' \cdot (\mathbf{u} \times \Omega) + \mathbf{u} \cdot (\mathbf{u}' \times \Omega) \right] - \nabla \cdot \left(p\mathbf{u}' + p'\mathbf{u} \right)$$

$$= -\nabla \cdot \left(p\mathbf{u}' + p'\mathbf{u} \right),$$

which integrates over a bounded domain to give

$$\frac{d}{dt} \int \mathbf{u} \cdot \mathbf{u}' dV = 0. \tag{7.3}$$

Now suppose that

$$\mathbf{u} = \Re\left\{\hat{\mathbf{u}}(\mathbf{x}) \exp(j\varpi t)\right\} = \Re\left\{\left[\hat{\mathbf{u}}_R + j\hat{\mathbf{u}}_I\right] \exp(j\varpi t)\right\},$$

and

$$\mathbf{u}' = \Re\left\{\hat{\mathbf{u}}'(\mathbf{x}) \exp(j\varpi' t)\right\} = \Re\left\{\left[\hat{\mathbf{u}}'_R + j\hat{\mathbf{u}}'_I\right] \exp(j\varpi' t)\right\},$$

where subscripts R and I indicate the real and imaginary parts of a complex quantity. Then

$$\mathbf{u} \cdot \mathbf{u}' = \hat{\mathbf{u}}_R \cdot \hat{\mathbf{u}}'_R \cos \varpi t \cos \varpi' t + \hat{\mathbf{u}}_I \cdot \hat{\mathbf{u}}'_I \sin \varpi t \sin \varpi' t$$
$$- \hat{\mathbf{u}}_R \cdot \hat{\mathbf{u}}'_I \cos \varpi t \sin \varpi' t - \hat{\mathbf{u}}_I \cdot \hat{\mathbf{u}}'_R \sin \varpi t \cos \varpi' t ,$$

from which we find

$$2\mathbf{u} \cdot \mathbf{u}' = \left[\hat{\mathbf{u}}_R \cdot \hat{\mathbf{u}}'_R - \hat{\mathbf{u}}_I \cdot \hat{\mathbf{u}}'_I\right] \cos\left((\varpi + \varpi')t\right)$$
$$- \left[\hat{\mathbf{u}}_R \cdot \hat{\mathbf{u}}'_I + \hat{\mathbf{u}}_I \cdot \hat{\mathbf{u}}'_R\right] \sin\left((\varpi + \varpi')t\right)$$
$$+ \left[\hat{\mathbf{u}}_R \cdot \hat{\mathbf{u}}'_R + \hat{\mathbf{u}}_I \cdot \hat{\mathbf{u}}'_I\right] \cos\left((\varpi - \varpi')t\right)$$
$$+ \left[\hat{\mathbf{u}}_R \cdot \hat{\mathbf{u}}'_I - \hat{\mathbf{u}}_I \cdot \hat{\mathbf{u}}'_R\right] \sin\left((\varpi - \varpi')t\right). \tag{7.4}$$

If $\mathbf{u}$ and $\mathbf{u}'$ are the same mode, *i.e.* $\varpi = \varpi'$ and $\hat{\mathbf{u}} = \hat{\mathbf{u}}'$, then (7.3) and (7.4) demand

$$\int \hat{\mathbf{u}}_R^2 dV = \int \hat{\mathbf{u}}_I^2 dV, \quad \int \hat{\mathbf{u}}_R \cdot \hat{\mathbf{u}}_I dV = 0, \tag{7.5}$$

which is a form of equipartition of energy. On the other hand, when $\varpi \neq \pm \varpi'$, then (7.3) and (7.4) require

$$\int \hat{\mathbf{u}}_R \cdot \hat{\mathbf{u}}'_R dV = \int \hat{\mathbf{u}}_I \cdot \hat{\mathbf{u}}'_I dV = \int \hat{\mathbf{u}}_R \cdot \hat{\mathbf{u}}'_I dV = 0, \tag{7.6}$$

which tells us that distinct inertial modes are orthogonal.

7.1.3 Angular Momentum Conservation and the Mean Circulation Theorem

We now consider the consequences of angular momentum conservation for inertial modes. To focus thoughts, we shall consider the case where the domain is a sphere of radius R, though the key results generalize in an obvious way to other axisymmetric domains. Let V_C be a cylinder control volume that spans the sphere and whose axis is aligned with $\mathbf{\Omega} = \Omega \hat{\mathbf{e}}_z$, passing through the centre of the sphere. The cylinder has radius R_C, a cylindrical surface S_C, and top and bottom surfaces S_{TC} and S_{BC}. Also, let S_T and S_B be the spherical caps that span the top and bottom of the cylinder, as shown in Figure 7.2.

The axial component of the vorticity equation can be written as

$$\frac{\partial \omega_z}{\partial t} = 2(\mathbf{\Omega} \cdot \nabla)u_z = 2\Omega \nabla \cdot (u_z \hat{\mathbf{e}}_z), \tag{7.7}$$

which integrates over V_C to give the important result

$$\frac{d}{dt} \int_{V_C} \omega_z dV = 2\Omega \left\{ \int_{S_{TC}} \mathbf{u} \cdot d\mathbf{S} - \int_{S_{BC}} \mathbf{u} \cdot d\mathbf{S} \right\} = 0. \tag{7.8}$$

It is natural to suspect that (7.8) is a manifestation of angular momentum conservation. That suspicion is fuelled by the fact that

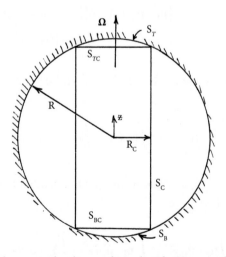

Figure 7.2 A cylinder control volume, aligned with $\mathbf{\Omega}$, spans the interior of a sphere.

$$\omega_z = \frac{1}{r}\frac{\partial}{\partial r}(ru_\theta) - \frac{1}{r}\frac{\partial u_r}{\partial \theta},$$

in (r, θ, z) coordinates, and so (7.8) can be rewritten as

$$\frac{d}{dt}\int_{S_C} u_\theta dS = 0. \tag{7.9}$$

Let us see if we can deduce (7.9) directly from angular momentum conservation. The angular momentum density of the fluid is governed by the cross product of $\mathbf{x}$ with (7.2),

$$\frac{\partial}{\partial t}(\mathbf{x} \times \mathbf{u}) = 2\mathbf{x} \times (\mathbf{u} \times \mathbf{\Omega}) + \nabla \times (p\mathbf{x}/\rho).$$

Moreover, the z component of this may be rewritten using the identity

$$\nabla \cdot \left[\left(x^2 - z^2\right)\mathbf{\Omega}\mathbf{u}\right] = 2\Omega(\mathbf{x} \cdot \mathbf{u} - zu_z) = -[2\mathbf{x} \times (\mathbf{u} \times \mathbf{\Omega})]_z,$$

which yields

$$\frac{\partial}{\partial t}(\mathbf{x} \times \mathbf{u})_z = -\nabla \cdot \left[r^2\mathbf{\Omega}\mathbf{u}\right] + \left[\nabla \times (p\mathbf{x}/\rho)\right]_z, \tag{7.10}$$

in (r, θ, z) coordinates. We now integrate (7.10) over the volume enclosed by S_C, S_T, and S_B, while noting that the pressure term converts to a surface integral which is readily shown to be zero. (Physically, there is no net axial torque acting on this volume arising from pressure.) The end result is

$$\frac{d}{dt}\int r u_\theta dV = -R_C^2\Omega\int_{S_C} \mathbf{u} \cdot d\mathbf{S} = 0, \tag{7.11}$$

where we have used conservation of mass to equate the surface integral to zero.

Now suppose that we apply (7.11) to the annulus created by two cylindrical surfaces of slightly different radii, say R_C and $R_C + \delta R_C$. Then the angular momentum within the annulus is clearly conserved and we have

$$\frac{d}{dt}\int_{S_C} r u_\theta dS = R_C\frac{d}{dt}\int_{S_C} u_\theta dS = 0,$$

which brings us back to (7.9).

Equation (7.9) is often couched in terms of circulation. Let us rewrite the equation as

$$\frac{d}{dt} \int \left\{ \oint_C \mathbf{u} \cdot d\mathbf{r} \right\} dz = 0, \tag{7.12}$$

where C is a circle of radius R_C that sits on the cylindrical surface. We can interpret (7.12) as saying that the depth-averaged circulation is conserved on R_C. Now suppose $\mathbf{u}$ is an inertial mode, *i.e.* $\mathbf{u} = \hat{\mathbf{u}}(\mathbf{x}) \exp(j\bar{\omega}t)$, where the real part is understood. Then we have

$$\int \left\{ \oint_C \hat{\mathbf{u}} \cdot d\mathbf{r} \right\} dz = 0, \ \bar{\omega} \neq 0, \tag{7.13}$$

and the depth-averaged circulation is necessarily zero.

This generalizes in an obvious way to other axisymmetric bodies (whose axis of symmetry is aligned with $\mathbf{\Omega}$), and the underlying principle is always angular momentum conservation. It also generalizes to certain non-axisymmetric bodies, where it applies to any cylindrical control volume of *constant depth* that is aligned with $\mathbf{\Omega}$ and spans the interior of the body from the bottom boundary to the top. In particular, the cylindrical control volume need not have a circular cross-section. To show this, we let V_C be *any* cylindrical control volume that spans the domain and is bounded top and bottom by the *horizontal* surfaces S_{TC} and S_{BC}. Then, irrespective of the cross-sectional shape of the cylinder, (7.8) yields

$$\frac{d}{dt} \int_{V_C} \omega_z dV = 2\Omega \left\{ \int_{S_{TC}} \mathbf{u} \cdot d\mathbf{S} - \int_{S_{BC}} \mathbf{u} \cdot d\mathbf{S} \right\} = 0, \tag{7.14}$$

because there is zero net mass flux through S_{TC} and S_{BC}. We now use Stokes theorem to rewrite this as

$$\frac{d}{dt} \int \left\{ \oint_C \mathbf{u} \cdot d\mathbf{r} \right\} dz = 0, \tag{7.15}$$

where C is a horizontal contour that lies on the surface of the cylinder. This is a particular example of a more general result known as the *mean circulation theorem.* The full significance of this theorem for inertial modes is discussed in Greenspan (1968).

7.2 Inertial Modes in a Cylinder: An Illustrative Example

The detailed mathematical structure of most confined inertial modes is rather complex, even for very simple geometries, such as a sphere. Luckily, however, modes in a circular cylinder turn out to be reasonably accessible. So we shall focus here on inertial modes in a circular cylinder of height H and radius R (see Figure 7.3), primarily as an illustrative example which provides some hints as to the structure of modes in less accessible geometries. We shall use cylindrical polar coordinates, (r, θ, z), throughout.

7.2.1 Axisymmetric Modes as Standing Waves

As noted in §6.4, any axisymmetric flow must satisfy the kinematic relationships

$$\mathbf{u}_p = \nabla \times [(\Psi/r)\hat{\mathbf{e}}_\theta], \quad \boldsymbol{\omega}_p = \nabla \times [(\Gamma/r)\,\hat{\mathbf{e}}_\theta], \quad \nabla_*^2\Psi = -r\omega_\theta, \qquad (7.16)$$

where $\Gamma = ru_\theta$, Ψ is the Stokes streamfunction, the subscript p stands for poloidal, and ∇_*^2 is the Stokes operator, (2.48). Moreover, the azimuthal component of (7.2), and of the corresponding vorticity equation, yields

$$\frac{\partial\Gamma}{\partial t} = -2\Omega ru_r = 2\Omega\frac{\partial\Psi}{\partial z}, \qquad (7.17)$$

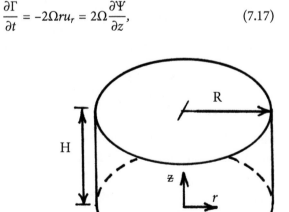

Figure 7.3 Geometry considered in this section.

and

$$\frac{\partial}{\partial t} r\omega_\theta = -\frac{\partial}{\partial t} \nabla_*^2 \Psi = 2\Omega \frac{\partial \Gamma}{\partial z}, \tag{7.18}$$

which combine to give the wave-like equations

$$\frac{\partial^2}{\partial t^2} \nabla_*^2 \Psi + (2\Omega)^2 \frac{\partial^2 \Psi}{\partial z^2} = 0, \tag{7.19}$$

$$\frac{\partial^2}{\partial t^2} \nabla_*^2 \Gamma + (2\Omega)^2 \frac{\partial^2 \Gamma}{\partial z^2} = 0. \tag{7.20}$$

It is easier to work with Ψ, rather than Γ, since the corresponding boundary conditions are simply $\Psi = 0$. So let us look for standing waves of the form

$$\Psi = \hat{\Psi}(r) \sin(n\pi z/H) \exp(j\varpi t), \quad n = 1, 2, 3..., \tag{7.21}$$

which clearly satisfies the boundary condition at the base, $z = 0$, and top, $z = H$, of the cylinder. The wave equation (7.19) now yields

$$r^2 \frac{d^2}{dr^2}\left(\frac{\hat{\Psi}}{r}\right) + r\frac{d}{dr}\left(\frac{\hat{\Psi}}{r}\right) + \left\{r^2\left(\frac{n\pi}{H}\right)^2\left[\left(\frac{2\Omega}{\varpi}\right)^2 - 1\right] - 1\right\}\frac{\hat{\Psi}}{r} = 0.$$

It is convenient to introduce

$$\beta = \frac{n\pi}{H}\sqrt{(2\Omega/\varpi)^2 - 1}, \tag{7.22}$$

and rescale r as $\eta = \beta r$. The governing equation for $\hat{\Psi}/r$ is then

$$\eta^2 \frac{d^2}{d\eta^2}\left(\frac{\hat{\Psi}}{r}\right) + \eta\frac{d}{d\eta}\left(\frac{\hat{\Psi}}{r}\right) + (\eta^2 - 1)\frac{\hat{\Psi}}{r} = 0, \tag{7.23}$$

whose solution is, to within an arbitrary pre-factor, $\hat{\Psi}/r \sim J_1(\eta)$. We conclude that the solution for Ψ, which has an arbitrary amplitude, can be written as

$$\Psi = r J_1(\beta r) \sin(n\pi z/H) \cos(\varpi t). \tag{7.24}$$

It remains to satisfy the boundary condition $\Psi = 0$ at $r = R$, and it is this that fixes the discrete frequencies associated with these standing waves. Let $\delta_{1,p}$ be the zeros of J_1, with $p = 1, 2, 3...$ and $\delta_{1,1} = 3.832$. Then $\beta R = \delta_{1,p}$, and so (7.22) can be rearranged to give the wave frequencies as a function of n and p,

$$\frac{\varpi_{n,p}}{2\Omega} = \pm\frac{n\pi/H}{\sqrt{\left(\delta_{1,p}/R\right)^2 + (n\pi/H)^2}} = \pm\frac{k_{z,n}}{\sqrt{k_{r,p}^2 + k_{z,n}^2}}, \tag{7.25}$$

where

$$k_{r,p} = \delta_{1,p}/R, \qquad k_{z,n} = n\pi/H, \tag{7.26}$$

are the discrete wavenumbers in the radial and axial directions. Note the similarity between (7.25) and (6.16) in the form $\varpi/2\Omega = \pm k_z/k$. In any event, Ψ can now be rewritten as

$$\Psi_{n,p} = rJ_1\left(\delta_{1,p}r/R\right)\sin\left(n\pi z/H\right)\cos\left(\varpi_{n,p}t\right). \tag{7.27}$$

Having found Ψ, the poloidal velocity can be obtained from $\mathbf{u}_p = \nabla \times [(\Psi/r)\hat{\mathbf{e}}_\theta]$, while the azimuthal velocity is dictated by (7.17). A little algebra now yields

$$u_r^{(n,p)} = -\frac{n\pi}{H}J_1\left(\delta_{1,p}r/R\right)\cos\left(n\pi z/H\right)\cos\left(\varpi_{n,p}t\right), \tag{7.28}$$

$$u_\theta^{(n,p)} = \pm\sqrt{\left(\frac{\delta_{1,p}}{R}\right)^2 + \left(\frac{n\pi}{H}\right)^2}J_1\left(\delta_{1,p}r/R\right)\cos\left(n\pi z/H\right)\sin\left(\varpi_{n,p}t\right), \tag{7.29}$$

$$u_z^{(n,p)} = \frac{\delta_{1,p}}{R}J_0\left(\delta_{1,p}r/R\right)\sin\left(n\pi z/H\right)\cos\left(\varpi_{n,p}t\right). \tag{7.30}$$

Alternatively, in terms of $\mathbf{u} = \hat{\mathbf{u}}(r, z)\exp\left(j\varpi t\right)$, we have

$$\hat{u}_r^{(n,p)} = -\frac{n\pi}{H}J_1\left(\delta_{1,p}r/R\right)\cos\left(n\pi z/H\right), \tag{7.31}$$

$$\hat{u}_\theta^{(n,p)} = \mp j\sqrt{\left(\frac{\delta_{1,p}}{R}\right)^2 + \left(\frac{n\pi}{H}\right)^2}J_1\left(\delta_{1,p}r/R\right)\cos\left(n\pi z/H\right), \tag{7.32}$$

$$\hat{u}_z^{(n,p)} = \frac{\delta_{1,p}}{R}J_0\left(\delta_{1,p}r/R\right)\sin\left(n\pi z/H\right). \tag{7.33}$$

It is readily confirmed that, for a single inertial mode,

$$\int \hat{\mathbf{u}}_R^2 dV = \int \hat{\mathbf{u}}_I^2 dV, \qquad \int \hat{\mathbf{u}}_R \cdot \hat{\mathbf{u}}_I dV = 0, \tag{7.34}$$

in accordance with (7.5). Moreover, since J_0 and J_1 satisfy the orthogonality conditions

$$\int_0^1 sJ_0\left(\delta_{1,p}s\right) J_0\left(\delta_{1,q}s\right) ds = 0, \quad p \neq q,$$

$$\int_0^1 sJ_1\left(\delta_{1,p}s\right) J_1\left(\delta_{1,q}s\right) ds = 0, \quad p \neq q,$$

where $s = r/R$, distinct axisymmetric modes satisfy

$$\int \hat{\mathbf{u}}_R \cdot \hat{\mathbf{u}}_R' dV = \int \hat{\mathbf{u}}_I \cdot \hat{\mathbf{u}}_I' dV = \int \hat{\mathbf{u}}_R \cdot \hat{\mathbf{u}}_I' dV = 0, \quad \varpi_{n,p} \neq \varpi'_{n,p}, \tag{7.35}$$

in accordance with (7.6).

So much for the axisymmetric modes, or standing waves. The analysis becomes somewhat more involved when we turn to non-axisymmetric modes, which drift in θ.

7.2.2 Non-Axisymmetric Modes as Azimuthally Drifting Waves

For non-axisymmetric modes, it is convenient to work directly with the three velocity components, whose axial and azimuthal forms can be fixed from the outset. For example, homogeneity in θ allows us to adopt the notation $\mathbf{u} = \hat{\mathbf{u}}(r, z) \exp\left(j(\varpi t + m\theta)\right)$ for the mth azimuthal mode, with $\varpi < 0$ representing a prograde wave and $\varpi > 0$ a retrograde wave. Moreover, since u_z vanishes on the top and bottom of the cylinder, we have $\hat{u}_z \sim \sin\left(n\pi z/H\right)$ for the nth axial mode. Also, continuity takes the form

$$\frac{1}{r}\frac{\partial}{\partial r}(r\hat{u}_r) + \frac{jm\hat{u}_\theta}{r} + \frac{\partial \hat{u}_z}{\partial z} = 0, \tag{7.36}$$

and so we might expect $\hat{u}_r$, $\hat{u}_\theta \sim \cos\left(n\pi z/H\right)$, just like for the axisymmetric modes. It is natural, therefore, to seek modes of the form

$$u_r = \hat{u}_r(r) \cos\left(n\pi z/H\right) \exp(jm\theta) \exp(j\varpi t), \tag{7.37}$$

$$u_\theta = \hat{u}_\theta(r) \cos\left(n\pi z/H\right) \exp(jm\theta) \exp(j\varpi t), \tag{7.38}$$

$$u_z = \hat{u}_z(r) \sin\left(n\pi z/H\right) \exp(jm\theta) \exp(j\varpi t). \tag{7.39}$$

In such a case, the axial vorticity,

$$\omega_z = \frac{1}{r}\frac{\partial}{\partial r}(ru_\theta) - \frac{1}{r}\frac{\partial u_r}{\partial \theta}, \tag{7.40}$$

becomes

$$\omega_z = \left[\frac{1}{r}\frac{d}{dr}(r\hat{u}_\theta) - \frac{jm\hat{u}_r}{r}\right] \cos(n\pi z/H) \exp(jm\theta) \exp(j\omega t). \tag{7.41}$$

We now need three equations to determine the three velocity components. We shall choose to work with continuity, (7.36), which becomes

$$\frac{1}{r}\frac{d}{dr}(r\hat{u}_r) + \frac{jm\hat{u}_\theta}{r} + \frac{n\pi}{H}\hat{u}_z = 0, \tag{7.42}$$

and with the axial component of the vorticity equation,

$$\frac{\partial \omega_z}{\partial t} = 2(\mathbf{\Omega} \cdot \nabla)u_z,$$

which yields

$$j\omega\left[\frac{1}{r}\frac{d}{dr}(r\hat{u}_\theta) - \frac{jm\hat{u}_r}{r}\right] = 2\Omega\frac{n\pi}{H}\hat{u}_z. \tag{7.43}$$

Finally, we shall use the axial component of the inertial wave equation (6.14),

$$\frac{\partial^2}{\partial t^2}\nabla^2 u_z + (2\Omega)^2\frac{\partial^2 u_z}{\partial z^2} = 0,$$

from which we obtain

$$\frac{1}{r}\frac{d}{dr}r\frac{d\hat{u}_z}{dr} - \frac{m^2\hat{u}_z}{r^2} - \left(\frac{n\pi}{H}\right)^2\hat{u}_z + \left(\frac{2\Omega}{\omega}\right)^2\left(\frac{n\pi}{H}\right)^2\hat{u}_z = 0. \tag{7.44}$$

These three equations can be tidied up somewhat. First, it is convenient to reintroduce the parameter

$$\beta = \frac{n\pi}{H}\sqrt{(2\Omega/\omega)^2 - 1}, \tag{7.45}$$

and rescale r as $\eta = \beta r$. Then (7.42), (7.43), and (7.44) become

$$\frac{d}{d\eta}(\eta \hat{u}_r) + jm\hat{u}_\theta + \left(\frac{n\pi}{\beta H}\right)\eta \hat{u}_z = 0, \tag{7.46}$$

$$j\frac{d}{d\eta}(\eta \hat{u}_\theta) + m\hat{u}_r = \pm\sqrt{1 + (n\pi/\beta H)^2}\ \eta \hat{u}_z, \tag{7.47}$$

$$\eta^2 \frac{d^2 \hat{u}_z}{d\eta^2} + \eta\frac{d\hat{u}_z}{d\eta} + (\eta^2 - m^2)\hat{u}_z = 0, \tag{7.48}$$

where the positive sign in (7.47) corresponds to $\hat{\omega} > 0$ (a retrograde wave). From (7.48) we see that $\hat{u}_z$, which has an arbitrary amplitude, can be we written as a Bessel function of order m, say

$$\hat{u}_z = \frac{\beta H}{n\pi}J_m(\eta). \tag{7.49}$$

Equations (7.46) and (7.47) now become

$$\frac{d}{d\eta}(\eta \hat{u}_r) + jm\hat{u}_\theta + \eta J_m(\eta) = 0, \tag{7.50}$$

$$j\frac{d}{d\eta}(\eta \hat{u}_\theta) + m\hat{u}_r = \pm\sqrt{1 + (\beta H/n\pi)^2}\ \eta J_m(\eta), \tag{7.51}$$

whose solution is readily shown to be

$$\hat{u}_r = J'_m(\eta) \pm \sqrt{1 + (\beta H/n\pi)^2}\frac{m}{\eta}J_m(\eta), \tag{7.52}$$

$$\hat{u}_\theta = \pm j\sqrt{1 + (\beta H/n\pi)^2}\ J'_m(\eta) + j\frac{m}{\eta}J_m(\eta), \tag{7.53}$$

where the prime indicates a derivative with respect to η.

It remains to satisfy the boundary condition $\hat{u}_r = 0$ at $r = R$, and it is this that fixes the discrete frequencies associated with these azimuthally drifting waves. Let $\delta^{\pm}_{m,n,p}$ be the real, positive solutions of the two transcendental equations

$$xJ'_m(x) \pm \sqrt{1 + x^2(H/n\pi R)^2}\ mJ_m(x) = 0, \tag{7.54}$$

where $p = 1, 2, 3 \ldots$. Here $p = 1$ is the smallest solution for a given m and n, and also for a given choice of $+$ or $-$. The solutions are then ordered such that

$0 < \delta^+_{m,n,1} < \delta^+_{m,n,2} < \delta^+_{m,n,3}\cdots$ and $0 < \delta^-_{m,n,1} < \delta^-_{m,n,2} < \delta^-_{m,n,3}\cdots$. Alternatively, using the recurrence relationship

$$x J'_m(x) = m J_m(x) - x J_{m+1}(x), \tag{7.55}$$

we can rewrite (7.54) as

$$x J_{m+1}(x) - \left\{ 1 \pm \sqrt{1 + x^2 (H/n\pi R)^2} \right\} m J_m(x) = 0, \tag{7.56}$$

and $\delta^{\pm}_{m,n,p}$ become the real, positive solutions of (7.56). Either way, (7.52) tells us that the radial boundary condition requires $\beta R = \delta^{\pm}_{m,n,p}$. Having determined that $\beta = \delta^{\pm}_{m,n,p}/R$, we now return to (7.45), which fixes the discrete frequencies for these waves as

$$\frac{\varpi^{\pm}_{m,n,p}}{2\Omega} = \pm \frac{n\pi/H}{\sqrt{\left(\delta^{\pm}_{m,n,p}/R\right)^2 + (n\pi/H)^2}}. \tag{7.57}$$

It is readily confirmed that, for the special case of $m = 0$ (*i.e.* axisymmetric standing waves), the frequencies predicted by (7.57) revert to those given by (7.25). Moreover, our expression for $\hat{u}_z$, (7.49), reverts to (7.33), at least to within a change in the (arbitrary) amplitude.

The triplet of positive integers, (m, n, p), acts as a proxy for the azimuthal, axial, and radial wavenumbers, with m the azimuthal wavenumber, $k_{z,n} = n\pi/H$ the axial wavenumber, and p a measure of the number of reversals in the radial direction. Indeed, it may be shown that there exist $p - 1$ zeros of $\hat{u}_r$ in the range $0 < r < R$ for the pth radial mode. Evidently, the radial complexity of the modes increases as p rises.

The velocity components for each mode may now be found by returning to equations (7.49), (7.52), and (7.53). It is necessary only to replace β with $\delta^{\pm}_{m,n,p}/R$, η with $\delta^{\pm}_{m,n,p}r/R$, and ϖ with $\varpi^{\pm}_{m,n,p}$. For example, the axial velocity for the (m, n, p) retrograde mode is

$$u_z^{(m,n,p)} = \frac{H}{n\pi} \frac{\delta^+_{m,n,p}}{R} J_m\left(\delta^+_{m,n,p}r/R\right) \sin\left(n\pi z/H\right) \cos\left(m\theta + \varpi^+_{m,n,p}t\right), \tag{7.58}$$

while that for the (m, n, p) prograde mode is

$$u_z^{(m,n,p)} = \frac{H}{n\pi} \frac{\delta^-_{m,n,p}}{R} J_m\left(\delta^-_{m,n,p}r/R\right) \sin\left(n\pi z/H\right) \cos\left(m\theta + \varpi^-_{m,n,p}t\right). \tag{7.59}$$

Table 7.1 The frequencies of some retrograde
($\bar{\omega} > 0$) and prograde ($\bar{\omega} < 0$) inertial modes in a
circular cylinder of unit aspect ratio, $R = H$, and
for $n = 1$

m	p	$\delta^+_{m,1,p}$	$\bar{\omega}^+_{m,1,p}/2\Omega$	$\delta^-_{m,1,p}$	$\bar{\omega}^-_{m,1,p}/2\Omega$
1	1	2.514	0.7808	4.970	−0.5343
1	2	5.703	0.4825	8.210	−0.3574
1	3	8.872	0.3338	11.39	−0.2659
1	4	12.03	0.2527	14.55	−0.2110
1	5	15.18	0.2026	17.70	−0.1747
2	1	4.107	0.6076	6.056	−0.4605
2	2	7.386	0.3914	9.372	−0.3178
2	3	10.59	0.2843	12.59	−0.2421
2	4	13.77	0.2224	15.77	−0.1953
2	5	16.94	0.1824	18.94	−0.1636

Of course, similar expressions for $\hat{u}_r$ and $\hat{u}_\theta$ may be obtained from equations
(7.52) and (7.53).

Note that the $m = 1$ modes have the surprising property that both $\hat{u}_r$ and $\hat{u}_\theta$
are non-zero on the axis of the cylinder, $r = 0$. This is evident from the last term
on the right of both (7.52) and (7.53), which takes the form $\hat{u}_r, \hat{u}_\theta \sim m J_m(\eta)/\eta$.
Clearly, this is zero for $\eta \to 0$ for all m other than $m = 1$, but it is non-zero for
$m = 1$. It turns out that this is not restricted to inertial waves. Rather, it is a
general property of the Fourier decomposition of vector fields in cylindrical
coordinates that the $m = 1$ modes, but only the $m = 1$ modes, need not vanish
on the axis. (See the discussion in Lewis & Bellan, 1990.)

The numerical values of $\delta^\pm_{m,n,p}$ and $\bar{\omega}^\pm_{m,n,p}/2\Omega$ are catalogued in Zhang & Liao
(2017) for a range of values of m and p, all-be-it restricted to cases where $n = 1$
and $R = H$. Some of these values are reproduced in Table 7.1, which shows the
retrograde ($\bar{\omega} > 0$) and prograde ($\bar{\omega} < 0$) inertial modes for a circular cylinder
of unit aspect ratio, $R = H$, and for $m = 1, 2$ and $p = 1, 2, 3, 4, 5$. In all cases,
$n = 1$.

7.3 Some Comments on Modes in a Sphere

Inertial modes in a sphere can be described in either cylindrical polar coordinates, as discussed by Greenspan (1968), or in spherical polar coordinates. An
analysis of the inviscid modes is simpler when cylindrical coordinates are used,

but even then the amount of mathematical detail is somewhat exhausting. In this section, we provide a brief, *qualitative* overview of inertial modes in a sphere of radius R, intended primarily as a steppingstone to more specialist reading. It is partly based on the description given in Zhang & Liao (2017), where a more complete account can be found.

7.3.1 Four Classes of Solutions

As for all axisymmetric domains (whose symmetry axis is aligned with $\boldsymbol{\Omega}$), the starting point is to look for solutions of the form

$$\mathbf{u} = \hat{\mathbf{u}}(r,z)\exp\left(j(\bar{\omega}t + m\theta)\right), \quad m = 0, 1, 2, 3..., \tag{7.60}$$

and it turns out that four broad classes of solutions can be distinguished. These are:

class 1: axisymmetric modes ($m = 0$), symmetric about the equator;
class 2: non-axisymmetric modes ($m \geq 1$), symmetric about the equator;
class 3: axisymmetric modes ($m = 0$), antisymmetric about the equator;
class 4: non-axisymmetric modes ($m \geq 1$), antisymmetric about the equator.

As always, the axisymmetric modes ($m = 0$) are standing waves, and the non-axisymmetric modes ($m \geq 1$) are azimuthally drifting modes which may be divided into prograde waves ($\bar{\omega} < 0$) and retrograde waves ($\bar{\omega} > 0$).

As with inertial modes in a cylinder, it is convenient to label the modes using a triplet of positive integers, (m, n, p), where the integers act, more or less, as proxies for the azimuthal, axial, and radial wavenumbers. However, the formal definitions of n and p for a sphere are messy, with different authors using different conventions. We shall follow the definitions adopted in Zhang & Liao (2017), where n is a rough measure of the axial complexity of a mode, with $n \geq 1$, and p reflects the radial structure (in a cylindrical polar sense) of the mode. Note, however, that unlike modes in a cylinder, the indices n and p do *not* correlate directly to the number of zeros in the axial and radial directions.

For a sphere, the solutions of governing equation (7.2) can be written in terms of the *Poincaré polynomials of degree p*, $P_{m,\delta}^{p}(r,z;\sigma)$, as defined in Zhang et al. (2004). Here $\sigma = \bar{\omega}/2\Omega$ and $\delta = 0$ for equatorially symmetric modes, or else $\delta = 1$ for antisymmetric modes. Each $P_{m,\delta}^{p}$ yields both the frequencies and the velocity fields for a range of inertial modes of given m and δ, that range varying with p. The three indices (m, n, p) are then assigned to the modes as follows. All of the allowable frequencies for a given $P_{m,\delta}^{p}$ are calculated, labelled $\varpi_{m,n,p}$, and ordered according to $|\varpi_{m,1,p}| < |\varpi_{m,2,p}| < |\varpi_{m,3,p}|$ ... That is to say, for a given m, δ, and p, the ordering of n is chosen to ensure that the frequency rises as n increases. Note that the ranges of n and p vary with the class of modes, as shown in Table 7.2.

We now consider the four classes of inertial modes one by one, starting with standing waves symmetric about the equator (class 1). The left-hand side of Table 7.3 shows the standing wave frequencies as a function of p and n. Note that, by construction, and for a given p, the frequency rises as n increases. Conversely, for a given n, the frequency *falls* as p increases. Both of these trends are consistent with modes in a cylinder.

In practice, the low mode numbers are the most important, as the high modes are prone to viscous damping. The simplest class 1 mode is $(m, n, p) = (0, 1, 2)$, and this has a simple analytical structure in cylindrical polar coordinates. For example, the spatial structure of the azimuthal velocity is

$$\hat{u}_{\theta}^{(0,1,2)} \sim \left(2z^2 + (r^2 + z^2 - R^2)\right) r, \tag{7.61}$$

Table 7.2 The ranges of the indices used to label inertial modes in a sphere, with $n \geq 1$

class 1 ($m = 0$)	class 2 ($m \geq 1$)	class 3 ($m = 0$)	class 4 ($m \geq 1$)
$p \geq 2, n \leq p - 1$	$p \geq 1, n \leq 2p$	$p \geq 1, n \leq p$	$p \geq 0, n \leq 2p + 1$

Table 7.3 The frequencies of equatorially symmetric inertial modes in a sphere

Equatorially symmetric standing waves (Class 1 modes) $m = 0$, $p \geq$ 2, $n \leq p - 1$			Equatorially symmetric drifting waves (Class 2 modes) $m = 1$, $p \geq 1$, $n \leq 2p$		
p	n	$\varpi_{0,n,p}/2\Omega$	p	n	$\varpi_{1,n,p}/2\Omega$
2	1	0.6547	1	1	−0.0883
3	1	0.4689	1	2	0.7550
3	2	0.8302	2	1	−0.0341
4	1	0.3631	2	2	0.5228
4	2	0.6772	2	3	−0.5917
4	3	0.8998	2	4	0.9030
5	1	0.2958	3	1	−0.0181
5	2	0.5652	3	2	0.3951
5	3	0.7845	3	3	−0.4314
5	4	0.9340	3	4	0.7369

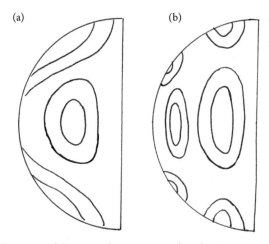

Figure 7.4 Schematic of the spatial structures of u_θ for two equatorially symmetric standing waves: (a) $(m, n, p) = (0, 1, 2)$; (b) $(m, n, p) = (0, 1, 3)$.

(Zhang & Liao, 2017), which is shown schematically in Figure 7.4(a). This is negative in the equatorial regions and positive near the poles. The next mode in Table 7.3 is $(m, n, p) = (0, 1, 3)$, whose spatial structure is also shown schematically in Figure 7.4. Note that, as we move from $(0, 1, 2)$ to $(0, 1, 3)$, we pick up a new radial node, as might have been expected since p has increased by 1. However, as we move from $(0, 1, 2)$ to $(0, 2, 3)$, which is the third mode in Table 7.3, we pick up a new axial node (per hemisphere), but *not* a new radial node (see Zhang & Liao, 2017). This emphasizes the fact that the indices n and p do not correlate directly to the number of zeros in the axial and radial directions.

The right-hand side of Table 7.3 shows the drifting wave frequencies for class 2 modes as a function of p and n, with m fixed at $m = 1$. Once again, for a given p, $|\hat{\omega}_{m,n,p}|$ is seen to rise as n increases, whereas, for a given n, the frequency falls as p increases. Note that the frequencies for all the $n = 1$ modes are particularly small, and so these modes drift very slowly in the azimuthal direction. Not surprisingly, the corresponding velocity fields are close to geostrophic, exhibiting almost no variation in z. Note also that these nearly geostrophic modes are all prograde. The same quasi-geostrophic behaviour is seen in the $n = 1$ modes at higher values of m, with the approximate frequencies

$$\frac{\hat{\omega}_{m,1,p}}{2\Omega} \approx -\frac{1}{m+2}\left[\sqrt{1 + \frac{m(m+2)}{p(2p+2m+1)}} - 1\right], \tag{7.62}$$

(see Zhang & Liao, 2017). Expression (7.62) is accurate to within a fraction of a per cent.

The two simplest class 2 modes in Table 7.3 are $(m, n, p) = (1, 1, 1)$, a prograde wave, and $(m, n, p) = (1, 2, 1)$, which is a retrograde wave. The spatial structure of the azimuthal velocity for either of these two drifting modes turns out to be

$$\hat{u}_\theta^{(1,n,1)} \sim \left\{ 4\left(\frac{\varpi_{1,n,1}}{2\Omega}\right)^2 z^2 + \left(1 - \frac{\varpi_{1,n,1}}{2\Omega}\right)\left(3 + \frac{\varpi_{1,n,1}}{2\Omega}\right) r^2 - \frac{4}{5} R^2 \right\} e^{j\theta}. \quad (7.63)$$

(Again, see Zhang & Liao, 2017.) Note that the low frequency of the (1, 1, 1) mode means that the pre-factor in front of z^2 is small, around 0.0312, whereas the pre-factor in front of r^2 is 3.17, a factor of 100 larger. This mode therefore exhibits only a relatively weak dependence on z, as noted above.

In addition to the nearly geostrophic ($n = 1$) modes, there is a second important subclass of the equatorially symmetric drifting waves. These are the $m \gg 1$ modes, which turn out to be equatorially trapped waves. In particular, the kinetic energy of these modes is concentrated close to the equatorial plane and to the outer boundary $|\mathbf{x}| \approx R$. Indeed, the amplitude of these modes falls off exponentially with distance from the point $(r, z) = (R, 0)$, with a characteristic decay length of R/m. Because these modes are spatially localized, their behaviour is insensitive to the boundary conditions in the polar regions, or to the existence of an inner sphere, concentric with the outer sphere.

Let us now consider those modes which are antisymmetric about the equator (classes 3 and 4). The left-hand side of Table 7.4 shows the antisymmetric standing wave frequencies as a function of p and n. As usual, for a given p, the frequency rises as n increases, and for a given n, the frequency falls as p increases. The simplest mode is now $(m, n, p) = (0, 1, 1)$, and this standing wave turns out to have a particularly simple spatial structure, with $\hat{u}_\theta^{(0,1,1)} \sim zr$.

The right-hand side of Table 7.4 shows the frequencies of the antisymmetric drifting waves as a function of p and n, with m fixed at $m = 1$. Unlike the symmetric drifting waves, there are no quasi-geostrophic modes. However, when $m \gg 1$, the antisymmetric drifting waves do exhibit equatorially trapped modes, not unlike their symmetric counterparts. The simplest class 4 modes correspond to $n = 1$, $p = 0$, whose frequencies are

$$\frac{\varpi_{m,1,0}}{2\Omega} = \frac{1}{m + 1}. \quad (7.64)$$

These are retrograde waves of relatively simple spatial structure.

Table 7.4 The frequencies of equatorially antisymmetric inertial modes in a sphere

Equatorially antisymmetric standing waves (Class 3 modes) $m = 0, p \geq 1, n \leq p$			Equatorially antisymmetric drifting waves (Class 4 modes) $m = 1, p \geq 0, n \leq 2p + 1$		
p	n	$\varpi_{0,n,p}/2\Omega$	p	n	$\varpi_{1,n,p}/2\Omega$
1	1	0.4472	0	1	0.5000
2	1	0.2852	1	1	0.3060
2	2	0.7651	1	2	−0.4100
3	1	0.2093	1	3	0.8540
3	2	0.5917	2	1	0.2202
3	3	0.8717	2	2	−0.2687
4	1	0.1653	2	3	0.6530
4	2	0.4779	2	4	−0.7021
4	3	0.7388	2	5	0.9308
4	4	0.9195	—	—	–

7.3.2 The Spin-Over Mode

Perhaps the most discussed antisymmetric drifting wave is the $(m, n, p) = (1, 1, 0)$ mode, which is often called the *spin-over mode*. This mode has uniform vorticity in the rotating frame, and is given by

$$\boldsymbol{\omega}^{(1,1,0)} = \boldsymbol{\omega}_0 \left[\cos(\theta + \Omega t)\hat{\mathbf{e}}_r - \sin(\theta + \Omega t)\hat{\mathbf{e}}_\theta\right], \qquad (7.65)$$

in cylindrical polar coordinates, (r, θ, z), or equivalently,

$$\boldsymbol{\omega}^{(1,1,0)} = \boldsymbol{\omega}_0 \left[\cos(\Omega t)\hat{\mathbf{e}}_x - \sin(\Omega t)\hat{\mathbf{e}}_y\right], \qquad (7.66)$$

in Cartesian coordinates. (Here $\boldsymbol{\omega}_0$ is the amplitude of $\boldsymbol{\omega}$.) Thus, the vorticity is uniform and, when viewed in a rotating frame of reference, it rotates in a retrograde sense at the rate Ω.

Crucially, however, in an *inertial* frame of reference, (7.66) represents a vector which is *fixed is space*. In an inertial frame, then, the total vorticity, $\boldsymbol{\omega}_{net} = 2\boldsymbol{\Omega} + \boldsymbol{\omega}_0$, is uniform and directed along a fixed axis, and so the fluid is in a state of rigid-body rotation about a fixed axis which is inclined to $\boldsymbol{\Omega}$ (see Figure 7.5). Clearly, in the inertial frame, there is *no* inertial oscillation driven by Newton's second law, and the apparent oscillation in the rotating frame is simply an artefact of the change in frame of reference. It would seem,

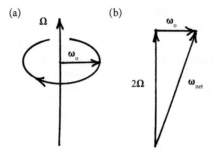

Figure 7.5 The spin-over mode vorticity in: (a) the rotating frame; (b) the inertial frame.

therefore, that terms such as *inertial oscillation* or *inertial mode* are entirely inappropriate for the inviscid spin-over mode, as the concept of inertia does not arise.

Evidently, the spin-over mode is not limited to small amplitudes, and indeed ω_0 can take any magnitude relative to $\mathbf{\Omega}$. Moreover, it is readily generated in the laboratory by first letting the fluid and sphere co-rotate, and then abruptly changing the direction of $\mathbf{\Omega}$ for the sphere. We shall return to the spin-over mode in Chapter 10, when we discuss precession.

7.3.3 A Footnote: The Completeness of Inertial Modes in a Sphere

The fact that distinct inertial modes are orthogonal gives one hope that, for a given geometry, the modes are also *complete*, in the sense that any smooth, incompressible velocity field can be represented as a sum over all m, n, and p of the various modes. In the case of a cylinder, it seems plausible (though not entirely certain) that (7.58) and (7.59) do indeed represent a complete set of basis functions, being Fourier in θ and z, and Bessel in r. The question mark, however, centres on the unusual definition of $\delta^{\pm}_{m,n,p}$ demanded by (7.54), which is not the usual condition required for the orthogonality of Bessel functions. So, (7.58) and (7.59) are *not* conventional Bessel series.

Although it has long been suspected that the inertial modes for a sphere are complete, a demonstration that this is indeed the case has proved surprisingly elusive. Recently, however, Ivers et al. (2015) formulated a completeness proof using the Weierstrass polynomial approximation theorem. This was quickly followed by Backus & Rieutord (2017), who formulated a proof in Hilbert space and managed to establish completeness of the Poincaré velocity polynomials in axisymmetric ellipsoids.

The observation that the inertial modes in a sphere represent a complete set facilitates, at least in principle, the study of inviscid spherical flows subject to slip boundaries. For example, it means that the weakly nonlinear evolution of an arbitrary initial condition can be reduced, using orthogonality, to a set of coupled ordinary differential equations for the time-dependant amplitudes of the various inertial modes, the coupling arising from weak inertial forces (assuming Ro is small but finite). In practice, however, the complexity of the inertial modes makes such a calculation far from straight forward.

$$* * *$$

That concludes our brief discussion of inertial modes in confined, axisymmetric domains. A more detailed account of such modes may be found in Greenspan (1968), while modes in a sphere are discussed at length in Chapter 5 of Zhang & Liao (2017).

References

Backus, G., & Rieutord, M., 2017, Completeness of inertial modes of an incompressible inviscid fluid in a co-rotating ellipsoid, *Phys, Rev. E*, **95**, 053116.

Greenspan, H.P., 1968, *The Theory of Rotating Fluids*, Cambridge University Press.

Ivers, D.J., Jackson, A., & Winch, D., 2015, Enumeration, orthogonality and completeness of the incompressible Coriolis modes in a sphere, *J. Fluid Mech.*, **766**, 468.

Lewis, R.H., & Bellan, P.M., 1990, Physical constraints on the coefficients of Fourier expansions in cylindrical coordinates, *J. Math. Phys.*, **31**(11), 2592.

Zhang, K., & Liao, X., 2017, *Theory and Modelling of Rotating Fluids*, Cambridge University Press.

Zhang, K., Liao, X., & Earnshaw, P., 2004, The Poincaré equation: A new polynomial and its unusual properties, *J. Maths Phys.*, **45** (12), 4777–90.

Chapter 8

Rossby Waves

An Example of Quasi-Geostrophic Motion

A strong background rotation can support a second type of wave motion, over and above inertial waves. These are known as *Rossby waves* and they are commonly observed in the atmosphere and in the oceans, typically at very large scales. Atmospheric Rossby waves, which are also called *planetary waves*, appear as giant meanders in high-altitude winds with wavelengths of up to 10^3 km, while oceanic Rossby waves are low-amplitude waves that may take months to cross an ocean basin.

Rossby waves are best thought of as weak, slow perturbations to a geostrophic base state, *i.e.* as a form of quasi-geostrophic wave motion whose frequency is much less than Ω. They occur in confined systems and indeed they have some connection to confined inertial waves, although it is simplest to treat them as an entirely new class of wave motion. Rossby waves arise naturally in rapidly rotating, shallow-water flows, as we shall see in Chapter 9. However, the simplest configuration in which Rossby waves can occur is a rotating fluid which is confined between two, non-parallel planes, one plane being normal to the rotation axis. We shall focus on that simpler geometry in this chapter.

8.1 Rossby Waves Between Plane, Non-Parallel Boundaries

Consider the geometry shown in Figure 8.1, where a rotating fluid is confined between two nearly parallel planes, $z = 0$ and $z = h(y) = h_0 - \alpha y$, where h_0 is the mean separation of the two planes and α is a small positive constant. As usual, we take $\mathbf{\Omega} = \Omega \hat{\mathbf{e}}_z$, ignore viscosity, and consider small-amplitude perturbations to a state of rest in the rotating frame. Note that, if λ is a typical horizontal wavelength, where $\lambda > h_0$, then we choose α such that $\alpha << h_0/\lambda < 1$. In such cases, any variation in $h(y)$ is much less than h_0.

We shall discuss the physical mechanisms responsible for Rossby waves towards the end of this chapter, after we have introduced the concept of *potential vorticity*. However, in the meantime, we might picture the waves

The Dynamics of Rotating Fluids. P. A. Davidson, Oxford University Press. © Peter A Davidson (2024).
DOI: 10.1093/9780191994272.003.0008

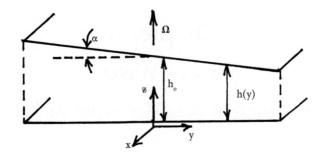

Figure 8.1 The simplest configuration in which Rossby waves can arise.

arising as follows. In a geostrophic flow the constraint $\partial u_z/\partial z = 0$ ensures that a fluid column cannot change its height, and so individual fluid particles must follow contours of constant depth. If a fluid column is displaced laterally in our configuration, then this geostrophic constraint is broken, and it turns out that the fluid tries to return to its original geostrophic height. Of course, inertia ensures an overshoot, and this gives rise to waves.

An elementary analysis of Rossby waves proceeds as follows. We start by assuming that the wave motion is predominantly planar, in the sense that $|\omega_z| \gg |\omega_x|, |\omega_y|$. (We shall check retrospectively that, when $\alpha \ll 1$, this is indeed the case.) Since ω_x and ω_y are small, the low-Ro vorticity equation,

$$\frac{\partial \boldsymbol{\omega}}{\partial t} = 2\boldsymbol{\Omega} \cdot \nabla \mathbf{u}, \tag{8.1}$$

gives us

$$\frac{\partial u_x}{\partial z} = \frac{1}{2\Omega} \frac{\partial \omega_x}{\partial t} \approx 0, \quad \frac{\partial u_y}{\partial z} = \frac{1}{2\Omega} \frac{\partial \omega_y}{\partial t} \approx 0, \tag{8.2}$$

$$\frac{\partial \omega_z}{\partial t} = 2\Omega \frac{\partial u_z}{\partial z}. \tag{8.3}$$

Now (8.2), combined with $\nabla \cdot \mathbf{u} = 0$, demands $\partial^2 u_z/\partial z^2 = 0$, and since $u_z = -\alpha u_y$ on $z = h$, we conclude that

$$\frac{\partial u_z}{\partial z} = -\frac{\alpha u_y}{h_0}. \tag{8.4}$$

Our governing equations are now

$$\frac{\partial u_x}{\partial x} + \frac{\partial u_y}{\partial y} = \frac{\alpha u_y}{h_0} \ll \frac{u_y}{\lambda}, \tag{8.5}$$

$$\frac{\partial \omega_z}{\partial t} = -\frac{2\Omega\alpha}{h_0} u_y, \tag{8.6}$$

where we have used $\alpha << h_0/\lambda$ in (8.5). Evidently, the horizontal motion, $\mathbf{u}_\perp$, is solenoidal to leading order in α. We may therefore introduce a streamfunction, $\psi(x, y, t)$, for this planar motion, defined by $\mathbf{u}_\perp = \nabla \times (\psi\hat{\mathbf{e}}_z)$. It follows that $\omega_z = -\nabla^2\psi$ and $u_y = -\partial\psi/\partial x$, so that (8.6) becomes

$$\frac{\partial}{\partial t}\nabla^2\psi = -\frac{2\Omega\alpha}{h_0}\frac{\partial\psi}{\partial x}. \tag{8.7}$$

This is our governing equation for Rossby waves between nearly parallel boundaries.

Equation (8.7) supports progressive waves of the form $\psi \sim \exp\left[j(\mathbf{k}_\perp \cdot \mathbf{x} - \bar{\omega}t)\right]$, where $\mathbf{k}_\perp$ is the horizontal wavevector. The corresponding dispersion relationship is

$$\bar{\omega} = -\frac{2\Omega\alpha}{h_0}\frac{k_x}{k_\perp^2}, \tag{8.8}$$

and since we insist that $\bar{\omega} > 0$, this requires $k_x < 0$. The phase and the group velocities are then

$$\mathbf{c}_p = -\frac{2\Omega\alpha}{h_0 k_\perp^4}k_x\mathbf{k}_\perp, \tag{8.9}$$

and

$$c_{g,x} = \frac{2\Omega\alpha}{h_0 k_\perp^4}\left(k_x^2 - k_y^2\right), \qquad c_{g,y} = \frac{2\Omega\alpha}{h_0 k_\perp^4}\left(2k_x k_y\right), \tag{8.10}$$

or equivalently,

$$\mathbf{c}_g = \frac{2\Omega\alpha}{h_0 k_\perp^2}\left(\cos 2\theta\hat{\mathbf{e}}_x + \sin 2\theta\hat{\mathbf{e}}_y\right), \tag{8.11}$$

where θ is the angle $\mathbf{k}_\perp$ makes to the x-axis. Evidently, Rossby waves are dispersive, with anisotropic dispersion characteristics. For the important special case of $k_y = 0$, these expressions simplify to

$$\mathbf{c}_p = -\frac{2\Omega\alpha}{h_0 k_x^2}\hat{\mathbf{e}}_x, \qquad \mathbf{c}_g = \frac{2\Omega\alpha}{h_0 k_x^2}\hat{\mathbf{e}}_x. \tag{8.12}$$

We must now check that, when $\alpha \ll 1$, we do indeed satisfy $|\omega_x|, |\omega_y| \ll |\omega_z|$. First, we note that $\omega_z \sim k_\perp u_y$, while

$$\omega_x = \frac{\partial u_z}{\partial y} - \frac{\partial u_y}{\partial z} \approx \frac{\partial u_z}{\partial y} \sim k_\perp u_z \sim \alpha k_\perp u_y \ll |\omega_z|, \tag{8.13}$$

and

$$\omega_y = \frac{\partial u_x}{\partial z} - \frac{\partial u_z}{\partial x} \approx -\frac{\partial u_z}{\partial x} \sim k_\perp u_z \sim \alpha k_\perp u_y \ll |\omega_z|, \tag{8.14}$$

ensure $|\boldsymbol{\omega}_\perp| \ll |\omega_z|$, exactly as required. Moreover, (8.1) combines with (8.8) to give us

$$\frac{\partial \mathbf{u}_\perp}{\partial z} = \frac{1}{2\Omega}\frac{\partial \boldsymbol{\omega}_\perp}{\partial t} \sim \frac{\varpi}{2\Omega}|\boldsymbol{\omega}_\perp| \sim \frac{\alpha}{k_\perp h_0}(\alpha k_\perp u_y) = \alpha\frac{\alpha u_y}{h_0} \ll \frac{\partial u_z}{\partial z}, \tag{8.15}$$

and so the z-derivatives of u_x and u_y are much smaller than $\partial u_z/\partial z$, which justifies (8.2).

Perhaps some comments are in order. First, we have $\varpi \sim \alpha\Omega \ll \Omega$, and so these are *low-frequency* waves. Second, since $c_g \sim \alpha\Omega/k_\perp \ll \Omega/k_\perp$, Rossby waves have a group velocity which is much smaller than that of inertial waves. Third, for the important special case of $k_y = 0$, $\mathbf{c}_g$ and $\mathbf{c}_p$ are equal and opposite, as shown in Figure 8.2. In such cases, a wave packet propagates in the opposite direction to the wave crests which sit within that packet. Fourth, the x-component of the phase velocity is always negative, in stark contrast to progressive inertial waves, which can disperse in any direction. (We shall see in Chapter 9 that terrestrial Rossby waves have a phase velocity that is always *westward*.) Finally, although we appear to have used α as a small parameter in a somewhat ad hoc way, the description above is readily rewritten as a formal perturbation analysis in α.

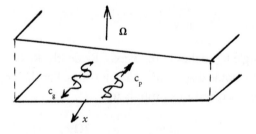

Figure 8.2 For Rossby waves in which $\mathbf{k}_\perp = k_x \hat{\mathbf{e}}_x$, $\mathbf{c}_g$, and $\mathbf{c}_p$ are equal and opposite.

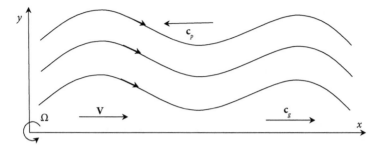

Figure 8.3 A Rossby wave is superimposed on a uniform geostrophic motion whose velocity is chosen to be equal and opposite to the phase velocity. The flow is then steady.

Now, suppose that a Rossby wave in which $\mathbf{k}_\perp = k_x \hat{\mathbf{e}}_x$ is superimposed on a uniform geostrophic motion whose velocity is chosen to be equal and opposite to the phase velocity, $\mathbf{V} = -\mathbf{c}_p$, as shown in Figure 8.3. Then the undulating wave pattern will appear steady, and indeed this is what tends to happen when Rossby waves are triggered by flow passing over an obstacle. That is to say, the waves generated by the obstacle tend to pick a wavelength such that $\mathbf{c}_p = -\mathbf{V}$, leaving the flow steady in a frame of reference attached to the obstacle. Since $\mathbf{c}_g = -\mathbf{c}_p = \mathbf{V}$, the energy of such a wave system propagates downstream with a velocity of $\mathbf{c}_g = \mathbf{V}$ relative to the fluid, but with a velocity of $2\mathbf{V}$ relative to the object that triggered the waves.

8.2 Rossby Waves in a Sliced Cylinder

So far, we have considered Rossby waves which are unbounded in the x–y plane, and so we now turn to bounded domains. Perhaps the simplest bounded geometry is a sliced cylinder of mean depth h_0 and radius R, as shown in Figure 8.4. We shall use cylindrical polar coordinates, (r, θ, z), centred on the base of the cylinder.

In the absence of a radial boundary, the discussion above shows that the simplest form of Rossby wave is $\psi \sim \exp\left[j(k_x x - \tilde{\omega}t)\right]$, where $k_x < 0$. So, in the presence of a radial boundary, it is natural to look for solutions of the form $\psi \sim f(r)\exp\left[j(k_x x - \tilde{\omega}t)\right]$, $k_x < 0$. However, our experience with bounded inertial waves suggests that we should allow for waves that drift in the azimuthal direction, and so we take as our trial solution

$$\psi \sim f(r)\exp\left[j(m\theta - kx - \tilde{\omega}t)\right], \quad f(R) = 0, \tag{8.16}$$

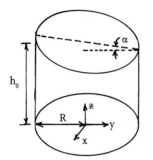

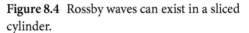

Figure 8.4 Rossby waves can exist in a sliced cylinder.

where $k = |k_x|$. Moreover, since we insist that $\hat{\omega} > 0$, the integer m must be allowed to take negative values, so as to admit retrograde waves. (A positive m represents a prograde wave.)

Our governing equation is, of course,

$$\frac{\partial}{\partial t} \nabla^2 \psi = -\frac{2\Omega\alpha}{h_0} \frac{\partial \psi}{\partial x}, \tag{8.17}$$

which becomes

$$\frac{\partial}{\partial t} \nabla^2 \psi = -\frac{2\Omega\alpha}{h_0} \left[\cos\theta \frac{\partial \psi}{\partial r} - \frac{\sin\theta}{r} \frac{\partial \psi}{\partial \theta} \right], \tag{8.18}$$

when written in polar coordinates. Substituting (8.16) into (8.18) yields, after some algebra,

$$\frac{1}{r}\frac{d}{dr}\left(rf'(r) \right) - \left[\frac{m^2}{r^2} + k^2 + \frac{2m}{r}k\sin\theta \right]f - 2jk\cos\theta f'(r)$$

$$= -\frac{\Omega\alpha}{\hat{\omega} k h_0} \left[2k^2 f + \frac{2m}{r}k\sin\theta f + 2jk\cos\theta f'(r) \right]. \tag{8.19}$$

The imaginary terms in (8.19) now demand that

$$\hat{\omega} = \alpha\Omega/k h_0, \tag{8.20}$$

which might be compared with (8.8), rewritten as

$$\hat{\omega} = \frac{\alpha\Omega}{|k_x| h_0} \cdot \frac{2k_x^2}{k_\perp^2}. \tag{8.21}$$

Given this frequency, the real terms in (8.19) yield

$$\frac{1}{r}\frac{d}{dr}\left(rf'(r)\right) + \left[k^2 - \frac{m^2}{r^2}\right]f = 0,\tag{8.22}$$

whose solution is the Bessel function $f \sim J_m(kr)$. The boundary condition $f(R) = 0$ now requires $k = \delta_{m,n}/R$, where $\delta_{m,n}$ is the nth zero of J_m.

In summary, then, a monochromatic Rossby wave in a sliced cylinder takes the form

$$\psi_{m,n} \sim J_m\left(\delta_{m,n}r/R\right)\exp\left[j\left(m\theta - kx - \hat{\omega}t\right)\right],\tag{8.23}$$

whose frequency is

$$\hat{\omega}_{m,n} = \frac{\alpha\Omega R}{\delta_{m,n}h_0}.\tag{8.24}$$

The simplest modes are

$$\psi_{0,n} \sim J_0\left(\delta_{0,n}r/R\right)\exp\left[j\left(k_x x - \hat{\omega}t\right)\right],\tag{8.25}$$

where $k_x = -\delta_{0,n}/R$. This resembles a Rossby wave in an unbounded domain, but with a radial envelope confining the wave pattern. The phase velocity is in the negative x direction, as shown in Figure 8.5, but there is no net energy flux in that direction.

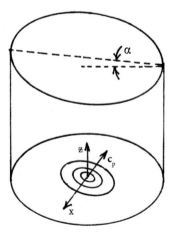

Figure 8.5 The simplest form of Rossby wave in a sliced cylinder.

8.3 Potential Vorticity and the Rossby Wave Equation

We shall now derive the governing equation for Rossby waves in a different way, one that generalizes more readily to rotating, shallow-water flow. Our starting point is to derive Helmholtz's second law in a rotating frame of reference. In §6.3.2 we showed that

$$\frac{D\boldsymbol{\omega}_{net}}{Dt} = (\boldsymbol{\omega}_{net} \cdot \nabla)\mathbf{u}, \quad \boldsymbol{\omega}_{net} = 2\boldsymbol{\Omega} + \boldsymbol{\omega}, \tag{8.26}$$

where D/Dt and $\mathbf{u}$ are both measured in the rotating frame and $\boldsymbol{\omega}_{net}$ is the net vorticity in an inertial frame. Equivalently, in the rotating frame we have

$$\frac{\partial \boldsymbol{\omega}_{net}}{\partial t} = \nabla \times (\mathbf{u} \times \boldsymbol{\omega}_{net}). \tag{8.27}$$

This then combines with (2.73) to give

$$\frac{d}{dt} \int_S \boldsymbol{\omega}_{net} \cdot d\mathbf{S} = 0, \tag{8.28}$$

where S is any surface that spans a closed *material* curve C_m.

Let us now apply (8.28) to a narrow column of fluid in Figure 8.1, of height h and cross-sectional area A. (Since $\partial \mathbf{u}_\perp/\partial z \approx 0$, we are at liberty to demand that the column is always composed of the same fluid particles.) The column conserves both $(2\Omega + \omega_z)A$ and $\rho h A$. It follows that

$$\frac{D}{Dt}\left(\frac{2\Omega + \omega_z}{h}\right) = \frac{DQ}{Dt} = 0, \tag{8.29}$$

where the materially conserved quantity, $Q = (2\Omega + \omega_z)/h$, is known as the *potential vorticity* in shallow-water theory. (The same name is used for a more general quantity in the theory of stratified fluids.) However, since $\omega_z \ll \Omega$ and $\alpha \ll 1$, we can rewrite Q as

$$Q = \frac{2\Omega + \omega_z}{h_0 - \alpha y} = \frac{2\Omega}{h_0}\left(1 + \frac{\omega_z}{2\Omega} + \frac{\alpha y}{h_0}\right) = \frac{2\Omega}{h_0} + \frac{1}{h_0}\left(\omega_z + \frac{2\Omega \alpha y}{h_0}\right), \tag{8.30}$$

and it follows that

$$\frac{D}{Dt}\left(\omega_z + \frac{2\Omega\alpha y}{h_0}\right) = \frac{Dq}{Dt} = 0, \tag{8.31}$$

where $q = \omega_z + (2\Omega\alpha y)/h_0$. For small-amplitude Rossby waves, this linearizes to give

$$\frac{\partial\omega_z}{\partial t} + u_y\frac{2\Omega\alpha}{h_0} = 0, \tag{8.32}$$

which is our governing equation for Rossby waves, (8.6). In Chapter 9 we shall see that this alternative derivation of (8.6) generalizes quite readily to rotating, shallow-water flow.

8.4 The Physical Mechanism of Rossby Waves

We close this chapter with a brief discussion of the physical mechanism behind Rossby waves. Consider an initial state of rest in the rotating frame and focus on three material columns, A, B, and C, which are initially positioned on the line y = constant, as shown in Figure 8.6. If column B is shifted to the right, the conservation of $q = \omega_z + 2\Omega\alpha y/h_0$ ensures that it acquires a negative ω_z. Moreover, the Biot–Savart law tells us that this negative ω_z induces a negative u_y at column A and a positive u_y at column C. Column A now moves to the

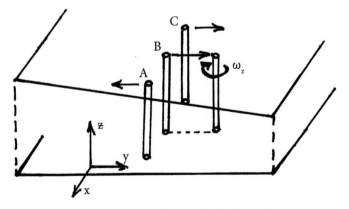

Figure 8.6 The physical mechanism behind Rossby waves.

left, acquiring a positive ω_z in the process, while column C moves to the right, acquiring a negative ω_z through the conservation of q. Both columns A and C now induce a velocity at column B which shifts it back towards its equilibrium position. Of course, inertia ensures that column B overshoots and this gives rise to an oscillation. This mechanism not only shows how oscillations can be sustained, but it also suggests that it is natural for the phase velocity of the resulting waves to be directed along the x-axis.

* * *

This is all we shall say about Rossby waves between nearly parallel boundaries. As noted at the beginning of this chapter, Rossby waves also occur in rotating, shallow-water flow. These are associated with a sloping bottom topography, and also with the variation with latitude of the component of Ω normal to the surface of the earth (the so-called β-effect). We shall discuss such cases in Chapter 9. There is a large literature on Rossby waves, reflecting their importance in meteorological and oceanic dynamics. Readers seeking more detail will find definitive accounts in Greenspan (1968), Pedlosky (1987), and Vallis (2017).

References

Greenspan, H.P., 1968, *The Theory of Rotating Fluids*, Cambridge University Press.
Pedlosky, J., 1987, *Geophysical Fluid Dynamics*, 2nd Ed., Springer-Verlag.
Vallis, G.K., 2017, *Atmospheric and Oceanic Fluid Dynamics*, 2nd Ed., Cambridge University Press.

Chapter 9

Rotating, Shallow-Water Flow

Inertia–Gravity Waves and Rossby Waves Revisited

The oceans and atmosphere are thin by comparison with the lateral dimensions of ocean gyres or large-scale atmospheric flows. Consequently, large-scale motions in the oceans and atmosphere are often modelled using a form of shallow-water theory which is not unlike that used by civil engineers to model open-channel flow. The main difference is that the rotation of the earth is often important for large-scale flows, and so the Coriolis force must be incorporated into the shallow-water equations. In this chapter we shall consider the various types of wave motion that are captured by a rotating, shallow-water framework, culminating in a return to the Rossby waves of Chapter 8. We start, however, with the shallow-water equations themselves.

9.1 The Shallow-Water Equations for an Inviscid Fluid

9.1.1 The Non-Rotating Shallow-Water Equations

Let us derive the conventional, non-rotating, shallow-water equations. For simplicity, we restrict ourselves to an inviscid fluid. Consider the flow shown in Figure 9.1. This is characterized by the fact that the fluid depth, h, is much less than the horizontal scale of the motion, $h \ll \ell$, which is the defining feature of shallow-water flow. The smallness of the ratio $\delta = h/\ell$ suggests that the flow is almost planar, in the sense that $|u_z| \ll |\mathbf{u}_\perp|$, where $\mathbf{u}_\perp$ is the horizontal motion. It is natural, therefore, to neglect the vertical acceleration of the fluid, and the horizontal and vertical components of the inviscid equation of motion are then

$$\frac{D\mathbf{u}_\perp}{Dt} = -\nabla_\perp (p/\rho), \qquad 0 = -\nabla_\parallel (p/\rho) + \mathbf{g}. \tag{9.1}$$

The second of these is simply a hydrostatic force balance and it integrates to give a gauge pressure of $p = \rho g(h - z)$. The horizontal equation of motion can now be rewritten as

The Dynamics of Rotating Fluids. P. A. Davidson, Oxford University Press. © Peter A Davidson (2024).
DOI: 10.1093/9780191994272.003.0009

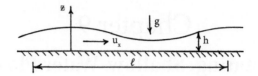

Figure 9.1 Shallow-water flow has $\delta = h/\ell \ll 1$.

$$\frac{D\mathbf{u}_\perp}{Dt} = -g\nabla h, \tag{9.2}$$

which is the usual shallow-water equation. Since $h = h(x, y, t)$, (9.2) tells us that, if $\mathbf{u}_\perp$ starts out as independent of z, then it stays that way, and we shall restrict ourselves to such cases.

Let us now estimate the error made in neglecting the vertical acceleration. Noting that (9.2) gives $u_\perp^2/\ell \sim gh/\ell$, the magnitude of the neglected vertical acceleration in (9.1) can be estimated as

$$\frac{u_\perp}{\ell} u_z \sim \frac{u_\perp}{\ell}\left(\frac{hu_\perp}{\ell}\right) \sim \frac{u_\perp^2}{\ell} \cdot \frac{h}{\ell} \sim \left(\frac{gh}{\ell}\right)\frac{h}{\ell} \sim \delta^2 g \ll g,$$

which is indeed negligible.

To complete the shallow-water equations, we must add to (9.2) conservation of mass. Consider the control volume shown in Figure 9.2, which has a fixed base and sides and whose upper surface is the free surface of the fluid. Let the curve C, which encloses the base A, be composed of line elements $d\mathbf{r}$ and have a horizontal unit normal $\mathbf{n}$. Since $\mathbf{u}_\perp = \mathbf{u}_\perp(x, y)$, mass conservation yields

$$\frac{d}{dt}\int_A (\rho h)\,dxdy = -\oint_C (\rho h\mathbf{u}_\perp)\cdot\mathbf{n}\,dr = -\int_A \nabla\cdot(\rho h\mathbf{u}_\perp)\,dxdy,$$

from which

$$\int_A \left[\frac{\partial h}{\partial t} + \nabla\cdot(h\mathbf{u}_\perp)\right]dxdy = 0.$$

The continuity equation appropriate to shallow-water flow is evidently

$$\frac{\partial h}{\partial t} + \nabla\cdot(h\mathbf{u}_\perp) = \frac{Dh}{Dt} + h\nabla\cdot\mathbf{u}_\perp = 0. \tag{9.3}$$

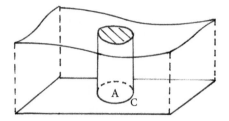

Figure 9.2 The control volume used to apply conservation of mass.

9.1.2 Adding Rotation: Potential Vorticity Revisited

In order to move closer to large-scale geophysical flows, we must embrace background rotation. So we now add rotation, $\boldsymbol{\Omega} = \Omega \hat{\mathbf{e}}_z$, move into the rotating frame of reference, and add the Coriolis force to the momentum equation. This gives

$$\frac{D\mathbf{u}_\perp}{Dt} = \frac{\partial \mathbf{u}_\perp}{\partial t} + (\mathbf{u}_\perp \cdot \nabla)\,\mathbf{u}_\perp = 2\mathbf{u}_\perp \times \boldsymbol{\Omega} - g\nabla h, \qquad (9.4)$$

which we rewrite as

$$\frac{\partial \mathbf{u}_\perp}{\partial t} = \mathbf{u}_\perp \times (\boldsymbol{\omega} + 2\boldsymbol{\Omega}) - \nabla\left(gh + \mathbf{u}_\perp^2/2\right), \qquad (9.5)$$

where $\boldsymbol{\omega} = \nabla \times \mathbf{u}_\perp$. Since $\mathbf{u}_\perp = \mathbf{u}_\perp(x, y)$, $\boldsymbol{\omega}$ points in the z direction and the curl of (9.5) is

$$\frac{\partial \boldsymbol{\omega}}{\partial t} = \nabla \times (\mathbf{u}_\perp \times (\boldsymbol{\omega} + 2\boldsymbol{\Omega})) = -\left[(\mathbf{u}_\perp \cdot \nabla)\,\omega + (\omega + 2\Omega)\nabla \cdot \mathbf{u}_\perp\right]\hat{\mathbf{e}}_z,$$

or equivalently,

$$\frac{D}{Dt}(\omega + 2\Omega) = -(\omega + 2\Omega)\nabla \cdot \mathbf{u}_\perp. \qquad (9.6)$$

This is more or less the usual scalar vorticity equation for an inviscid, two-dimensional flow, except for unexpected term on the right. This arises because the flow is not strictly two-dimensional, but rather quasi-two-dimensional, and $\mathbf{u}_\perp$ is therefore not solenoidal. In any event, we now combine (9.6) with mass conservation in the form of

$$\frac{Dh}{Dt} + h\nabla \cdot \mathbf{u}_\perp = 0,$$

to give

$$\frac{1}{h}\frac{D}{Dt}(\omega + 2\Omega) = (\omega + 2\Omega)\frac{1}{h^2}\frac{Dh}{Dt} = -(\omega + 2\Omega)\frac{D}{Dt}\left(\frac{1}{h}\right).$$

This yields the key vorticity equation for inviscid shallow-water flow,

$$\frac{D}{Dt}\left(\frac{2\Omega + \omega}{h}\right) = \frac{DQ}{Dt} = 0, \tag{9.7}$$

where the materially conserved quantity, $Q = (2\Omega + \omega)/h$, is the called the *potential vorticity* (a rather obscure name).

We have been here before. In §8.3 we deduced (9.7), not from the rotating shallow-water equations, but from Helmholtz's second law applied in the rotating frame, *i.e.*

$$\frac{d}{dt}\int_S \boldsymbol{\omega}_{net} \cdot d\mathbf{S} = 0, \ \ \boldsymbol{\omega}_{net} = 2\boldsymbol{\Omega} + \boldsymbol{\omega}, \tag{9.8}$$

where S is any surface that spans a closed *material* curve C_m. In particular, we applied (9.8) to a material volume which consists of a narrow column of fluid of height h and cross-sectional area A. This gives

$$(2\Omega + \omega)A = \text{constant.} \tag{9.9}$$

Since the column is always composed of the same fluid particles, it conserves of both $(2\Omega + \omega)A$ and $m = \rho h A$, and (9.7) follows immediately.

The conservation of potential vorticity, often abbreviated to PV, is one of the most important concepts in rotating, shallow-water flow, and we shall return to it time and again in this chapter.

9.1.3 An Energy Equation

Let us now return to the momentum and continuity equations,

$$\frac{D\mathbf{u}_\perp}{Dt} = 2\mathbf{u}_\perp \times \boldsymbol{\Omega} - \nabla(gh), \ \ \frac{Dh}{Dt} + h\nabla \cdot \mathbf{u}_\perp = 0, \tag{9.10}$$

and see if we can derive a mechanical energy equation from them. In particular, we wish to establish the equation governing the energy density

$$e = \tfrac{1}{2}\rho h \mathbf{u}_\perp^2 + \tfrac{1}{2}\rho g h^2, \tag{9.11}$$

which is the sum of the kinetic and potential energies per unit volume integrated through the depth of fluid.

The first step is to take the dot product of the momentum equation with $\mathbf{u}_\perp$ and then eliminate $\nabla \cdot \mathbf{u}_\perp$ using mass conservation. This yields

$$\frac{D}{Dt}\left(\tfrac{1}{2}\mathbf{u}_\perp^2\right) = -\,\mathbf{u}_\perp \cdot \nabla\left(gh\right) = -\,\nabla \cdot \left(gh\mathbf{u}_\perp\right) + gh\nabla \cdot \mathbf{u}_\perp = -\,\nabla \cdot \left(gh\mathbf{u}_\perp\right) - g\frac{Dh}{Dt}.$$

Next, we multiplying through by h and rearrange terms to give

$$\frac{D}{Dt}\left(\tfrac{1}{2}h\mathbf{u}_\perp^2 + \tfrac{1}{2}gh^2\right) - \tfrac{1}{2}\mathbf{u}_\perp^2\frac{Dh}{Dt} = -\,h\nabla \cdot \left(gh\mathbf{u}_\perp\right),$$

which is more conveniently written as

$$\frac{D}{Dt}\left(\tfrac{1}{2}h\mathbf{u}_\perp^2 + \tfrac{1}{2}gh^2\right) = \tfrac{1}{2}\mathbf{u}_\perp^2\frac{Dh}{Dt} - \left[\tfrac{1}{2}gh^2\nabla \cdot \mathbf{u}_\perp + \nabla \cdot \left(\tfrac{1}{2}gh^2\mathbf{u}_\perp\right)\right].$$

Using continuity once again now yields

$$\frac{D}{Dt}\left(\tfrac{1}{2}h\mathbf{u}_\perp^2 + \tfrac{1}{2}gh^2\right) = -\left(\tfrac{1}{2}h\mathbf{u}_\perp^2 + \tfrac{1}{2}gh^2\right)\nabla \cdot \mathbf{u}_\perp - \nabla \cdot \left(\tfrac{1}{2}gh^2\mathbf{u}_\perp\right).$$

Finally, this can be rewritten in terms of the energy density, e, as

$$\frac{De}{Dt} = -\,e\nabla \cdot \mathbf{u}_\perp - \nabla \cdot \left(\tfrac{1}{2}\rho g h^2\mathbf{u}_\perp\right), \tag{9.12}$$

which yields the energy equation for shallow-water flow,

$$\frac{\partial e}{\partial t} = -\nabla \cdot \left(e\mathbf{u}_\perp\right) - \nabla \cdot \left(\tfrac{1}{2}\rho g h^2\mathbf{u}_\perp\right). \tag{9.13}$$

The physical interpretation of (9.13) requires the use of a control volume, such as that shown in Figure 9.3. When integrated over such a volume, the first term on the right of (9.13) tells us that the mechanical energy within the volume declines due to the convective flux of energy out through the sides of the control volume. This is certainly consistent with energy conservation, but it does not explain the final divergence on the right of (9.13). It is readily confirmed that the second divergence represents the work done on the fluid within the control volume by the horizontal pressure forces, $\frac{1}{2}\rho g h^2$, acting on the sides of the control volume, as indicated in Figure 9.3. Note that, for steady flows, we have $\nabla \cdot (h\mathbf{u}_\perp) = 0$, which allows us to rewrite (9.13) as

$$\nabla \cdot \left[\left(\tfrac{1}{2}\rho \mathbf{u}_\perp^2 + \rho g h \right) h\mathbf{u}_\perp \right] = h\mathbf{u}_\perp \cdot \nabla \left(\tfrac{1}{2}\rho \mathbf{u}_\perp^2 + \rho g h \right) = 0. \qquad (9.14)$$

Of course, this is just Bernoulli's equation.

9.2 Small-Amplitude Perturbations in a Rotating, Shallow-Water System

The shallow-water equations discussed above are fully nonlinear, with no restriction on the magnitude of the motion. The only restrictions we have imposed so far are $\delta = h/\ell \ll 1$, which defines the term 'shallow-water', and the fact that the motion is inviscid. We now introduce a second small parameter and place an additional limitation on the flow. If h_0 is the mean depth, we write $h = h_0 + \eta$, define $\varepsilon = \eta/h_0$, and demand that $\varepsilon \ll 1$. In short, we consider small-amplitude disturbances to a quiescent state. This allows us to explore the various types of surface gravity waves which are supported by the rotating, shallow-water equations.

9.2.1 Linearized Dynamics, Geostrophic Flow, and the Rossby Deformation Radius

We now assume that η, and hence $|\mathbf{u}_\perp|$, is small. We therefore restrict ourselves to small perturbations about a state of rest. This means that we can linearize the shallow-water equations to give

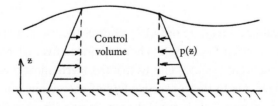

Figure 9.3 The pressure forces acting on the sides of a control volume.

$$\frac{\partial \mathbf{u}_\perp}{\partial t} = 2\mathbf{u}_\perp \times \mathbf{\Omega} - g\nabla\eta, \qquad \frac{\partial \eta}{\partial t} + h_0 \nabla \cdot \mathbf{u}_\perp = 0. \tag{9.15}$$

The corresponding vorticity equation can be obtained by taking the curl of (9.15), or else by linearizing (9.6). Either way, we find

$$\frac{\partial \omega}{\partial t} = -2\Omega \nabla \cdot \mathbf{u}_\perp = \frac{2\Omega}{h_0}\frac{\partial \eta}{\partial t}, \tag{9.16}$$

which is conveniently rewritten as

$$\frac{\partial}{\partial t}\left(\omega - \frac{2\Omega\eta}{h_0}\right) = \frac{\partial q_0}{\partial t} = 0, \tag{9.17}$$

where $q_0 = \omega - (2\Omega\eta)/h_0$. Note that (9.17) demands $q_0 \neq q_0(t)$, and so $q_0(x,y)$ is set by the initial condition. Note also the close similarity to q in the Rossby wave equation (8.31), although in (9.17) q_0 is independent of time, whereas in (8.31) q is materially conserved.

As in (8.31), q_0 is simply a linearized version of the potential vorticity. That is, when $\varepsilon = \eta/h_0$ and $|\mathbf{u}_\perp|$ are both small,

$$Q = \frac{2\Omega + \omega}{h_0 + \eta} \approx \frac{2\Omega}{h_0}\left(1 + \frac{\omega}{2\Omega} - \frac{\eta}{h_0}\right) = \frac{2\Omega}{h_0} + \frac{1}{h_0}\left(\omega - \frac{2\Omega\eta}{h_0}\right) = \frac{2\Omega}{h_0} + \frac{q_0}{h_0},$$

and so the linearized version of (9.7) is $\partial q_0/\partial t = 0$.

These linearized equations support weak, steady flows as well as small-amplitude waves. Let us start with the steady flows, where (9.15) simplifies to $\nabla \cdot \mathbf{u}_\perp = 0$ and

$$2\mathbf{u}_\perp \times \mathbf{\Omega} = g\nabla\eta. \tag{9.18}$$

This balance between the Coriolis and pressure forces is, of course, a geostrophic force balance. As $\mathbf{u}_\perp$ is solenoidal in a steady flow, we can introduce a streamfunction defined by $\mathbf{u}_\perp = \nabla \times (\psi\hat{\mathbf{e}}_z)$, from which $2\mathbf{u}_\perp \times \mathbf{\Omega} = -2\Omega\nabla\psi$. Substituting this into (9.18) yields

$$\eta = -\frac{2\Omega}{g}\psi, \tag{9.19}$$

and so the streamlines are aligned with contours of constant depth. Of course, this is exactly what the Taylor–Proudman theorem requires, since $\partial u_z/\partial z = 0$ demands that vertical columns of fluid cannot change their height as they move around.

The governing equation for these weak, steady flows can be obtained by combining $q_0 = \omega - (2\Omega\eta)/h_0$ with (9.19) to give

$$\omega + \frac{(2\Omega)^2}{gh_0}\psi = q_0(x, y),$$

or equivalently,

$$\nabla^2\psi - \frac{(2\Omega)^2}{gh_0}\psi = -q_0(x, y). \tag{9.20}$$

Of course, this can equally be written in terms of η, as

$$\nabla^2\eta - \frac{(2\Omega)^2}{gh_0}\eta = \frac{2\Omega}{g}q_0(x, y). \tag{9.21}$$

Equation (9.20) is usually rewritten in the form

$$\nabla^2\psi - \frac{\psi}{R_d^2} = -q_0(x, y), \qquad R_d = \sqrt{gh_0/2\Omega}, \tag{9.22}$$

where R_d is called the *Rossby deformation radius*. Given q_0, this is readily solved for ψ, with R_d setting the scale of the motion. In the oceans, R_d is of the order of 10^3 km.

9.2.2 Inertia–Gravity Waves and Geostrophic Adjustment

We now consider wave-like solutions of the linearized shallow-water equations, (9.15).

If we take the divergence of the linearized momentum equation, and substitute for $\nabla \cdot \mathbf{u}_\perp$ using continuity, we obtain

$$\frac{\partial}{\partial t}(\nabla \cdot \mathbf{u}_\perp) = \frac{\partial}{\partial t}\left(-\frac{1}{h_0}\frac{\partial\eta}{\partial t}\right) = -g\nabla^2\eta + 2\Omega\omega.$$

On substituting for ω using $q_0 = \omega - (2\Omega\eta)/h_0$, and rearranging terms, we obtain the wave-like equation

$$\frac{\partial^2\eta}{\partial t^2} - (gh_0)\nabla^2\eta + (2\Omega)^2\eta = -2\Omega h_0 q_0. \tag{9.23}$$

Waves governed by this equation are called *inertia–gravity* (or *Poincaré*) waves. Note that (9.21) is a special case of (9.23). Note also that q_0 is set by the initial conditions, according to $q_0 = (\omega - 2\Omega\eta/h_0)_{t=0}$, with a quiescent initial state giving $q_0 h_0 = -2\Omega\eta_{t=0}$.

There are two possibilities, $q_0 = 0$ and $q_0 \neq 0$. When $q_0 = 0$ we have a homogeneous wave-like equation which supports progressive waves of the form $\eta \sim \exp(\mathbf{k} \cdot \mathbf{x} - \tilde{\omega}t)$. The corresponding dispersion relationship is

$$\tilde{\omega}^2 = gh_0 k^2 + (2\Omega)^2, \tag{9.24}$$

where k is the magnitude of the two-dimensional wavevector, *i.e.* $k^2 = k_x^2 + k_y^2$. Note that, when $\Omega \to 0$ (or $R_d k \gg 1$), (9.24) recovers the frequency of conventional surface gravity waves on shallow water, which are non-dispersive and have a phase speed of $c = \pm\sqrt{gh_0}$. Conversely, when $gh_0 \to 0$ (or $R_d k \ll 1$), (9.24) gives the frequency of inertial waves whose phase velocity is vertical, such waves having zero group velocity. Equation (9.24) also tells us that the phase and group velocities of inertia–gravity waves are

$$\mathbf{c}_p = \frac{gh_0 + (2\Omega/k)^2}{\tilde{\omega}}\mathbf{k}, \qquad \mathbf{c}_g = \frac{gh_0}{\tilde{\omega}}\mathbf{k}, \tag{9.25}$$

with $\mathbf{c}_g \to 0$ as $gh_0 \to 0$, as must be the case.

When $q_0 \neq 0$, we have an inhomogeneous wave-like equation. An informative special case is an unbounded domain in which the initial condition is quiescent and the initial surface profile is everywhere flat, except close to the origin. In such a case, waves are triggered near the origin which then disperse off to infinity leaving behind a steady geostrophic flow governed by

$$\nabla^2\psi - \frac{\psi}{R_d^2} = -q_0(x,y), \qquad \eta = -\frac{2\Omega}{g}\psi. \tag{9.26}$$

This kind of spontaneous emission of inertia–gravity waves, which leaves behind a residual geostrophic flow, is referred to as *geostrophic adjustment*.

A simple example illustrates the process. Consider an initial surface profile which is one-dimensional and given by

$$\eta(x < 0) = \eta_0, \qquad \eta(x > 0) = -\eta_0, \tag{9.27}$$

where η_0 is a constant. This is shown in Figure 9.4(a). If the initial state is quiescent, then

$$q_0(x < 0) = -2\Omega\eta_0/h_0, \qquad q_0(x > 0) = 2\Omega\eta_0/h_0,$$

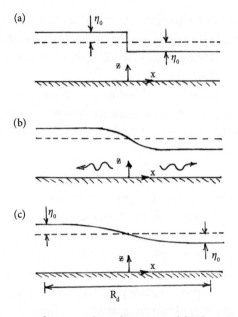

Figure 9.4 The process of geostrophic adjustment. (a) The initial surface profile. (b) Waves disperse off to infinity. (c) The final geostrophic state.

and (9.26) has the solution

$$\eta(x < 0, t \to \infty) = \eta_0 \left[1 - \exp\left(-|x|/R_d\right)\right],$$
$$\eta(x > 0, t \to \infty) = -\eta_0 \left[1 - \exp\left(-x/R_d\right)\right],$$

$$(9.28)$$

as shown in Figure 9.4(c). Thus, the initial discontinuity in η is smoothed out over a distance of order R_d.

The distribution of ψ now follows from $\psi = -g\eta/2\Omega$, and the residual geostrophic velocity field is readily shown to be

$$\mathbf{u}(x, t \to \infty) = -\frac{g\eta_0}{2\Omega R_d} \exp\left(-|x|/R_d\right) \hat{\mathbf{e}}_y.$$

$$(9.29)$$

This is illustrated in Figure 9.5. The final state is evidently a localized jet in the negative y-direction whose width is $\sim R_d$, which is of the order of 10^3 km in the oceans.

9.2.3 Kelvin Waves at a Boundary

A particularly interesting form of inertia–gravity wave occurs when waves are excited close to a sidewall (or a coastline), as shown in Figure 9.6. These are

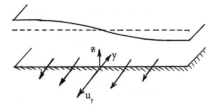

Figure 9.5 The residual geostrophic velocity distribution after geostrophic adjustment.

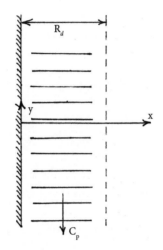

Figure 9.6 A Kelvin wave propagates down a western coastline.

called *Kelvin waves* in the oceanographic literature, and the oceans are full of them.

Suppose that, because of the boundary on the left, $\mathbf{u}_\perp$ has no x component, i.e. $\mathbf{u}_\perp = u_y(x, y, t)\hat{\mathbf{e}}_y$. Then the linearized equations (9.15) reduce to

$$\text{momentum}: \quad 0 = 2\Omega u_y - g\frac{\partial \eta}{\partial x}, \quad \frac{\partial u_y}{\partial t} = -g\frac{\partial \eta}{\partial y}, \quad (9.30)$$

$$\text{continuity}: \quad \frac{\partial \eta}{\partial t} = -h_0\frac{\partial u_y}{\partial y}. \quad (9.31)$$

Combining the second of (9.30) with (9.31) yields

$$\frac{\partial^2 u_y}{\partial t^2} = (gh_0)\frac{\partial^2 u_y}{\partial y^2}, \quad (9.32)$$

and so the wave propagates along the boundary with a wave speed equal to that of a conventional surface gravity wave, $c = \sqrt{gh_0}$. Thus, we have d'Alembert's solution,

$$u_y \sim f(y \pm ct), \tag{9.33}$$

where f is an arbitrary function of its argument.

If we now assume that we are dealing with the homogeneous wave equation, in which $q_0 = 0$, then we also have

$$\omega = \frac{\partial u_y}{\partial x} = \frac{2\Omega}{h_0}\eta. \tag{9.34}$$

This combines with the first of (9.30) to give

$$\frac{\partial^2 u_y}{\partial x^2} = \frac{(2\Omega)^2}{gh_0}u_y = \frac{u_y}{R_d^2}, \tag{9.35}$$

and so u_y decays exponentially with distance from the boundary, as $u_y \sim \exp(-x/R_d)$. Combining these results yields the *evanescent* wave

$$u_y = u_0 \exp(-x/R_d)f(y \pm ct), \tag{9.36}$$

$$\eta = \eta_0 \exp(-x/R_d)f(y \pm ct), \tag{9.37}$$

which is indeed a solution of the homogeneous wave-like equation (9.23).

However, this is not the end of the story. Substituting (9.36) and (9.37) back into (9.34), and also into the second of (9.30), gives us

$$\eta_0/h_o = -u_0/c, \qquad \eta_0/h_o = \mp u_0/c, \tag{9.38}$$

respectively, and so only the top sign is acceptable. We conclude that Kelvin waves take the general form

$$u_y = -\left(\eta_0 c/h_0\right)\exp(-x/R_d)f(y + ct), \tag{9.39}$$

$$\eta = \eta_0 \exp(-x/R_d)f(y + ct), \tag{9.40}$$

which propagate in the negative y direction only, as shown in Figure 9.6.

More generally, as long as $\Omega > 0$, which is exactly what we have assumed throughout, a Kelvin wave propagates in a direction such that an observer moving with the wave sees the boundary on their right. In the northern hemisphere, then, Kelvin waves propagate in an anticlockwise (or cyclonic) sense around an ocean basin. If we now repeat the analysis for $\Omega < 0$, which requires that we redefine the deformation radius as $R_d = -\sqrt{gh_0}/2\Omega$, it turns out that an observer moving with the wave sees the boundary on their left. Thus, in the southern hemisphere, Kevin waves propagate in a clockwise sense around an ocean basin, which is also in a cyclonic direction.

Kelvin waves are generated by changing wind patterns. One notable example occurs when an El Niño event impacts on an eastern oceanic boundary in the equatorial regions. Kelvin waves then propagate poleward along the coast, both north and south of the equator.

9.3 The Shallow-Water Equations in the Quasi-Geostrophic Limit

9.3.1 The Quasi-Geostrophic Shallow-Water (QGSW) Equations

So far we have assumed that:

(i) the flow is inviscid;
(ii) we have shallow water, $\delta = h/\ell << 1$;
(iii) changes in the depth are small, $\varepsilon = \eta/h_0 << 1$;
(iv) as a consequence of (iii) above, the horizontal velocity, $\mathbf{u}_\perp$, is small.

This allowed us to linearize the equations of motion and obtain solutions in terms of 'fast waves', *i.e.* waves whose frequency exceeds 2Ω, and hence is much higher than that of quasi-geostrophic Rossby waves.

We shall now shift emphasis and refocus on slow, weak changes to a geostrophic state, in which the fast waves discussed above are filtered out. We shall retain all four restrictions listed above, while formalizing (iv) to read

$$\mathrm{Ro} = u_\perp/\Omega\ell << 1. \qquad (9.41)$$

Crucially, however, we shall now demand that time derivatives are restricted to the *convective timescale* $\ell/u_\perp$, so that $\partial \mathbf{u}_\perp/\partial t \sim u_\perp^2/\ell$. Since Ro $<< 1$, this requires that time derivatives are weak, and we shall see that this excludes (or filters out) the fast inertia–gravity waves discussed above. We shall also assume that the horizontal scales of motion do not significantly exceed the deformation radius, so that $(\Omega\ell)^2/gh_0$ is of order unity, or smaller. Finally, we shall refrain from automatically linearizing all equations on the assumption that wave amplitudes are infinitesimally small. Rather, we start with the nonlinear equations and progressively simplify those using the restrictions listed above.

Let us return to the nonlinear, shallow-water equations (9.3), (9.4), and (9.6),

$$\frac{D\eta}{Dt} + h_0 \nabla \cdot \mathbf{u}_\perp = 0, \quad \frac{D\mathbf{u}_\perp}{Dt} = 2\mathbf{u}_\perp \times \mathbf{\Omega} - g\nabla\eta, \tag{9.42}$$

$$\frac{D\omega}{Dt} = -(\omega + 2\Omega)\nabla \cdot \mathbf{u}_\perp, \tag{9.43}$$

where we have used (iii) above to replaced h by h_0 in the continuity equation. To leading order in Ro, the momentum equation reduces to the geostrophic balance

$$2\mathbf{u}_\perp^{(0)} \times \mathbf{\Omega} = g\nabla\eta^{(0)}, \tag{9.44}$$

where the superscript (0) indicates that these are the leading-order contributions (in Ro) to $\mathbf{u}_\perp$ and η. This yields $\eta^{(0)} \sim u_\perp \Omega \ell / g$, and so our restriction on time derivatives requires

$$\frac{1}{h_0}\frac{D\eta}{Dt} \sim \frac{u_\perp^2 \Omega}{g h_0} \sim \frac{u_\perp}{\Omega \ell}\frac{u_\perp}{\ell} = \text{Ro}\frac{u_\perp}{\ell},$$

where we have used $g h_0 \sim (\Omega \ell)^2$. Evidently, to leading order in Ro, mass conservation reduces to $\nabla \cdot \mathbf{u}_\perp^{(0)} = 0$. Alternatively, we could have established $\nabla \cdot \mathbf{u}_\perp^{(0)} = 0$ by taking the curl of (9.44). Either way, our leading-order equations are identical to the steady, linearized, geostrophic equations of §9.2.1. Thus, we have

$$\mathbf{u}_\perp^{(0)} = \nabla \times (\psi \hat{\mathbf{e}}_z), \quad \eta^{(0)} = -\frac{2\Omega}{g}\psi, \quad \omega^{(0)} = -\nabla^2\psi. \tag{9.45}$$

To allow for slow, weak departures from geostrophic motion we must now go to the next order in Ro, and write

$$\eta = \eta^{(0)} + \text{Ro}\,\eta^{(1)} + ..., \quad \mathbf{u}_\perp = \mathbf{u}_\perp^{(0)} + \text{Ro}\,\mathbf{u}_\perp^{(1)} + ... \tag{9.46}$$

Mass conservation and the vorticity equation now yield

$$\frac{D\eta^{(0)}}{Dt} = -h_0 \nabla \cdot \left(\text{Ro}\,\mathbf{u}_\perp^{(1)}\right),$$

$$\frac{D\omega^{(0)}}{Dt} = -2\Omega \nabla \cdot \left(\mathrm{Ro}\, \mathbf{u}_\perp^{(1)}\right),$$

where the convective derivatives are based on $\mathbf{u}_\perp^{(0)}$. Eliminating $\nabla \cdot \mathbf{u}_\perp^{(1)}$ from these equations yields

$$\frac{D}{Dt}\left(\omega^{(0)} - \frac{2\Omega}{h_0}\eta^{(0)}\right) = 0, \tag{9.47}$$

which combines with (9.45) to give

$$\frac{D}{Dt}\left(\omega^{(0)} + \frac{\psi}{R_d^2}\right) = \frac{Dq}{Dt} = 0, \tag{9.48}$$

where

$$q = \omega^{(0)} - \frac{2\Omega\eta^{(0)}}{h_0} = \omega^{(0)} + \frac{\psi}{R_d^2} = -\nabla^2\psi + \frac{\psi}{R_d^2}. \tag{9.49}$$

Equation (9.48) is the key result of this section. Of course, it is a form of potential vorticity conservation. That is, when $\varepsilon = \eta/h_0$ and Ro are both small,

$$Q^{(0)} = \frac{2\Omega + \omega^{(0)}}{h_0 + \eta^{(0)}} \approx \frac{2\Omega}{h_0}\left(1 + \frac{\omega^{(0)}}{2\Omega} - \frac{\eta^{(0)}}{h_0}\right) = \frac{2\Omega}{h_0} + \frac{1}{h_0}\left(\omega^{(0)} - \frac{2\Omega\eta^{(0)}}{h_0}\right)$$

$$= \frac{2\Omega}{h_0} + \frac{q}{h_0},$$

and so (9.48) follows directly from the leading-order contribution (in Ro) to (9.7). Indeed, q is called the *quasi-geostrophic shallow-water potential vorticity*.

Equation (9.48) is often called the *quasi-geostrophic shallow-water equation*, or the QGSW equation for short. It is central to quasi-geostrophic shallow-water theory. As we shall see, it is readily generalized to include bottom topography, friction, and so on. It is also very similar to the Rossby wave equation (8.31), and so it should come as no surprise that (9.48) supports low-frequency Rossby waves. Finally, we note that, in the limit of $R_d \to \infty$, (9.48) reverts to the conventional vorticity equation for inviscid, two-dimensional flow. This limit is equivalent to insisting that the upper surface is rigid, i.e. $\eta^{(0)}/h_0 = -\psi/(2\Omega R_d^2) \to 0$, and is commonly referred to as the *rigid-lid approximation*.

9.3.2 Potential Vorticity Inversion

Given that q is materially conserved, it is natural to interpret the development of a quasi-geostrophic shallow-water flow in terms of the evolution of
the potential vorticity field, $q(\mathbf{x}, t)$. It is important, therefore, that we establish
the shallow water equivalent of the Biot–Savart law. That is to say, we would
like to find a way of inverting (or solving)

$$\nabla^2\psi - \psi/R_d^2 = -q(\mathbf{x}) \tag{9.50}$$

in an infinite domain, so that, given the instantaneous distribution of q, we can
find the associated velocity field. We can then follow the evolution of a QGSW
flow from given initial conditions by advecting q, inverting (9.50) to update ψ,
advecting q again, and so on.

To establish such an inversion formula, it is natural to consider first the simple case where q takes the form of a delta function located at the origin. Having
found the corresponding distribution of ψ, we can then use superposition to
determine the flow for an arbitrary distribution of $q(\mathbf{x})$. So let us start with
the case $q = \delta(\mathbf{x})$, where δ is the two-dimensional delta function. Clearly, when
$q = \delta(\mathbf{x})$, the flow is axisymmetric and so, in polar coordinates, (9.50) becomes

$$r^2\frac{d^2\psi}{dr^2} + r\frac{d\psi}{dr} - \frac{r^2}{R_d^2}\psi = -r^2\delta(\mathbf{x}),$$

or equivalently,

$$s^2\frac{d^2\psi}{ds^2} + s\frac{d\psi}{ds} - s^2\psi = -r^2\delta(\mathbf{x}), \tag{9.51}$$

where $s = r/R_d$. For $s \neq 0$, this has the solution $\psi = aK_0(s)$, where K_0 is the usual
modified Bessel function and a is a constant to be determined. To find a we
integrate (9.50) over a circle of radius r_0 centred on the origin. Gauss' theorem
then gives us

$$\oint_{r=r_0} \nabla\psi \cdot d\mathbf{S} - \frac{1}{R_d^2}\int_0^{r_0} 2\pi r\psi dr = -1, \tag{9.52}$$

and since $K'_0(s) = -K_1(s)$, (9.52) becomes

$$(2\pi a) \left[s_0 K_1(s_0) + \int_0^{s_0} s K_0(s) ds \right] = 1.$$

Finally, we note that

$$\int_0^{s_0} s K_0(s) ds = - [s K_1(s)]_0^{s_0} = 1 - s_0 K_1(s_0),$$

and so we conclude that $a = 1/(2\pi)$. Hence, the solution of (9.50) for $q = \delta(\mathbf{x})$ is

$$\psi = \frac{1}{2\pi} K_0 (r/R_d). \tag{9.53}$$

Next we note that, if q is a delta function of strength q' located at position $\mathbf{x}'$, and we interpret r in (9.53) as the distance from the source, then (9.53) generalizes to

$$\psi(\mathbf{x}) = \frac{q'}{2\pi} K_0 (|\mathbf{x} - \mathbf{x}'|/R_d). \tag{9.54}$$

The final step is to combine (9.54) with superposition. This tells us that, if $q(\mathbf{x})$ is a continuous distribution of potential vorticity, then the associated streamfunction in an infinite domain is

$$\psi(\mathbf{x}, t) = \frac{1}{2\pi} \int K_0 (|\mathbf{x} - \mathbf{x}'|/R_d) q(\mathbf{x}', t) d\mathbf{x}'. \tag{9.55}$$

This is the equivalent of the Biot–Savart law for quasi-geostrophic shallow-water flow and the application of equation (9.55) is sometimes called *PV inversion*.

When combined with (9.48), this inversion formula offers a complete description of QGSW flow. That is, starting with some specified distribution of q, (9.55) allows us to calculate ψ, which in turn enables us to advance q in time using $Dq/Dt = 0$. This then yields a new velocity field, and hence a new distribution of q, and so it goes on. We shall make use of PV inversion in Chapter 18 when we discuss Rossby wave turbulence.

Note that equation (9.55) has an unexpected and useful property which makes it somewhat different to the Biot–Savart law. The modified Bessel function K_0 decays as an exponential for large argument. It follows that a blob of potential vorticity has a finite domain of influence, of the order of R_d, with the induced velocity exponentially small outside that domain. This is different to the Biot–Savart law, where a blob of vorticity casts a long shadow, causing the velocity to fall off slowly as a power law.

9.3.3 An Energy Equation for QGSW Flow

We now turn to energy. In §9.1.3 we introduced the energy density for shallow-water flow,

$$e = \tfrac{1}{2}\rho h \mathbf{u}_\perp^2 + \tfrac{1}{2}\rho g h^2, \tag{9.56}$$

which is the sum of the kinetic and potential energies per unit volume integrated through the depth of fluid. For a QGSW flow subject to the constraints listed in §9.3.1, we might expect this to simplify to

$$e_{QG} = \tfrac{1}{2}\rho h_0 \mathbf{u}_\perp^2 + \tfrac{1}{2}\rho g \eta^2, \tag{9.57}$$

where, for simplicity, we have omitted the superscript (0) on $\mathbf{u}_\perp$ and η. We shall now confirm that the global conservation of e_{QG} follows directly from the QGSW equations. We assume that the domain is bounded, with $\psi = 0$ at the boundary, or else the motion is unbounded but localized, so that $\psi \to 0$ in the far field.

We start by noting that, in the light of definition (9.49), we have

$$\tfrac{1}{2}q\psi = \tfrac{1}{2}(\nabla\psi)^2 + \frac{\psi^2}{2R_d^2} - \tfrac{1}{2}\nabla \cdot (\psi\nabla\psi), \tag{9.58}$$

which combines with (9.45) to give

$$\tfrac{1}{2}\rho h_0 q\psi = \tfrac{1}{2}\rho h_0 \mathbf{u}_\perp^2 + \tfrac{1}{2}\rho g \eta^2 - \nabla \cdot \left(\tfrac{1}{2}\rho h_0 \psi\nabla\psi\right), \tag{9.59}$$

where again we have omitted the superscript (0) on $\mathbf{u}_\perp$ and η. Evidently, e_{QG} and $\tfrac{1}{2}\rho h_0 q\psi$ differ by a divergence, which integrates to zero. We therefore transfer attention to $\tfrac{1}{2}q\psi$.

Now q is materially conserved and so we have

$$\frac{D(q\psi)}{Dt} = \left(-\nabla^2\psi + \frac{\psi}{R_d^2}\right)\frac{\partial\psi}{\partial t} = \frac{\partial}{\partial t}\left(\frac{\mathbf{u}_\perp^2}{2} + \frac{\psi^2}{2R_d^2}\right) - \nabla \cdot \left(\frac{\partial\psi}{\partial t}\nabla\psi\right), \tag{9.60}$$

where the convective derivative is based on $\mathbf{u}_\perp^{(0)}$. Moreover, (9.58) differentiates to give

$$\frac{D}{Dt}(q\psi) - \mathbf{u}_\perp^{(0)} \cdot \nabla(q\psi) = 2\frac{\partial}{\partial t}\left(\frac{\mathbf{u}_\perp^2}{2} + \frac{\psi^2}{2R_d^2}\right) - \nabla \cdot \left[\frac{\partial}{\partial t}(\psi\nabla\psi)\right]. \qquad (9.61)$$

Crucially, subtracting (9.60) from (9.61) yields

$$\frac{\partial}{\partial t}\left(\frac{\mathbf{u}_\perp^2}{2} + \frac{\psi^2}{2R_d^2}\right) + \nabla \cdot \left[\psi\left(q\mathbf{u}_\perp^{(0)} - \nabla\frac{\partial\psi}{\partial t}\right)\right] = 0, \qquad (9.62)$$

or equivalently,

$$\frac{\partial e_{QG}}{\partial t} + \nabla \cdot \left[\rho h_0 \psi\left(q\mathbf{u}_\perp^{(0)} - \nabla(\partial\psi/\partial t)\right)\right] = 0.$$

Finally, integrating (9.59) and (9.62) over the entire domain yields the energy equation

$$\rho h_0 \frac{d}{dt}\int \left(\tfrac{1}{2}q\psi\right) d\mathbf{x} = \frac{d}{dt}\int \left(\tfrac{1}{2}\rho h_0 \mathbf{u}_\perp^2 + \tfrac{1}{2}\rho g\eta^2\right) d\mathbf{x} = 0, \qquad (9.63)$$

which confirms that e_{QG} is indeed globally conserved.

9.4 Adding Topography and the β-Effect to the QGSW Equations: Rossby Waves II

9.4.1 Generalized Potential Vorticity Conservation

The shallow-water systems we have considered so far are too idealized to apply to the atmosphere or to an ocean. We shall now show how a more realistic set of equations are readily developed through a generalization of the potential vorticity, although we continue to neglect friction. The first thing to note is that the Ω which appears in our equations is the component of rotation normal to the free surface. For the earth, this varies with latitude, being zero at the equator, positive in the north, negative in the south, and maximal at the poles. In order to allow for this spatial variation in Ω it is common to consider a plane, called the β-plane, which is locally tangent to the surface of the earth, with x pointing to the east, y directed to the north, and z normal to the surface, as shown in Figure 9.7. The 2Ω that appears in the shallow-water equations is

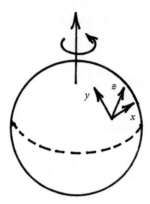

Figure 9.7 The β-plane.

then written as $2\Omega = 2\Omega_0 + \beta y$, where β is *small* and positive. This is called the β-plane approximation.

The second point is that the bottom surface is rarely flat, but rather undulates. We can allow for slow, gentle undulations, as shown in Figure 9.8, without violating the restrictions of QGSW theory, provided that the bed height, h_b, is much less than h_0. The water depth is now $h = h_0 - h_b + \eta$. When we include the β-effect and bottom topography, the potential vorticity remains a material invariant, because (9.7) continues to hold, but it takes the more general form

$$Q^{(0)} = \frac{2\Omega + \omega^{(0)}}{h} = \frac{2\Omega_0 + \beta y + \omega^{(0)}}{h_0 - h_b + \eta^{(0)}}, \tag{9.64}$$

to leading order in Ro. This may be expanded when β, η/h_0, h_b/h_0, and Ro $\sim \omega^{(0)}/2\Omega_0$ are all small, to give

$$Q^{(0)} \approx \frac{2\Omega_0}{h_0}\left(1 + \frac{\omega^{(0)} + \beta y}{2\Omega_0} + \frac{h_b - \eta^{(0)}}{h_0}\right)$$

$$= \frac{2\Omega_0}{h_0} + \frac{1}{h_0}\left(\omega^{(0)} + \beta y - \frac{2\Omega_0\eta^{(0)}}{h_0} + \frac{2\Omega_0 h_b}{h_0}\right).$$

This, in turn, leads to a generalized version of the shallow-water quasi-geostrophic potential vorticity,

$$q = \omega^{(0)} + \beta y - \frac{2\Omega_0\eta^{(0)}}{h_0} + \frac{2\Omega_0 h_b}{h_0} = -\nabla^2\psi + \frac{\psi}{R_d^2} + \beta y + \frac{2\Omega_0 h_b}{h_0}, \tag{9.65}$$

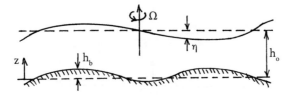

Figure 9.8 Shallow-water flow over bottom topography.

where (9.45) now reads $\eta^{(0)} = -2\Omega_0\psi/g$ and R_d is defined as $R_d = \sqrt{gh_0}/2\Omega_0$. Since $Q^{(0)}$ is a material invariant, so is q,

$$\frac{D}{Dt}\left[-\nabla^2\psi + \frac{\psi}{R_d^2} + \beta y + \frac{2\Omega_0 h_b}{h_0}\right] = 0, \tag{9.66}$$

which is the governing equation for QGSW flow with bottom topography and a β-effect.

9.4.2 Rossby Waves on the β-plane

Now consider the case of $R_d \to \infty$, or equivalently $\ell \ll R_d$, where the upper surface is rigid in the sense that $\eta^{(0)} \to 0$, as discussed in §9.3.1. Additionally, we take the bottom surface to be flat but slightly inclined, say $h_b = \alpha y$, as shown in Figure 9.9. Then our QGSW equation becomes

$$\frac{D}{Dt}\left[-\nabla^2\psi + \left(\frac{2\Omega_0\alpha}{h_0} + \beta\right)y\right] = 0, \tag{9.67}$$

which, for small-amplitude motion, may be linearized to give

$$\frac{\partial}{\partial t}\nabla^2\psi = u_y\left(\frac{2\Omega_0\alpha}{h_0} + \beta\right). \tag{9.68}$$

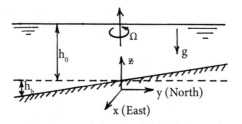

Figure 9.9 QGSW flow in the rigid-lid approximation and with $h_b = \alpha y$.

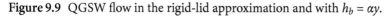

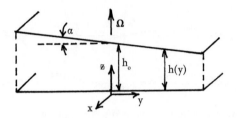

Figure 9.10 The geometry considered in Chapter 8 when discussing Rossby waves.

We have been here before. In Chapter 8 we considered quasi-geostrophic disturbances in a rotating fluid bounded by two almost parallel surfaces, as shown in Figure 9.10. In particular, we showed that the conservation of potential vorticity yields

$$\frac{D}{Dt}\left(-\nabla^2\psi + \frac{2\Omega\alpha y}{h_0}\right) = 0. \tag{9.69}$$

Moreover, we noted that, for small-amplitude disturbances, (9.69) linearizes to give

$$\frac{\partial}{\partial t}\nabla^2\psi = u_y\frac{2\Omega\alpha}{h_0},$$

which is the governing equation for linear Rossby waves between two boundaries.

It follows immediately that our QGSW equation (9.68) supports linear Rossby waves whose properties are identical to those described in Chapter 8. It also follows that the β-effect is exactly equivalent to a sloping lower boundary, which is a remarkably convenient result. Rossby waves driven by the β-effect are often called *planetary waves*, while those driven by the bottom profile are called *topographic Rossby waves*.

Since (9.67) supports linear Rossby waves, we might expect that (9.66) does also, and that is indeed the case. When $h_b = \alpha y$, their governing equation is

$$\frac{\partial}{\partial t}\left(\nabla^2\psi - \frac{\psi}{R_d^2}\right) = u_y\left(\frac{2\Omega_0\alpha}{h_0} + \beta\right),$$

which is a direct generalization of (9.68), whereas linear waves associated with an arbitrary bottom profile are governed by

$$\frac{\partial}{\partial t}\left(\nabla^2\psi - \frac{\psi}{R_d^2}\right) = \beta u_y + \frac{2\Omega_0}{h_0}\mathbf{u}\cdot\nabla h_b. \qquad (9.70)$$

We now focus on Rossby waves associated exclusively with the β-effect. In such cases (9.70) simplifies to

$$\frac{\partial}{\partial t}\left(\nabla^2\psi - \frac{\psi}{R_d^2}\right) = -\beta\frac{\partial\psi}{\partial x}, \qquad (9.71)$$

which supports progressive waves of the form $\psi \sim \exp\left[j\left(\mathbf{k}_\perp\cdot\mathbf{x} - \bar{\omega}t\right)\right]$, where $\mathbf{k}_\perp$ is the horizontal wavevector. The corresponding dispersion relationship is clearly

$$\bar{\omega} = -\frac{\beta k_x}{k_\perp^2 + R_d^{-2}}, \qquad k_\perp = |\mathbf{k}_\perp|, \qquad (9.72)$$

and since we insist that $\bar{\omega} > 0$, this requires $k_x < 0$. The phase and the group velocities, on the other hand, are readily shown to be

$$\mathbf{c}_p = -\frac{\beta k_x}{k_\perp^2\left(k_\perp^2 + R_d^{-2}\right)}\mathbf{k}_\perp, \qquad (9.73)$$

and

$$c_{gx} = \frac{\beta\left(k_x^2 - k_y^2 - R_d^{-2}\right)}{\left(k_\perp^2 + R_d^{-2}\right)^2}, \qquad c_{gy} = \frac{2\beta k_x k_y}{\left(k_\perp^2 + R_d^{-2}\right)^2}. \qquad (9.74)$$

If we recall that x points to the east on the β-plane, then we see that the x-component of the phase velocity is always westward. However, the x-component of the group velocity can be in either direction, though it is also to the west for wavelengths much larger than R_d.

For the important special case of $k_y = 0$ and $R_d \to \infty$, the expressions above simplify to

$$\mathbf{c}_p = -\frac{\beta}{k_x^2}\hat{\mathbf{e}}_x, \qquad \mathbf{c}_g = \frac{\beta}{k_x^2}\hat{\mathbf{e}}_x, \qquad (9.75)$$

and so $\mathbf{c}_p$ and $\mathbf{c}_g$ are equal and opposite, with the phase velocity always westward and the group velocity to the east. We can now reinterpret Figure 8.3, which shows Rossby waves generated by flow over an obstacle, in terms of

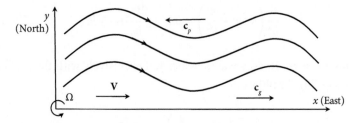

Figure 9.11 A Rossby wave whose phase velocity is westward is superimposed on a uniform geostrophic motion, **V**, which is equal and opposite to $\mathbf{c}_p$. The resulting flow is steady.

motion on the β-plane. Consider a wave in which $k_y = 0$ and $R_d \gg \ell$, and suppose that it is superimposed on a uniform geostrophic motion whose velocity, **V**, is to the east and equal but opposite to the phase velocity. Then we have the situation shown in Figure 9.11, where the wave pattern appears steady. (A steady wave pattern often occurs when waves are triggered by flow over an obstacle.) Since $\mathbf{c}_g = -\mathbf{c}_p$, the energy of such a wave propagates eastward with a velocity of $\mathbf{c}_g = \mathbf{V}$ relative to the fluid, but with the higher velocity of $2\mathbf{V}$ relative to the object that triggered the waves.

* * *

That concludes our brief excursion into rotating, shallow-water flow. The topic is central to both oceanography and meteorology, and so it is covered in some detail in texts on geophysical fluid dynamics. Excellent accounts may be found in, for example, McWilliams (2006), Salmon (1998), and Vallis (2017).

References

McWilliams, J.C., 2006, *Geophysical Fluid Dynamics*, Cambridge University Press.
Salmon, R., 1998, *Lectures on Geophysical Fluid Dynamics*, Oxford University Press.
Vallis, G.K., 2017, *Atmospheric and Oceanic Fluid Dynamics*, 2nd Ed., Cambridge University Press.

Chapter 10
Precession

10.1 An Example of Precessing Flow: Motion in Planetary Cores

The symmetry axis of an inclined, spinning gyroscope slowly orbits, or cones, around a vertical axis through its pivot point, as shown in Figure 10.1(a). The symmetry axis of the gyroscope is said to *precess* about the vertical axis. The motion of a fluid within a precessing container is an interesting problem that finds applications in the fuel tanks of manoeuvring spacecraft and in the liquid iron cores of the terrestrial planets. If a container simply spins without precession, then viscous stresses ensure that, after some transient, the fluid co-rotates with the container. However, if the angular velocity of the container continually precesses about a fixed inertial axis, then the container is constantly accelerating in the inertial frame (Figure 10.1b). The viscous and pressure torques exerted by the container wall on the fluid, which try to enforce co-rotation on the fluid, now have to compete with the inertia of the fluid. In such cases, the instantaneous orientation of the mean vorticity of the fluid tends to lag behind that of the spin axis of the container.

In this chapter, we focus on the precession of the earth and its liquid iron core. We first determine the precessional motion of the mantle by treating the

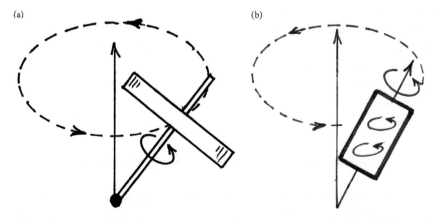

(a)   (b)

Figure 10.1 (a) A precessing gyroscope. (b) Precession of a container filled with fluid.

The Dynamics of Rotating Fluids. P. A. Davidson, Oxford University Press. © Peter A Davidson (2024).
DOI: 10.1093/9780191994272.003.0010

earth as a rigid body. Subsequently, we consider the response of the fluid core to the precession of the more massive mantle. Although we focus on the earth and its liquid core, much of what we discuss is relevant to other precessional flows.

Perhaps we should start by saying something about the structure of the earth. The earth, which is almost (but not quite) spherical, has a mean outer radius of 6370 km. It comprises an iron core, a rocky mantle, and a thin outer crust (Figure 10.2a). The iron core, which contains roughly one third of the earth's total mass, is divided into two parts. The liquid outer core has a radius of approximately 3480 km, while the inner core is solid and has a radius of 1220 km. (The inner part of the core is solid because of the high pressure at the centre of the earth.) Crucially, the earth is not exactly spherical. Rather, because the mantle is not rigid, the equatorial radius is slightly larger than the polar radius, by around 21 km. Thus, the earth has the geometry of a spinning top. Finally, the spin axis of the earth, which is (almost) fixed in direction, is inclined to the axis of the orbital plane by 23.4°, as shown in Figure 10.2(b). This combination of an inclined spin axis and an equatorial bulge means that the sun (and indeed the moon) exerts a weak *gravitational torque* on the earth.

As we shall see, the earth experiences two distinct forms of precession. First, if we ignore the weak solar and lunar gravitational torques, the difference in the equatorial and polar radii means that the earth can undergo a form of *free precession*. This was known to Newton and Euler, but is now called the *Chandler wobble*, after the astronomer who first detected it. In free precession, the earth's rotation vector cones around the geometric north pole with a period of around 430 days. However, the amplitude of the precession is small, with the point where the rotation vector pierces the earth's crust being a mere 9 m from the geometric north pole. The second kind of precession, *forced precession*, arises directly from the solar and lunar torques and is called *equinox precession*. Here the earth's spin axis precesses about an axis aligned with the solar spin axis. This is a slow process, with a period of around 26,000 years. Equinox precession has been much studied, as has the response of the earth's fluid core to the equinox precession of the mantle.

10.2 A Crash Course on Rigid-Body Precession

The primary focus of this chapter is precession within planetary cores. However, it seems prudent to first take a step back and review the classical theory of rigid-body precession. In part, this is to establish some key physical concepts,

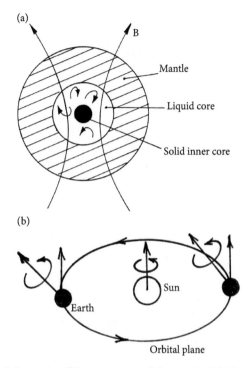

Figure 10.2 (a) Schematic of the structure of the earth. (b) The spin axis of the earth is inclined to the orbital plane.

such as the distinction between free and forced precession. However, it is also to agree the precise meaning of certain basic terms (which can vary somewhat from author to author), and to introduce a system of notation.

10.2.1 Euler's Equations

The starting point for rigid-body precession is Euler's equations for a rotating body. However, before introducing Euler's equations, let us briefly recall the discussion in Chapter 4 regarding changing frames of reference. In §4.1, we considered an inertial frame with unit vectors $(\mathbf{i}, \mathbf{j}, \mathbf{k})$, and a non-inertial frame with unit vectors $(\mathbf{i}^*, \mathbf{j}^*, \mathbf{k}^*)$. The starred system shares a common origin with the inertial frame, but rotates relative to the inertial frame with angular velocity $\boldsymbol{\Omega}$, as shown in Figure 10.3. Of course, any vector $\mathbf{A}$ can be represented in either system, as

$$\mathbf{A} = A_x\mathbf{i} + A_y\mathbf{j} + A_z\mathbf{k} = A_x^*\mathbf{i}^* + A_y^*\mathbf{j}^* + A_z^*\mathbf{k}^*.$$

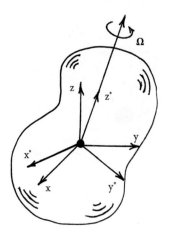

Figure 10.3 Two coordinate systems, one inertial, (x, y, z), and one non-inertial, (x^*, y^*, z^*).

Since $(\mathbf{i}, \mathbf{j}, \mathbf{k})$ is fixed in the inertial frame, while $(\mathbf{i}^*, \mathbf{j}^*, \mathbf{k}^*)$ is fixed in the rotating frame, the time derivatives of $\mathbf{A}$, as measured by different observers in the two frames, are

$$\frac{d\mathbf{A}}{dt} = \dot{A}_x \mathbf{i} + \dot{A}_y \mathbf{j} + \dot{A}_z \mathbf{k}, \qquad \frac{d^*\mathbf{A}}{dt} = \dot{A}_x^* \mathbf{i}^* + \dot{A}_y^* \mathbf{j}^* + \dot{A}_z^* \mathbf{k}^*,$$

where a dot indicates the time derivative of a scalar. Of course, $d\mathbf{A}/dt$ and $d^*\mathbf{A}/dt$ are not the same thing. For example, a vector $\mathbf{A}$ which is fixed in the starred system, $d^*\mathbf{A}/dt = 0$, rotates in the inertial frame, and simple kinematics tells us that

$$\frac{d\mathbf{A}}{dt} = \mathbf{\Omega} \times \mathbf{A}.$$

More generally, the two time derivatives, $d\mathbf{A}\,dt$ and $d^*\mathbf{A}\,dt$, are related by

$$\frac{d\mathbf{A}}{dt} = \frac{d^*\mathbf{A}}{dt} + \mathbf{\Omega} \times \mathbf{A}, \tag{10.1}$$

(see (4.3)). Note that (10.1) holds even if $\mathbf{\Omega}$ is a function of time. Note also that

$$\frac{d\mathbf{\Omega}}{dt} = \frac{d^*\mathbf{\Omega}}{dt}. \tag{10.2}$$

Let us now turn from kinematics to dynamics. If we wish to describe the rotation of a rigid body, we must first agree on a reference point within the body about which to measure its angular momentum, $\mathbf{H}$, angular velocity, $\mathbf{\Omega}$, moment of inertia tensor, I_{ij}, and any external torque applied to the body, $\mathbf{T}$. If there is a fixed point in an inertial frame about which the body rotates, then that is the obvious choice. If there is not, then we choose a coordinate system whose origin is located at the centre of mass of the body. This allows us to separate the motion of the body into a possible translation of its centre of mass and a rotation about that centre of mass. In either case, fixed point or centre of mass, Newtonian mechanics tells us that

$$\frac{dH_i}{dt} = \frac{d}{dt}(I_{ij}\Omega_j) = T_i. \tag{10.3}$$

When our reference point is the centre of mass, we shall restrict ourselves to cases in which there is no *net* external force acting on the body, and so the centre of mass is stationary in a suitable inertial frame.

Unfortunately, (10.3) is inconvenient because I_{ij} is constantly changing in the inertial frame, whereas I_{ij} is constant in a frame of reference attached to the body. So we introduce a second, non-inertial coordinate system, which shares a common origin with the inertial space frame, but rotates with the body. This is called *the body frame*. Equation (10.1) now tells us that

$$\frac{d\mathbf{H}}{dt} = \frac{d^*\mathbf{H}}{dt} + \mathbf{\Omega} \times \mathbf{H} = \mathbf{T}, \tag{10.4}$$

from which

$$I_{ij}\frac{d\Omega_j}{dt} + (\mathbf{\Omega} \times \mathbf{H})_i = T_i, \tag{10.5}$$

where I_{ij} is now measured in the rotating frame and we have used (10.2). Since I_{ij} is a symmetric tensor, we can always find an orientation of the body-frame coordinates which makes I_{ij} diagonal, *i.e.* the principal axes of the body. In such a coordinate system we can represent I_{ij} by the vector $\mathbf{I} = (I_1, I_2, I_3)$, where $(1, 2, 3)$ represents the principal axes which are fixed in the body.

Equation (10.5) now simplifies to

$$I_1 \frac{d\Omega_1}{dt} + (I_3 - I_2)\, \Omega_2 \Omega_3 = T_1, \tag{10.6}$$

$$I_2 \frac{d\Omega_2}{dt} + (I_1 - I_3)\, \Omega_3 \Omega_1 = T_2, \tag{10.7}$$

$$I_3 \frac{d\Omega_3}{dt} + (I_2 - I_1)\, \Omega_1 \Omega_2 = T_3, \tag{10.8}$$

which are known as *Euler's equations*.

If Ω is initially aligned with I_1, and $\mathbf{T} = 0$, these equations admit the steady solution $\Omega_1 = $ constant and $\Omega_2 = \Omega_3 = 0$. It turns out that, in the absence of friction, such a solution is linearly stable if I_1 is the largest or smallest principal moment of inertia, but unstable if I_1 is the intermediate one (see Goldstein, 1980, §5.6).

Note that we can obtain an energy equation from (10.5) by taking the dot product of that equation with Ω. In principal coordinates this yields

$$\frac{d}{dt}\left(\frac{1}{2}I_1\Omega_1^2 + \frac{1}{2}I_2\Omega_2^2 + \frac{1}{2}I_3\Omega_3^2 \right) = \mathbf{T} \cdot \Omega, \tag{10.9}$$

and when there is no external torque, we have the invariant,

$$\text{K.E.} = \frac{1}{2}\Omega \cdot \mathbf{H} = \frac{1}{2}I_1\Omega_1^2 + \frac{1}{2}I_2\Omega_2^2 + \frac{1}{2}I_3\Omega_3^2 = \text{constant.} \tag{10.10}$$

10.2.2 Torque-Free Precession of an Axisymmetric Body

We now consider the case where there is no external torque and the body is axisymmetric with respect to principal axis 3, so that $I_1 = I_2$. In such cases Euler's equations simplify to

$$\frac{d\Omega_1}{dt} + (\hat\beta \Omega_3)\Omega_2 = 0, \qquad \frac{d\Omega_2}{dt} - (\hat\beta \Omega_3)\Omega_1 = 0, \tag{10.11}$$

$$\Omega_3 = \text{constant}, \tag{10.12}$$

where $\hat\beta = (I_3 - I_1)/I_1$. Evidently, Ω_1 and Ω_2 take the form

$$\Omega_1 = \Omega_\perp \cos\left(\hat\beta\Omega_3 t + \phi\right), \qquad \Omega_2 = \Omega_\perp \sin\left(\hat\beta\Omega_3 t + \phi\right), \tag{10.13}$$

or equivalently,

$$\mathbf{\Omega}_\perp = \mathbf{\Omega} - \Omega_3\hat{\mathbf{e}}_3 = \Omega_\perp\left[\cos(\hat{\beta}\Omega_3 t), \sin(\hat{\beta}\Omega_3 t), 0\right], \qquad (10.14)$$

where $\Omega_\perp$ is a constant and ϕ has been eliminated through a suitable choice of $t = 0$.

We conclude that $\mathbf{\Omega}_\perp$ rotates about $\hat{\mathbf{e}}_3$ in the body frame, with an angular velocity of $\hat{\beta}\Omega_3$. For a disc-like object (known as an *oblate* symmetrical body), we have $I_3 > I_1$ and $\hat{\beta} > 0$, and so $\mathbf{\Omega}_\perp$ rotates in the same sense as Ω_3. Conversely, a cigar-like object (a *prolate* symmetrical body) has $I_3 < I_1$ and hence $\hat{\beta} < 0$, and so $\mathbf{\Omega}_\perp$ rotates in the opposite sense to Ω_3. This is shown in Figure 10.4. For a sphere, where $\hat{\beta} = 0$, $\mathbf{\Omega}$ is a constant.

In summary, then, the vector $\mathbf{\Omega}$ traces out a cone, called the *body cone*, as it *precesses* around the symmetry axis. Since Ω_3 and $\Omega_\perp$ are both constant, so is $|\mathbf{\Omega}|$, as is the angle between $\mathbf{\Omega}$ and $\hat{\mathbf{e}}_3$, $\alpha_b = \tan^{-1}(\Omega_\perp/\Omega_3)$. (The subscript b stands for 'body'.) The situation is as shown in Figure 10.5(a).

So far we have worked only in the non-inertial body frame. It is natural to ask what the motion looks like when viewed from an inertial frame of reference, and to that end we must now choose a suitable inertial (or space) frame. Since $d\mathbf{H}/dt = 0$ when $\mathbf{T} = 0$, it is natural to adopt an inertial frame in which the z-axis is aligned with $\mathbf{H}$. The angle between $\mathbf{H}$ and $\mathbf{\Omega}$, which we shall label α_s (with subscript s for 'space'), is then given by

$$\cos\alpha_s = \frac{\mathbf{\Omega}\cdot\mathbf{H}}{|\mathbf{\Omega}|\,|\mathbf{H}|} = \frac{2(\text{K.E.})}{|\mathbf{\Omega}|\,|\mathbf{H}|}. \qquad (10.15)$$

(a)   (b)

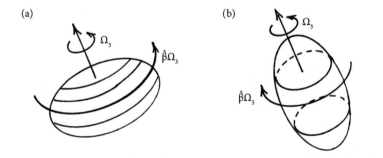

Figure 10.4 (a) For an oblate (disc-like) object, $\mathbf{\Omega}_\perp$ rotates about $\hat{\mathbf{e}}_3$ in the same sense as Ω_3. (b) For a prolate (cigar-like) body, $\mathbf{\Omega}_\perp$ rotates about $\hat{\mathbf{e}}_3$ in the opposite sense to Ω_3.

However, $\mathbf{H}$ and $|\mathbf{\Omega}|$ are both conserved in an inertial frame, while (10.10) tells us that the kinetic energy is also constant. It follows that α_s, like α_b, is an invariant of the motion. When viewed in the space frame, then, the situations is as shown in Figure 10.5(b), with $\mathbf{\Omega}$ precessing around $\mathbf{H}$ at a constant angle, α_s. This is called *the space cone*.

We now consider the relationship between the body cone and the space cone. The first point to note is that $\mathbf{H}$, $\mathbf{\Omega}$, $\mathbf{\Omega}_\perp$, and the symmetry axis, $\hat{\mathbf{e}}_3$, are all coplanar. This follows from

$$\mathbf{H} = I_1\mathbf{\Omega}_\perp + I_3\Omega_3\hat{\mathbf{e}}_3 = I_1\mathbf{\Omega} + (I_3 - I_1)\Omega_3\hat{\mathbf{e}}_3 = I_1\left[\mathbf{\Omega} + \hat{\beta}\Omega_3\hat{\mathbf{e}}_3\right]. \qquad (10.16)$$

The second point is that the two precession angles, α_s and α_b, are uniquely related through $\hat{\beta}$. To see why, we note that (10.16) combined with $\cos\alpha_b = \Omega_3/|\mathbf{\Omega}|$ gives us

$$2(\text{K.E.}) = \mathbf{\Omega} \cdot \mathbf{H} = I_1\left[\mathbf{\Omega}^2 + \hat{\beta}\Omega_3^2\right] = I_1\mathbf{\Omega}^2\left[1 + \hat{\beta}\cos^2\alpha_b\right], \qquad (10.17)$$

as well as

$$\mathbf{H}^2 = I_1^2\left[\mathbf{\Omega} + \hat{\beta}\Omega_3\hat{\mathbf{e}}_3\right]^2 = I_1^2\mathbf{\Omega}^2\left[1 + (2\hat{\beta} + \hat{\beta}^2)\cos^2\alpha_b\right]. \qquad (10.18)$$

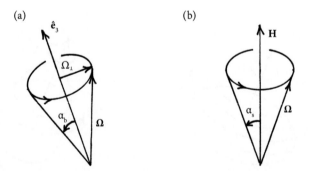

Figure 10.5 (a) In the body frame, $\mathbf{\Omega}$ traces out a cone (the body cone) as it precesses about the symmetry axis. $|\mathbf{\Omega}|$ and α_b are both conserved. (b) In an inertial space frame, $\mathbf{\Omega}$ traces out a cone (the space cone) as it precesses about $\mathbf{H}$. $|\mathbf{\Omega}|$, α_s, and $\mathbf{H}$ are all conserved.

These combine with (10.15) to give

$$\cos^2\alpha_s = \frac{4(\text{K.E.})^2}{\Omega^2 H^2} = \frac{\left(1 + \hat{\beta}\cos^2\alpha_b\right)^2}{1 + (2\hat{\beta} + \hat{\beta}^2)\cos^2\alpha_b},$$ (10.19)

which, after some algebra, may be rearranged as

$$\frac{\sin^2\alpha_s}{\sin^2\alpha_b} = \frac{\hat{\beta}^2\cos^2\alpha_b}{\sin^2\alpha_b + (1 + \hat{\beta})^2\cos^2\alpha_b}.$$ (10.20)

Equation (10.20) demands that a disc-like body, for which $\hat{\beta} > 0$, always has $\alpha_s < \alpha_b$.

Now $\mathbf{H}$, $\mathbf{\Omega}$, and $\hat{\mathbf{e}}_3$ must remain coplanar as $\mathbf{\Omega}$ precesses around $\mathbf{H}$ in the space cone. It follows that, as the plane containing $\mathbf{\Omega}$ and $\hat{\mathbf{e}}_3$ rotates about $\mathbf{H}$, the symmetry axis precesses around $\mathbf{H}$ at a constant angle. Moreover, the space cone and the body cone touch along the line defined by $\mathbf{\Omega}$, and so the body cone 'rolls' (without slipping) on the surface of the space cone. For an oblate body, where $\alpha_s < \alpha_b$, the space cone lies inside the body cone, as shown in Figure 10.6(a). However, for a prolate body, the body cone rolls on the outside surface of the space cone (see Figure 10.6b). Note that, according to Figure 10.6, an oblate body has $\mathbf{\Omega}_\perp$ rotating in the same sense as Ω_3, whereas a prolate body has $\mathbf{\Omega}_\perp$ rotating in the opposite sense to Ω_3, which is consistent with (10.14).

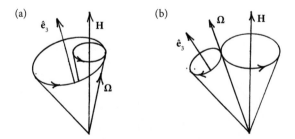

Figure 10.6 The body cone rolls (without slipping) on the surface of the space cone. (a) A disc-like (oblate) body. (b) A cigar-like (prolate) body.

10.2.3 Euler Angles

We now return to kinematics in order to introduce a particularly convenient way of describing the orientation and motion of a rigid body in three dimensions. The *Euler angles* are almost universally used for this purpose, as they provide an efficient way of relating the principal-axis body frame (1, 2, 3) to the inertial space frame (x, y, z).

Consider Figure 10.7, which shows the instantaneous position of the body frame, (1, 2, 3), and the inertial space frame, (x, y, z). The body need not be axisymmetric, but if it is, then the symmetry axis is taken as 3. Also, if the body is torque-free, we take the z-axis to be aligned with **H**, as in §10.2.2. The tilted semi-disc in Figure 10.7 represents the plane defined by the principal body axes 1 and 2, and the intersection of this plane with the x–y plane is called the *line of nodes* (an obscure astronomical name). A third right-handed Cartesian system is now introduced, denoted (α, β, γ), where the γ-axis is aligned with the body axis 3, the α-axis sits on the line of nodes, and the β-axis lies on the 1–2 plane. The Euler angles θ, φ, and ψ are now defined as follows:

- the angle θ is the polar angle measured from the z-axis to the body axis 3;
- the angle φ is the azimuthal angle measured from the x-axis to the line of nodes;
- the angle ψ is measured from the line of nodes, or α-axis, to the body axis 1.

Now, an arbitrary rotation may be expressed as the product of three successive rotations about three different axes. So we can use the Euler angles to rotate from the space frame to the body frame using the following successive rotations:

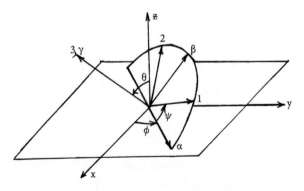

Figure 10.7 The definition of the Euler angles.

(i)   rotate about the z-axis by the angle ϕ to establish the line of nodes;

(ii) rotate about the line of nodes by the angle θ to establish the body axis 3;

(iii) rotate about the body axis 3 by the angle ψ to establish the final orientation.

Note that, for an axisymmetric body, $\mathbf{H}$, $\hat{\mathbf{e}}_3$, and $\mathbf{\Omega}_\perp = \mathbf{\Omega} - \Omega_3\hat{\mathbf{e}}_3$ are all coplanar (see (10.16)). It follows that, if z is aligned with $\mathbf{H}$, then $\mathbf{\Omega}_\perp$ is aligned with the β-axis.

Now an incremental change in the orientation of the body can be accomplished by incremental changes in θ, ϕ, and ψ, and if those changes occur in a time dt, it follows from (i)→(iii) above that the angular velocity of the body can be expressed as

$$\mathbf{\Omega} = \dot{\phi}\hat{\mathbf{e}}_z + \dot{\theta}\hat{\mathbf{e}}_\alpha + \dot{\psi}\hat{\mathbf{e}}_3. \qquad (10.21)$$

When dealing with axisymmetric bodies it is conventional to refer to $\dot{\phi}\hat{\mathbf{e}}_z$ as *precession* and to $\dot{\theta}\hat{\mathbf{e}}_\alpha$ as *nutation* (after the Latin word for 'nodding'). This terminology, which we embrace, is illustrated in Figure 10.8. (Beware, however, that different communities have adopted different naming conventions.) Note that the term precession is now being applied to the coning of the symmetry axis about the z-axis in the space frame. By contrast, in §10.2.2, the term precession was applied to the coning of $\mathbf{\Omega}$, either about the symmetry axis in the body frame, or about $\mathbf{H}$ in the space frame.

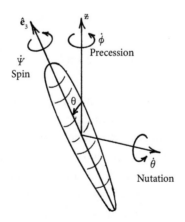

Figure 10.8 Precession and nutation of an axisymmetric body in terms of Euler angles.

We shall now show how to express the body-frame components of $\boldsymbol{\Omega}$ in terms of the Euler angles. First, we note that

$$\hat{\mathbf{e}}_\alpha = \hat{\mathbf{e}}_1 \cos \psi - \hat{\mathbf{e}}_2 \sin \psi, \quad \hat{\mathbf{e}}_\beta = \hat{\mathbf{e}}_1 \sin \psi + \hat{\mathbf{e}}_2 \cos \psi, \tag{10.22}$$

combines with

$$\hat{\mathbf{e}}_z = \hat{\mathbf{e}}_\beta \sin \theta + \hat{\mathbf{e}}_3 \cos \theta,$$

to give

$$\hat{\mathbf{e}}_z = (\hat{\mathbf{e}}_1 \sin \psi + \hat{\mathbf{e}}_2 \cos \psi) \sin \theta + \hat{\mathbf{e}}_3 \cos \theta. \tag{10.23}$$

Next, (10.21) can be rewritten as

$$\boldsymbol{\Omega} = \dot{\theta}(\hat{\mathbf{e}}_1 \cos \psi - \hat{\mathbf{e}}_2 \sin \psi) + \dot{\phi}\hat{\mathbf{e}}_z + \dot{\psi}\hat{\mathbf{e}}_3, \tag{10.24}$$

from which we obtain

$$\Omega_1 = \dot{\phi} \sin \psi \sin \theta + \dot{\theta} \cos \psi, \tag{10.25}$$

$$\Omega_2 = \dot{\phi} \cos \psi \sin \theta - \dot{\theta} \sin \psi, \tag{10.26}$$

$$\Omega_3 = \dot{\psi} + \dot{\phi} \cos \theta. \tag{10.27}$$

Finally, we note that (10.25)→(10.27) combine with (10.22) to yield

$$\left(\Omega_\alpha, \Omega_\beta, \Omega_\gamma\right) = \left(\dot{\theta}, \ \dot{\phi} \sin \theta, \ \dot{\psi} + \dot{\phi} \cos \theta\right), \tag{10.28}$$

which is more or less self-evident from an inspection of Figure 10.8.

We now have a fairly complete description of the motion in terms of Euler angles. We shall use this framework to investigate the axisymmetric 'heavy top' in §10.2.5, and also reconsider the axisymmetric 'free top' in §10.2.4. First, however, we close this section with a comment on the special case of axisymmetric, torque-free bodies. In such cases (10.16) requires $\boldsymbol{\Omega}_\perp$ to be aligned with the β-axis, Ω_3 and $\Omega_\perp$ are both constant, and the angles between $\mathbf{H}$, $\boldsymbol{\Omega}$, and $\hat{\mathbf{e}}_3$ remain fixed (see §10.2.2). This last point demands $\dot{\theta} = 0$, while the constancy of Ω_3 and Ω_β combines with (10.28) to ensure that $\dot{\phi}$ and $\dot{\psi}$ are both constant.

In short, for an axisymmetric, torque-free body, we have

$$\dot{\theta} = 0, \quad \dot{\phi} = \text{constant}, \quad \dot{\psi} = \text{constant}. \tag{10.29}$$

10.2.4 Torque-Free Precession of an Axisymmetric Body Revisited

In §10.2.2 we considered the precession of $\boldsymbol{\Omega}$, either about the symmetry axis in the body frame, or about $\mathbf{H}$ in the space frame. However, the Euler angles focus our attention on the rotation of the *symmetry axis* about $\mathbf{H}$ in the space frame. This is, perhaps, a more intuitive approach, since it is the precession of the symmetry axis in an inertial frame that one primarily observes when watching a spinning top. So, let us return to the torque-free, axisymmetric body of §10.2.2 and reconsider its motion in terms of Euler angles.

We have already seen that θ, $\dot{\phi}$, and $\dot{\psi}$ are all constant. It is also clear that $\dot{\psi} = -\hat{\beta}\Omega_3$. This is because $\boldsymbol{\Omega}_\perp$, which is aligned with the β-axis, rotates in the body frame at the rate $\hat{\beta}\Omega_3$ (see (10.14)), and it follows that the α-axis also rotates in the body frame at the rate $\hat{\beta}\Omega_3$. The expression $\hat{\beta}\Omega_3 = -\dot{\psi}$ then follows from an inspection of Figure 10.7. We can therefore rewrite (10.27) as

$$I_1\Omega_3 = I_1\dot{\phi}\cos\theta - (I_3 - I_1)\,\Omega_3,$$

which fixes the precession rate for the symmetry axis at

$$\dot{\phi} = \frac{I_3\Omega_3}{I_1\cos\theta}. \tag{10.30}$$

This confirms that the precession of the symmetry axis about $\mathbf{H}$ is always in the same sense as Ω_3, as expected. The situation is summarized in Figure 10.9, which shows the relative orientations of the body cone, space cone, and symmetry axis, as well as the directions of precession of both $\hat{\mathbf{e}}_3$ and $\boldsymbol{\Omega}$ about $\mathbf{H}$ in the inertial frame, and the direction of precession of $\boldsymbol{\Omega}$ about $\hat{\mathbf{e}}_3$ in the body frame. An oblate body is on the left and a prolate body on the right.

Equation (10.30) is sometimes rewritten in terms of $\dot{\phi}$ and $\dot{\psi}$. Using (10.27) to eliminate Ω_3 from (10.30) yields

$$I_3\dot{\psi} = (I_1 - I_3)\,\dot{\phi}\cos\theta. \tag{10.31}$$

This form can be useful when considering the unforced precession of a planet, whose nearly spherical shape ensures $I_3 \gg (I_3 - I_1)$. In such cases $\dot{\psi}$ is tiny,

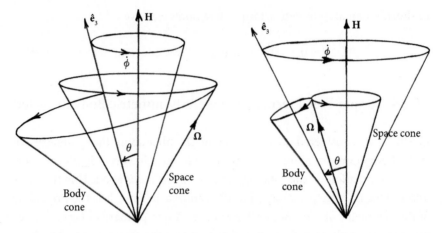

Figure 10.9 The relative orientations of the body cone, space cone, and symmetry axis for (a) an oblate (disc-like) body and (b) a prolate (cigar-like) body.

i.e. $\dot{\psi} \ll \dot{\phi} \cos \theta$, and hence Ω_3 is dominated by the precession, $\dot{\phi}$, with $\Omega_3 \approx \dot{\phi} \cos \theta$. We shall discuss the free precession of the Earth in §10.3.1.

10.2.5 Forced-Precession: The Heavy Top

So far we have ignored external torques, an omission we shall now remedy. The most commonly discussed example of forced precession is the 'heavy top' (see Figure 10.10). This is an axisymmetric body which is pivoted about a fixed point P and subject to the gravitational force mg which acts through the centre of mass. As always, we use (1,2,3) to represent the principal axes in the body frame, and θ, ϕ, and ψ for the Euler angles. We take P to be our reference point about which **H**, **Ω**, and **T** are calculated, and since **H** is not conserved in this problem, we align the z-axis of our inertial frame with $-\mathbf{g}$.

It is conventional to analyse the heavy top using Lagrange's equations, which are indeed well suited to that job. However, for our purposes, Euler's equations provide a more direct route. The first step is to determine the components of the gravitational torque in the body frame of reference, which we now do. If ℓ is the distance from the centre of mass to P, then the magnitude of the gravitational torque is clearly

$$T = |\mathbf{T}| = mg\ell \sin \theta. \tag{10.32}$$

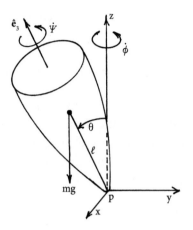

Figure 10.10 The heavy top. The Euler angles are θ, ϕ, and ψ, as usual.

Moreover, a comparison of Figure 10.10 with Figure 10.7 confirms that the axis about which this torque acts is the line of nodes, $\hat{\mathbf{e}}_\alpha$. Since (10.22) requires

$$\hat{\mathbf{e}}_\alpha = \hat{\mathbf{e}}_1 \cos\psi - \hat{\mathbf{e}}_2 \sin\psi,$$

we conclude that

$$\mathbf{T} = mg\ell \sin\theta \,(\hat{\mathbf{e}}_1 \cos\psi - \hat{\mathbf{e}}_2 \sin\psi). \tag{10.33}$$

We now return to Euler's equations, (10.6)→(10.8), and incorporate these components of $\mathbf{T}$ into them. This yields

$$I_1 \frac{d\Omega_1}{dt} + (I_3 - I_1)\,\Omega_2\Omega_3 = mg\ell \sin\theta \cos\psi, \tag{10.34}$$

$$I_1 \frac{d\Omega_2}{dt} - (I_3 - I_1)\,\Omega_3\Omega_1 = -mg\ell \sin\theta \sin\psi, \tag{10.35}$$

$$\frac{d\Omega_3}{dt} = 0. \tag{10.36}$$

A technical difficulty now arises in that the left-hand side of these equations is expressed in terms of the body frame, whereas the right-hand side uses Euler angles. We therefore need to add to Euler's equations the kinematic relationships (10.25)→(10.27):

$$\Omega_1 = \dot{\phi}\sin\psi\sin\theta + \dot{\theta}\cos\psi, \tag{10.37}$$

$$\Omega_2 = \dot{\phi}\cos\psi\sin\theta - \dot{\theta}\sin\psi, \tag{10.38}$$

$$\Omega_3 = \dot{\psi} + \dot{\phi}\cos\theta. \tag{10.39}$$

Equations (10.34)→(10.39) support solutions in which the polar angle θ is a constant (pure precession), and also those in which θ oscillates, that is the top *nutates* as it precesses. However, in the interests of brevity, we shall restrict ourselves to solutions in which θ is constant. In particular, we seek solutions of the form

$$\dot{\theta} = 0, \quad \dot{\phi} = \text{constant}, \quad \dot{\psi} = \text{constant}, \tag{10.40}$$

which clearly satisfies (10.36) and also characterizes the free precession of §10.2.4. (Motion governed by (10.40) is called *regular precession*.) Two of our kinematic expressions now simplify to

$$\Omega_1 = \dot{\phi}\sin\psi\sin\theta, \quad \Omega_2 = \dot{\phi}\cos\psi\sin\theta, \tag{10.41}$$

which we use to substitute for Ω_1 and Ω_2 in (10.34) and (10.35). *Either* of these two Euler equations yields, after a little algebra,

$$\dot{\phi}\left[I_1\dot{\psi} + (I_3 - I_1)\Omega_3\right] = mg\ell, \tag{10.42}$$

which may be combined with (10.39) in the form

$$\dot{\psi} + \dot{\phi}\cos\theta = \Omega_3 = \text{constant}. \tag{10.43}$$

The fact that *both* (10.34) and (10.35) lead to (10.42) points to a redundancy in our equations, which arises from axial symmetry. In any event, expressions (10.42) and (10.43) are readily solved for $\dot{\phi}$ and $\dot{\psi}$ in terms of the two constants of the motion, Ω_3 and $mg\ell$.

We are almost there. We now eliminate $\dot{\psi}$ from (10.42) and (10.43) to yield a quadratic equation for $\dot{\phi}$,

$$\dot{\phi}^2 I_1\cos\theta - I_3\Omega_3\dot{\phi} + mg\ell = 0,$$

whose roots are

$$\dot{\phi} = \frac{I_3\Omega_3}{2I_1\cos\theta}\left[1 \pm \sqrt{1 - 4\frac{(mg\ell)(I_1\cos\theta)}{(I_3\Omega_3)^2}}\,\right]. \tag{10.44}$$

Note that, for $mg\ell = 0$, (10.44) reduces to the precession rate for an unforced top, as given by (10.30). Evidently, nutation-free precession is possible only if

$$\Omega_3 > \Omega_{min} = \frac{2}{I_3}\sqrt{(mg\ell)(I_1 \cos \theta)}, \qquad (10.45)$$

and so steady precession cannot be maintained if the top spins too slowly.

The regime in which $\Omega_3 \gg \Omega_{min}$, sometimes called a *fast top*, is an interesting one. It can be thought of as either the fast rotation limit or else the weak torque limit. In this regime the two solutions of (10.44), which are known as the *fast and slow* precession solutions, are

$$\dot{\phi} = \frac{I_3\Omega_3}{I_1 \cos \theta}, \text{ (fast precession)}, \qquad (10.46)$$

$$\dot{\phi} = \frac{mg\ell}{I_3\Omega_3}, \text{ (slow precession)}. \qquad (10.47)$$

The first of these is not unexpected, and is simply the unforced precession rate, (10.30). The second is more interesting and indeed it is slow precession which is normally observed in a rapidly spinning top. Combining the slow precession solution with (10.42) yields $\dot{\psi} \approx \Omega_3$, which in turn gives us

$$\frac{\dot{\phi}}{\dot{\psi}} = \frac{(mg\ell)(I_1 \cos \theta)}{(I_3\Omega_3)^2} \frac{I_3}{I_1 \cos \theta} \ll \frac{I_3}{I_1 \cos \theta}. \qquad (10.48)$$

Evidently, the slow precession solution is characterized by

$$\dot{\phi} = \frac{mg\ell}{I_3\Omega_3} \ll \dot{\psi}, \qquad \Omega_3 \approx \dot{\psi}. \qquad (10.49)$$

Note that, because the precession is tiny here, *i.e.* $\dot{\phi} \ll \dot{\psi}$, Ω_3 is dominated by $\dot{\psi}$. This is exactly the opposite to the free precession of a nearly spherical body, where (10.31) demands that $\dot{\psi}$ is tiny, and hence Ω_3 is dominated by the precession, *i.e.* $\Omega_3 \approx \dot{\phi} \cos \theta$.

Perhaps we should close this section by noting that the neglect of nutation in our discussion of the heavy top is somewhat artificial, since special initial conditions are required in order to eliminate it. Indeed, in practice, nutation is nearly always observed immediately after the release of a top. Consider, for example, an initial condition in which

$$\dot{\theta} = 0, \quad \dot{\phi} = 0, \quad \dot{\psi} = \Omega_3,$$

so that, initially, the symmetry axis has no velocity and all of the rotation is around $\hat{e}_3$. On release, the symmetry axis falls under gravity, but as it falls it picks up a precessional velocity which eventually halts the fall and reverses it (see Figure 10.11). This is the start of an oscillation of the symmetry axis in θ, *i.e.* a nutation, which accompanies the precession. It turns out that, for a fast top, the amplitude of the nutation is of the order of $mg\ell I_1/(I_3\Omega_3)^2$, which is small. So, typically, the initial nutation is quickly damped out by friction at the pivot and cannot be observed after a short time. However, nutation is usually observed immediately following the release of a top. The theory of nutation is well developed and interested readers will find an extended discussion in Goldstein (1980).

10.3 Free and Forced Precession of the Earth

10.3.1 The Chandler Wobble

The rapid rotation of the earth causes the mantle, which is not entirely rigid, to bulge slightly in the equatorial regions. Hence, to leading order, the earth is an oblate spheroid, with

$$\hat{\beta} = (I_3 - I_1)/I_1 \approx 1/300.$$

It follows that, if perturbed, the earth can exhibit free precession of the type described in §10.2.2, with a period of around 300 days. Of course, the earth is not free of external torques. For example, the moon and sun exert gravitational torques on the equatorial bulge, and this leads to a form of forced precession, as

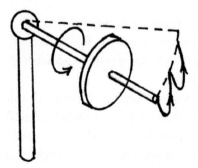

Figure 10.11 Nutation (an oscillation in θ) often accompanies precession of a heavy top.

discussed below in §10.3.2. However, these external gravitational torques are extremely weak, and the resulting *equinox precession* is very slow, measured in hundreds of centuries. So the forced and free precessions of the earth are effectively decoupled and distinct phenomena.

It was Euler who, following an earlier suggestion by Newton, first predicted in 1749 that the earth might exhibit free precession with a period of ~300 days. However, it was not until Chandler, an American astronomer, carefully analysed the available data in 1891 that just such a phenomenon was confirmed, with Chandler estimating a period of ~430 days. That is to say, the *celestial north pole*, which is where the rotation vector of the earth pierces the earth's surface, is observed to circle around the *geometric north pole* every 430 days, as shown in Figure 10.12. Chandler's observations have since been confirmed many times, with the dominant rotation period being 433 days. The amplitude of the motion is, however, very small, with the celestial north pole deviating from the geometric north pole by no more than 9 or 10 m, and indeed the deviation is often less.

The phenomenon is, however, altogether more complicated than the simple free precession envisaged by Euler. There are three main differences. First, the observed motion of the celestial north pole is not a simple circle surrounding the geometric north pole. Rather, the motion is more erratic, as clearly shown in Figure 10.12(b). Perhaps this is why the name *Chandler's wobble* has stuck, whereas *Euler's precession* has not. Second, the dominant frequency (of 433 days) is significantly longer than the precessional frequency of 300 days. Third, frictional forces will naturally dampen any free precession, with typical estimates of the decay time being around several decades to one century. So there must be some form of random excitation that continually triggers the precession.

Simon Newcomb was the first to provide an explanation for the discrepancy in frequency, noting that the earth is not a rigid body and that the elasticity of the mantle allows the wobble to induce elastic deformations that decrease $I_3 - I_1$, and so increase the period of the precession. Likewise, the wobble induces small displacements in the surface of the oceans (the so-called *pole tide*), which could reduce the effective value of $I_3 - I_1$, also increasing the period of precession.

The source of the random excitation is still a matter of debate, with the traditional explanation of earthquakes progressively giving way to theories involving motion in the atmosphere and the oceans. For example, fluctuations in the density distribution within the oceans can cause small changes in the pressure forces exerted on the seabed. All-in-all, it would seem that this is a story which is far from complete.

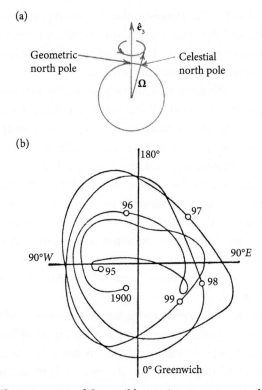

Figure 10.12 The migration of the earth's rotation vector around the geometric north pole. (a) Euler's ideal precession. (b) The actual motion from 1895 to 1900 is more erratic.

10.3.2 Equinox Precession

We now turn to the slow, forced precession of the earth, whose origins lie in the gravitational torques exerted by the sun and the moon on the equatorial bulge of the earth. Historically, this has been known by the astronomical name *precession of the equinoxes.*

The spin axis of the earth is inclined to the axis of its orbital plane (or ecliptic plane), by an angle of $\theta = 23.4°$ (see Figure 10.13a). This combination of an inclined spin axis and an equatorial bulge ensures that the sun exerts a net gravitational torque on the earth, as shown schematically in Figure 10.13(b). If we choose a space frame whose z-axis is aligned with the axis of the orbital plane, then this solar torque tends to align the symmetry axis of the earth with the z-axis. It also acts about an axis aligned with the line of nodes, just like the heavy top of §10.2.5. We might anticipate, therefore, that the solar torque

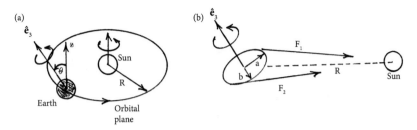

Figure 10.13 (a) The spin axis of the earth is inclined to its orbital plane. (b) An inclined spin axis combined with an equatorial bulge ensures that the sun exerts a torque on the earth.

induces a similar form of forced precession. Note, however, that the gravitational torques in the two problems are of opposite signs, and this will prove to be significant.

Determining the solar torque is a rather detailed calculation. However, a very crude estimate (accurate to within an unknown pre-factor of order unity) can be established using a simple toy problem. Since the torque arises purely from the asymmetry of the earth's mass distribution, we replace the earth by a dumbbell aligned with the equatorial plane and composed of two masses of magnitude $\hat{\beta} m_e$ separated by a distance of $2a$, where m_e is the mass of the earth, a is the equatorial radius, and $\hat{\beta} = (I_3 - I_1)/I_1$, as in §10.3.1. Also, treating the earth as an oblate spheroid of uniform density gives us $I_3 = 2m_e a^2/5$ and $I_1 = m_e(a^2 + b^2)/5$, where b is the polar radius.

Now, suppose that R is the orbital radius, with $R \gg a$, and θ is the inclination of the spin axis to the z-axis of the space frame. Then it is clear from Figure 10.14 that the distances from the sun to the upper (closer) and lower (more distant) masses are given by

$$\ell_1^2 = R^2 - 2aR\cos\theta + a^2 \approx R^2\left[1 - 2\left(a/R\right)\cos\theta\right],$$

$$\ell_2^2 = R^2 + 2aR\cos\theta + a^2 \approx R^2\left[1 + 2\left(a/R\right)\cos\theta\right].$$

Also, the gravitational forces on the upper and lower masses in this toy problem are

$$F_1 = G\frac{\hat{\beta} m_e M_\odot}{\ell_1^2}, \qquad F_2 = G\frac{\hat{\beta} m_e M_\odot}{\ell_2^2},$$

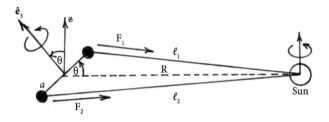

Figure 10.14 A toy problem that gives a crude estimate of the solar torque.

where G is the universal gravitational constant and M_Θ the solar mass. So the corresponding torque is approximately

$$T_{\text{sun}} \sim (F_1 - F_2)\, a \sin \theta = \frac{G\hat{\beta} m_e M_\Theta a \sin \theta}{R^2}\, [4\,(a/R) \cos \theta],$$

and substituting for $\hat{\beta}$ gives us

$$T_{\text{sun}} \sim 10\frac{GM_\Theta(I_3 - I_1)}{R^3} \sin \theta \cos \theta.$$

However, the solar torque varies as the earth orbits around the sun, being a maximum at solstice (the orientation shown in Figure 10.13b) and zero at the spring equinox. This suggests the revised estimate

$$T_{\text{sun}} \sim 5\frac{GM_\Theta(I_3 - I_1)}{R^3} \sin \theta \cos \theta.$$

(Note that averaging the torque over an orbit is a reasonable thing to do as the period of precession is measured in hundreds of centuries.) In fact, when averaged over an orbit, the actual solar torque turns out to be

$$T_{\text{sun}} = \frac{3GM_\Theta(I_3 - I_1)}{2R^3} \sin \theta \cos \theta. \tag{10.50}$$

We now return to our expression for the slow precession of a fast top, (10.47), and replace the torque on the top, $-mg\ell \sin \theta$, by the solar torque, (10.50). Note the minus sign. This arises because the gravitational torques in the two problems are in opposite directions, so that equinox precession is like a heavy top whose centre of mass lies *below* the pivot point. In any event, combining

Table 10.1 Properties of the sun, earth, and moon

	Sun	Earth	Moon
Mean radius (km)	6.96×10^5	6.37×10^3	1.74×10^3
Mass (kg)	1.99×10^{30}	5.97×10^{24}	7.35×10^{22}
Mean orbital radius (km)	–	1.50×10^8	3.84×10^5

(10.47) with (10.50) gives the precession rate as

$$\dot{\phi} = -\frac{3GM_\Theta}{2\Omega_3 R^3} \frac{I_3 - I_1}{I_3} \cos\theta, \tag{10.51}$$

(see Goldstein, 1980). Using the data from Table 10.1, (10.51) gives a precessional period of around 81,000 years.

A similar calculation for the lunar torque yields a value which is 2.2 times that of the solar torque, and hence a precessional period roughly halve of that associated with the sun. (Although the moon is far less massive, its proximity to the earth wins out, and so the torque is larger.) Of course, these two torques do not act in isolation and so there is the altogether more complicated problem of calculating the precession of the earth under the simultaneous influence of both torques (called *lunisolar precession*). This is simplified somewhat by the fact that the lunar orbit lies almost, but not quite, in the same plane as the earth's orbital plane. Thus, to a first approximation, the torques may be simply added algebraically, giving a net torque roughly 3.2 times that of the solar torque, and a period of precession around 31% of the solar precessional period, *i.e.* approximately 25,000 years. This is reasonably consistent with the observed precession rate of around 26,000 years.

We close this section by repeating a point first made in our discussion of the heavy top. In slow, forced precession of this type, $\dot{\phi}$ is tiny and so Ω_3 is dominated by $\dot{\psi}$, *i.e.* $\Omega_3 \approx \dot{\psi}$. This is exactly the opposite to the free precession of a nearly spherical body, such as the Chandler wobble, where (10.31) demands that $\dot{\psi}$ is tiny, and so Ω_3 is dominated by $\dot{\phi}\cos\theta$.

10.4 Forced Precession of the Earth's Fluid Core

We now consider the motion induced in the liquid core of a terrestrial planet in response to the forced precession of its mantle. Here the mindset changes. We no longer wish to determine the rate of precession of the mantle resulting from given external torques. Rather, we take Ω_3, θ, and $\dot{\phi}$ to be *prescribed* for the mantle, and seek to determine the fluid response to that motion. To simplify matters, we shall assume that there is no nutation of the mantle, and that the precession is uniform, so that θ, $\dot{\phi}$, and $\dot{\psi}$ are all constant. We also introduce the notation Ω_0 and Ω_p for $\dot{\psi}\hat{e}_3$ and $\dot{\phi}\hat{e}_z$ (see Figure 10.15), so the mantle has the angular velocity

$$\Omega = \Omega_0 + \Omega_p = \dot{\psi}\hat{e}_3 + \dot{\phi}\hat{e}_z. \tag{10.52}$$

We have in mind situations where $\dot{\phi} \ll \dot{\psi}$, as in equinox precession, so that $|\Omega_p| \ll |\Omega_0|$.

10.4.1 A Third Frame of Reference: The Precession Frame

We have already introduced two frames of reference: the body frame, which is the natural vantage point if you are standing on the surface of the earth, watching the stars, and the space frame, which is the more intuitive frame if you are watching a spinning top on your desk. When it comes to fluid motion, however, it turns out that a third frame is often more useful, called the *precession frame*.

To understand the precession frame, it is probably easiest to imagine that you are setting up a precession experiment in the laboratory. Suppose that you have a nearly spherical, axisymmetric container constructed from perspex and filled with water. It is mounted on a spinning table (the spin table) whose axis is aligned with the symmetry axis of the body and tilted away from the vertical. The spin table rotates rapidly at the rate Ω_0. In order to generate some

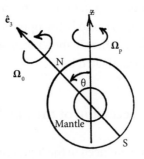

Figure 10.15 The motion of the mantle is prescribed and we wish to determine the resulting flow in the liquid core.

precession, the spin table is itself mounted on a second, horizontal turntable (the precession table) which spins slowly about a vertical axis at the rate Ω_p. We can view the contents of the container from three different vantage points, as shown in Figure 10.16. As before, there is the space frame (observer 1) and the body frame (observer 3). However, there is now the option of adopting a frame of reference that rotates with the precession table (observer 2). This is the precession frame, and it is the reference frame that would normally be used for making measurements in such an experiment, perhaps using cameras mounted on the precession table. Crucially, just as measurements are most readily made in the precession frame, so an analysis of the induced flow is often most easily carried out in that frame. Let us try to understand why.

Suppose we decided to analyse the flow in the body frame. Then the angular velocity of that frame, $\Omega = \Omega_0 + \Omega_p$, is unsteady when seen from an inertial frame, because it rotates around the z-axis in a space cone. Indeed, using (10.1) and (10.2), and noting that Ω_0 is constant in the body frame, while Ω_p is constant in the inertial frame, we have

$$\frac{d\Omega}{dt} = \frac{d^*\Omega}{dt} = \frac{d^*\Omega_p}{dt} = \left(\frac{d}{dt} - \Omega\times\right)\Omega_p = -\Omega \times \Omega_p = \Omega_p \times \Omega_0. \quad (10.53)$$

Because Ω is time dependent, we must include two fictitious body forces when we write down the equation of motion for the fluid in the body frame, as shown by equation (4.6). These are:

$$\text{Coriolis force,} \quad 2\mathbf{u} \times \Omega(t), \quad (10.54)$$

$$\text{Poincaré force,} \quad \mathbf{x} \times \frac{d\Omega}{dt} = \mathbf{x} \times (\Omega_p \times \Omega_0). \quad (10.55)$$

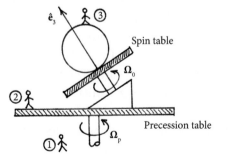

Figure 10.16 Definition of the precession frame. Observer 2 is in the precession frame.

(The centrifugal force, by contrast, is simply absorbed into the pressure gradient.) Both of these forces are time dependent in the body frame, and so the induced flow will also be time dependent in that frame.

Contrast (10.54) and (10.55) with what happens in the precession frame. The angular velocity of that frame, $\mathbf{\Omega}_p$, is steady when seen from an inertial frame. Consequently, there is no Poincaré force and the Coriolis force, $2\mathbf{u} \times \mathbf{\Omega}_p$, is steady. We now have the possibility of a steady flow and hence a simpler solution. There are, however, two drawbacks to the use of the precession frame. First, in such a frame, the velocity of the inside surface of the container is not zero, but rather given by $\mathbf{u} = \mathbf{\Omega}_0 \times \mathbf{x}$, and the no-slip condition demands that this is imposed on the fluid as a boundary condition. Second, because of the boundary condition $\mathbf{u} = \mathbf{\Omega}_0 \times \mathbf{x}$, the Rossby number is never small in this frame, even when the rotation $\mathbf{\Omega}_0$ is very rapid, and so the Navier–Stokes equation cannot be linearized by dispensing with the nonlinear inertial term. Some authors view this second drawback as a particularly heavy price to pay, and so they prefer to stay in the body frame (which we should now call the *mantle frame*). However, depending on the exact problem at hand, it often pays to adopt the precession frame.

Finally, we note that a comparison of Figures 10.7 and 10.16 confirms that the right-handed Cartesian system (α, β, γ) shown in Figure 10.7 is, in fact, fixed in the precession frame, with $\mathbf{\Omega}_p$ lying in the β–γ plane. So our coordinate system in the precession frame has an origin at the centre of the planet, has axes aligned with (α, β, γ), and has coordinates (x', y', z') along those axes.

10.4.2 Poincaré's Inviscid Analysis

Before discussing viscous effects, it is informative to consider an ingenious inviscid solution to this problem which was developed by Poincaré (1910). We take the fluid domain to be an oblate spheroid of equatorial radius a and polar radius b, so that $I_3 = 2m_c a^2/5$ and $I_1 = m_c (a^2 + b^2)/5$, where m_c is the mass of the core. (We shall ignore the solid inner core.) The essence of Poincaré's solution is to assume that the fluid has uniform vorticity and so is in a state of uniform rotation, except to the extent that a potential flow is added to $\mathbf{u}$ in order to satisfy the non-spherical boundary conditions. In short, in the *precession frame*, we have

$$\mathbf{u} = \mathbf{\Omega}_f \times \mathbf{x} + \nabla \hat{\phi}, \qquad \boldsymbol{\omega} = 2\mathbf{\Omega}_f. \tag{10.56}$$

Here $\boldsymbol{\Omega}_f$ is spatially uniform but a function of time, the subscript f stands for *fluid*, and $\hat{\phi}$ is chosen to satisfy the boundary condition $\mathbf{u} \cdot \mathbf{n} = 0$ on the boundary S. The problem now falls into two parts: kinematic and dynamic. That is to say, for a given $\boldsymbol{\Omega}_f$, the determination of $\mathbf{u}$ is a matter of kinematics. The temporal variation of $\boldsymbol{\Omega}_f$, on the other hand, is a question of dynamics. Note that (10.56) is well posed in a kinematic sense, since its divergence yields

$$\nabla^2 \hat{\phi} = 0, \quad \nabla \hat{\phi} \cdot \mathbf{n} = -(\boldsymbol{\Omega}_f \times \mathbf{x}) \cdot \mathbf{n} \quad \text{on } S, \tag{10.57}$$

which has a unique solution for a given $\boldsymbol{\Omega}_f$. Note also that, if the domain is close to spherical, then (10.57) demands that $\hat{\phi}$ is small and so $\mathbf{u}$ is close to rigid-body rotation.

Physically, the position is this. If there is no precession, *i.e.* $\boldsymbol{\Omega}_p = 0$, then the fluid can simply co-rotate with the mantle, and so $\boldsymbol{\Omega}_f = \boldsymbol{\Omega}_0$ is one option, and indeed a small amount of friction ensures that it is *the* solution. However, for a finite precession, the orientation of $\boldsymbol{\Omega}$ in an inertial frame is constantly changing and the inertia of the fluid ensures that changes in $\boldsymbol{\Omega}_f$ lag behind changes in $\boldsymbol{\Omega}$. Moreover, if $|\boldsymbol{\Omega}_p| \ll |\boldsymbol{\Omega}_0|$, that lag will be small.

Let us start with the kinematic half of the problem. To make the flow fit into the ellipsoid

$$\frac{x'^2}{a^2} + \frac{y'^2}{a^2} + \frac{z'^2}{b^2} = 1,$$

we stretch the coordinates to transform the ellipsoid into a unit sphere, assume rigid-body rotation in the sphere, and then transform back. This yields

$$\mathbf{u}_{\text{stretched}} = \left(\frac{u_\alpha}{a}, \frac{u_\beta}{a}, \frac{u_\gamma}{b} \right) = \tilde{\boldsymbol{\Omega}} \times \left(\frac{x'}{a}, \frac{y'}{a}, \frac{z'}{b} \right)$$

for some $\tilde{\boldsymbol{\Omega}}$, which we shall relate to $\boldsymbol{\Omega}_f$. Our velocity components are then

$$u_\alpha = \frac{a}{b} \tilde{\Omega}_\beta z' - \tilde{\Omega}_\gamma y', \tag{10.58}$$

$$u_\beta = \tilde{\Omega}_\gamma x' - \frac{a}{b} \tilde{\Omega}_\alpha z', \tag{10.59}$$

$$u_\gamma = \frac{b}{a} \tilde{\Omega}_\alpha y' - \frac{b}{a} \tilde{\Omega}_\beta x', \tag{10.60}$$

which clearly satisfy $\nabla \cdot \mathbf{u} = 0$ and may be shown to also ensure $\mathbf{u} \cdot \mathbf{n} = 0$ on S. The vorticity is evidently

$$\boldsymbol{\omega} = 2\boldsymbol{\Omega}_f = \left(\left(\frac{b}{a} + \frac{a}{b}\right)\tilde{\Omega}_\alpha, \left(\frac{b}{a} + \frac{a}{b}\right)\tilde{\Omega}_\beta, 2\tilde{\Omega}_\gamma\right) = 2\left(e\tilde{\Omega}_\alpha, e\tilde{\Omega}_\beta, \tilde{\Omega}_\gamma\right),$$

where $e = (a^2 + b^2)/2ab$. We may now eliminate $\tilde{\boldsymbol{\Omega}}$ from our expressions for $\mathbf{u}$ by writing

$$\tilde{\boldsymbol{\Omega}} = \left(e^{-1}\Omega_{f,\alpha}, e^{-1}\Omega_{f,\beta}, \Omega_{f,\gamma}\right),$$

which yields

$$u_\alpha = \frac{2a^2}{a^2 + b^2}\Omega_{f,\beta}z' - \Omega_{f,\gamma}y', \tag{10.61}$$

$$u_\beta = \Omega_{f,\gamma}x' - \frac{2a^2}{a^2 + b^2}\Omega_{f,\alpha}z', \tag{10.62}$$

$$u_\gamma = \frac{2b^2}{a^2 + b^2}\left(\Omega_{f,\alpha}y' - \Omega_{f,\beta}x'\right). \tag{10.63}$$

That completes the kinematic part of the problem.

We now turn to dynamics, which determines $\boldsymbol{\Omega}_f(t)$. In the precession frame, which incorporates the Coriolis force $2\mathbf{u}\times\boldsymbol{\Omega}_p$, the inviscid vorticity equation is,

$$\frac{\partial\boldsymbol{\omega}}{\partial t} + \mathbf{u} \cdot \nabla\boldsymbol{\omega} = \boldsymbol{\omega} \cdot \nabla\mathbf{u} + \nabla \times \left(2\mathbf{u} \times \boldsymbol{\Omega}_p\right) = \left(\boldsymbol{\omega} + 2\boldsymbol{\Omega}_p\right) \cdot \nabla\mathbf{u}.$$

Since $\boldsymbol{\omega} = 2\boldsymbol{\Omega}_f$, this becomes

$$\frac{d\boldsymbol{\Omega}_f}{dt} = \left(\boldsymbol{\Omega}_f + \boldsymbol{\Omega}_p\right) \cdot \nabla\mathbf{u}. \tag{10.64}$$

We now substitute for $\mathbf{u}$ using (10.61)→(10.63), while noting that $\Omega_{p,\alpha} = 0$ because $\boldsymbol{\Omega}_p$ lies in the β–γ plane. This yields evolution equations for the three components of $\boldsymbol{\Omega}_f$:

$$\frac{d\Omega_{f,\alpha}}{dt} = \frac{a^2 - b^2}{a^2 + b^2}\Omega_{f,\beta}\,\Omega_{f,\gamma} + \frac{2a^2}{a^2 + b^2}\Omega_{f,\beta}\,\Omega_{p,\gamma} - \Omega_{f,\gamma}\,\Omega_{p,\beta}, \tag{10.65}$$

$$\frac{d\Omega_{f,\beta}}{dt} = -\frac{a^2 - b^2}{a^2 + b^2}\Omega_{f,\alpha}\,\Omega_{f,\gamma} - \frac{2a^2}{a^2 + b^2}\Omega_{f,\alpha}\,\Omega_{p,\gamma}, \tag{10.66}$$

$$\frac{d\Omega_{f,\gamma}}{dt} = \frac{2b^2}{a^2 + b^2}\Omega_{f,\alpha}\,\Omega_{p,\beta}. \tag{10.67}$$

These expressions may be simplified by introducing the parameter

$$\tilde{\beta} = \frac{a^2 - b^2}{a^2 + b^2} + \frac{2a^2}{a^2 + b^2}\frac{\Omega_{p,\gamma}}{\Omega_{f,\gamma}} = \frac{I_3 - I_1}{I_1} + \frac{I_3}{I_1}\frac{\Omega_{p,\gamma}}{\Omega_{f,\gamma}}, \tag{10.68}$$

which allows us to rewrite our governing equations in the form

$$\frac{d\Omega_{f,\alpha}}{dt} = \left(\tilde{\beta}\Omega_{f,\beta} - \Omega_{p,\beta}\right)\Omega_{f,\gamma}, \tag{10.69}$$

$$\frac{d\Omega_{f,\beta}}{dt} = -\tilde{\beta}\Omega_{f,\alpha}\,\Omega_{f,\gamma}, \tag{10.70}$$

$$\frac{d\Omega_{f,\gamma}}{dt} = \frac{2b^2}{a^2 + b^2}\Omega_{f,\alpha}\,\Omega_{p,\beta}. \tag{10.71}$$

We now take advantage of the fact that, for equinox precession of the earth, we have $\Omega_p/\Omega_0 \approx 1.06 \times 10^{-7}$ and $(I_3 - I_1)/I_1 \approx 1/393$, and so

$$\Omega_p/\Omega_0 \ll (I_3 - I_1)/I_1 \ll 1.$$

Because $\Omega_p \ll \Omega_0$, we expect $\boldsymbol{\Omega}_f$ to be almost aligned with $\boldsymbol{\Omega}_0$, with $\Omega_{f,\gamma} \approx |\boldsymbol{\Omega}_0|$ and $(\Omega_{f,\alpha}, \Omega_{f,\beta}) \ll \Omega_0$. Assuming that $d/dt \sim \tilde{\beta}\Omega_0$ (see below), (10.69)–(10.71) then suggest

$$(\Omega_{f,\alpha}, \Omega_{f,\beta}) \sim \Omega_p/\tilde{\beta}, \quad \Omega_{f,\gamma} = \Omega_0\left[1 + O\left(\left(\Omega_p/\tilde{\beta}\Omega_0\right)^2\right)\right]. \tag{10.72}$$

Indeed, we shall see shortly that (10.72) is exactly what the viscous equations of motion yield. So let us replace $\Omega_{f,\gamma}$ by Ω_0 in (10.68)→(10.70) to give

$$\tilde{\beta} = \frac{I_3 - I_1}{I_1} + \frac{I_3}{I_1}\frac{\Omega_{p,\gamma}}{\Omega_0} = \text{constant}, \tag{10.73}$$

and

$$\frac{d\Omega_{f,\alpha}}{dt} = \Omega_0 \left(\tilde{\beta}\Omega_{f,\beta} - \Omega_{p,\beta} \right), \quad \frac{d\Omega_{f,\beta}}{dt} = -\tilde{\beta}\Omega_0\Omega_{f,\alpha}, \quad (10.74)$$

where the constant $\tilde{\beta}$ is a generalization of $\hat{\beta} = (I_3 - I_1)/I_1$. (Note that the second contribution to $\tilde{\beta}$ in (10.73) is very small, and so $\tilde{\beta} \approx \hat{\beta}$.) Crucially, for a constant $\tilde{\beta}$, (10.74) yields un-damped inertial oscillations at the frequency $\tilde{\beta}\Omega_0$. Moreover, approximation (10.74) admits the *steady* solution,

$$\Omega_{f,\beta} = \Omega_{p,\beta}/\tilde{\beta}, \quad \Omega_{f,\alpha} = 0. \quad (10.75)$$

In such cases, $\mathbf{\Omega}_0$, $\mathbf{\Omega}_f$, and $\mathbf{\Omega}_p$ are co-planar, all lying in the β–γ plane. Finally, invoking the approximation $\Omega_{f,\gamma} = \Omega_0$, our steady solution becomes

$$\mathbf{\Omega}_f = \mathbf{\Omega}_0 + \left(\Omega_{p,\beta}/\tilde{\beta} \right) \hat{\mathbf{e}}_\beta. \quad (10.76)$$

Consider now the effects of a small but finite amount of viscous dissipation. In particular, consider a weakly damped initial value problem in which $\Omega_{f,\gamma} \approx |\mathbf{\Omega}_0|$ at $t = 0$. Then we might reasonably suppose that (10.74), and the associated inertial oscillation, remains a good approximation, despite the viscous stresses, except to the extent that energy is slowly drained from the oscillatory core flow. In such a situation, it seems probable that the flow asymptotes to the steady solution (10.76). (See the discussion in Tilgner, 2009.) Thus, when the precession is weak and the fluid domain is close to spherical, the fluid tends to a state of steady, rigid-body rotation in the precession frame. This is reminiscent of the spin-over mode introduced in §7.3.2 and it is shown in Figure 10.17.

For the liquid core of the earth we have $\hat{\beta} \approx 1/393$, while equinox precession dictates $\Omega_{p,\beta}/\Omega_0 \approx 1.06 \times 10^{-7} \sin 23.4$. In such a case we have $\Omega_{f,\beta}/\Omega_0 \approx 1.65 \times 10^{-5}$, which makes the angle between $\mathbf{\Omega}_0$ and $\mathbf{\Omega}_f$ tiny, of order 10^{-5}. Perhaps this is hardly a surprise, given how small the solar and lunar torques are, and hence how weak the equinox precession is. Nevertheless, this angle is larger, by a factor of ~11, than that between $\mathbf{\Omega}_0$ and $\hat{\mathbf{e}}_3$ induced by the Chandler wobble of the mantle.

We need to make one final comment about Poincaré's inviscid solution. It is readily confirmed that (10.69)→(10.71) yield the energy equation

$$\frac{d}{dt} \frac{1}{2} \left(I_1\Omega_{f,\alpha}^2 + I_1\Omega_{f,\beta}^2 + I_3\Omega_{f,\gamma}^2 \right) = -\left(\frac{I_3 - I_1}{I_1} \right)^2 I_1\Omega_{f,\alpha}\Omega_{p,\beta}\Omega_{f,\gamma}, \quad (10.77)$$

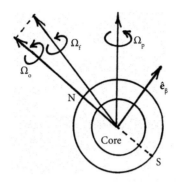

Figure 10.17 The rotation of the mantle and liquid core as seen in the precession frame.

where the inertial term on the right is (minus) the rate of change of kinetic energy associated with the irrotational velocity $\nabla\hat{\phi}$ in (10.56). We shall return to this shortly.

10.4.3 Viscous Boundary Layers

The fact that the fluid and mantle have slightly different velocities means that Ekman layers form on the mantle, and this gives rise to a frictional torque acting on the fluid. Such a torque is absent from (10.69)→(10.71), and this, in turn, means that (10.76) requires a viscous correction. The nature of these Ekman layers, and of the viscous correction to (10.76), has been studied analytically by Busse (1968), heuristically by Vanyo & Likins (1972), and experimentally by Noir et al. (2003), among others.

Let $\mathbf{T}_v$ be the viscous torque exerted by the mantle on the fluid. Then our governing equations in the precession frame, (10.69)→(10.71), become

$$\frac{d\Omega_{f,\alpha}}{dt} = \left(\tilde{\beta}\Omega_{f,\beta} - \Omega_{p,\beta}\right)\Omega_{f,\gamma} + T_{v,\alpha}/I_1, \tag{10.78}$$

$$\frac{d\Omega_{f,\beta}}{dt} = -\tilde{\beta}\Omega_{f,\alpha}\,\Omega_{f,\gamma} + T_{v,\beta}/I_1, \tag{10.79}$$

$$\frac{d\Omega_{f,\gamma}}{dt} = \frac{2b^2}{a^2+b^2}\Omega_{f,\alpha}\,\Omega_{p,\beta} + T_{v,\gamma}/I_3, \tag{10.80}$$

which yield the energy equation

$$\frac{d}{dt}\frac{1}{2}\left(I_1\Omega_{f,\alpha}^2 + I_1\Omega_{f,\beta}^2 + I_3\Omega_{f,\gamma}^2\right) = -\left(\frac{I_3-I_1}{I_1}\right)^2 I_1\Omega_{f,\alpha}\Omega_{p,\beta}\Omega_{f,\gamma} + \mathbf{T}_v \cdot \mathbf{\Omega}_f. \tag{10.81}$$

In the steady state, which must ultimately be reached through viscous dissipation, it is readily confirmed that (10.78) and (10.79) require

$$\mathbf{T}_v \cdot \left(\Omega_{f,\alpha}\hat{\mathbf{e}}_\alpha + \Omega_{f,\beta}\hat{\mathbf{e}}_\beta\right) = I_1 \Omega_{f,\alpha}\Omega_{p,\beta}\Omega_{f,\gamma},$$

and so our steady energy equation yields

$$\mathbf{T}_v \cdot \boldsymbol{\Omega}_f = \left(\frac{I_3 - I_1}{I_1}\right)^2 I_1 \Omega_{f,\alpha}\Omega_{p,\beta}\Omega_{f,\gamma} = \left(\frac{I_3 - I_1}{I_1}\right)^2 \mathbf{T}_v \cdot \left(\Omega_{f,\alpha}\hat{\mathbf{e}}_\alpha + \Omega_{f,\beta}\hat{\mathbf{e}}_\beta\right). \tag{10.82}$$

Given that $\hat{\beta} = (I_3 - I_1)/I_1$ is small for the earth's core, (10.82) tells us that, in the steady state, the viscous torque is almost perpendicular to $\boldsymbol{\Omega}_f$. Indeed, the viscous torque does no work when the domain is spherical.

To make progress, we need to say something about the nature of $\mathbf{T}_v$. So, let us follow the heuristic arguments of Vanyo & Likins (1972) and suppose that the frictional torque acting on the fluid is proportional to $\boldsymbol{\Omega}_0 - \boldsymbol{\Omega}_f$, say $\mathbf{T}_v = \kappa I_1 \left(\boldsymbol{\Omega}_0 - \boldsymbol{\Omega}_f\right)$, where κ is some form of friction coefficient. (For a laminar Ekman layer, κ is proportional to the product of ν with the inverse of the Ekman-layer thickness.) Equation (10.82) then yields

$$\left(\boldsymbol{\Omega}_0 - \boldsymbol{\Omega}_f\right) \cdot \boldsymbol{\Omega}_f = \left(\frac{I_3 - I_1}{I_1}\right)^2 \left(\boldsymbol{\Omega}_0 - \boldsymbol{\Omega}_f\right) \cdot \left(\boldsymbol{\Omega}_f - \Omega_{f,\gamma}\hat{\mathbf{e}}_\gamma\right),$$

and on expanding the product on the right, we obtain the simple yet important result

$$\left(\boldsymbol{\Omega}_0 - \boldsymbol{\Omega}_f\right) \cdot \boldsymbol{\Omega}_f = -\left(\frac{I_3 - I_1}{I_1}\right)^2 \left(\Omega_{f,\alpha}^2 + \Omega_{f,\beta}^2\right). \tag{10.83}$$

Remarkably, (10.83) is independent of κ, and hence independent of how we model the Ekman layers. It rests exclusively on the assumption that the viscous torque is proportional to $\boldsymbol{\Omega}_0 - \boldsymbol{\Omega}_f$. Given that $(I_3 - I_1)/I_1 \approx 1/393$ for the earth's core, this suggests that $\boldsymbol{\Omega}_f \cdot \boldsymbol{\Omega}_0 \approx \Omega_f^2$, and indeed this simple model would have $\boldsymbol{\Omega}_f \cdot \boldsymbol{\Omega}_0 = \Omega_f^2$ to leading order in the small parameter $\hat{\beta}$. Physically, this is because the viscous torque does very little work when the fluid domain is close to spherical, and so $\left(\boldsymbol{\Omega}_0 - \boldsymbol{\Omega}_f\right) \cdot \boldsymbol{\Omega}_f \approx 0$.

We note in passing that incorporating $\mathbf{T}_\nu = \kappa I_1 (\mathbf{\Omega}_0 - \mathbf{\Omega}_f)$ into the steady versions of (10.78) and (10.79) gives us

$$\left(\tilde{\beta}\Omega_{f,\beta} - \Omega_{p,\beta}\right)\Omega_{f,\gamma} = \kappa\Omega_{f,\alpha}, \quad \tilde{\beta}\Omega_{f,\alpha}\,\Omega_{f,\gamma} = -\kappa\Omega_{f,\beta},$$

from which we find

$$\Omega_{f,\beta} = \frac{1}{1 + \left(\kappa/\tilde{\beta}\Omega_{f,\gamma}\right)^2}\frac{\Omega_{p,\beta}}{\tilde{\beta}}, \quad \Omega_{f,\alpha} = -\left(\frac{\kappa}{\tilde{\beta}\Omega_{f,\gamma}}\right)\Omega_{f,\beta}.$$

These revert to (10.75) for $\kappa \to 0$, as they must. Moreover, for $\Omega_{f,\beta} \sim \Omega_p/\tilde{\beta} \ll \Omega_0$, (10.83) yields

$$\Omega_{f,\gamma} = \Omega_0\left[1 - \frac{(2ab)^2}{(a^2 + b^2)^2}\frac{\left(\Omega_{f,\alpha}^2 + \Omega_{f,\beta}^2\right)}{\Omega_0^2}\right].$$

This brings us back to (10.72), which we had assumed in §10.4.2 based on simple physical arguments.

Noir et al. (2003), following on from Busse (1968), provide a more formal analysis for this geometry, including an asymptotic description of the Ekman layers. They show that the frictional torque acting on the fluid is not of the form $\mathbf{T}_\nu \sim (\mathbf{\Omega}_0 - \mathbf{\Omega}_f)$, but rather has one term proportional to $\mathbf{\Omega}_0 - \mathbf{\Omega}_f$ and a second proportional to $(\mathbf{\Omega}_0 - \mathbf{\Omega}_f) \times \mathbf{\Omega}_f$. Crucially, however, this extra term does no work. Nor does it contribute to $\mathbf{T}_\nu \cdot \left(\Omega_{f,\alpha}\hat{\mathbf{e}}_\alpha + \Omega_{f,\beta}\hat{\mathbf{e}}_\beta\right)$ in (10.82), and so our derivation of (10.83) remains valid. Thus, (10.83) is not merely an artefact of the assumption that $\mathbf{T}_\nu \sim (\mathbf{\Omega}_0 - \mathbf{\Omega}_f)$ in Vanyo & Likins (1972). Rather, it has much wider validity. Not surprisingly, then, both Busse (1968) and Noir et al. (2003) conclude that $\mathbf{\Omega}_f \cdot \mathbf{\Omega}_0 = \Omega_f^2$, at least to leading order in $\hat{\beta}$.

Actually, for earth-like values of $\hat{\beta}$, it turns out that the frictional torque associated with the Ekman layers does not substantially change the inviscid results of the previous section, particularly (10.76), provided that the Ekman number, $Ek = \nu/\Omega_0 a^2$, is smaller than $Ek \sim 10^{-7}$. (See, for example, the discussion in Tilgner, 2009.) This might be compared with an estimate of $Ek \sim 10^{-15}$ for the core of the earth. In this sense, it might be supposed that the precessional Ekman layers are of limited geophysical significance.

However, there are good reasons for wanting to know, or at least have an estimate of, how much energy is dissipated in the earth's core due to precession, and this requires a detailed knowledge of the structure of the Ekman layers. In the case of the earth, the angle between $\mathbf{\Omega}_0$ and $\mathbf{\Omega}_f$ is of the order of 10^{-5}, which suggests boundary-layer velocities of around 4 mm/s. Tilgner (2009) suggests that the associated precessional dissipation may be as high as 10^6 MW.

Moreover, much of our understanding of precessional flows comes from laboratory experiments, and the Ekman number in such experiments is rarely so small that frictional torques can be ignored. Such experiments display a variety of intriguing phenomena, many of which were predicted theoretically, such as critical latitudes at which the Ekman layers break down, internal cylindrical shear layers (some, but not all, associated with the breakdown of the Ekman layers), and a transition to turbulence at higher precession rates. All of these phenomena are discussed in the reviews of Tilgner (2009) and Le Bars et al. (2015).

10.4.4 The Breakdown of Ekman Layers at Critical Latitudes

We close this section with a brief discussion of the breakdown of the Ekman layers at critical latitudes. Here we shall find it convenient to adopt a frame of reference attached to the mantle, in which $\mathbf{\Omega}_f$ precesses around $\hat{\mathbf{e}}_3$ at the rate $-\mathbf{\Omega}_0$. In order to simplify the discussion, we shall ignore many of the complications associated with a precessing mantle (though not the fact that $\mathbf{\Omega}_f$ is forced to cone around $\mathbf{\Omega}_0$) by approximating the inertial-frame angular velocity of the mantle by $\mathbf{\Omega}_0$, rather than $\mathbf{\Omega}_0 + \mathbf{\Omega}_p$.

Let us return to the derivation of the inertial wave equation (6.14), only this time we retain viscosity in the *linearized* equation of motion,

$$\frac{\partial \mathbf{u}}{\partial t} = 2\mathbf{u} \times \mathbf{\Omega}_0 - \nabla\left(p\,\rho\right) + \nu\nabla^2\mathbf{u}. \tag{10.84}$$

The equivalent vorticity equation is

$$\left(\frac{\partial}{\partial t} - \nu\nabla^2\right)\boldsymbol{\omega} = 2(\mathbf{\Omega}_0 \cdot \nabla)\mathbf{u}, \tag{10.85}$$

and applying the operator $(\partial/\partial t - \nu\nabla^2)\,\nabla\times$ to (10.85), while noting that $\nabla\times$ commutes with ∇^2, yields

$$\left(\frac{\partial}{\partial t} - \nu\nabla^2\right)^2 \nabla^2\mathbf{u} + (2\mathbf{\Omega}_0 \cdot \nabla)^2\mathbf{u} = 0. \tag{10.86}$$

We now think of (10.86) as describing not just a viscous inertial wave, but also an oscillating Ekman layer.

Let ϑ be the polar angle measured from the spin axis $\hat{\mathbf{e}}_3$ to a location on the core-mantle boundary (CMB), and ξ a coordinate that points radially inward from the CMB. Within an Ekman layer the Laplacians in (10.86) are dominated by derivatives in ξ, and so (10.86) becomes

$$\left(\frac{\partial}{\partial t} - \nu\frac{\partial^2}{\partial\xi^2}\right)^2\frac{\partial^2\mathbf{u}}{\partial\xi^2} + (2\Omega_0\cos\vartheta)^2\frac{\partial^2\mathbf{u}}{\partial\xi^2} = 0. \tag{10.87}$$

Note that, when the time derivative in (10.87) is set to zero, we recover the governing equation for steady Ekman layers, (5.16).

Now, in the mantle frame, $\boldsymbol{\Omega}_f$ precesses around $\hat{\mathbf{e}}_3$ at the rate $-\Omega_0$. So let us look for a solution of the form $(\mathbf{u} - \mathbf{u}_f) \sim \exp\left[j\left(\xi/\delta - \Omega_0 t\right)\right]$, where δ is complex. Equation (10.87) then yields

$$\delta^2 = \frac{\nu}{j\Omega_0\left(1 \pm 2\cos\vartheta\right)}. \tag{10.88}$$

This is the key result. There are singularities at $2\cos\vartheta = \mp 1$, that is at colatitudes of $\vartheta = 60°$ and $\vartheta = 120°$. This identifies the locations at which the conventional Ekman scaling breaks down. In the vicinity of these critical latitudes, the Ekman layer thickness rescales from $O(\mathrm{Ek}^{1/2})$ to $O(\mathrm{Ek}^{2/5})$, i.e. it thickens, and this occurs over a narrow range of latitudes, of lateral extent $\mathrm{Ek}^{1/5}$ (see, for example, Le Bars et al., 2015).

In these critical regions, the Ekman pumping is considerably more intense than that normally associated with an Ekman layer. This, in turn, triggers a cylindrical shear layer within the inviscid interior which is aligned with the rotation axis of the fluid and spans the critical latitudes, from $\vartheta = 60°$ to $\vartheta = 120°$. (See, for example, Noir et al., 2001b.) This vertical shear layer has a width of order $\mathrm{Ek}^{1/5}$, which matches the lateral extent of the rescaled Ekman layers. It consists of a radially confined jet-like motion in the azimuthal direction. Interestingly, certain precession experiments reveal additional, concentric shear layers, with the direction of the azimuthal jets alternating between successive layers (Vanyo, 2004).

Of course, it is important to understand the physical origin of these critical latitudes. In this respect, it is interesting to note that the critical latitudes are precisely the locations at which incident inertial waves of frequency Ω_0

have difficulty reflecting off the mantle, because the reflected wave has a group velocity which is *locally* parallel to the surface. The implication is that some of the energy of an incident inertial wave is absorbed into the Ekman layer at the critical latitudes. (The reflection of inviscid inertial waves off of plane surfaces, and the critical angles at which reflection cannot occur, is discussed in §6.8.) It seems plausible, therefore, that the breakdown of the Ekman layers at critical latitudes is intimately connected to the absorption of inertial-wave energy into the boundary layer at those locations. Certainly, experiments and numerical simulations show that, in a sphere, inertial waves can establish an oblique, conical ray pattern in which the rays ricochet off the mantle at the poles and then reflect back onto the mantle at the critical latitudes (see Figure 10.18). This is discussed in, for example, Noir et al. (2001a, b), and Le Bars et al. (2015), and supports the hypothesis that breakdown of the Ekman layers is caused by *incoming* inertial waves. However, the width of these conical rays is of the order of $\mathrm{Ek}^{1/5}$, which matches the lateral extent of the rescaled Ekman layers and is thicker than might otherwise be expected. This suggests that the rays also contain *outgoing* inertial waves. In practice, these conical rays are in fact *inertial modes*, which can be thought of as a superposition of both incoming and outgoing waves.

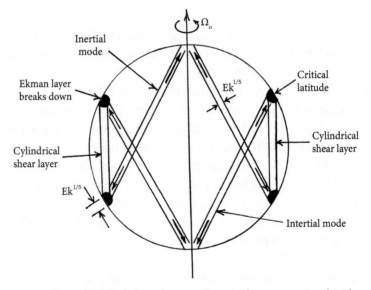

Figure 10.18 The cylindrical shear layer and conical rays associated with critical latitudes.

The relationship between the breakdown of Ekman layers at critical latitudes, the cylindrical shear layers observed in experiments, and the conical ray patterns seen in numerical simulations, is an ongoing area of research. For a recent overview of this topic see, for example, Kida (2021).

* * *

This concludes our introduction to precession. The topic is a subtle one, but an excellent introduction to rigid-body precession may be found in Goldstein (1980), while Tilgner (2009) and Le Bars et al. (2015) offer extensive reviews of precession in rotating fluids.

References

Busse, F.H., 1968, Steady fluid flow in a precessing spheroidal shell, *J. Fluid Mech.*, **33**, 739–51.

Goldstein, H., 1980, *Classical Mechanics*, 2nd Ed., Addison-Wesley.

Kida, S., 2021, Steady flow in a rapidly rotating spheroid with weak precession, part 2, *Fluid Dyn. Res.*, 53(2), 025501.

Le Bars, M., Cebron, D., & Le Gal, P., 2015, Flows driven by libration, precession, and tides, *Ann. Rev. Fluid Mech.*, **47**, 163–93.

Noir, J., Brito, D., Aldridge, K., & Cardin, P., 2001a, Experimental evidence of inertial waves in a precessing spheroidal cavity, *Geophys. Res. Lett.*, **28**(19), 3785–8.

Noir, J., Cardin, P., Jault, D., & Masson, J.-P., 2003, Experimental evidence of nonlinear resonance effects between retrograde precession and the tilt-over mode within a spheroid. *Geophys. J. Int.*, **154**, 407–16.

Noir, J., Jault, D., & Cardin, P., 2001b, Numerical study of the motions within a slowly precessing sphere at low Ekman number, *J. Fluid Mech.*, **437**, 283–99.

Poincaré, H., 1910, Sur la précession des corps déformables, *Bulletin Astronomique*, **27**, 321–56.

Tilgner, A., 2009, Rotational dynamics of the core, In: *Treatise on Geophysics*, vol. 8. Peter Olson, ed. Elsevier, 207–44.

Vanyo, J.P., 2004, Core–mantle relative motion and coupling, *Geophys. J. Int.*, **158**, 470–78.

Vanyo, J.P., & Likins, P.W., 1972, Rigid-body approximations to turbulent motion in a liquid-filled, precessing spherical cavity, *J. Appl. Mech.*, **39**, 18–24.

Chapter 11

Instability I

Taylor–Couette Flow

We now return to an inertial frame of reference and consider the stability of various categories of swirling flow. We start, in §11.1, with the simple case where the base flow is $\mathbf{u}_0 = u_\theta(r)\hat{\mathbf{e}}_\theta$. (We shall use cylindrical polar coordinates throughout this chapter.) This is then extended, in §11.2, to the helical flow $\mathbf{u}_0(r) = u_\theta\hat{\mathbf{e}}_\theta + u_z\hat{\mathbf{e}}_z$. However, we postpone our discussion of the stability of rotating, Rayleigh–Bénard convection until Chapter 12.

11.1 The Centrifugal Instability of Rayleigh and Taylor

11.1.1 Rayleigh's Criterion for Inviscid, Axisymmetric Disturbances

We have already discussed Rayleigh's seminal 1916 paper in §2.12.1. In this paper, Rayleigh considered the stability of the inviscid flow $\mathbf{u}_0 = u_\theta(r)\hat{\mathbf{e}}_\theta$ using a formal analogy between swirling and stratified fluids. Famously, he used the analogy to show that this flow is stable to axisymmetric disturbances if and only if

$$\Phi(r) = \frac{1}{r^3}\frac{d\Gamma^2}{dr} \geq 0, \tag{11.1}$$

where $\Gamma = ru_\theta$. The function $\Phi(r)$ is known as *Rayleigh's discriminant*.

Perhaps it is worth taking a moment to recall Rayleigh's simple physical argument, if only to contrast it with the more formal perturbation analysis commonly used to deduce (11.1). The inviscid, axisymmetric flow of a fluid of uniform density is governed by,

$$\frac{D\Gamma}{Dt} = 0, \quad \nabla \cdot \mathbf{u}_p = 0, \tag{11.2}$$

$$\frac{D\mathbf{u}_p}{Dt} = -\nabla\left(p/\rho\right) + \frac{\Gamma^2}{r^3}\hat{\mathbf{e}}_r, \tag{11.3}$$

The Dynamics of Rotating Fluids. P. A. Davidson, Oxford University Press. © Peter A Davidson (2024).
DOI: 10.1093/9780191994272.003.0011

where $\mathbf{u}_p(r, z) = (u_r, 0, u_z)$ is the poloidal velocity. Rayleigh noticed that there is an exact analogy between these equations and the axisymmetric (non-rotating) motion of an incompressible Boussinesq fluid of mean density $\bar{\rho}$ driven by density gradients in a *radial* gravitational field. To see how this analogy arises, we write $\rho = \bar{\rho} + \rho'$ for the density of the Boussinesq fluid, whose axisymmetric motion is then governed by

$$\frac{D\rho'}{Dt} = 0, \quad \nabla \cdot \mathbf{u}_p = 0, \tag{11.4}$$

$$\frac{D\mathbf{u}_p}{Dt} = -\nabla\left(p/\bar{\rho}\right) + \left(\rho'/\bar{\rho}\right)\mathbf{g}. \tag{11.5}$$

If we take $\mathbf{g} = \left(g^*/r^3\right)\hat{\mathbf{e}}_r$, where g^* is a constant, the analogy between the two systems is complete, with Γ^2 replacing $g^*\rho'/\bar{\rho}$ in (11.4) and (11.5). It is readily confirmed that, under this analogy, the azimuthal contribution to the kinetic energy of the swirling flow, $\rho u_\theta^2/2$, corresponds to the potential energy of the radially stratified fluid (see §2.12.1).

Now, since $\mathbf{g}$ is radial, our Boussinesq fluid admits static equilibria of the form $\bar{\rho} + \rho' = \rho_0(r)$, and those equilibria are stable if and only if $\rho_0(r)$ is an increasing function of r, for all r. It follows from Rayleigh's analogy that a necessary and sufficient condition for the stability of the flow $\mathbf{u}_0 = u_\theta(r)\hat{\mathbf{e}}_\theta$ to inviscid, axisymmetric disturbances is that Γ^2 increases monotonically with radius. This is Rayleigh's celebrated centrifugal stability criterion.

Rayleigh also used an energy argument to interpret his stability criterion. Consider two, thin, circular rings of fluid of volume δV. One ring has radius r_1, angular momentum Γ_1 and kinetic energy $\frac{1}{2}\rho\delta V\left(\Gamma_1^2/r_1^2\right)$, while the other has radius $r_2 = r_1 + \delta r$ and angular momentum $\Gamma_2 = \Gamma_1 + \delta\Gamma$. If the rings exchange position while conserving their angular momenta, the change in kinetic energy is

$$\frac{1}{2}\rho\delta V\left(\Gamma_2^2 - \Gamma_1^2\right)\left(\frac{1}{r_1^2} - \frac{1}{r_2^2}\right) = \rho\delta V\left(\Gamma_2^2 - \Gamma_1^2\right)\frac{\delta r}{r_1^3} = \Phi(r)\rho\delta V(\delta r)^2. \tag{11.6}$$

Evidently, if $\Phi(r) < 0$, the kinetic energy associated with the azimuthal motion falls as a result of the perturbation and this releases energy to the disturbance, driving an instability. In the analogous stratified problem, this is equivalent to releasing potential energy by exchanging hoops of fluid in a region where $\rho_0(r)$ is a decreasing function of r.

The ease with which Rayleigh obtained his stability criterion, with a minimum of mathematical detail, is masterful. However, in order to progress to more complicated cases, such as viscous flows, we need to engage with a more formal perturbation analysis. So let us now derive Rayleigh's centrifugal criterion using a more conventional approach.

As before, we restrict ourselves to inviscid, axisymmetric disturbances. Let the perturbed flow be

$$\mathbf{u}(r, z, t) = (\Gamma(r)/r)\,\hat{\mathbf{e}}_\theta + \mathbf{u}'(r, z, t), \tag{11.7}$$

with $|\mathbf{u}'| \ll \Gamma/r$. Since the disturbance is axisymmetric, we have

$$\left(u_r', 0, u_z'\right) = \left(-\frac{1}{r}\frac{\partial \Psi'}{\partial z},\ 0,\ \frac{1}{r}\frac{\partial \Psi'}{\partial r}\right), \tag{11.8}$$

where Ψ is the Stokes streamfunction, (2.49). Moreover, our perturbed flow is governed by the inviscid equations (2.88), rewritten in the form

$$\frac{D}{Dt}(\Gamma + \Gamma') = 0, \qquad \frac{D}{Dt}\left(\frac{\omega_\theta'}{r}\right) = \frac{\partial}{\partial z}\left(\frac{(\Gamma + \Gamma')^2}{r^4}\right). \tag{11.9}$$

Note that the azimuthal vorticity, ω_θ', is related to the Stokes streamfunction by

$$\omega_\theta' = \frac{\partial u_r'}{\partial z} - \frac{\partial u_z'}{\partial r} = -\frac{1}{r}\nabla_*^2 \Psi', \tag{11.10}$$

where, as usual, ∇_*^2 is the Stokes operator, (2.48).

We now linearize (11.9) about the base flow $\mathbf{u}_0(r) = (\Gamma/r)\,\hat{\mathbf{e}}_\theta$. This gives us

$$\frac{\partial \Gamma'}{\partial t} + u_r'\frac{d\Gamma}{dr} = 0, \qquad \frac{\partial}{\partial t}\left(\frac{\omega_\theta'}{r}\right) = \frac{2\Gamma}{r^4}\frac{\partial \Gamma'}{\partial z}, \tag{11.11}$$

which we rewrite as

$$\frac{\partial \Gamma'}{\partial t} - \frac{1}{r}\frac{\partial \Psi'}{\partial z}\frac{d\Gamma}{dr} = 0, \tag{11.12}$$

$$\frac{\partial}{\partial t} \nabla_*^2 \Psi' = -\frac{2\Gamma}{r^2} \frac{\partial \Gamma'}{\partial z}.$$ (11.13)

We are almost there. The final step is to eliminate Γ' from these equations, which yields the governing equation for an inviscid, axisymmetric disturbance of small amplitude,

$$\frac{\partial^2}{\partial t^2} \nabla_*^2 \Psi' + \Phi(r) \frac{\partial^2 \Psi'}{\partial z^2} = 0.$$ (11.14)

Note that, when the base flow consists of uniform rotation, $\Gamma = \Omega r^2$, then $\Phi = (2\Omega)^2$ and (11.14) yields the governing equation for axisymmetric inertial waves, (6.27).

Now suppose that the fluid is confined to the annulus $R_1 < r < R_2$, but unbounded in z, so that $\Psi'(R_1) = \Psi'(R_2) = 0$. Then we have homogeneity in z and we are free to look for modes of the form

$$\Psi' = \hat{\Psi}(r) \exp\left[j(k_z z - \hat{\omega} t)\right].$$ (11.15)

We take k_z to be real but allow $\hat{\omega}$ to be complex. Evidently, if $\hat{\omega}$ is real the flow is stable, whereas a positive imaginary part of $\hat{\omega}$ indicates instability. Substituting (11.15) into (11.14) leads to the eigenvalue problem

$$r\frac{d}{dr}\frac{1}{r}\frac{d\hat{\Psi}}{dr} + \left(\frac{\Phi(r)}{\hat{\omega}^2} - 1\right) k_z^2 \hat{\Psi} = 0, \quad \hat{\Psi}(R_1) = \hat{\Psi}(R_2) = 0.$$ (11.16)

We now multiply through (11.16) by $\hat{\Psi}/r$ and rewrite the result as

$$\frac{d}{dr}\left(\frac{\hat{\Psi}}{r}\frac{d\hat{\Psi}}{dr}\right) - \frac{1}{r}\left(\frac{d\hat{\Psi}}{dr}\right)^2 + \left(\frac{\Phi(r)}{\hat{\omega}^2} - 1\right) k_z^2 \frac{\hat{\Psi}^2}{r} = 0.$$ (11.17)

Integrating across the annulus $R_1 < r < R_2$, and invoking the boundary conditions for $\hat{\Psi}$, now yields

$$\hat{\omega}^2 \int_{R_1}^{R_2} \left[\frac{1}{r}\left(\frac{d\hat{\Psi}}{dr}\right)^2 + k_z^2 \frac{\hat{\Psi}^2}{r}\right] dr = k_z^2 \int_{R_1}^{R_2} \left[\Phi(r)\frac{\hat{\Psi}^2}{r}\right] dr.$$ (11.18)

This tells us that ϖ^2 is positive, and the flow stable, if Φ is everywhere positive. Conversely, ϖ^2 is negative, and the flow unstable, if Φ is everywhere negative. More generally, standard Sturm–Liouville theory tells us that ϖ^2 takes on negative values, and the flow is unstable, if Φ is negative anywhere in the range $R_1 < r < R_2$ (see Drazin & Reid, 1981). We have recovered Rayleigh's centrifugal criterion.

11.1.2 A Counter-Example: Two-Dimensional, Inviscid Disturbances

Often, axisymmetric modes are the most dangerous for the velocity profile $u_\theta(r)$. However, we shall now show that this is not always the case. In particular, we shall consider inviscid, two-dimensional disturbances and show that such a perturbation can destabilize certain flows that are otherwise stable by Rayleigh's criterion.

Before discussing the two-dimensional stability of swirling flows, let us digress and recall the conditions under which a parallel shear flow, $\mathbf{u}_0 = u_x(y)\hat{\mathbf{e}}_x$, may succumb to an inviscid, two-dimensional instability. Here Rayleigh's celebrated *inflection point theorem* states that a necessary condition for instability of the base flow $\mathbf{u}_0 = u_x(y)\hat{\mathbf{e}}_x$ is that the vorticity, $\omega_z(y) = -du_x/dy$, exhibits a local minimum or a local maximum (see, for example, Drazin & Reid, 1981). Put another way, an inviscid instability can occur only if $d\omega_z/dy$ changes sign somewhere in the flow.

The equivalent result for the swirling flow $\mathbf{u}_0(r) = (\Gamma/r)\hat{\mathbf{e}}_\theta$ was also obtained by Rayleigh (1880). This states that a necessary condition for the instability of the inviscid flow $\mathbf{u}_0(r) = (\Gamma/r)\hat{\mathbf{e}}_\theta$ to two-dimensional disturbances is that the $\omega_z(r)$ exhibits a local minimum or a local maximum, i.e. $d\omega_z/dr$ changes sign somewhere in the flow.

Noting that $r\omega_z = d\Gamma/dr$, it is relatively easy to identify velocity profiles that are unstable to planar disturbances, yet stable to axisymmetric perturbations. Consider the velocity profile shown in Figure 11.1. Here $d\Gamma/dr > 0$ and so both $\omega_z(r)$ and $\Phi(r)$ are everywhere positive. The flow is therefore stable to axisymmetric disturbances. However, $\omega_z(r)$ is a maximum near the centre of the annulus, and so the flow is potentially unstable to planar perturbations. Now suppose that $d\Gamma/dr$ is large near the centre of the annulus, but weak elsewhere. Then the annular vortex sheet associated with the sharp rise in $\Gamma(r)$ will succumb to a Kelvin–Helmholtz instability, provided that the vortex sheet is sharp enough. So, this is an example where planar perturbations are more dangerous than axisymmetric ones.

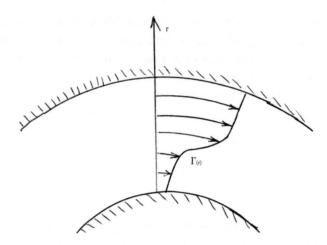

Figure 11.1 An angular momentum distribution that is stable to axisymmetric disturbances but unstable to planar perturbations.

11.1.3 Viscous Instability: Taylor's Analysis of Flow in an Annulus

Rayleigh's analysis is restricted to an inviscid fluid. We now turn to Taylor (1923), who extended Rayleigh's ideas into the viscous regime. As in §11.1.1, the flow is confined to the annular gap $R_1 < r < R_2$, this time driven by the rotation of one or both boundaries (so-called *Taylor–Couette flow*). Once again, we restrict the discussion to axisymmetric disturbances and our base flow is $\mathbf{u}_0(r) = (\Gamma/r)\hat{\mathbf{e}}_\theta$. However, for a viscous fluid, Γ is no longer an arbitrary function of r, but rather constrained to satisfy the azimuthal component of the Navier–Stokes equation, (2.87), which reduces here to $\nu \nabla_*^2 \Gamma = 0$. This represents a balance between the viscous torques exerted on a fluid annulus, and can be written as

$$\rho \nu r \frac{d}{dr}\frac{1}{r}\frac{d\Gamma}{dr} = \frac{1}{r}\frac{d}{dr}\left(r^2 \tau_{r\theta}\right) = 0. \tag{11.19}$$

Now suppose that Ω_1 and Ω_2 are the rotation rates of the inner and outer cylinders, as shown in Figure 11.2. Then (11.19) integrates to give

$$\Gamma = \frac{\Omega_2 R_2^2 - \Omega_1 R_1^2}{R_2^2 - R_1^2} r^2 - \frac{(\Omega_2 - \Omega_1) R_1^2 R_2^2}{R_2^2 - R_1^2} = A r^2 + B. \tag{11.20}$$

We focus on the case where Ω_1 and Ω_2 are both positive, and in which $\Omega_2 R_2^2 < \Omega_1 R_1^2$, as this is centrifugally unstable by Rayleigh's inviscid criterion. Not

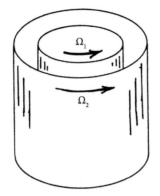

Figure 11.2 Taylor–Couette flow.

unsurprisingly, such a flow is stable provided that a suitably defined Reynolds number is small enough. However, it is observed that the same flow becomes unstable to an axisymmetric mode if the Reynolds number is sufficiently large. We seek to predict the onset of this instability.

The equations governing axisymmetric disturbances to Taylor–Couette flow are (2.86) and (2.87) which, in the current context, take the form

$$\frac{D}{Dt}(\Gamma + \Gamma') = \nu \nabla_*^2 \Gamma', \tag{11.21}$$

$$\frac{D}{Dt}\left(\frac{\omega_\theta'}{r}\right) = \frac{\partial}{\partial z}\left(\frac{(\Gamma + \Gamma')^2}{r^4}\right) + \frac{\nu}{r^2}\nabla_*^2(r\omega_\theta'). \tag{11.22}$$

These are the viscous counterparts of (11.9). Next, assuming the disturbance is small, and linearizing about the base flow, we obtain the viscous versions of (11.12) and (11.13),

$$\left(\frac{\partial}{\partial t} - \nu \nabla_*^2\right)\Gamma' - \frac{1}{r}\frac{\partial \Psi'}{\partial z}\frac{d\Gamma}{dr} = 0, \tag{11.23}$$

$$\left(\frac{\partial}{\partial t} - \nu \nabla_*^2\right)\nabla_*^2 \Psi' = -\frac{2\Gamma}{r^2}\frac{\partial \Gamma'}{\partial z}. \tag{11.24}$$

Finally, we apply the operator $(\partial/\partial t - \nu \nabla_*^2)\nabla_*^2$ to (11.23), taking advantage of the fact that $d\Gamma/dr$ is linear in r in Taylor–Couette flow. This gives us

$$\left(\frac{\partial}{\partial t} - \nu \nabla_*^2\right)^2 \nabla_*^2 \Gamma' = \frac{1}{r}\frac{d\Gamma}{dr}\frac{\partial}{\partial z}\left(\frac{\partial}{\partial t} - \nu \nabla_*^2\right)\nabla_*^2 \Psi',$$

and substituting for Ψ' using (11.24) yields the viscous counterpart of (11.14),

$$\left(\frac{\partial}{\partial t} - \nu \nabla_*^2\right)^2 \nabla_*^2 \Gamma' + \Phi(r)\frac{\partial^2 \Gamma'}{\partial z^2} = 0. \tag{11.25}$$

This is the governing equation for axisymmetric disturbances. Note that $\Phi(r)$ is prescribed in Taylor–Couette flow, and given by $\Phi = 4A\Gamma/r^2$.

To help find solutions of (11.25), we now follow Taylor (1923) and adopt the narrow-gap approximation, which requires $d = R_2 - R_1 \ll R_1$. Also, to focus thoughts, we shall take $\Omega_2 = 0$, so that

$$\Gamma = \frac{\Omega_1 R_1^2 \left(R_2^2 - r^2\right)}{(R_2 + R_1)d} = -A\left(R_2^2 - r^2\right).$$

It is convenient at this point to introduce the new radial coordinate $s = r - R_1$, so that

$$\Gamma = \Omega_1 R_1^2 \frac{R_2 + r}{R_2 + R_1}\left(1 - s/d\right) \approx \Omega_1 R_1^2 \left(1 - s/d\right), \tag{11.26}$$

from which we obtain, in the narrow-gap limit,

$$\Phi(s) = 4A\frac{\Gamma}{r^2} \approx -\frac{2\Omega_1^2 R_1}{d}\left(1 - \frac{s}{d}\right). \tag{11.27}$$

The narrow-gap approximation also allows us to replace ∇_*^2 by a conventional Laplacian, and so (11.25) becomes

$$\left(\frac{\partial}{\partial t} - \nu \nabla^2\right)^2 \nabla^2 \Gamma' = \frac{2\Omega_1^2 R_1}{d}\left(1 - \frac{s}{d}\right)\frac{\partial^2 \Gamma'}{\partial z^2}. \tag{11.28}$$

We now note that both theory and experiment confirm that the axisymmetric instability sets in as a non-oscillatory mode, so that, at onset, we may ignore the time derivatives in (11.28). Thus, for marginal stability, and in the narrow-gap limit, we have

$$(\nabla^2)^3 \Gamma' = \frac{2\Omega_1^2 R_1}{\nu^2 d}\left(1 - \frac{s}{d}\right)\frac{\partial^2 \Gamma'}{\partial z^2}. \tag{11.29}$$

Now all spatial derivatives scale on d in the narrow-gap approximation, which tells us that the appropriate dimensionless control parameter for this instability is

$$\text{Ta} = \frac{\Omega_1^2 R_1 d^3}{\nu^2}. \tag{11.30}$$

This is known as the *Taylor number*, although different authors define Ta in slightly different ways. Of course, Ta is effectively the square of a Reynolds number.

The determination of the critical value of Ta is now relatively straightforward, if somewhat tiresome. However, a good approximation can be obtained by replacing $\Phi(s)$ in (11.29) by its average value, $\bar{\Phi}$, (see Drazin & Reid, 1981). Taking $\Gamma' = \hat{\Gamma}'(s) \cos(k_z z)$ then yields the more accessible expression

$$\left(\frac{d^2}{ds^2} - k_z^2 \right)^3 \hat{\Gamma}' + \frac{\text{Ta}}{d^4} k_z^2 \hat{\Gamma}' = 0. \tag{11.31}$$

Now, for a given k_z, there are solutions of (11.31) for certain values of Ta only, and there will be a minimum value, $(\text{Ta})_{\min}$, for each k_z. The smallest of these values of $(\text{Ta})_{\min}$ yields the critical Taylor number, $(\text{Ta})_{\text{crit}}$, and associated critical wavenumber, k_z. The final result of such a calculation is that (11.31) predicts the most unstable axisymmetric mode to be $k_z d = 3.117$, with an associated Taylor number of $(\text{Ta})_{\text{crit}} = 1708$. A similar analysis of the exact equation (11.29) yields $k_z d = 3.127$ and $(\text{Ta})_{\text{crit}} = 1695$.

Let us now relax the assumption that $\Omega_2 = 0$, requiring only that $0 < \Omega_2 < \Omega_1$. Then

$$\Gamma = \frac{\Omega_1 R_1^2 \left(R_2^2 - r^2 \right) + \Omega_2 R_2^2 \left(r^2 - R_1^2 \right)}{(R_2 + R_1) d} \approx \Omega_1 R_1^2 \left(1 - \frac{s}{d} \right) + \Omega_2 R_2^2 \frac{s}{d},$$

from which, in the narrow-gap limit,

$$\Phi(s) = 4A \frac{\Gamma}{r^2} \approx -\frac{2 \left(\Omega_1 R_1^2 - \Omega_2 R_2^2 \right)}{R_1 d} \left[\Omega_1 \left(1 - \frac{s}{d} \right) + \Omega_2 \frac{s}{d} \right].$$

Our approximate eigenvalue problem (in which $\Phi(s)$ is replaced by its mean value, $\bar{\Phi}$), and the associated Taylor number, then generalize to

$$\left(\frac{d^2}{ds^2} - k_z^2 \right)^3 \hat{\Gamma}' + \frac{\text{Ta}}{d^4} k_z^2 \hat{\Gamma}' = 0, \tag{11.32}$$

$$\text{Ta} = -\frac{\bar{\Phi}d^4}{\nu^2} = \frac{(\Omega_1 + \Omega_2)(\Omega_1 R_1^2 - \Omega_2 R_2^2)\, d^3}{\nu^2 R_1}. \tag{11.33}$$

Since the eigenvalue problem is unchanged, we still have $(\text{Ta})_{\text{crit}} = 1708$ at $k_z d = 3.117$.

These predictions are in excellent agreement with the experimental evidence, which shows that the instability takes the form of a cellular motion. Interestingly, the instability quickly saturates to yield a new laminar flow consisting of a regular pattern of vortices, called *Taylor vortices*, as shown in Figure 11.3(a). However, if the value of Ta is further increased, the Taylor vortices themselves go unstable, yielding an array of non-axisymmetric vortices that migrate around the inner cylinder (see Figure 11.3b).

11.1.4 The Experimental Evidence for Taylor–Couette Flow

Let us consider the experimental evidence in a little more detail, restricting the discussion to the case where $\Omega_2 = 0$. For low to modest Ta, the flow is azimuthal and given by (11.20), except near the top and bottom of the cylinders where Ekman pumping sets up a secondary flow in the r–z plane. However, if the cylinders are long, this Ekman pumping is localized around

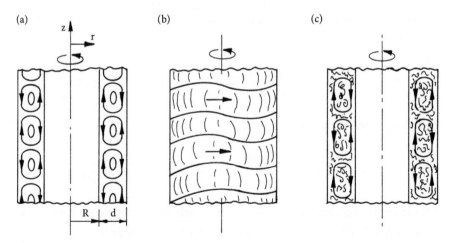

Figure 11.3 Flow between concentric cylinders. Only the inner cylinder rotates. (a) Taylor vortices. (b) Wavy Taylor vortices. (c) Turbulent Taylor vortices. Reproduced from Davidson (2015).

the ends of the apparatus. Now suppose we slowly increase Ta by increasing Ω_1. As Ta approaches $(\text{Ta})_{\text{crit}}$, steady Taylor vortices appear in the form of an axisymmetric, cellular pattern, as shown in Figures 11.4(a). These vortices have an aspect ratio close to unity and are of alternating sense in the r–z plane. Within them, a poloidal flow is superimposed on the primary azimuthal motion and so fluid particles follow helical paths on toroidal surfaces. Note that the laminar flow in Figures 11.4(a) is steady, indicating that the exponential growth predicted by linear stability theory has saturated through nonlinear interactions. Despite this nonlinearity, the axial wavelength of the Taylor vortices is reasonably close to that predicted by linear stability theory, i.e. $k_z d \approx 3.1$.

It turns out that, because any apparatus has a finite length, the onset of this instability is a little more complex than our theoretical analysis would suggest. Specifically, the Taylor vortices do not suddenly appear at $\text{Ta} = (\text{Ta})_{\text{crit}}$, as predicted by idealized theory. Rather, they emerge gradually as Ta approaches $(\text{Ta})_{\text{crit}}$ from below. In particular, the poloidal vortices established by Ekman pumping near the ends of the apparatus, which exist even at modest values of Ta, start to multiply and extend into the interior of the flow as Ta approaches $(\text{Ta})_{\text{crit}}$. By the time $\text{Ta} = (\text{Ta})_{\text{crit}}$, the vortices which have spread inward from the two ends of the apparatus are able to link up near the centre, establishing the flow shown in Figure 11.4(a). In short, the appearance of

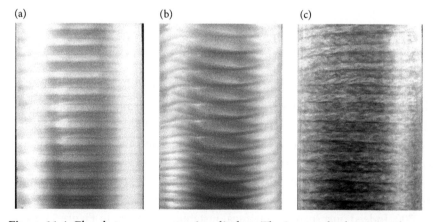

Figure 11.4 Flow between concentric cylinders. The inner cylinder rotates but the outer one does not. (a) Taylor vortices at $\text{Ta} \approx (\text{Ta})_{\text{crit}}$. (b) Wavy Taylor vortices at $\text{Ta} = 4(\text{Ta})_{\text{crit}}$ with an azimuthal wavenumber of $m = 3$. (c) Turbulent Taylor vortices at $\text{Ta} = 676(\text{Ta})_{\text{crit}}$. Courtesy of Q. Xiao, T.T. Lim, and Y.T. Chew, National University of Singapore.

Taylor vortices is a continuous, rather than an abrupt, process. The fact that the Taylor vortices creep in from the ends of the apparatus has a number of nontrivial consequences, such as opening up the possibility of non-uniqueness and hysteresis. That is to say, the exact size and sense of the Taylor vortices in a given experiment can be influenced by the history of how the final state is approached. This is particularly the case if rapid changes in Ta are made.

Now suppose that Ta is increased beyond $(Ta)_{crit}$. It is observed that, at a few multiples of $(Ta)_{crit}$, the Taylor vortices themselves become unstable to a non-axisymmetric disturbance. The flow then bifurcates to an unsteady, non-axisymmetric motion, as shown in Figure 11.4(b). This new flow pattern is said to consist of *wavy Taylor vortices* and it comprises non-axisymmetric Taylor vortices that migrate around the inner cylinder with an azimuthal wavenumber that depends on the value of Ta. Note that, although this unsteady flow is more complex than the original steady Taylor vortex pattern, it is still laminar (i.e. non-chaotic). Yet further increases in Ta triggers additional bifurcations to ever more complex states until eventually, when Ta is large enough, the flow becomes fully turbulent. However, embedded within the turbulence there is a time-averaged component of the motion which resembles the laminar Taylor vortices of Figure 11.4(a), as shown in Figures 11.4(c). This mean, cellular motion is said to consist of *turbulent Taylor vortices*. When Ta becomes very large, however, the turbulent Taylor vortices disappear, leaving only turbulence superimposed on the mean azimuthal motion.

11.1.5 The Stability of the Boundary Layer on a Rotating Cylinder

Consider a long cylinder of radius R immersed in a viscous fluid. At $t = 0$ the cylinder is made to rotate at the speed Ω. A thin, laminar boundary layer starts to develop on the surface of the cylinder, of thickness $\delta(t)$, as vorticity slowly diffuses away from the surface and into the fluid. Within this boundary layer the angular momentum drops from $\Gamma = \Omega R^2$ at $r = R$ to $\Gamma \approx 0$ at $r = R + \delta$. From the discussion above, we might expect this boundary layer to become centrifugally unstable at a critical value of Ta $= |\bar{\Phi}| \, \delta^4 / \nu^2$ ~ 1700. Indeed, just such a centrifugal instability is shown in Figure 11.5(b). Let us see if we can estimate how long it takes for the instability to first appear.

As long as the boundary layer remains thin, the curvature of the surface should not influence the rate of diffusion of vorticity from the surface of the

(a)   (b)

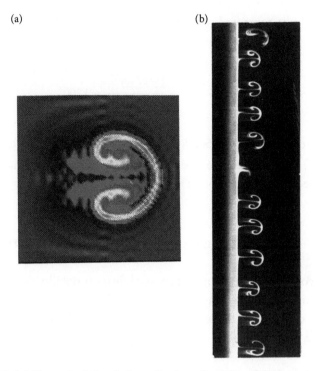

Figure 11.5 (a) Numerical simulation of a ring of swirling fluid bursting radially outward under the action of the centrifugal force. The image is a cross-section in the r–z plane and is coloured by $\Gamma = r u_\theta$. (b) Essentially the same type of structure is generated by the centrifugal instability of a boundary layer on a rotating cylinder. Reproduced from Davidson (2015).

cylinder. So we can adapt the well-known solution for the diffusion of vorticity from an impulsively started plate to give

$$\omega_z = \frac{1}{r}\frac{\partial \Gamma}{\partial r} = -\frac{\Omega R}{\sqrt{\pi v t}}\exp\left(-\frac{s^2}{4vt}\right), \quad \Gamma = \Omega R^2 \text{erfc}\left(\frac{s}{\sqrt{4vt}}\right), \tag{11.34}$$

where $s = r - R$ and erfc is the complimentary error function. Adjacent to the surface of the cylinder, this yields

$$\Phi(s \to 0) = -\frac{2\Omega^2 R}{\delta}(1 - s/\delta), \quad \delta = \sqrt{\pi v t}, \tag{11.35}$$

which might be compared with (11.27). This suggests $|\bar{\Phi}| \sim \Omega^2 R/\sqrt{\pi v t}$, with instability at

$$\text{Ta} = -\frac{\bar{\Phi}\delta^4}{v^2} \sim \frac{\Omega^2 R(\pi t)^{3/2}}{v^{1/2}} \sim 1700, \qquad (11.36)$$

or equivalently, $\Omega t_{\text{crit}} \sim 45\left(\Omega R^2/v\right)^{-1/3}$. In fact, experiments give $\Omega t_{\text{crit}} \approx 98\left(\Omega R^2/v\right)^{-1/3}$, the difference arising largely from our crude estimate of $(\text{Ta})_{\text{crit}}$, but also because the base flow is not steady.

11.2 The Influence of Axial Flow on Stability

We now consider the influence of an axial flow on stability. In particular, we consider an extension of Rayleigh's centrifugal criterion which asserts that a *sufficient* condition for the axisymmetric stability of the inviscid flow $\mathbf{u}_0(r) = u_\theta\hat{\mathbf{e}}_\theta + u_z\hat{\mathbf{e}}_z$ is that, for all r,

$$\Phi(r) = \frac{1}{r^3}\frac{d\Gamma^2}{dr} > \frac{1}{4}\left(\frac{du_z}{dr}\right)^2. \qquad (11.37)$$

The physical idea behind (11.37) is shown in Figure 11.6. Suppose that $\Phi > 0$ and consider two, thin rings of fluid, A and B, located at r and $r + dr$, and each of volume δV. If they exchange radial position while conserving angular momentum, then the rise in the kinetic energy of the azimuthal motion is, according to (11.6), $\rho\delta V\Phi(r)(dr)^2$. Such an exchange is possible only if this rise in azimuthal kinetic energy is compensated for by a corresponding fall in the kinetic energy associated with the axial flow. However, it turns out that the fall in axial kinetic energy resulting from this exchange of rings cannot exceed $\frac{1}{4}\rho\delta V(du_z/dr)^2(dr)^2$. Stability is therefore assured provided $\Phi(r) > \frac{1}{4}(du_z/dr)^2$, for all r.

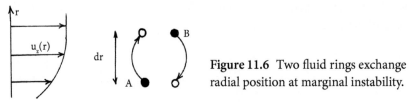

Figure 11.6 Two fluid rings exchange radial position at marginal instability.

11.2.1 A Heuristic Energy Argument

Various authors have offered heuristic explanations as to why the fall in axial kinetic energy cannot exceed $\frac{1}{4}(du_z/dr)^2\rho\delta V(dr)^2$. Typically, these proceed as follows. It is plausible that, on arrival at their new positions, the two rings have an axial velocity which is somewhere between their original speed and the speed at their new locations. Thus,

$$u_z^A(r+dr) = u_z^A(r) + \alpha_A\frac{du_z}{dr}dr, \quad u_z^B(r) = u_z^B(r+dr) - \alpha_B\frac{du_z}{dr}dr,$$

for two constants α_A and α_B which are expected to take values in the range $0\rightarrow 1$. However, the need to retain the same axial flow rate suggests we take $\alpha_A = \alpha_B$, and hence

$$u_z^A(r+dr) = u_z^A(r) + \alpha\frac{du_z}{dr}dr, \quad u_z^B(r) = u_z^B(r+dr) - \alpha\frac{du_z}{dr}dr. \qquad (11.38)$$

The change in the axial kinetic energy of ring A is then

$$\Delta\left(\tfrac{1}{2}mu_z^2\right)_A = \frac{1}{2}\rho\delta V\left[2\alpha u_z(r)\frac{du_z}{dr}dr + \alpha^2\left(\frac{du_z}{dr}\right)^2(dr)^2\right], \qquad (11.39)$$

while that of ring B is

$$\Delta\left(\tfrac{1}{2}mu_z^2\right)_B = \frac{1}{2}\rho\delta V\left[-2\alpha u_z(r+dr)\frac{du_z}{dr}dr + \alpha^2\left(\frac{du_z}{dr}\right)^2(dr)^2\right], \qquad (11.40)$$

giving a total change of

$$\Delta\left(\tfrac{1}{2}mu_z^2\right) = -\alpha(1-\alpha)\rho\delta V\left(\frac{du_z}{dr}\right)^2(dr)^2. \qquad (11.41)$$

Since we expect $0 < \alpha < 1$, this represents a reduction in kinetic energy. Although we do not know the value of α, we can at least place an upper bound on $\left|\Delta\left(\tfrac{1}{2}mu_z^2\right)\right|$ as a function of α, and this maximum clearly corresponds to $\alpha = \frac{1}{2}$. We conclude that

$$\left|\Delta\left(\tfrac{1}{2}mu_z^2\right)\right| \leq \frac{1}{4}\rho\delta V\left(\frac{du_z}{dr}\right)^2(dr)^2. \qquad (11.42)$$

Now, marginal instability is energetically permissible if the fall in axial kinetic energy compensates for the rise in azimuthal kinetic energy, whereas it is not realizable if $\left|\Delta\left(\frac{1}{2}u_z^2\right)\right| < \Delta\left(\frac{1}{2}u_\theta^2\right)$. We conclude that, as claimed above, stability is guaranteed whenever

$$\Phi(r) > \frac{1}{4}\left(\frac{du_z}{dr}\right)^2. \tag{11.43}$$

11.2.2 Rayleigh's Analogy Between Swirl and Buoyancy Revisited

The argument above is physically plausible but not rigorous, as we have no right to assume (11.38). In fact, this stability criterion was first established in a rigorous fashion by Howard & Gupta (1962). However, given Rayleigh's analogy between swirl and buoyancy, the criterion comes as no surprise. The point is this. It is well known that an inviscid shear flow, $\mathbf{u}_0 = u_x(z)\hat{\mathbf{e}}_x$, located in a stably stratified Boussinesq fluid with a smooth density profile, $\rho = \rho_0(z)$, is linearly stable provided the *local Richardson number*, Ri, exceeds ¼, i.e.

$$\text{Ri} = \frac{g\,|d\rho_0/dz|}{\bar{\rho}(du_x/dz)^2} > \frac{1}{4}, \tag{11.44}$$

where $\bar{\rho}$ is a mean density. This was proposed by Taylor in 1931, though a rigorous proof (by Howard again) had to wait until 1961. Physically, this criterion arises because, if we exchange two adjacent parcels of fluid, A and B, of equal volume and located at z and $z + dz$, then the *increase* in potential energy (per unit volume) resulting from the exchange is

$$\Delta(\text{P.E.}) = \rho_A g(dz) - \rho_B g(dz) = -\frac{d\rho_0}{dz}g(dz)^2 = \left|\frac{d\rho_0}{dz}\right|g(dz)^2. \tag{11.45}$$

This rise in potential energy must be compensated for by a corresponding drop in kinetic energy, and we have already seen that the fall in kinetic energy cannot exceed

$$\left|\Delta\left(\frac{1}{2}\bar{\rho}u_x^2\right)\right|_{\text{max}} = \frac{1}{4}\bar{\rho}\left(\frac{du_x}{dz}\right)^2(dz)^2. \tag{11.46}$$

Thus, the exchange cannot take place, and the flow is stable, provided

$$\frac{1}{4}\bar{\rho}\left(\frac{du_x}{dz}\right)^2 (dz)^2 < \left|\frac{d\rho_0}{dz}\right| g(dz)^2, \tag{11.47}$$

which brings us back to (11.44).

Given Rayleigh's analogy between swirl and buoyancy, the step from (11.44) to (11.37) is not large, and indeed the formal proof of (11.37) in Howard & Gupta (1962) is remarkably similar to that of (11.44) in Howard (1961).

* * *

This is all we have to say about the stability of swirling flows of uniform density. Readers keen for more details will find an extensive discussion in Drazin & Reid (1981). Also, Rayleigh (1916) is always a delight to read.

References

Davidson, P.A., 2015, *Turbulence: An Introduction for Scientists and Engineers*, 2nd Ed., Oxford University Press.

Drazin, P.G. & Reid, W.H., 1981, *Hydrodynamic Stability*, Cambridge University Press.

Howard, L.N., 1961, Note on a paper of John W. Miles, *J. Fluid Mech.*, **10**, 509–12.

Howard, L.N., & Gupta, A.S., 1962, On the hydrodynamic and hydromagnetic stability of swirling flows, *J. Fluid Mech.*, **14**(3), 463–76.

Rayleigh, Lord, 1880, On the stability, or instability, of certain fluid motions. *Proc. Lond. Math. Soc.*, **11**, 57–70.

Rayleigh, Lord, 1916, On the dynamics of revolving fluids, *Proc. Royal Soc., A*, **93**, 148.

Taylor, G.I., 1923, Stability of a viscous liquid contained between two rotating cylinders, *Phil. Trans. Roy. Soc. Lond. A*, **223**, 289–343.

Taylor, G.I., 1931, Effect of variation in density on the stability of superposed streams of fluid, *Proc. Roy. Soc. A*, **132**, 499–523.

Chapter 12
Instability II
Rotating Convection

We now turn to buoyant, rotating flows, starting with an introduction to convection.

12.1 Convection Without Rotation: The Rayleigh–Bénard Problem

12.1.1 The Experiments of James Thomson and Henri Bénard

Consider a layer of liquid of depth d which sits on a heated metal plate whose temperature is uniform and steady. The upper surface may be free, or else bounded by a second, cooler metal plate. A uniform temperature difference, ΔT, is maintained between the lower and upper surfaces, and if ΔT is large enough, natural convection occurs in which buoyant fluid rises, loses its heat to the cooler upper surface, and then sinks back down. Natural convection in such layers was first studied by James Thomson (Kelvin's elder brother) in 1882, and then in more detail by Henri Bénard in 1900, Thomson using soapy water and Bénard a type of molten wax. Both noted that the upper surface became tessellated (tilled), reflecting the cellular nature of the convection below. In Thomson's case, the pattern was angular, irregular, and constantly evolving, while Bénard observed a stable, hexagonal pattern (see Figure 12.1). Ironically, although Bénard's experiments are widely cited in the context of buoyant convection, they were dominated by the Marangoni effect (spatial variations in surface tension arising from horizontal temperature gradients), as noted by Pearson (1958). In any event, in experiments where buoyancy dominates, the liquid is observed to rise at the centre of the cells and fall at the edges. Moreover, Bénard noticed that the convection died out if ΔT is too low, although he attributed this to the wax solidifying.

The idea that there is a critical value of ΔT, say ΔT_{crit}, below which convection cannot occur was first proposed by Rayleigh (1916), who considered the

The Dynamics of Rotating Fluids. P. A. Davidson, Oxford University Press. © Peter A Davidson (2024).
DOI: 10.1093/9780191994272.003.0012

(a)   (b)

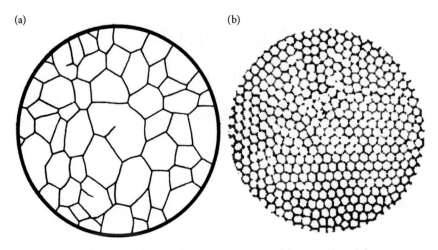

Figure 12.1 Tessellated free surface in convection. (a) Reproduced from
Thomson (1882). (b) Reproduced from Bénard (1900).

onset of convection to be a consequence of an instability. Bénard was reluctant
to accept this, but Rayleigh's point was that buoyant convection is opposed by
viscous forces and a simple energy argument says that convection can occur
only if the rate of working of the buoyancy forces exceeds the rate of viscous
dissipation. When the temperature difference between the two surfaces is too
low, or the fluid viscosity is too large, no such motion can occur. Rather, heat
is transferred by thermal conduction alone, as shown in Figure 12.2(a).

As Rayleigh showed, the transition from static conduction to steady convection is controlled by the *Rayleigh number*,

$$\mathrm{Ra} = \frac{g\beta\Delta T d^3}{\nu\alpha},\tag{12.1}$$

where α is the thermal diffusivity of the fluid, $\beta = -(\partial\rho/\partial T)/\bar{\rho}$ the expansion
coefficient, and $\bar{\rho}$ the mean fluid density. It is only when Ra exceeds a certain
critical threshold, $(\mathrm{Ra})_{\mathrm{crit}}$, that the static equilibrium is destabilized and convection sets in. This is analogous to the sudden appearance of Taylor vortices
in a rotating fluid at $\mathrm{Ta} = (\mathrm{Ta})_{\mathrm{crit}}$, and indeed we shall see that there are close
connections between these two flows. Note that large values of both ν and α
are stabilizing, the former because of energy dissipation, and the latter because
it allows hot rising fluid to shed its excess temperature by diffusion.

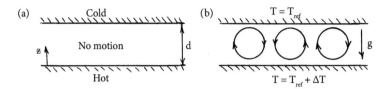

Figure 12.2 Rayleigh–Bénard convection in a layer of fluid of depth d driven by a temperature difference, ΔT. (a) $\Delta T < \Delta T_{\text{crit}}$, conduction only. (b) $\Delta T > \Delta T_{\text{crit}}$, convection.

12.1.2 Rayleigh's Stability Analysis I: Framing the Eigenvalue Problem

We now describe Rayleigh's stability analysis, where the equilibrium whose stability is in question is $\mathbf{u}_0 = 0$ and $T = T_0(z)$, T_0 being linear in z. Rayleigh adopted the Boussinesq approximation, in which variations in density are small and may be ignored, except to the extent that they introduce a buoyancy force $\delta \rho \mathbf{g}$ (see §2.11). In Rayleigh–Bénard convection, it is convenient to rewrite this force as $-\bar{\rho}\beta(T - T_{\text{ref}})\mathbf{g}$, where $\mathbf{g} = -g\hat{\mathbf{e}}_z$ and T_{ref} is a reference temperature. Our governing equations are then continuity, $\nabla \cdot \mathbf{u} = 0$, and

$$\frac{D\mathbf{u}}{Dt} = -\nabla\left(p/\bar{\rho}\right) + \nu\nabla^2\mathbf{u} - \beta(T - T_{\text{ref}})\mathbf{g}, \tag{12.2}$$

$$\frac{DT}{Dt} = \alpha\nabla^2 T, \tag{12.3}$$

where p is the departure from a *linear* (in z) hydrostatic pressure distribution.

We now write the perturbation in temperature as $\vartheta = T - T_0(z)$ and consider small-amplitude disturbances. Linearizing (12.2) and (12.3) about the equilibrium state gives us

$$\frac{\partial\mathbf{u}}{\partial t} = -\nabla\left(\delta p/\bar{\rho}\right) + \nu\nabla^2\mathbf{u} - \beta\vartheta\mathbf{g}, \tag{12.4}$$

$$\frac{\partial\vartheta}{\partial t} + u_z\frac{dT_0}{dz} = \alpha\nabla^2\vartheta, \tag{12.5}$$

with the vorticity equation corresponding to (12.4) being

$$\left[\frac{\partial}{\partial t} - \nu\nabla^2\right]\boldsymbol{\omega} = g\beta(\nabla\vartheta) \times \hat{\mathbf{e}}_z. \tag{12.6}$$

Note that ω_z obeys a simple diffusion equation. Taking the curl a second time yields

$$\left[\frac{\partial}{\partial t} - \nu \nabla^2\right] \nabla^2 \mathbf{u} = g\beta \left[\hat{\mathbf{e}}_z \nabla^2 \vartheta - \frac{\partial}{\partial z} \nabla \vartheta\right],$$

whose z component is

$$\left[\frac{\partial}{\partial t} - \nu \nabla^2\right] \nabla^2 u_z = g\beta \nabla_\perp^2 \vartheta, \tag{12.7}$$

where

$$\nabla_\perp^2 = \partial^2/\partial^2 x + \partial^2/\partial^2 y.$$

If we rewrite (12.5) in the form

$$\left[\frac{\partial}{\partial t} - \alpha \nabla^2\right] \vartheta = u_z \frac{\Delta T}{d}, \tag{12.8}$$

we can eliminate u_z from (12.7) and (12.8) to give the governing equation for ϑ,

$$\left[\frac{\partial}{\partial t} - \nu \nabla^2\right]\left[\frac{\partial}{\partial t} - \alpha \nabla^2\right] \nabla^2 \vartheta = \frac{g\beta \Delta T}{d} \nabla_\perp^2 \vartheta. \tag{12.9}$$

Now, it is possible to show that at marginal stability we have exponential growth without oscillation, as in the viscous centrifugal instability. Consequently, at onset we have

$$\alpha \nabla^2 \vartheta = -(\Delta T/d)\, u_z, \tag{12.10}$$

and

$$(\nabla^2)^3 u_z = \frac{g\beta \Delta T}{\nu \alpha d} \nabla_\perp^2 u_z, \quad (\nabla^2)^3 \vartheta = \frac{g\beta \Delta T}{\nu \alpha d} \nabla_\perp^2 \vartheta. \tag{12.11}$$

Moreover, since d is the only geometric length scale, (12.11) tells us that the stability threshold is determined exclusively by $\mathrm{Ra} = g\beta \Delta T d^3/\nu\alpha$, and by the boundary conditions.

Let us now turn to the boundary conditions, which depend on whether the upper boundary is a solid plate or a free surface, which we treat as a flat, stress-free boundary. In either case, we have $\vartheta = u_z = 0$ at $z = 0$, d. If the boundaries at $z = 0, d$ are no-slip, then $\omega_z = 0$ and continuity demands that $\partial u_z/\partial z = 0$. On the other hand, if the upper boundary is stress free, then the vertical gradients in both u_x and u_y vanish at $z = d$ and $\nabla \cdot \mathbf{u} = 0$ now requires $\partial^2 u_z/\partial z^2 = 0$. In summary, the boundary conditions are:

$$\text{no-slip surface,} \quad \vartheta = u_z = \partial u_z/\partial z = \omega_z = 0,$$

$$\text{stress-free surface,} \quad \vartheta = u_z = \partial^2 u_z/\partial z^2 = \partial \omega_z/\partial z = 0.$$

In order to find the critical value of Ra, Rayleigh noted that (12.11) admits separable solutions of the form

$$\vartheta = \hat{\vartheta}(z)f(x, y), \quad u_z = \hat{u}_z(z)f(x, y), \tag{12.12}$$

provided that f satisfies the Helmholtz equation,

$$\nabla_{\perp}^2 f + k^2 f = 0. \tag{12.13}$$

The constant k is a form of horizontal wavenumber, which has yet to be determined. Assuming solutions of this form, (12.11) now yields, for marginally unstable modes, the eigenvalue problem

$$\left(\frac{d^2}{dz^2} - k^2\right)^3 \hat{u}_z + \frac{\text{Ra}}{d^4}k^2\hat{u}_z = 0. \tag{12.14}$$

It is of interest to compare (12.14) with (11.31), which governs (at least approximately) the axisymmetric Taylor instability in the narrow-gap approximation. The equations are identical, with Ra replacing Ta. Clearly, there are close links between these two problems.

12.1.3 Rayleigh's Stability Analysis II: Slip Boundaries Top and Bottom

The most important configurations are no-slip at $z = 0$ and no-slip or stress free at $z = d$. Nevertheless, it is instructive to follow Rayleigh and consider the somewhat artificial case where the top and bottom surfaces are both stress free, as this captures the essence of the instability with minimal effort. That is, by inspection, $\hat{\vartheta} \sim \hat{u}_z \sim \sin(n\pi z/d)$, $n = 1, 2, 3, \ldots$, satisfies all the boundary conditions and (12.14) then yields

$$\text{Ra} = \frac{\left((n\pi)^2 + a^2\right)^3}{a^2}, \quad a = kd. \tag{12.15}$$

The most unstable mode is clearly $n = 1$ and so the critical value of a is the one that minimizes Ra in (12.15) subject to $n = 1$. This gives us

$$a_{\text{crit}} = \pi/\sqrt{2}, \quad (\text{Ra})_{\text{crit}} = 27\pi^4/4 = 657.5. \tag{12.16}$$

Now, there is a second way of getting the same result which is physically more revealing. We start by noting that (12.4) yields the energy equation

$$\frac{\partial}{\partial t}\left(\frac{1}{2}\mathbf{u}^2\right) = \nabla \cdot \left[\nu\mathbf{u} \times \boldsymbol{\omega} - (\delta p/\bar{\rho})\,\mathbf{u}\right] - \nu\omega^2 + g\beta\vartheta u_z. \tag{12.17}$$

Moreover, it is readily confirmed that, on integrating over a single convection cell and invoking Gauss' theorem, the divergence in (12.17) vanishes. This leaves us with

$$\frac{d}{dt}\int \frac{1}{2}\mathbf{u}^2 dV = -\nu\int \omega^2 dV + g\beta\int \vartheta u_z dV. \tag{12.18}$$

The integrals on the right of (12.18) are the rate of viscous dissipation and the rate of working of the buoyancy force. Evidently, the criterion for instability is that the rate of working of the buoyancy force exceeds the viscous dissipation, and at marginal stability we have

$$g\beta\int \vartheta u_z dV = \nu\int \omega^2 dV. \tag{12.19}$$

Let us now try to estimate these two integrals. Suppose that the instability takes the form of two-dimensional rolls in the x-z plane. The simplest form for u_z and ϑ which satisfies (12.13) and the boundary conditions at $z = 0$ and $z = d$ is then

$$(u_z, \vartheta) = (u_0, \vartheta_0) \sin(\pi z/d) \sin(kx). \tag{12.20}$$

Since the instability sets in as an exponential growth without oscillation, at marginal stability ϑ is determined by (12.10), which fixes the relationship between u_0 and ϑ_0 as

$$\vartheta_0 = \frac{\Delta T}{\alpha d}\left[(\pi/d)^2 + k^2\right]^{-1}u_0.$$

Moreover, we can use continuity to find u_x, and hence $\boldsymbol{\omega}$, from u_z. We are now in a position to estimate both integrals in (12.19). Integrating over a single cell, our buoyancy and dissipation integrals are

$$g\beta \int \vartheta u_z dV = \frac{g\beta\Delta T}{4\alpha d}\left[(\pi/d)^2 + k^2\right]^{-1}\frac{\pi du_0^2}{k}, \tag{12.21}$$

$$\nu \int \boldsymbol{\omega}^2 dV = \frac{\nu}{4}\left[(\pi/d)^2 + k^2\right]^2\frac{\pi du_0^2}{k^3}. \tag{12.22}$$

Equating these integrals, we find that the *marginal stability curve* is

$$(\mathrm{Ra})_{\mathrm{marginal}} = \frac{\left(\pi^2 + a^2\right)^3}{a^2}, \quad a = kd, \tag{12.23}$$

which brings us back to (12.15), but with $n = 1$. If we now choose a to maximize the ratio of the rate of working of the buoyancy force to the rate of dissipation, then we arrive back at (12.16). In short, the eigenvalue problem (12.14) is equivalent to the energy balance (12.19). We shall find this useful in §12.1.5 when estimating $(\mathrm{Ra})_{\mathrm{crit}}$ for the more difficult problem of no-slip boundaries. Finally, we note that, because ω_z is governed by a diffusion equation with no source term, the boundary conditions ensure that $\omega_z = 0$.

12.1.4 More Realistic Cases: No-slip Boundaries

We now consider the more important cases of rigid-rigid and rigid-free boundaries. In such cases, analytical progress is awkward (though see the energy argument in §12.1.5) and (12.14) is most readily solved numerically. The critical values for Ra and $a = kd$ are given in Table 12.1 for all three combinations of boundary conditions.

For no-slip boundaries top and bottom, it turns out that $(\mathrm{Ra})_{\mathrm{crit}} = 1708$ at $a_{\mathrm{crit}} = 3.117$, and the corresponding marginal stability curve is shown in

Table 12.1 Critical values of Ra and $a = kd$ in Rayleigh–Bénard convection

Boundary conditions	Free-free	Rigid-free	Rigid-rigid
$(\mathrm{Ra})_{\mathrm{crit}}$	657.5	1,101	1,708
a_{crit}	2.221	2.682	3.117

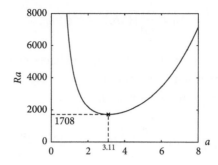

Figure 12.3 The marginal stability curve of Ra versus $a = kd$ for convection between two rigid boundaries. Above the curve, the equilibrium is unstable to modes of a given k.

Figure 12.3. However, we have seen these numbers before. The critical threshold for the appearance of Taylor vortices in Couette flow is $(Ta)_{crit} = 1695$ at $kd = 3.127$, and when we use the approximate eigenvalue equation (11.31), these change to $(Ta)_{crit} = 1708$ at $kd = 3.117$. Of course, this similarity is no accident and could have been anticipated by comparing (12.14) with (11.31).

The value of $(Ra)_{crit}$ for a free upper surface, $(Ra)_{crit} = 1101$, is somewhat lower than that for two rigid boundaries, and the value of $(Ra)_{crit} = 657.5$ for two free surfaces is lower still. This ranking of the values of $(Ra)_{crit}$ is because a no-slip boundary condition induces more shear, and hence more dissipation, than a slip boundary condition.

Note that our analysis has determined the critical value of Ra, but not the form of $f(x, y)$ in (12.12). Indeed, the only restrictions on f are that it must satisfy Helmholtz's equation and the lateral boundary conditions. Note also that, when Ra is slightly larger than $(Ra)_{crit}$, the initial instability quickly saturates to give way to steady convection, and this occurs through nonlinear interactions. Consequently, linear theory has little to say about the pattern of steady convection cells, other than to set the dominant length scale through the critical value of k. In practice, a variety of patterns are observed, as discussed in §12.1.6. Finally, we note that, if Ra is significantly larger than $(Ra)_{crit}$, then there is a point at which the Bénard cells themselves become unstable, just as Taylor vortices eventually become unstable when Ta is large enough.

12.1.5 An Approximate Energy Analysis for No-slip Boundaries Top and Bottom

We shall now show that the energy argument of §12.1.3, which rests on the fact that

$$g\beta \int \vartheta u_z dV = v \int \omega^2 dV \tag{12.24}$$

at onset, yields a surprisingly accurate estimate of $(\text{Ra})_{\text{crit}}$ for the important case of rigid boundaries top and bottom. Suppose that the instability takes the form of two-dimensional rolls in the x-z plane. Then the vertical velocity field

$$u_z = \frac{u_0}{d^4}z^2(d-z)^2 \sin(kx), \tag{12.25}$$

where k is a horizontal wavenumber and u_0 a constant, is consistent with the boundary conditions

$$\vartheta = u_z = \partial u_z / \partial z = 0, \quad \text{at} \quad z = 0, \quad d. \tag{12.26}$$

So let us adopt (12.25) as a kinematically admissible guess and see where it takes us.

The first step is to note that, for marginal stability, combining (12.25) and (12.26) with (12.10) yields, after a certain amount of algebra,

$$\vartheta = \frac{\Gamma \sin kx}{a^4}\left[a^4\eta^2(1-\eta)^2 + 2a^2(1-6\eta+6\eta^2) + 24\right]$$
$$-\frac{2(a^2+12)\Gamma \sin kx}{a^4 \cosh(a/2)}\cosh(a\eta - a/2).$$

where

$$a = kd, \quad \eta = z/d, \quad \Gamma = \Delta T u_0 d/\alpha a^2.$$

Next, u_x can be determined from u_z using continuity and the associated vorticity field is readily shown to be

$$\omega_y = \frac{u_0 \cos kx}{da}\left[2(1-6\eta+6\eta^2) - a^2\eta^2(1-\eta)^2\right].$$

Finally, we integrate over a single convection cell to estimate the dissipation and buoyancy integrals in (12.24). This yields

$$v \int \omega^2 dV = \frac{2\pi v u_0^2}{a^3}\left[\frac{1}{5} + \frac{a^2}{105} + \frac{a^4/4}{630}\right], \tag{12.27}$$

for the dissipation integral. The buoyancy integral is messier, but using the definite integral

$$
\int_{0}^{1} (1 - p^2)^2 \cosh\left(\frac{ap}{2}\right) dp = \frac{8\left[(a/2)^2 + 3\right]}{(a/2)^5} \sinh(a/2) - \frac{24}{(a/2)^4} \cosh(a/2),
$$

we (eventually) obtain the somewhat cumbersome expression

$$
g\beta \int \vartheta u_z dV = \frac{\pi g \beta \Delta T d^3 u_0^2}{2\alpha a^{11}} \left[\frac{a^8}{630} - \frac{2a^6}{105} + \frac{4a^4}{5} + 48a^2 + 576 \right.
$$
$$
\left. -(a^2 + 12)^2 \frac{\tanh(a/2)}{a/8} \right]. \tag{12.28}
$$

It follows from (12.24) that, at marginal stability, we have the following approximate relationship between Ra and a:

$$
\left[\frac{4}{5} + \frac{4a^2}{105} + \frac{a^4}{630} \right] = \frac{\text{Ra}}{a^8} \left[\frac{a^8}{630} - \frac{2a^6}{105} + \frac{4a^4}{5} + 48a^2 + 576 \right.
$$
$$
\left. -(a^2 + 12)^2 \frac{\tanh(a/2)}{a/8} \right]. \tag{12.29}
$$

The corresponding marginal stability curve is shown in Figure 12.3.

Of course, (12.29) is only an estimate, based on the guess (12.25). However, it turns out to be astonishingly accurate, with (12.29) predicting a critical value of $(\text{Ra})_{\text{crit}} = 1708.1$ corresponding to $a_{\text{crit}} = 3.113$. The true values are $(\text{Ra})_{\text{crit}} = 1707.8$ and $a_{\text{crit}} = 3.117$, as shown in Table 12.1. Thus, the error in $(\text{Ra})_{\text{crit}}$ is less than one part in 5000.

12.1.6 Nonlinear Saturation

So far, we have said nothing about the mechanisms by which the nonlinear interactions cause the linear instability to saturate. Nor have we discussed the role of nonlinearity in determining the form of the steady convection pattern that emerges from the instability. Let us start with saturation of the linear instability.

The saturation mechanism for convection rolls between parallel plates can be understood as follows. Suppose that Ra > $(\text{Ra})_{\text{crit}}$ and consider an initial value problem in which the instability first starts to grow. Once convection begins, heat is transported predominantly by convection within the interior, but by conduction close to the two plates. Convection is more efficient and so the horizontally averaged temperature profile progressively evolves, with $T(z)$ becoming increasingly uniform within the interior, and with most of the temperature drop, ΔT, occurring within thin thermal boundary layers at the top and bottom plates. Likewise, the vertical gradients in horizontal velocity become increasingly concentrated within thin mechanical boundary layers adjacent to the plates. Thus, as the instability grows, the buoyant source of convection is increasingly diminished within the interior, while the viscous dissipation of energy is enhanced within thin mechanical boundary layers. The end result is that the flow eventually stabilizes, provided, of course, that Ra is not much greater than $(\text{Ra})_{\text{crit}}$.

It turns out that two-dimensional rolls, square cells, and hexagonal cells are all permissible forms of steady convection, although two-dimensional rolls are the most common pattern. Moreover, the values of $\text{Ra}/(\text{Ra})_{\text{crit}}$ and ν/α, as well as the boundary conditions, are all important in determining the observed pattern and its relative robustness. For rolls, the bifurcation to steady convection is *supercritical*, which means that, for Ra slightly larger than $(\text{Ra})_{\text{crit}}$, an initial exponential growth will quickly saturate to yield steady convection. However, the emergence of hexagons, which occurs in thin layers where Marangoni effects are dominant, is typically *subcritical*, which means that a finite-amplitude disturbance can gain traction and trigger convection for $\text{Ra} < (\text{Ra})_{\text{crit}}$. Subcritical hexagons can also appear in a non-Boussinesq fluid subject to a large ΔT. Here hysteresis effects are typically important, with the observed pattern being sensitive to the history of variations in ΔT.

All of this assumes Ra is close to $(\text{Ra})_{\text{crit}}$, and the behaviour is more complex when Ra is significantly larger than $(\text{Ra})_{\text{crit}}$. Just as laminar Taylor vortices become unstable when Ta is large enough, so too do convection rolls at large Ra. When Ra is sufficiently supercritical, a secondary instability leads to three-dimensional rolls. This requires only a modest increase in Ra when ν/α is small, but a more substantial increase for large ν/α. Further bifurcations, perhaps only three or four in total, eventually lead to turbulent convection, which is not inconsistent with the Ruelle–Takens modification to Landau's classical theory of transition to turbulence. (See Drazin, 1992, for a discussion of the Ruelle–Takens route to turbulence.) There is a rich literature on the development of bifurcations and turbulence in Rayleigh–Bénard convection, although we shall not pause to discuss it here. However, useful reviews are Bodenschatz et al. (2000) and Manneville (2006).

12.2 Rotating Rayleigh–Bénard Convection I: Non-Oscillatory Instability

We now turn to the main topic of this chapter, which is rotating convection. Since the Taylor–Proudman theorem tells us that a rapidly rotating fluid tends to suppress three-dimensional motion, it should come as no surprise that rotation acts to inhibit the instability of a fluid heated from below. Moreover, a rotating fluid can sustain inertial waves, and so we might anticipate that oscillations at the onset of convection are a possibility. We shall see that this is indeed the case, though *only* when the Ekman and Prandtl numbers,

$$\mathrm{Ek} = \nu/\Omega d^2, \quad \mathrm{Pr} = \nu/\alpha, \tag{12.30}$$

are small, so that the absolute vortex lines are almost frozen into the fluid and the wave mechanism of §6.3.2 can be realized. In any event, we start with the simpler problem of exponential growth without oscillation, leaving oscillatory growth for §12.3. To keep the algebra simple, we shall focus primarily on the case where the top and bottom boundaries are stress free, so that the boundary conditions are

$$\vartheta = u_z = \partial^2 u_z/\partial z^2 = \partial \omega_z/\partial z = 0\,, \quad z = 0, d. \tag{12.31}$$

12.2.1 A Linear Stability Analysis

Once again, we consider a layer of fluid of depth d which is heated from below and subject to a uniform temperature difference, ΔT. The only change is that we now impose a background rotation on the fluid (and the boundaries), of the form $\mathbf{\Omega} = \Omega \hat{\mathbf{e}}_z$. Our governing equations in a rotating frame are then

$$\frac{D\mathbf{u}}{Dt} = 2\mathbf{u} \times \mathbf{\Omega} - \nabla\left(p/\bar{\rho}\right) + \nu\nabla^2\mathbf{u} - \beta(T - T_{\mathrm{ref}})\mathbf{g}, \tag{12.32}$$

$$\frac{DT}{Dt} = \alpha\nabla^2 T, \tag{12.33}$$

which generalize (12.2). As in §12.1.2, we linearize these about the equilibrium state $\mathbf{u}_0 = 0$ and $T = T_0(z)$ to give

$$\left(\frac{\partial}{\partial t} - \nu\nabla^2\right)\mathbf{u} = 2\mathbf{u} \times \mathbf{\Omega} - \nabla\left(\delta p/\bar{\rho}\right) - \beta\vartheta\mathbf{g}, \tag{12.34}$$

$$\frac{\partial\vartheta}{\partial t} + u_z\frac{dT_0}{dz} = \alpha\nabla^2\vartheta, \tag{12.35}$$

where $\vartheta = T - T_0(z)$. The vorticity equation corresponding to (12.34) is evidently

$$\left[\frac{\partial}{\partial t} - \nu\nabla^2\right]\boldsymbol{\omega} = 2\boldsymbol{\Omega} \cdot \nabla\mathbf{u} + g\beta(\nabla\vartheta) \times \hat{\mathbf{e}}_z, \qquad (12.36)$$

and taking the curl a second time yields

$$\left[\frac{\partial}{\partial t} - \nu\nabla^2\right]\nabla^2\mathbf{u} = g\beta\left[\hat{\mathbf{e}}_z\nabla^2\vartheta - \frac{\partial}{\partial z}\nabla\vartheta\right] - 2\boldsymbol{\Omega} \cdot \nabla\boldsymbol{\omega}. \qquad (12.37)$$

Finally, we note that the z components of (12.36) and (12.37) are

$$\left[\frac{\partial}{\partial t} - \nu\nabla^2\right]\omega_z = 2\Omega\frac{\partial u_z}{\partial z}, \qquad (12.38)$$

$$\left[\frac{\partial}{\partial t} - \nu\nabla^2\right]\nabla^2 u_z = g\beta\,\nabla_\perp^2\vartheta - 2\Omega\frac{\partial\omega_z}{\partial z}, \qquad (12.39)$$

which constitute our governing equations, along with (12.35), which we now rewrite as

$$\left[\frac{\partial}{\partial t} - \alpha\nabla^2\right]\vartheta = \frac{\Delta T}{d}u_z. \qquad (12.40)$$

We might compare (12.38)→(12.40) with the governing equations for the non-rotating case, (12.6)→(12.8). We see immediately that, in stark contrast to non-rotating convection, the introduction of rotation demands that ω_z (and hence the helicity, $h = \mathbf{u} \cdot \boldsymbol{\omega}$) is non-zero. Thus, as fluid particles rise and fall, they follow helical trajectories. Simple, two-dimensional rolls are therefore not a permissible unstable mode.

Now, experiments show that, for most liquids and gases, the most unstable disturbance grows exponentially without oscillation. In fact, we shall see shortly that, for stress-free boundaries top and bottom, oscillatory growth is the preferred mode *only* when Pr < 0.677. Thus, oscillatory growth is largely, but not exclusively, limited to liquid metals. So, for the time being, let us assume that, at marginal stability, time derivatives are zero. Our governing equations then reduce to

$$\nu \nabla^2 \omega_z = -2\Omega \frac{\partial u_z}{\partial z}, \tag{12.41}$$

$$\nu \left(\nabla^2\right)^2 u_z = 2\Omega \frac{\partial \omega_z}{\partial z} - g\beta \, \nabla_\perp^2 \vartheta, \tag{12.42}$$

$$\alpha \nabla^2 \vartheta = -\frac{\Delta T}{d} u_z. \tag{12.43}$$

As before, we look for separable solutions of the form

$$\vartheta = \hat{\vartheta}(z) f(x, y), \quad u_z = \hat{u}_z(z) f(x, y), \quad \omega_z = \hat{\omega}_z(z) f(x, y), \tag{12.44}$$

where f satisfies the Helmholtz equation,

$$\nabla_\perp^2 f + k^2 f = 0. \tag{12.45}$$

Assuming solutions of this type, the marginally unstable modes are governed by

$$\alpha \left(\frac{d^2}{dz^2} - k^2\right) \hat{\vartheta} = -\frac{\Delta T}{d} \hat{u}_z, \tag{12.46}$$

$$\nu \left(\frac{d^2}{dz^2} - k^2\right) \hat{\omega}_z = -2\Omega \frac{d\hat{u}_z}{dz}, \tag{12.47}$$

$$\nu \left(\frac{d^2}{dz^2} - k^2\right)^2 \hat{u}_z = 2\Omega \frac{d\hat{\omega}_z}{dz} + g\beta \, k^2 \hat{\vartheta}. \tag{12.48}$$

These combine to yield the eigenvalue problem

$$d^4 \left(\frac{d^2}{dz^2} - k^2\right)^3 \hat{u}_z + \left(\frac{2\Omega d^2}{\nu}\right)^2 \frac{d^2 \hat{u}_z}{dz^2} + \frac{g\beta \Delta T d^3}{\nu \alpha} k^2 \hat{u}_z = 0, \tag{12.49}$$

which is a generalization of (12.14). Following Chandrasekhar (1961), it has become conventional to regard the dimensionless pre-factor in front of the second term in (12.49) as a form of Taylor number, rather than the inverse square of the Ekman number. So let us introduce the Taylor-like number

$$\hat{\text{Ta}} = 4\Omega^2 d^4 / \nu^2, \tag{12.50}$$

and rewrite (12.49) as

$$\left(d^2 \frac{d^2}{dz^2} - a^2\right)^3 \hat{u}_z + \hat{T}a \, d^2 \frac{d^2 \hat{u}_z}{dz^2} + Ra \, a^2 \hat{u}_z = 0, \tag{12.51}$$

where $a = kd$, as usual.

So far, we have assumed exponential growth without oscillation at onset, but made no assumption about the nature of the boundary conditions at the top and bottom of the fluid layer. We now restrict ourselves to the simple (if somewhat artificial) case of stress-free boundaries top and bottom. We are then free to take

$$\left(\hat{u}_z, \hat{\vartheta}\right) = (u_0, \vartheta_0) \sin(\pi z/d), \tag{12.52}$$

which satisfies and the boundary conditions (12.31) at $z = 0, d$. Our marginal stability curve is then

$$(Ra)_{marginal} = \frac{\left(\pi^2 + a^2\right)^3 + \pi^2 \hat{T}a}{a^2}, \tag{12.53}$$

which is a generalization of (12.23). The critical wavenumber that minimizes $(Ra)_{marginal}$, and hence sets the critical value of Ra, is governed by

$$2\left(a^2_{crit}/\pi^2\right)^3 + 3\left(a^2_{crit}/\pi^2\right)^2 = 1 + \hat{T}a/\pi^4, \tag{12.54}$$

from which we have

$$a^2_{crit} = \frac{\pi^2}{2}\left(1 + \frac{\hat{T}a}{\left(\pi^2 + a^2_{crit}\right)^2}\right), \quad (Ra)_{crit} = 3\left(\pi^2 + a^2_{crit}\right)^2. \tag{12.55}$$

The cubic equation (12.54) is readily solved and the values of a_{crit} and $(Ra)_{crit}$, as functions of $\hat{T}a$, are tabulated in Table 12.2. It can be seen that $(Ra)_{crit}$ rises monotonically with $\hat{T}a$, so that rotation is indeed stabilizing.

Let us now consider the case of strong rotation. For $\hat{T}a \gg \pi^4$, (12.55) gives us

$$a_{crit} = \left(\pi^2 \hat{T}a/2\right)^{1/6}, \quad (Ra)_{crit} = 3\left(\pi^2 \hat{T}a/2\right)^{2/3}, \tag{12.56}$$

Table 12.2 Critical values of Ra and $a = kd$ in rotating, Rayleigh–Bénard convection

$\hat{T}a = 4\Omega^2 d^4/\nu^2$	Stress-free boundaries		No-slip boundaries	
	a_{crit}	$(Ra)_{crit}$	a_{crit}	$(Ra)_{crit}$
10	2.27	6.77×10^2	3.12	1.71×10^3
100	2.59	8.26×10^2	3.15	1.76×10^3
500	3.28	1.28×10^3	3.30	1.94×10^3
1,000	3.71	1.68×10^3	3.50	2.15×10^3
2,000	4.22	2.30×10^3	3.75	2.53×10^3
5,000	5.01	3.67×10^3	4.25	3.47×10^3
10,000	5.70	5.38×10^3	4.80	4.71×10^3
30,000	6.96	1.02×10^4	5.80	8.33×10^3
100,000	8.63	2.13×10^4	7.20	1.68×10^4
1,000,000	12.86	9.22×10^4	10.80	7.11×10^4

and so strong rotation leads to $kd \gg 1$, *i.e.* a horizontal wavelength significantly less than d. It turns out that the same $(Ra)_{crit} \sim \hat{T}a^{2/3}$ law holds for no-slip boundaries top and bottom, and also for a combination of no-slip at the lower boundary and stress free at the upper surface. It is necessary only to change the pre-factor in the scaling law. (See, for example, Chandrasekhar, 1961.) This law is sometimes rewritten in the form

$$\frac{g\beta d(\Delta T)_{crit}}{\alpha\Omega} \sim \left(\frac{\Omega d^2}{\nu}\right)^{1/3}, \tag{12.57}$$

and it is tempting to interpret (12.57) as saying that a low viscosity is stabilizing, in the sense that $\Delta T_{crit} \to \infty$ as $\nu \to 0$ for a given α. However, some caution is required in drawing such a conclusion, since we could equally write the law as

$$\frac{g\beta(\Delta T)_{crit}}{\Omega^2 d} \sim Pr^{-1}\left(\frac{\Omega d^2}{\nu}\right)^{-2/3}, \tag{12.58}$$

which gives $\Delta T_{crit} \to 0$ as $\nu \to 0$ for fixed Pr.

12.2.2 The Origin of the Criterion $(Ra)_{crit} \sim \hat{T}a^{2/3}$

The simplest way to understand the origins of the $(Ra)_{crit} \sim \hat{T}a^{2/3}$ law for stress-free boundaries is to note that horizontal gradients dominate over vertical gradients for large $\hat{\text{T}}$a. Thus, $(12.46) \rightarrow (12.48)$ simplify to

$$\alpha k^2 \hat{\vartheta} = \frac{\Delta T}{d} \hat{u}_z, \tag{12.59}$$

$$\nu k^2 \hat{\omega}_z = 2\Omega \frac{d\hat{u}_z}{dz}, \tag{12.60}$$

$$\nu k^4 \hat{u}_z = 2\Omega \frac{d\hat{\omega}_z}{dz} + g\beta k^2 \hat{\vartheta} = \frac{(2\Omega)^2}{\nu k^2} \frac{d^2 \hat{u}_z}{dz^2} + \frac{g\beta \Delta T}{\alpha d} \hat{u}_z. \tag{12.61}$$

Note that the structure of these equations is such that specifying $u_z = 0$ on $z = 0$, d ensures that all other boundary conditions in (12.31) are automatically satisfied. Note also that (12.61) can be written as

$$\hat{\text{T}}ad^2 \frac{d^2 \hat{u}_z}{dz^2} + \left(\text{Ra} - a^4 \right) a^2 \hat{u}_z = 0, \tag{12.62}$$

so we have moved from a sixth-order to a second-order equation, consistent with the reduction in the number of independent boundary conditions.

For stress-free boundaries top and bottom, we may take $\hat{u}_z \sim \sin (\pi z / d)$, and (12.62) then yields the marginal stability curve

$$(\text{Ra})_{\text{marginal}} = a^4 + \frac{\pi^2 \hat{\text{T}}a}{a^2}, \tag{12.63}$$

consistent with (12.53) at large a. This, in turn, yields a critical wavenumber, and a critical value of Ra, of $a_{\text{crit}} \sim \hat{\text{T}}a^{1/6}$ and $(\text{Ra})_{\text{crit}} \sim \hat{\text{T}}a^{2/3}$, which brings us back to (12.56).

Note that some authors use $\text{Ek} = \nu / \Omega d^2$ in preference to $\hat{\text{T}}a$, and also $\lambda = 2\pi / k$ instead of k. The scaling laws (12.56) then become

$$\lambda_{\text{crit}} / d \sim \text{Ek}^{1/3}, \quad (\text{Ra})_{\text{crit}} \sim \text{Ek}^{-4/3}. \tag{12.64}$$

The first of these, $\lambda_{\text{crit}} / d \sim \text{Ek}^{1/3}$, follows more or less directly from the viscous-Coriolis force balance (12.60), rewritten as

$$\nu \nabla_\perp^2 \omega_z = -2\Omega \frac{\partial u_z}{\partial z}, \tag{12.65}$$

and from the assumption that the helicity is strong, so that $u_\perp \sim u_z$. By way of contrast, the scaling $(\text{Ra})_{\text{crit}} \sim \text{Ek}^{-4/3}$ has its roots in (12.61), rewritten as

$$\nu\left(\nabla_\perp^2\right)^2 u_z = 2\Omega\frac{\partial \omega_z}{\partial z} - g\beta\,\nabla_\perp^2 \vartheta. \tag{12.66}$$

Equating the viscous and buoyancy terms, and using $\alpha\nabla_\perp^2\vartheta = -(\Delta T/d)u_z$ to estimate ϑ, leads directly to $(\text{Ra})_{\text{crit}} \sim a_{\text{crit}}^4$. This combines with $a_{\text{crit}} \sim \text{Ek}^{-1/3}$ to give $(\text{Ra})_{\text{crit}} \sim \text{Ek}^{-4/3}$.

When the boundary conditions are no-slip, rather than stress free, (12.61) must break down at $z = 0, d$, because $u_z = \vartheta = 0$ there, yet $\partial\omega_z/\partial z$ is non-zero. The problem is that Ekman layers form, and so we are not free to ignore vertical gradients close to the boundaries. Nevertheless, provided that $\text{Ek} \ll 1$, the asymptotic laws $\lambda_{\text{crit}}/d \sim \text{Ek}^{1/3}$ and $(\text{Ra})_{\text{crit}} \sim \text{Ek}^{-4/3}$ continue to hold (Chandrasekhar, 1961). Presumably, this is because the force balances (12.65) and (12.66) remain valid throughout the bulk of the fluid at low Ek.

In fact, leaving aside the entire question of stability, equations (12.65) and (12.66) govern *steady*, viscous convection at *low Rossby number* in which axial gradients are much weaker than horizontal gradients. It follows that the transverse scale of the columnar convection cells in such a flow, say δ, must also follow the scaling law $\delta/d \sim \text{Ek}^{1/3}$.

12.2.3 An Interpretation of the Stability Criterion in Terms of Energy

We shall now deduce the marginal stability curve (12.53) using the energy equation

$$g\beta\int \vartheta u_z dV = \nu\int \boldsymbol{\omega}^2 dV. \tag{12.67}$$

Our motivation is twofold. On the one hand, this second derivation will show explicitly how the introduction of helicity increases the viscous dissipation and so helps stabilize the equilibrium state. On the other hand, the energy method requires a full knowledge of the vorticity field (and hence the velocity field), rather than just u_z and ω_z. So an energy analysis will help expose the underlying structure of the unstable mode. As before, we take

$$(u_z, \vartheta) = (u_0, \vartheta_0) \sin (\pi z/d) f(x, y), \quad \omega_z = \omega_0 \cos (\pi z/d) f(x, y), \qquad (12.68)$$

where

$$\nabla_\perp^2 f + k^2 f = 0, \qquad (12.69)$$

while (12.46) and (12.47) require

$$(\pi^2 + a^2) \omega_0 = \frac{2\pi\Omega d}{\nu} u_0, \qquad (12.70)$$

$$(\pi^2 + a^2) \vartheta_0 = \frac{\Delta T d}{\alpha} u_0. \qquad (12.71)$$

Our primary task is to determine u_x and u_y as a function of $f(x, y)$, so that we may evaluate ω^2 in the dissipation integral. To that end, we write the horizontal components of $\mathbf{u}$ as

$$\mathbf{u}_\perp = \cos (\pi z/d) \, \mathbf{h}(x, y), \qquad (12.72)$$

where the divergence and curl of $\mathbf{h}$ are evidently

$$\nabla \cdot \mathbf{h} = - (\pi u_0/d) f, \quad \nabla \times \mathbf{h} = \omega_0 f \hat{\mathbf{e}}_z. \qquad (12.73)$$

Since the divergence and curl of $\mathbf{h}$ are uniquely determined by f, $\mathbf{h}$ itself is also uniquely determined by f. It is readily confirmed, by direct substitution, that $\mathbf{h}$, and hence $\mathbf{u}_\perp$, are given by

$$\mathbf{u}_\perp = \cos (\pi z/d) \, \mathbf{h} = \cos (\pi z/d) \left[\frac{\pi u_0}{ka} \nabla f - \frac{\omega_0}{k^2} \hat{\mathbf{e}}_z \times \nabla f \right]. \qquad (12.74)$$

The horizontal components of the vorticity field are then

$$\boldsymbol{\omega}_\perp = - \sin (\pi z/d) \left[\frac{\pi \omega_0}{ka} \nabla f + \left(1 + \frac{\pi^2}{a^2} \right) u_0 \hat{\mathbf{e}}_z \times \nabla f \right], \qquad (12.75)$$

from which

$$\nu \omega_\perp^2 = \nu \left[\left(1 + \pi^2/a^2 \right)^2 u_0^2 + \frac{\pi^2 \omega_0^2}{k^2 a^2} \right] \sin^2 (\pi z/d) \left(\nabla f \right)^2. \qquad (12.76)$$

We shall return to (12.74) and (12.75) shortly, when we discuss the spatial structure of the unstable mode. In the meantime, we focus on the integrals in (12.67), which we take to be integrals over a single convection cell. In addition to (12.76), we require

$$\nu\omega_z^2 = \nu\omega_0^2\cos^2(\pi z/d)f^2,$$
$$g\beta\vartheta u_z = g\beta\vartheta_0 u_0\sin^2(\pi z/d)f^2.$$

We now substituting for $(\nabla f)^2$ in (12.76) using $(\nabla f)^2 = k^2 f^2 + \nabla\cdot(f\nabla f)$, noting that symmetry requires the divergence to integrate to zero. Our energy balance now yields

$$g\beta\vartheta_0 u_0 = \nu\omega_0^2 + \nu k^2\left[\left(1 + \pi^2/a^2\right)^2 u_0^2 + \frac{\pi^2\omega_0^2}{k^2 a^2}\right], \tag{12.77}$$

or equivalently,

$$g\beta\vartheta_0 u_0 = \nu k^2\left(1 + \pi^2/a^2\right)^2 u_0^2 + \nu\left(1 + \pi^2/a^2\right)\omega_0^2. \tag{12.78}$$

Note that the axial vorticity, which is absent in the non-rotating case, increases the viscous dissipation. The final step is to substitute for ϑ_0 and ω_0 using (12.70) and (12.71). This yields

$$\frac{g\beta\Delta T d}{\alpha(\pi^2 + a^2)} = \frac{\nu k^2}{a^4}\left(\pi^2 + a^2\right)^2 + \frac{4\pi^2\Omega^2 d^2}{\nu a^2(\pi^2 + a^2)}, \tag{12.79}$$

which returns us to (12.53),

$$(\mathrm{Ra})_{\mathrm{marginal}} = \frac{\left(\pi^2 + a^2\right)^3 + \pi^2\hat{T}a}{a^2}. \tag{12.80}$$

The important point, however, is (12.77). Evidently, the introduction of ω_z, i.e. the introduction of helicity, has increased the viscous dissipation, and indeed $\nu\omega_z^2$ dominates the dissipation at large $\hat{T}a$. Physically, the helical path of a rising fluid particle in a rotating flow is longer than its non-rotating counterpart, and so a fluid particle in rotating convection experiences more viscous dissipation for a given release of potential energy.

12.2.4 The Spatial Structure of the Unstable Mode

We now consider the spatial structure of the unstable mode with stress-free boundaries. Let us start by calculating the helicity. From (12.68) we have

$$u_z \omega_z = u_0 \omega_0 f^2 \sin\left(\frac{\pi z}{d}\right) \cos\left(\frac{\pi z}{d}\right) = \frac{\pi \Omega d u_0^2 f^2}{\nu \left(\pi^2 + a_{\text{crit}}^2\right)} \sin\left(\frac{2\pi z}{d}\right), \qquad (12.81)$$

while (12.74) and (12.75) yield

$$\mathbf{u}_\perp \cdot \boldsymbol{\omega}_\perp = \frac{u_0 \omega_0 (\nabla f)^2}{k^2} \sin\left(\frac{\pi z}{d}\right) \cos\left(\frac{\pi z}{d}\right) = \frac{\pi \Omega d u_0^2 (\nabla f)^2}{\nu k^2 \left(\pi^2 + a_{\text{crit}}^2\right)} \sin\left(\frac{2\pi z}{d}\right).$$
$$(12.82)$$

We conclude that, when $\boldsymbol{\Omega}$ is antiparallel to $\mathbf{g}$, the helicity is negative (left-handed spirals) in the upper half of the fluid layer, and positive (right-handed spirals) in the lower half. Moreover, if $\boldsymbol{\Omega}$ is reversed, then so is the helicity distribution. Finally, when horizontally averaged, $u_z \omega_z$ and $\mathbf{u}_\perp \cdot \boldsymbol{\omega}_\perp$ make identical contributions to the net helicity.

Turning next to kinetic energy, (12.74) yields

$$\mathbf{u}_\perp^2 = \left[\frac{\pi^2 u_0^2}{k^2 a_{\text{crit}}^2} + \frac{\omega_0^2}{k^4}\right] \cos^2 (\pi z/d)\, (\nabla f)^2.$$

or equivalently, using (12.70),

$$\mathbf{u}_\perp^2 = \frac{\pi^2 u_0^2}{k^2 a_{\text{crit}}^2} \left[1 + \frac{\hat{\mathrm{T}}\mathrm{a}}{\left(\pi^2 + a_{\text{crit}}^2\right)^2}\right] \cos^2 (\pi z/d)\, (\nabla f)^2.$$

This combines with (12.55) to give the remarkably simple result

$$\mathbf{u}_\perp^2 = \frac{2 u_0^2}{k^2} \cos^2 (\pi z/d)\, (\nabla f)^2, \qquad (12.83)$$

which is independent of the rotation rate. When compared with

$$u_z^2 = u_0^2 \sin^2 (\pi z/d) f^2, \qquad (12.84)$$

we see that, when horizontally averaged, $\mathbf{u}_\perp^2$ and u_z^2 make similar contributions to the net kinetic energy.

Let us now consider the convection pattern at onset. As with non-rotating Rayleigh–Bénard convection, rolls, square cells, and hexagonal cells are all permissible patterns. Let us start with rolls, in which f is a function of x only, say $f = \sin(kx)$. Then (12.74) simplifies to

$$\mathbf{u}_\perp = \left[\frac{\pi u_0}{a_{\text{crit}}} \hat{\mathbf{e}}_x - \frac{\omega_0}{k} \hat{\mathbf{e}}_y \right] \cos(\pi z/d) \cos(kx), \tag{12.85}$$

which combines with (12.70) to give

$$\frac{u_y}{u_x} = -\frac{\omega_0 d}{\pi u_0} = -\frac{\sqrt{\hat{\text{T}}\text{a}}}{\pi^2 + a_{\text{crit}}^2} = \text{constant}. \tag{12.86}$$

Since this ratio is independent of $\mathbf{x}$, the streamlines are confined to planes, just like in non-rotating convection, only these planes are no longer normal to the axis of the rolls. If we adopt a second coordinate system aligned with the planes containing the streamlines, then the effective wavelength, λ_{eff}, is increased relative to λ_x (the wavelength in the x direction) by a factor of

$$\lambda_{\text{eff}}/\lambda_x = \sqrt{1 + u_y^2/u_x^2} = \sqrt{1 + \hat{\text{T}}\text{a}/\left(\pi^2 + a_{\text{crit}}^2\right)^2} = \sqrt{2} a_{\text{crit}}/\pi,$$

where we have used (12.55). In terms of wavenumber, this becomes

$$k_{\text{eff}}/k = \pi \Big/ \left(\sqrt{2} a_{\text{crit}}\right) = \pi \Big/ \left(\sqrt{2} kd\right),$$

and we conclude that $k_{\text{eff}} d = \pi/\sqrt{2}$. Remarkably, this is independent of the rotation rate and equal to the critical wavenumber for non-rotating convection (Veronis, 1959). We conclude that, for the case of convective rolls, the primary effect of rotation is to twist the plane containing the streamlines away from the x-z plane. The wavelength measured in this rotated plane is then independent of the rotation rate. This is illustrated in Figure 12.4.

The case of square and hexagonal cells is discussed in detail in Veronis (1959) and Chandrasekhar (1961). The structure of the flow for square cells is shown in Figure 12.5. Note that the fluid paths are left-handed spirals in the upper half of the cell, and right-handed spirals in the lower half, consistent

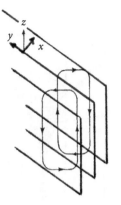

Figure 12.4 Convection pattern for rotating, Rayleigh–Bénard convection when the convection cells are rolls. Adapted from Veronis (1959).

(a)   (b)

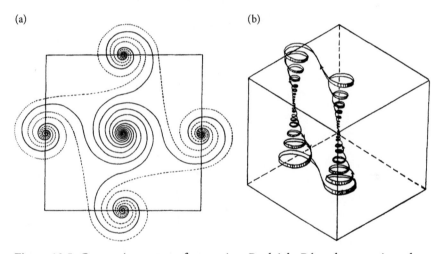

Figure 12.5 Convection pattern for rotating, Rayleigh–Bénard convection when the convection cells are square. (a) Plan view. (b) Perspective view. Reproduced from Veronis (1959).

with (12.81) and (12.82). The corresponding case of hexagonal cells is shown in Figure 12.6.

We shall not pause to discuss the nonlinear development of these various instabilities, but rather refer the reader to the review by Bodenschatz et al. (2000).

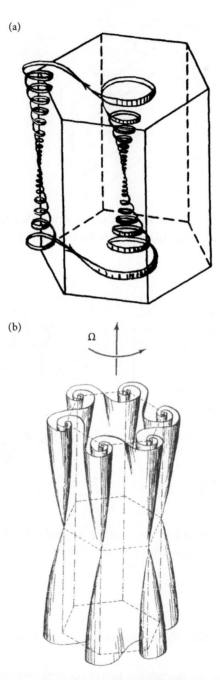

(a)

(b)

Ω

Figure 12.6 Convection pattern for rotating, Rayleigh–Bénard convection when the convection cells are hexagonal. (a) The path of a fluid particle. (b) Streamline surfaces. Reproduced from Veronis (1959).

12.3 Rotating Rayleigh–Bénard Convection II: Oscillatory Instability

We now turn to the question of oscillatory instability, where a marginally unstable mode oscillates as it grows. For simplicity, we focus on the case of stress-free boundaries top and bottom. We shall see that the Prandtl number, $Pr = \nu/\alpha$, now plays a key role.

12.3.1 A Linear Stability Analysis for Slip Boundaries

Suppose we look for modes of the form $(\vartheta, u_z, \omega_z) \sim \exp(pt)$, where p is complex. Then (12.38)→(12.40) simplify to

$$\left[p - \alpha\nabla^2\right]\vartheta = \frac{\Delta T}{d}u_z, \tag{12.87}$$

$$\left[p - \nu\nabla^2\right]\omega_z = 2\Omega\frac{\partial u_z}{\partial z}, \tag{12.88}$$

$$\left[p - \nu\nabla^2\right]\nabla^2 u_z = g\beta\,\nabla_\perp^2\vartheta - 2\Omega\frac{\partial\omega_z}{\partial z}. \tag{12.89}$$

As before, we look for separable solutions of the form

$$(u_z, \vartheta) \sim (u_0, \vartheta_0)\sin(\pi z/d)f(x,y), \quad \omega_z \sim \omega_0\cos(\pi z/d)f(x,y),$$

where f satisfies Helmholtz's equation, (12.69). Our governing equations now become

$$\left[p + \alpha\left((\pi/d)^2 + k^2\right)\right]\vartheta_0 = \frac{\Delta T}{d}u_0, \tag{12.90}$$

$$\left[p + \nu\left((\pi/d)^2 + k^2\right)\right]\omega_0 = \frac{2\pi\Omega}{d}u_0, \tag{12.91}$$

$$\left[p + \nu\left(\left(\frac{\pi}{d}\right)^2 + k^2\right)\right]\left(\left(\frac{\pi}{d}\right)^2 + k^2\right)u_0 = g\beta\,k^2\vartheta_0 - \frac{2\pi\Omega}{d}\omega_0. \tag{12.92}$$

Eliminating $(\vartheta_0, u_0, \omega_0)$ from these expressions yields a relationship between the Rayleigh number, Ra, and the Prandtl and Taylor numbers, $Pr = \nu/\alpha$ and $\hat{Ta} = 4\Omega^2 d^4/\nu^2$. Specifically, (12.90)→(12.92) give us

$$\frac{a^2 \text{Ra}}{(\sigma \Pr + \pi^2 + a^2)} = (\pi^2 + a^2)(\sigma + \pi^2 + a^2) + \frac{\pi^2 \hat{\text{T}} \text{a}}{(\sigma + \pi^2 + a^2)}, \qquad (12.93)$$

where $\sigma = pd^2/\nu$ and $a = kd$. Note that this is a generalization of the marginal stability curve (12.53), in the sense that we recover (12.53) if we put $\sigma = 0$ in (12.93).

For a marginally unstable mode σ is purely imaginary, so our next task is to split (12.93) into its real and imaginary parts. To minimize the algebra, it is convenient to introduce the auxiliary variables $\chi = a^2/\pi^2$ and $j\hat{\sigma} = \sigma/\pi^2$, where we take $\hat{\sigma}$ to be real, indicative of marginal stability. Rewriting (12.93) in terms of χ and $\hat{\sigma}$ gives

$$\frac{\chi(\text{Ra})_{\text{marginal}}}{(j\hat{\sigma} \Pr + 1 + \chi)} = \pi^4 (1 + \chi)(j\hat{\sigma} + 1 + \chi) + \frac{\hat{\text{T}}\text{a}}{(j\hat{\sigma} + 1 + \chi)}.$$

We now multiplying through by $j\hat{\sigma} \Pr + 1 + \chi$ and split the resulting expression into real and imaginary parts. This yields

$$\chi(\text{Ra})_{\text{marginal}} = \pi^4 (1 + \chi)\left[(1 + \chi)^2 - \Pr \hat{\sigma}^2\right] + \frac{(1 + \chi)^2 + \Pr \hat{\sigma}^2}{(1 + \chi)^2 + \hat{\sigma}^2}\hat{\text{T}}\text{a}, \qquad (12.94)$$

$$0 = \pi^4 (1 + \chi)(\Pr + 1) + \frac{(\Pr - 1)\hat{\text{T}}\text{a}}{(1 + \chi)^2 + \hat{\sigma}^2}. \qquad (12.95)$$

Finally, we use (12.95) to eliminate $\hat{\sigma}$ from (12.94), and after a little algebra we obtain

$$\chi(\text{Ra})_{\text{marginal}} = 2(1 + \Pr)\left[\pi^4(1 + \chi)^3 + \frac{\Pr^2}{(1 + \Pr)^2}\hat{\text{T}}\text{a}\right]. \qquad (12.96)$$

It is also useful to rewrite (12.95) in the form

$$\frac{(1 - \Pr)\hat{\text{T}}\text{a}}{(1 + \Pr)} = \pi^4(1 + \chi)\left((1 + \chi)^2 + \hat{\sigma}^2\right) \geq \pi^4(1 + \chi)^3. \qquad (12.97)$$

Equations (12.96) and (12.97) are the key to understanding the onset of an oscillatory mode. Note that, if we put $\hat{\sigma} = 0$ in (12.97) and substitute the result back into (12.96), we find

$$\chi(\text{Ra})_{\text{marginal}} = \pi^4(1 + \chi)^3 + \hat{\text{T}}\text{a},$$

which is the marginal stability curve for steady convection. Thus, $\hat{\sigma} = 0$ marks the intersection of the marginal stability curves for oscillatory and steady convection.

12.3.2 Marginal Stability: The Importance of Prandtl Number and Role of Inertial Waves

We now consider the conditions under which an oscillatory mode might be observed at onset. There are two issues to consider. First, we need to establish when an oscillatory mode is dynamically achievable. Second, even if such a mode is permissible, we need to know if it will be observed in practice. That is to say, for a given Pr and $\hat{\mathrm{Ta}}$, is Ra$_{\mathrm{crit}}$ for an oscillatory mode lower than that for a stationary mode? We start with the former question.

The first point to note is that (12.97) forbids oscillations whenever Pr > 1. In fact, we shall see that oscillatory modes require Pr < 0.67. Moreover, (12.97) may be rearranged as

$$\mathrm{Pr} \le \frac{\hat{\mathrm{Ta}} - \pi^4(1 + \chi)^3}{\hat{\mathrm{Ta}} + \pi^4(1 + \chi)^3}. \tag{12.98}$$

Evidently, the lower the value of $\hat{\mathrm{Ta}}$, the lower the value of Pr required for an oscillatory instability, and indeed we might expect a lower bound on $\hat{\mathrm{Ta}}$ below which oscillations are not permissible. We shall see that this is $\hat{\mathrm{Ta}} = 548$. Thus, to obtain an oscillatory mode, both the Prandtl number *and* the Ekman number, Ek = $\nu/\Omega d^2$, must be sufficiently small.

So why do oscillatory modes require a small Pr and Ek? The point is that, physically, these oscillations can be regarded as a form of inertial wave, and if the viscosity is too large, such waves will not be realized. That is, (12.38)→(12.40) combine to give

$$\left[\frac{\partial}{\partial t} - \alpha\nabla^2\right]\left\{\left[\frac{\partial}{\partial t} - \nu\nabla^2\right]^2 \nabla^2 u_z + (2\Omega)^2\frac{\partial^2 u_z}{\partial z^2}\right\} = \frac{g\beta\Delta T}{d}\left[\frac{\partial}{\partial t} - \nu\nabla^2\right]\nabla_\perp^2 u_z, \tag{12.99}$$

where the expression within the braces on the left is the viscous version of the inertial wave operator (6.14). Thus, for $\Delta T \to 0$, we obtain viscously damped inertial waves. If we now look for solutions of the form $\exp(j(\mathbf{k} \cdot \mathbf{x}_\perp - \hat{\omega}t)\sin(\pi z/d)$, then we obtain

$$\left(\hat{\omega} + j\alpha\kappa^2\right)\left\{\left(\hat{\omega} + j\nu\kappa^2\right)^2 - \hat{\omega}_{IW}^2\right\} = -\frac{g\beta\Delta T}{d}\frac{k^2}{\kappa^2}\left(\hat{\omega} + j\nu\kappa^2\right), \qquad (12.100)$$

where $\kappa^2 = k^2 + (\pi/d)^2$ is the square of the magnitude of the three-dimensional wave-vector, and $\hat{\omega}_{IW} = 2\pi\Omega/\kappa d$ is the inviscid inertial wave frequency, (6.16). Note that, for Pr = 1, this reduces to

$$\left(\hat{\omega} + j\nu\kappa^2\right)^2 = \hat{\omega}_{IW}^2 - \frac{g\beta\Delta T}{d}\frac{k^2}{\kappa^2},$$

which, for Ta $\gg$ Ra, is a thermally modified inertial wave that decays on the viscous timescale of $\tau_d = 1/\nu\kappa^2$. Now suppose we demand that $\hat{\omega}$ is real, so we are looking for neutral oscillations, as in the previous section. Then dividing through (12.100) by $\hat{\omega} + j\nu\kappa^2$, and taking the imaginary part, yields

$$\hat{\omega}^2 + \nu^2\left(k^2 + (\pi/d)^2\right)^2 = \frac{\alpha - \nu}{\alpha + \nu}\hat{\omega}_{IW}^2, \qquad (12.101)$$

which is equivalent to (12.97). Notice that, for $\nu = 0$, we recover the inertial wave frequency. In any event, (12.101) shows that, once again, neutral oscillations are possible only when Pr < 1, and also when Ek is small enough, i.e. $\nu\kappa^2 < \sqrt{(\alpha - \nu)/(\alpha + \nu)}\,\hat{\omega}_{IW}$.

12.3.3 When is a Marginal Oscillatory Mode Preferred Over Stationary Convection?

Let us now consider the question of when an oscillatory mode is preferred over its non-oscillatory counterpart, i.e. it has a lower value of $(\text{Ra})_{\text{crit}}$. The relationship between $\hat{\text{Ta}}$ and Pr at the crossover from oscillatory to non-oscillatory instability is discussed in Chandrasekhar (1961). It is shown there that no oscillatory mode can be observed for Pr > 0.677, and for a given value of Pr < 0.677, there is a threshold value of $\hat{\text{Ta}}$ *above* which the oscillatory mode is preferred. Moreover, for a given $\hat{\text{Ta}} > 548$, there is a crossover value of Pr *below* which the oscillatory mode takes precedence, as shown in Figure 12.7.

This last point is illustrated in Figure 12.8, which is taken from Chandrasekhar (1961) and shows the marginal stability curves as a function of χ for the case of $\hat{\text{Ta}} = 10^4$. The upper curve is the marginal stability line for steady convection, while the lower curves represent marginal stability to oscillatory convection at various values of Pr, as given by (12.96). For Pr < 0.513 (curves A and B), $(\text{Ra})_{\text{crit}}$ is lower for the oscillatory motion, and so that mode

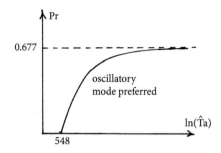

Figure 12.7 The threshold value of $\hat{T}a$ required for an oscillatory mode as a function of Pr.

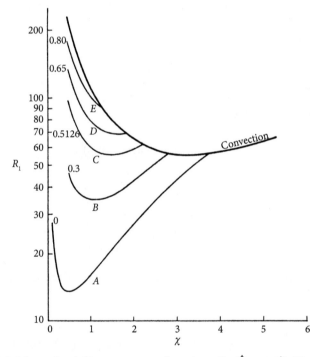

Figure 12.8 Neutral stability curves as a function of χ ($\hat{T}a = 10^4$). The upper curve is the marginal stability line for steady convection. The lower curves represent marginal stability to oscillatory convection at various values of Pr. $R_1 = Ra/\pi^4$. Reproduced from Chandrasekhar (1961).

takes precedence. However, for $Pr > 0.513$ (curves D and E), $(Ra)_{crit}$ is higher for the oscillatory motion and so that mode will not be observed. The point at which the oscillatory curves intersect with the steady convection line corresponds to $\hat{\sigma} = 0$, and above that line no oscillatory modes are dynamically admissible, let alone observable.

By constructing similar plots for a range of values of $\hat{\mathrm{T}}\mathrm{a}$, it is possible to determine the crossover values of Pr as a function of $\hat{\mathrm{T}}\mathrm{a}$, or equivalently, the threshold values of $\hat{\mathrm{T}}\mathrm{a}$ as a function of Pr. The results are shown in Table 12.3.

To understand where the values of $(\mathrm{Ra})_{\mathrm{crit}}$ in Table 12.3 come from, let us return to the marginal stability curve (12.96). The critical wavenumber that minimizes $(\mathrm{Ra})_{\mathrm{marginal}}$, and hence determines $(\mathrm{Ra})_{\mathrm{crit}}$, is governed by

$$2\left(a_{\mathrm{crit}}^2/\pi^2\right)^3 + 3\left(a_{\mathrm{crit}}^2/\pi^2\right)^2 = 1 + \frac{\mathrm{Pr}^2\hat{\mathrm{T}}\mathrm{a}}{(1+\mathrm{Pr})^2\pi^4}. \tag{12.102}$$

This is identical to (12.54) for steady convection, except that $\hat{\mathrm{T}}\mathrm{a}/\pi^4$ has been replaced by

$$\tilde{T} = \frac{\mathrm{Pr}^2\hat{\mathrm{T}}\mathrm{a}}{(1+\mathrm{Pr})^2\pi^4}.$$

From (12.96) and (12.102) we have

$$(\mathrm{Ra})_{\mathrm{crit}} = 6(1+\mathrm{Pr})\left(\pi^2 + a_{\mathrm{crit}}^2\right)^2, \tag{12.103}$$

and

$$a_{\mathrm{crit}}^2 = \frac{\pi^2}{2}\left(1 + \frac{\pi^4\tilde{T}}{\left(\pi^2 + a_{\mathrm{crit}}^2\right)^2}\right), \tag{12.104}$$

which correspond to the steady solutions (12.55). Moreover, solving the cubic equation (12.102) yields

$$\frac{2a_{\mathrm{crit}}^2}{\pi^2} = \left(1 + 2\tilde{T} + 2\sqrt{\tilde{T}(1+\tilde{T})}\right)^{1/3} + \left(1 + 2\tilde{T} - 2\sqrt{\tilde{T}(1+\tilde{T})}\right)^{1/3} - 1, \tag{12.105}$$

Table 12.3 The threshold value of $\hat{\mathrm{T}}\mathrm{a}$ above which an oscillatory mode is preferred

Pr	0	0.1	0.2	0.4	0.5	0.6	0.6766
Threshold $\hat{\mathrm{T}}\mathrm{a}$	548	728	990	3,163	8,505	68,150	∞
$(\mathrm{Ra})_{\mathrm{crit}}$ at crossover	1,315	1,471	1,669	2,890	4,910	16,790	∞

which applies also to steady convection, but with $\tilde{T}$ replaced by $\hat{T}a/\pi^4$. Thus, for a given $\tilde{T}$, or $\hat{T}a/\pi^4$, (12.105) gives a_{crit} for both oscillatory and steady convection. The corresponding values of $(Ra)_{crit}$ are then given by (12.103) and (12.55), respectively. A comparison of the critical Rayleigh numbers then dictates the preferred mode.

For strong rotation, say $Pr^2\hat{T}a \gg \pi^4$, (12.103) and (12.104) yield the simple results

$$a_{crit} = \left(\frac{Pr}{1 + Pr}\right)^{1/3}\left(\frac{\pi^2\hat{T}a}{2}\right)^{1/6}, \quad (Ra)_{crit} = \frac{6Pr^{4/3}}{(1 + Pr)^{1/3}}\left(\frac{\pi^2\hat{T}a}{2}\right)^{2/3}, \quad (12.106)$$

and so, once again, we obtain a $(Ra)_{crit} \sim \hat{T}a^{2/3}$ law. Comparing (12.106) with (12.56), we see that, for large $\hat{T}a$, $(Ra)_{crit}$ for oscillatory instability is less than that for non-oscillatory instability only when $2Pr^{4/3} < (1 + Pr)^{1/3}$, which requires $Pr < 0.6766$. This explains the final entry in Table 12.3, as well as the asymptote in Figure 12.7.

12.4 The Busse Annulus and Thermal Rossby Waves

12.4.1 The Busse Annulus

We close this chapter with a brief discussion of a useful model problem that has become known as the 'Busse annulus' (Busse, 1970). The motivation is that mildly supercritical convection in a rotating spherical shell takes the form of nearly two-dimensional convection rolls aligned with the rotation axis and surrounding the inner sphere, as shown in Figure 12.9.

One version of the Busse annulus is shown in Figure 12.10(a). It comprises a thin, rotating, annular gap of width d and length L which is heated at the inner surface, with the outer surface at T_{ref} and the inner one at $T_{ref} + \Delta T$. In order to allow for the possibility of linear Rossby waves, the top and bottom of the annulus have *small* slopes, s, and so the geometry is rather like an annular version of that used to study conventional Rossby waves, *i.e.* Figure 8.1. Since the gap is thin, we may ignore curvature and it is convenient to introduce local Cartesian coordinates, with x pointing eastward, y radially inward, and z aligned with the rotation axis. Gravity is radially inward, so that $\mathbf{g} = g\hat{\mathbf{e}}_y$. This coordinate system might be compared with that used to study Rossby waves on the β-plane in Chapter 9, as shown in Figure 12.10(b). In both cases, x points eastward.

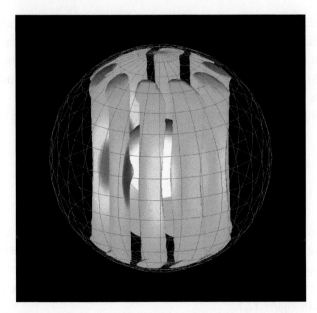

Figure 12.9 Convection in a rotating spherical shell. Courtesy of Emmanuel Dormy.

As with conventional Rossby waves, the motion is assumed to be quasi-geostrophic. Thus, as discussed in §8.1, we have $\omega_x, \omega_y \sim s|\omega_z|$, and hence $|\omega_z| \gg \omega_x, \omega_y$. Also, the velocity in the x-y plane, $\mathbf{u}_\perp$, is independent of z, at least to leading order in s. (Again, see §8.1.) The origin of coordinates is located on the outer surface, so the undisturbed temperature is $T_0 = T_{\text{ref}} + y\Delta T/d$, while the perturbed temperature is $T = T_0 + \vartheta(\mathbf{x}, t)$. Since the motion is (almost) planar, and the thermal boundary conditions are independent of z, we seek solutions of the form $\vartheta = \vartheta(x, y)$, to leading order in s. Finally, for simplicity, we (somewhat artificially) adopt slip boundary conditions.

12.4.2 A Linear Stability Analysis and Thermal Rossby Waves

We now perform a linear stability analysis on the Busse annulus. The resulting oscillations are known as 'thermal Rossby waves'. The perturbed equation of motion, (12.34), is

$$\left(\frac{\partial}{\partial t} - \nu\nabla^2\right)\mathbf{u} = 2\mathbf{u} \times \boldsymbol{\omega} - \nabla\left(\delta p/\bar{\rho}\right) - \beta\vartheta\mathbf{g},$$

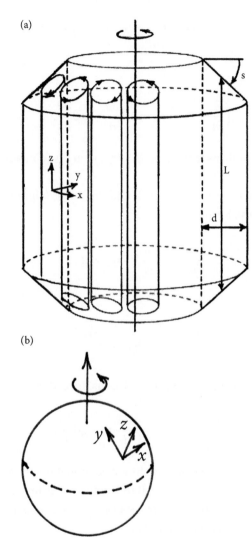

(a)

(b)

Figure 12.10 (a) The Busse annulus. (b) Coordinate system used for the β-plane in Chapter 9.

which gives the vorticity equation

$$\left(\frac{\partial}{\partial t} - \nu\nabla^2\right)\boldsymbol{\omega} = 2\Omega\frac{\partial \mathbf{u}}{\partial z} - g\beta(\nabla\vartheta) \times \hat{\mathbf{e}}_y. \tag{12.107}$$

Since $\omega_x, \omega_y \ll |\omega_z|$, and ϑ is assumed to be independent of z, we conclude that

$$2\Omega\frac{\partial \mathbf{u}_\perp}{\partial z} = \left(\frac{\partial}{\partial t} - \nu\nabla^2\right)\boldsymbol{\omega}_\perp \approx 0, \tag{12.108}$$

and

$$\left(\frac{\partial}{\partial t} - \nu\nabla^2\right)\omega_z = 2\Omega\frac{\partial u_z}{\partial z} - g\beta\frac{\partial \vartheta}{\partial x}, \tag{12.109}$$

the first of which confirms that, given $\vartheta = \vartheta(x, y)$, $\mathbf{u}_\perp$ is (almost) independent of z.

Next we note that $u_z = \pm s u_y$ at $z = \pm L/2$, and since $\mathbf{u}_\perp$ is (almost) independent of z, continuity in the form of

$$\frac{\partial^2 u_z}{\partial z^2} = -\frac{\partial}{\partial z}\nabla \cdot \mathbf{u}_\perp \approx 0,$$

yields

$$\frac{\partial u_z}{\partial z} = -\nabla \cdot \mathbf{u}_\perp \approx \frac{2su_y}{L}. \tag{12.110}$$

Substituting for $\partial u_z/\partial z$ in (12.109) then gives us, to leading order is s,

$$\left(\frac{\partial}{\partial t} - \nu\nabla^2\right)\omega_z = \frac{4\Omega s}{L}u_y - g\beta\frac{\partial \vartheta}{\partial x}, \tag{12.111}$$

which is a thermally forced version of the conventional Rossby wave equation, (8.6). In addition, we have $\nabla \cdot \mathbf{u}_\perp = 0$ to leading order in s, and so we may introduce a streamfunction for the transverse velocity, defined by $\mathbf{u}_\perp = \nabla \times (\psi\hat{\mathbf{e}}_z)$. Our vorticity equation now yields

$$\left(\frac{\partial}{\partial t} - \nu\nabla^2\right)\nabla^2\psi = \frac{4\Omega s}{L}\frac{\partial \psi}{\partial x} + g\beta\frac{\partial \vartheta}{\partial x}, \tag{12.112}$$

while the perturbed heat equation,

$$\left(\frac{\partial}{\partial t} - \alpha\nabla^2\right)\vartheta = -u_y\frac{dT_0}{dy} = \frac{\Delta T}{d}\frac{\partial \psi}{\partial x}, \tag{12.113}$$

also relates ϑ to ψ.

Note that our governing equations and boundary conditions contain between them six parameters: d, ΔT, $g\beta$, $\Omega s/L$, ν, and α. They also contain three dimensions (length, time, and temperature), which means that there are three independent dimensionless groups. We take these to be,

$$\text{Ra} = \frac{g\beta\Delta T d^3}{\nu\alpha}, \quad \text{Pr} = \frac{\nu}{\alpha}, \quad \text{Bu} = \frac{4\Omega s d^3}{\nu L}. \tag{12.114}$$

We now look for neutral oscillations in which ψ and ϑ are of the form

$$(\psi, \vartheta) \sim (\hat{\psi}, \hat{\vartheta}) \sin(\pi y/d) \exp\left[j(k_x x - \varpi t)\right],$$

where k_x and ϖ are real and, by convention, $\varpi > 0$. Equations (12.112) and (12.113) then give

$$\left[(\varpi + j\nu k^2) k^2 - \frac{4\Omega s k_x}{L}\right] \hat{\psi} = (g\beta k_x) \hat{\vartheta},$$

$$(\varpi + j\alpha k^2) \hat{\vartheta} = -\frac{\Delta T k_x}{d} \hat{\psi},$$

which yield

$$(\varpi + j\alpha k^2)\left[(\varpi + j\nu k^2) k^2 - \frac{4\Omega s k_x}{L}\right] = -\frac{g\beta\Delta T k_x^2}{d}, \tag{12.115}$$

where $k^2 = k_x^2 + (\pi/d)^2$. The real and imaginary parts of (12.115) are evidently

$$\varpi\left[\varpi k^2 - \frac{4\Omega s k_x}{L}\right] - \alpha\nu k^6 = -\frac{g\beta\Delta T k_x^2}{d}, \tag{12.116}$$

$$(1 + \text{Pr})k^2\varpi = \frac{4\Omega s}{L}k_x, \tag{12.117}$$

the second of which demands $k_x > 0$. Eliminating ϖ now gives us the marginal stability curve

$$(\text{Ra})_{\text{marginal}} = \frac{\left(\pi^2 + (k_x d)^2\right)^3}{(k_x d)^2} + \left(\frac{\text{Pr Bu}}{1 + \text{Pr}}\right)^2 \frac{1}{\pi^2 + (k_x d)^2}. \tag{12.118}$$

Note that, since $k_x > 0$, the phase velocity is in the positive x direction, *i.e.* to the east. This is the opposite of Rossby waves on the β-plane, as discussed in Chapter 9, whose phase velocity is to the west. Note also that, for Bu = 0, we have $\varpi = 0$ and (12.118) reverts to (12.23) for conventional Rayleigh–Bénard convection.

Finally, the wavenumber which minimizes Ra, and so sets the critical Rayleigh number, obeys

$$\frac{\left(\pi^2 + (k_x d)^2\right)^4 \left(2(k_x d)^2 - \pi^2\right)}{(k_x d)^4} = \left(\frac{\mathrm{Pr\,Bu}}{1 + \mathrm{Pr}}\right)^2,$$

and so for large Pr.Bu we find

$$(k_x d)_{\mathrm{crit}} = \left(\frac{\mathrm{Pr\,Bu}}{\sqrt{2}(1 + \mathrm{Pr})}\right)^{1/3}, \quad (\mathrm{Ra})_{\mathrm{crit}} = \frac{3}{2^{2/3}}\left(\frac{\mathrm{Pr\,Bu}}{1 + \mathrm{Pr}}\right)^{4/3}. \tag{12.119}$$

Thus, for strong rotation, we get tall, thin columns, which comes as no surprise given the $(kd)_{\mathrm{crit}} \sim \hat{\mathrm{Ta}}^{1/6}$ law for rotating, Rayleigh–Bénard convection.

* * *

This concludes our brief survey of rotating, Rayleigh–Bénard convection. Many more details of the linear theory may be found in Chandrasekhar (1961), which remains an authoritative source on this topic.

References

Bénard, H., 1900, Les tourbillons cellulaires dans une nappe liquide, *Revue Générale des Sciences Pures et Appliquées*, **11**, 1261–71, 1309–28.

Bodenschatz, E., Pesch, W., & Ahlers, G., 2000, Recent developments in Rayleigh–Bénard convection, *Ann. Rev. Fluid Mech.*, **32**, 709–78.

Busse, F.H., 1970, Thermal instabilities in rapidly rotating systems, *J. Fluid Mech.*, **44**, 441–60.

Chandrasekhar, S., 1961, *Hydrodynamic and Hydromagnetic Stability*, Oxford University Press.

Drazin, P.G., 1992, *Nonlinear Systems*, Cambridge University Press.

Manneville, P., 2006, Rayleigh–Bénard convection: thirty years of experimental, theoretical and modelling work. In: *Dynamics of Spatio-Temporal Cellular Structures: Henri Bénard Centenary Review*. I. Mutabazi, J.E. Wesfreid & E. Guyon, eds., Springer, 41–65.

Pearson, J.R.A., 1958, On convection cells induced by surface tension, *J. Fluid Mech.*, **4**, 489–500.

Rayleigh, Lord, 1916, On the convective currents in a horizontal layer of fluid when the higher temperature is on the underside, *Phil. Mag.*, **32**(6), 529–46.

Thomson, J., 1882, On a changing tessellated structure in certain liquids, *Proc. Phil. Soc. Glasgow*, **8**(2), 464–8.

Veronis, G., 1959, Cellular convection with finite amplitude in a rotating fluid, *J. Fluid Mech.*, **5**(3), 401–35.

Chapter 13

Vortex Breakdown

13.1 Observations of Vortex Breakdown in Pipes and on Delta Wings

Vortex breakdown is a phenomenon in which a slowly evolving vortex tube, within which the axial and swirling components of velocity are of similar magnitudes, suddenly and unexpectedly undergoes a rapid transition involving a sharp divergence of the flow away from the axis of the vortex. This is frequently observed in flow within a slowly diverging pipe, and on the upper surface of a highly swept wing, such as a delta wing.

In the case of a slowly diverging pipe, breakdown is usually, but not always, associated with the sudden appearance of a stagnation point on the axis of the pipe, followed by an approximately axisymmetric recirculation bubble within which the axial velocity reverses. This is often followed by an unsteady, helical wake, which eventually becomes turbulent. Typical examples are shown in Figure 13.1, which is taken from Sarpkaya (1971), where the angle of divergence of the pipe is around 3°. However, vortex breakdown in a pipe need not always be axisymmetric. For example, sometimes it adopts a twisting, helical structure that eventually gives way to turbulent mixing.

An example of vortex breakdown on a delta wing is shown in Figure 13.2. Here trailing vortices develop on the upper surface of the wing near the front, which then undergo vortex breakdown about a third of the way along the wing. Notice that, ahead of vortex breakdown, the core of the vortex appears to be slowly diverging, which suggests that the axial velocity in the vortex core is decreasing. This, in turn, suggests that the vortex is developing within a positive axial pressure gradient, which is indeed the case.

The circumstances under which vortex breakdown occurs vary, but the conditions which favour breakdown appear to be a strong component of axial vorticity (swirl), a slow divergence of the flow ahead of breakdown, and a positive axial pressure gradient. Of course, the last two points are closely coupled. The phenomenon also appears to be essentially an inviscid one. We shall discuss the various theories which have been proposed to explain vortex breakdown in §13.4. First, however, we shall consider two idealized model problems,

The Dynamics of Rotating Fluids. P. A. Davidson, Oxford University Press. © Peter A Davidson (2024).
DOI: 10.1093/9780191994272.003.0013

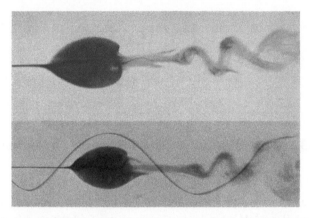

Figure 13.1 Axisymmetric vortex breakdown in a slowly diverging pipe. Reproduced from Sarpkaya (1971).

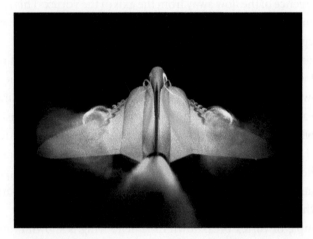

Figure 13.2 Scroll vortices on the upper surface of a delta wing showing vortex breakdown partway along the wing. (H. Werlé of ONERA, courtesy of B. Chanetz.)

the first of which is suggestive of breakdown in a diverging pipe, and the second of breakdown on a delta wing.

13.2 A Suggestive Model Problem of Flow in a Diverging Pipe

Consider inviscid, axisymmetric, helical flow in a circular tube of radius $R(z)$, described in cylindrical polar coordinates, (r, θ, z). Upstream, the tube is cylindrical and of radius $R = a$, as shown in Figure 13.3. The tube then expands

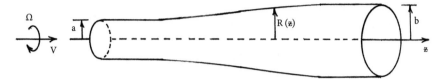

Figure 13.3 The flow domain under consideration.

before becoming cylindrical again, this time with a radius of $R = b$. In §2.10.2 we saw that such a flow is governed by the Squire–Long equation,

$$\nabla_*^2 \Psi = r^2 H'(\Psi) - \Gamma\Gamma'(\Psi), \tag{13.1}$$

where Ψ is the Stokes streamfunction, (2.9), ∇_*^2 is the Stokes operator, (2.48), $\Gamma = ru_\theta$ is the angular momentum density, and $H = \frac{1}{2}\mathbf{u}^2 + p/\rho$ is Bernoulli's function. Both Γ and H are constant along a streamline, and so (13.1) can be solved for Ψ provided that $\Gamma(\Psi)$ and $H(\Psi)$ are both specified at some upstream location.

We now consider the simplest of all possible inlet conditions, which is $u_\theta = \Omega r$ and $u_z = V$, i.e. rigid-body rotation plus uniform translation. In such a case the upstream distributions of Ψ, Γ, and p/ρ are $\Psi = \frac{1}{2}Vr^2$, $\Gamma = \Omega r^2$, and $p/\rho = \frac{1}{2}\Omega^2 r^2$, from which

$$\Gamma = (2\Omega/V)\,\Psi = k\Psi, \qquad H'(\Psi) = 2\Omega^2/V, \tag{13.2}$$

where $k = 2\Omega/V$. Moreover, the Squire–Long equation now becomes

$$\nabla_*^2 \Psi + k^2\left(\Psi - \tfrac{1}{2}Vr^2\right) = 0. \tag{13.3}$$

Finally, if F represents the departure of Ψ from its upstream value, defined through the expression $\Psi = \frac{1}{2}Vr^2 + rF(r, z)$, then (13.3) yields the *linear* equation,

$$\nabla^2 F + \left[k^2 - r^{-2}\right]F = 0. \tag{13.4}$$

This can be integrated to find the downstream flow, provided there is no reversal in u_z.

In §2.10.2 we noted that, when $b \to a$, so the entire tube is cylindrical, (13.4) supports stationary inertial waves of the form

$$F = AJ_1(k_r r) \cos(k_z z),$$ (13.5)

where A is a *finite* amplitude, k_r and k_z are related by $k_r^2 + k_z^2 = k^2$, and J_1 is the usual Bessel function. In a frame of reference moving with the fluid this is a progressive wave moving upstream with the phase velocity $V = 2\Omega/k$ and frequency $\hat{\omega} = k_z V = 2\Omega k_z/k$. However, (13.5) represents a standing wave in the radial direction, with the discrete values of k_r set by the requirement that $u_r = 0$ at $r = b$. This, in turn, requires $k_r b = \gamma_n$, where γ_n are the zeros of J_1. It follows that k_z is constrained to satisfy $k_z^2 b^2 + \gamma_n^2 = (kb)^2$, and so such waves are permissible only if $kb > \gamma_1 = 3.832$. Note that the axial component of the group velocity for such waves, now assumed to be of small amplitude, can be found by differentiating the expression $\hat{\omega} = 2\Omega k_z/k$. It is readily confirmed that

$$c_{g,z} = -\frac{2\Omega}{k}\frac{k_r^2}{k^2} = -\frac{k_r^2}{k^2}V,$$ (13.6)

which is also directed upstream, but is smaller in magnitude than the phase velocity. It follows that, in the laboratory frame, all wave energy gets swept downstream.

The linearity of (13.4) allows us to generalize the analysis of §2.10.2 to cases where $b > a$, provided that solutions of the Squire–Long equation can be extended to the downstream cylindrical section, i.e. there is no flow reversal. (If there is a reversal in u_z, then the analysis breaks down because we have assumed that $\Gamma(\Psi)$ and $H(\Psi)$ are prescribed at the inlet for *all* streamlines.) We now write

$$F(r, z) = F_0(r) + f(r, z), \quad f(b, z) = 0,$$ (13.7)

for the downstream cylindrical section, so that (13.4) plus $\Psi(r = b) = \Psi(r = a)$ yield

$$\nabla^2 F_0 + \left[k^2 - r^{-2}\right]F_0 = 0, \quad \tfrac{1}{2}Vb^2 + bF_0(b) = \tfrac{1}{2}Va^2,$$ (13.8)

$$\nabla^2 f + \left[k^2 - r^{-2}\right]f = 0, \quad f(b, z) = 0.$$ (13.9)

The solutions of (13.8) and (13.9) are

$$F_0 = AJ_1(kr), \quad \tfrac{1}{2}Vb^2 + bAJ_1(kb) = \tfrac{1}{2}Va^2,$$ (13.10)

$$f = CJ_1(k_r r)\cos(k_z z), \qquad J_1(k_r b) = 0, \tag{13.11}$$

where $k_r^2 + k_z^2 = k^2$. This allows the non-oscillatory base flow to evolve from one cylindrical section to the next, with finite-amplitude inertial waves downstream. As before, we have $k_z^2 b^2 + \gamma_n^2 = (kb)^2$, and so waves are permissible only when $kb > \gamma_1 = 3.832$.

We now follow Batchelor (1967) and consider cases where $kb < \gamma_1 = 3.832$, so the flow in the downstream cylindrical section does not oscillate in z: i.e. $f = 0$. (This is often called a *cylindrical flow*.) In such a case (13.10) yields, for the downstream section,

$$\Psi = \tfrac{1}{2}Vr^2 + ArJ_1(kr), \qquad \Gamma = k\Psi, \tag{13.12}$$

$$u_z = \frac{1}{r}\frac{\partial \Psi}{\partial r} = V + kAJ_0(kr), \qquad u_\theta = \Omega r + kAJ_1(kr), \tag{13.13}$$

$$A = \frac{V(a^2 - b^2)}{2bJ_1(kb)} = \frac{\Omega(a^2 - b^2)}{kbJ_1(kb)}. \tag{13.14}$$

We now eliminate A to give

$$\frac{u_z}{V} = 1 - \frac{b^2 - a^2}{b^2}\frac{\tfrac{1}{2}kb}{J_1(kb)}J_0(kr), \tag{13.15}$$

$$\frac{u_\theta}{\Omega r} = 1 - \frac{b^2 - a^2}{b^2}\frac{bJ_1(kr)}{rJ_1(kb)}, \tag{13.16}$$

from which we find

$$\frac{\partial u_z}{\partial r} = -\omega_\theta = \frac{b^2 - a^2}{b^2}\frac{\tfrac{1}{2}k^2 bV}{J_1(kb)}J_1(kr) > 0, \tag{13.17}$$

$$\frac{u_z(r=0)}{V} = \frac{u_\theta(r \to 0)}{\Omega r} = 1 - \frac{b^2 - a^2}{b^2}\frac{\tfrac{1}{2}kb}{J_1(kb)}. \tag{13.18}$$

We shall now unpick the physical content of (13.15)→(13.18). The shapes of $J_0(x)$ and $J_1(x)$ are shown in Figure 13.4. The first zero of $J_0(x)$ is at $x = 2.405$, and so for $0 < kb < 2.405$ the axial velocity in (13.15) is less than V for all r. However, for $2.405 < kb < 3.832$, u_z in the downstream cylindrical section is greater than V at the tube wall. Of course, continuity demands that the mean axial velocity in the downstream section is less than V, and so any acceleration of the fluid near the wall will accentuate the decline of u_z on the axis. For 0

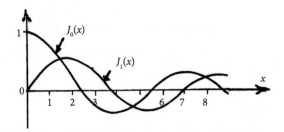

Figure 13.4 The shapes of $J_0(x)$ and $J_1(x)$.

$< kb < 3.832$, (13.16) tells us that $u_\theta < \Omega r$ for all r in the downstream section, consistent with Γ being conserved in all diverging streamlines.

Turning to (13.17), we see that, in the downstream section, u_z increases monotonically with r from the axis to the wall. Thus, while continuity demands that the mean axial velocity falls as we move from the upstream to the downstream section, the deceleration of the flow is more pronounced on the axis. This can be understood in one of two ways; in terms of vortex dynamics or in terms of pressure. The vorticity argument proceeds as follows. The streamlines diverge while passing through the expansion in the tube, and since Γ is constant on a streamline, this tells us that $\partial\Gamma/\partial z < 0$ throughout the expansion. The inviscid evolution equation for the azimuthal vorticity,

$$\frac{D}{Dt}\left(\frac{\omega_\theta}{r}\right) = \frac{\partial}{\partial z}\left(\frac{\Gamma^2}{r^4}\right),$$

(see 2.88), now demands that $\omega_\theta < 0$ in the downstream section, which, in turn, requires $\partial u_z/\partial r > 0$. Alternatively, suppose that Δp is the difference in pressure between the tube wall and the axis at any one cross-section in the expansion. Then, provided the expansion is gradual, so that the radial velocity is small, we have

$$\Delta p = \rho \int_0^R \left(\Gamma^2/r^3\right) dr.$$

The fact that $\partial\Gamma/\partial z < 0$ in the expansion now ensures that $\partial(\Delta p)/\partial z < 0$. Moreover, as the mean velocity falls across the expansion, so the mean pressure rises, and the fact that $\partial(\Delta p)/\partial z < 0$ tells us that the pressure rises faster on the axis than at the tube wall. Hence, the axial deceleration of the fluid within the expansion is highest on the axis, consistent with $\partial u_z/\partial r > 0$ in the downstream cylindrical section.

Finally, we turn to (13.18), the first part of which can be rewritten in the form

$$\frac{2\pi\Gamma(r \to 0)}{u_z(r = 0)\pi r^2} = \frac{2\Omega}{V},$$

reflecting the conservation of vorticity flux, $2\pi\Gamma$, and mass flux, $\rho u_z \pi r^2$, along a thin flux tube on the axis. More importantly, (13.18) tells us that a stagnation point forms whenever

$$\frac{\frac{1}{2}kb}{J_1(kb)} \geq \frac{b^2}{b^2 - a^2}. \tag{13.19}$$

Moreover, since Rayleigh's discriminant for this cylindrical flow can be written as

$$\Phi(r) = \frac{1}{r^3}\frac{d\Gamma^2}{dr} = \frac{2k\Gamma}{r^3}\frac{d\Psi}{dr} = \frac{2k\Gamma}{r^2}u_z, \tag{13.20}$$

it is likely that the flow ceases to be centrifugally stable when $u_z \leq 0$.

Of course, the solution ceases to be valid once reverse flow occurs, but nevertheless the equality sign in (13.19) suggests a breakdown of the solution at a certain critical value of kb. Now $J_1(x) \approx \frac{1}{2}x$ for small x, and so $\frac{1}{2}x/J_1(x)$ increases monotonically from unity to infinity throughout the range $0 < x < 3.832$. It follows that, for a given b/a, a stagnation point will inevitably appear on the downstream axis as kb increases, *i.e.* as Ω is increased at the inlet. Moreover, for $x < 1$, we have $J_1(x)/\frac{1}{2}x \approx 1 - \frac{1}{8}x^2$ to within 1%. So, for $kb < 1$, the stagnation point forms at $kb \approx \sqrt{8}\,(a/b)$. More generally, we can approximate (13.19) by

$$kb = \frac{2\Omega b}{V} \geq \sqrt{8}\frac{a}{b}\left[1 + ((\gamma_1^2/8) - 1)(a/b)^4\right]^{1/2}, \quad 0 < kb < \gamma_1. \tag{13.21}$$

The suggestion, of course, is that the appearance of a stagnation point on the axis signals the onset of vortex breakdown. If correct, this suggests that the prerequisites for vortex breakdown in a tube are a positive axial pressure gradient and (roughly)

$$\sqrt{8}\frac{a}{b}\left[1 + ((\gamma_1^2/8) - 1)(a/b)^4\right]^{1/2} < \frac{2\Omega b}{V} < \gamma_1. \tag{13.22}$$

We shall return to this idea in §13.4, where we shall see that there is some truth in this claim, but that more realistic inlet conditions lead to a more complex form of behaviour. In particular, vortex breakdown in a tube is sometimes associated with the appearance of a stagnation point, and sometimes with the loss of all possible solutions of (13.1), prior to the formation of a stagnation point.

13.3 A Toy Model of Vortex Breakdown on the Surface of a Delta Wing

We now consider a closely related problem, which serves as a simple model of vortex breakdown on the surface of a delta wing. Once again, this was first introduced by Batchelor (1967), though it was subsequently extended by Saffman (1992).

Consider an isolated, axisymmetric, inviscid vortex tube embedded within an irrotational flow. The circular vortex tube has radius $R(z)$ and we use cylindrical polar coordinates, (r, θ, z), centred on the tube. As in §13.2, the tube is initially cylindrical and of radius $R = a$. It then expands before becoming cylindrical again, with a radius of $R = b$. The velocity at the upstream end of the vortex tube is the same as for the confined flow of §13.2, i.e. $u_\theta = \Omega r$ and $u_z = V_1$, and the motion in the downstream section is taken to be steady and independent of z. The flux of vorticity along the vortex tube is $\Phi = 2\Omega\pi a^2$ and the external azimuthal velocity is evidently $u_\theta = \Phi/2\pi r$. The velocity at the edge of the vortex tube is taken to be continuous, with an external velocity of $u_z = V_1$ and $u_\theta = \Phi/2\pi a$ at the edge of the upstream cylindrical section, and $u_z = V_2$ and $u_\theta = \Phi/2\pi b$, or equivalently $u_\theta = \Omega a^2/b$, at the downstream section. The geometry is as shown in Figure 13.5.

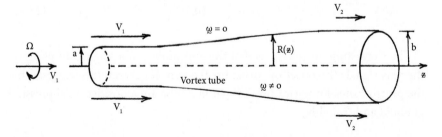

Figure 13.5 An isolated, inviscid, axisymmetric vortex tube undergoing an expansion.

At first sight, this looks to be identical to the previous problem. However, the key difference is that the external axial velocity is now prescribed, rather than the downstream radius of the tube. Thus, the downstream velocity $u_z(r = b) = V_2$ is given, and b has to be inferred. Of course, since we require the vortex tube to expand, we shall take $V_2 < V_1$. As before, the velocity components at the downstream cylindrical section are

$$\frac{u_z}{V_1} = 1 - \frac{b^2 - a^2}{b^2} \frac{\frac{1}{2}kb}{J_1(kb)} J_0(kr), \tag{13.23}$$

$$\frac{u_\theta}{\Omega r} = 1 - \frac{b^2 - a^2}{b^2} \frac{bJ_1(kr)}{rJ_1(kb)}, \tag{13.24}$$

where $k = 2\Omega/V_1$. This then yields

$$\frac{\partial u_z}{\partial r} = -\omega_\theta = \frac{b^2 - a^2}{b^2} \frac{\frac{1}{2}k^2 bV_1}{J_1(kb)} J_1(kr) > 0, \tag{13.25}$$

$$\frac{u_z(r = 0)}{V_1} = \frac{u_\theta(r \to 0)}{\Omega r} = 1 - \frac{b^2 - a^2}{b^2} \frac{\frac{1}{2}kb}{J_1(kb)}, \tag{13.26}$$

and

$$\frac{V_2}{V_1} = 1 - \frac{b^2 - a^2}{b^2} \frac{\frac{1}{2}kbJ_0(kb)}{J_1(kb)}. \tag{13.27}$$

Note that, to ensure that $b > a$ when $V_2 < V_1$, we must restrict ourselves to $ka < kb < 2.405$, the first zero of J_0. However, this encompasses most cases of practical interest and so is not overly restrictive. Note also that, as with the confined flow of §13.2, $\partial u_z/\partial r > 0$ in the downstream section, and so the fluid in the expansion decelerates fastest on the axis. The reason for this preferential deceleration of the fluid on the axis is discussed in §13.2.

Since V_2 is prescribed and b unknown, the key is to invert (13.27) to find b. For a given V_1/V_2 and ka, (13.27) can indeed be solved for kb, and the result is shown schematically in Figure 13.6. Here the upstream conditions are held fixed (in particular, ka is fixed), while V_1/V_2 is slowly increased from unity. Note that (13.27) requires that $V_2 = V_1$ not only when $kb = ka$, but also when $kb = 2.405$. Thus, the curves in Figure 13.6 bend back on themselves to intercept the vertical axis twice. It follows that, for given values of

V_1/V_2 and $ka = 2\Omega a/V_1$, there are two possible solutions for kb. Of course, the non-uniqueness of kb also implies a non-unique axial velocity profile in the downstream section.

There is a second surprise in Figure 13.6. For a given upstream condition, *i.e.* for a given $ka = 2\Omega a/V_1$, there is a maximum allowable value of V_1/V_2, beyond which there are no solutions of the Squire–Long equation. That is to say, for a given upstream rotation rate, solutions exist only for a finite deceleration of the external flow, and the stronger the upstream rotation, the smaller the allowable deceleration. The non-uniqueness of the downstream flow, and the upper limit on V_1/V_2, are both central features of vortex breakdown of an isolated vortex tube.

Figure 13.6 shows how b changes when the upstream conditions are held fixed and V_2/V_1 is slowly decreased. However, we might equally ask how b and u_z vary if V_2/V_1 is held fixed and we slowly increase the upstream rotation rate, that is we increase ka. To that end, we first note that, for weak rotation (*i.e.* small kb), (13.27) requires $b = a\sqrt{V_1/V_2}$, which we label as b_0. Following Batchelor (1967), it is convenient to normalize b by $b_0 = a\sqrt{V_1/V_2}$ and then

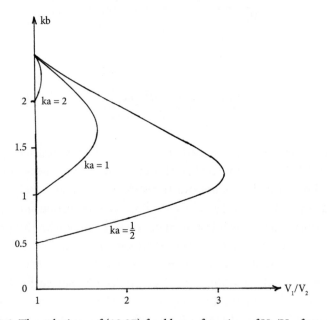

Figure 13.6 The solutions of (13.27) for kb as a function of V_1/V_2, for various values of ka.

consider how b/b_0 varies with increasing ka. Next, noting that

$$kb = \left(ka\sqrt{V_1/V_2}\right)\frac{b}{b_0},\tag{13.28}$$

we rewrite (13.27) as

$$\frac{(1 - V_2/V_1)(b/b_0)^2}{(b/b_0)^2 - V_2/V_1} = \frac{\frac{1}{2}kbJ_0(kb)}{J_1(kb)}, \quad kb = \left(ka\sqrt{V_1/V_2}\right)\frac{b}{b_0}.\tag{13.29}$$

For a fixed V_2/V_1, this can be used to show how b/b_0 varies with increasing ka. This is shown schematically in Figure 13.7 for the particular case of $V_1/V_2 = 2$. It is immediately apparent that there is a *fold* in the curve where it bends back on itself, and this occurs at $ka = 0.790$. We shall discuss the significance of this fold shortly.

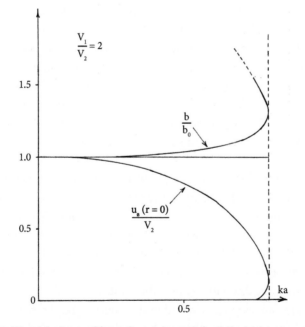

Figure 13.7 The solutions of (13.29) and (13.31) for b/b_0 and $u_z(r = 0)/V_2$ as a function of ka. Both curves are for the case of $V_1/V_2 = 2$.

Turning now to the downstream axial velocity, (13.26) and (13.27) combine to give

$$\frac{V_2 - u_z(r = 0)}{(V_1 - V_2)} = \left[\frac{1}{J_0(kb)} - 1\right] \geq 0, \tag{13.30}$$

which confirms that $u_z(r = 0) < V_2$. Equally, (13.30) may be written as

$$\frac{u_z(r = 0)}{V_2} = \frac{V_1}{V_2} - \left(\frac{V_1}{V_2} - 1\right)\frac{1}{J_0(kb)}, \tag{13.31}$$

which shows that a stagnation point will inevitably appear on the axis of the vortex tube when the rotation is sufficiently strong; in particular, when kb reaches a value such that $J_0(kb) = 1 - V_2/V_1$. Moreover, for a fixed V_2/V_1, (13.31) can be used to show how $u_z(r = 0)/V_2$ varies as kb is increased, and since (13.27) relates kb to ka (for a given V_2/V_1), we can also determine the variation of $u_z(r = 0)/V_2$ with ka.

To focus thoughts, let us return to the case of $V_1/V_2 = 2$ and to Figure 13.7, which shows schematically the variation of $u_z(r = 0)/V_2$ with ka. Evidently, u_z also exhibits a fold at $ka = 0.790$, and beyond this value of ka, there are no solutions of the Squire–Long equation (for $V_1/V_2 = 2$). Note that the axial velocity on the axis is still finite at the fold, with a value of $u_z(r = 0)/V_2 = 0.175$. If we follow the curve for $u_z(r = 0)$ past its fold, then there is a stagnation point on the axis at the slightly lower value of $ka = 0.780$. Thus, in the narrow range of $0.780 < ka < 0.790$, there are two possible solutions for both b and $u_z(r = 0)$. As in §13.2, we cannot follow the curves for b/b_0 or $u_z(r = 0)$ past the stagnation point, because our solution ceases to be valid once reverse flow occurs.

It seems probable that the appearance of a fold in these curves signals vortex breakdown, as suggested in Saffman (1992). As the upstream rotation (or ka) is increased, we follow the main branch of the curves for b/b_0 and $u_z(r = 0)/V_2$. Once ka exceeds 0.780, two downstream solutions are available, and it is conceivable that the flow might flip from the main branch to the secondary one. However, by the time ka reaches 0.790, the Squire–Long solution has nowhere to go, and something dramatic is inevitable. Note that, unlike the confined flow of §13.2, it is the appearance of a fold in the solution, rather than a stagnation point, that signals vortex breakdown here. However, we shall see that folds can also occur in pipe flow, depending on the shape of the inlet velocity profile. It just so happens that there is no fold for the inlet conditions considered in §13.2.

13.4 Theories of Vortex Breakdown

In the model problems of §13.2 (pipe flow) and §13.3 (a free vortex) we supposed the edge of the tube or vortex to be cylindrical both upstream and downstream, with a divergence in between. However, in the case of pipe flow, we could equally imagine the tube wall to slowly yet continually diverge, so that at any one cross-section the Squire–Long equation can be approximated by its quasi-cylindrical form, in which axial derivatives are ignored. Vortex breakdown then occurs (according to the hypothesis of §13.2) when kb is large enough to initiate a stagnation point on the axis, in accordance with (13.19). Similarly, for a free vortex, we might suppose that V_2/V_1 is slowly yet continually reduced with distance along the vortex (see Figure 13.2) so that, once again, we may use the quasi-cylindrical approximation to the Squire–Long equation at any one cross-section. Vortex breakdown then occurs, for a given ka, when we reach the fold in Figure 13.6. Note that these model problems seek only to establish the conditions under which vortex breakdown might occur, and say nothing about the structure of the flow following breakdown, as these solutions of the Squire–Long equation assume that all streamlines originate at the inlet.

The inviscid model problem of §13.2 has been extended by Buntine & Saffman (1995), who considered more complicated inlet conditions, thought to be more representative of experimental studies of swirling pipe flow. In particular, the axial components of $\mathbf{u}$ and $\boldsymbol{\omega}$ at the inlet were taken to be maximal on the axis of the tube. When the inlet velocity and vorticity profiles are relatively flat, the behaviour resembles that of §13.2, with a stagnation point appearing on the axis for a sufficiently strong inlet rotation, and this stagnation point is taken to signal vortex breakdown. However, when the axial components of $\mathbf{u}$ and $\boldsymbol{\omega}$ at the inlet have strong peaks on the axis, u_z ($r = 0$) at the exit exhibits a behaviour similar to that of a free vortex, as shown in Figure 13.8. That is, as the upstream rotation rate is increased, u_z ($r = 0$) falls at the exit, but a fold in the solution curve precedes the appearance of a stagnation point. It is now the appearance of a fold, rather than the formation of a stagnation point, that is associated with vortex breakdown. This suggests that vortex breakdown in a tube and in a free vortex is not so different.

One of the more striking observations of Buntine & Saffman (1995), which had been noted several times before, is that, as the fold in the solution curve for u_z ($r = 0$) is approached (by increasing the upstream rotation), the flow at the exit can support small-amplitude standing waves. Conversely, prior to the fold, all such inertial waves are swept downstream. A similar situation arises when there is no fold, but rather a stagnation point forms. Once again, small-amplitude standing waves can be (locally) supported at the exit as the stagnation point is approached (see Figure 13.8). The existence of standing waves means that, locally, the phase velocity, c_p, is equal and opposite to the flow speed, u_z, or equivalently $c = u_z - |c_p| = 0$, where c is the wave speed measured in the laboratory frame. The existence of standing waves at a fold or stagnation point is significant, because there have been several studies which compare vortex breakdown to a hydraulic jump in open-channel flow (see Squire, 1960 and Benjamin, 1962, 1967). In particular, it has been suggested that, as with a hydraulic jump, vortex breakdown represents a transition from a *supercritical* flow (defined as $u_z > |c_p|$) to a *subcritical* flow ($u_z < |c_p|$). This is an attractive idea, particularly as there are two possible solutions for the flow just ahead of a fold, analogous to the conjugate states either side of a hydraulic jump.

The analogy to a hydraulic jump is, however, not entirely straightforward, as there are important differences between surface gravity waves and inertial waves. For example, shallow-water surface gravity waves are non-dispersive, and so their phase and group velocities are equal. Thus, since subcritical shallow-water flow is defined in terms of the flow speed being less than the phase speed, downstream influences can always propagate upstream in such a subcritical flow. However, as noted in §13.2, and as pointed out by Benjamin (1962), the group velocity of inertial waves in a cylindrical tube is less than the

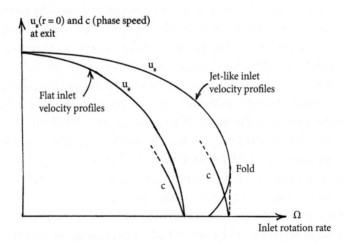

Figure 13.8 $u_z(r = 0)$ at the exit of a diverging pipe for flat and jet-like inlet conditions. c is the stream-wise phase speed (in the laboratory frame) of infinitesimal waves at the exit.

phase velocity (except when $k_z = 0$). Thus, if a standing wave is maintained in a swirling pipe flow, with the flow speed equal and opposite to the phase speed, then all of the wave energy is swept downstream. Now, it is normal to define criticality in such a pipe flow in terms of the appearance of standing waves (by analogy with shallow-water flow), and if that definition is adopted, down-stream disturbances with a finite k_z *cannot* propagate upstream in a marginally subcritical flow. In this sense, the analogy breaks down.

Another important difference is that the transition from a supercritical flow to a subcritical flow in a hydraulic jump demands a local loss of mechanical energy. Thus, intense energy dissipation is an essential feature of hydraulic jumps (except in very mild jumps, where the excess energy can be carried off by cnoidal waves). However, no such violent dissipation of energy is observed in an axisymmetric vortex breakdown bubble.

There is also a technical difficulty associated with the definition of the criti-cal state (in terms of a train of standing waves) in a diverging pipe flow or free vortex. This is because kb continually evolves with axial position, and so it is difficult to associate a spatially extended wave train with a particular cross-section of the vortex tube, unless the rate of divergence is very small and the flow quasi-cylindrical. However, despite these various difficulties, the analogy of vortex breakdown to the supercritical–subcritical transition in a hydraulic jump has received experimental and numerical support. (See, for example, the experiments of Sarpkaya, 1971, and the simulations of Ruith et al., 2003.)

There are other interpretations of vortex breakdown, over and above the folds and stagnation points of Batchelor and Saffman, and the hydraulic jump analogy of Squire and Benjamin. For example, some have suggested that it is a form of non-axisymmetric hydrodynamic instability, though this is disputed by Hall (1972). Yet another interpretation comes from integrating the quasi-cylindrical Navier–Stokes equation downstream to the point where rapid variations in z are predicted, which violates the initial assumption of a quasi-cylindrical flow (see Hall, 1972). Vortex breakdown is then interpreted as a collapse of the quasi-cylindrical assumption. Of course, this is not unlike the occurrence of a fold in a solution of the Squire–Long equation, and indeed it is quite possibly equivalent, at least in some sense.

There has been much discussion over which interpretation (or theory) of vortex breakdown best fits a given set of experiments, and this is not helped by the fact that vortex breakdown can mean different things to different people. The early reviews by Hall (1972) and Leibovich (1978) capture this dia-logue rather well. However, the fact that criticality (in the sense of Benjamin, 1962) coincides with the appearance of stagnations points and folds, suggests that these various interpretations of vortex breakdown are all closely related, perhaps even different sides of the same coin.

* * *

That concludes our brief excursion into vortex breakdown. We shall return to the topic in Chapter 15, when we discuss tornadoes.

References

Batchelor, G.K., 1967, *An Introduction to Fluid Dynamics*, Cambridge University Press.

Benjamin, T.B., 1962, Theory of the vortex breakdown phenomenon, *J. Fluid Mech.*, **14**, 593–629.

Benjamin, T.B., 1967, Some developments in the theory of vortex breakdown, *J. Fluid Mech.*, **28**, 65–84.

Buntine, J.D., & Saffman, P.G., 1995, Inviscid swirling flows and vortex breakdown, *Proc. Royal Soc.*, **449**, 139–53.

Hall, M.G., 1972, Vortex breakdown, *Ann. Rev. Fluid Mech.*, **4**, 195–218.

Leibovich, S., 1978, The structure of vortex breakdown, *Ann. Rev. Fluid Mech.*, **10**, 221–46.

Ruith, M.R., Chen, P., Meiburg, E., & Maxworthy, T., 2003, Three-dimensional vortex breakdown in swirling jets and wakes: direct numerical simulations. *J. Fluid Mech.*, **486**, 331–78.

Saffman, P.G., 1992, *Vortex Dynamics*, Cambridge University Press.

Sarpkaya, T., 1971, On stationary and travelling vortex breakdowns, *J. Fluid Mech.*, **45**(3), 545–59.

Squire, H.B., 1960, Analysis of the 'vortex breakdown' phenomenon. In: *Miszallaneen der Ange-wandten Mechanik Festschrift W. Tollmein*. M. Schaefer, ed. Akad. Verlag, 306.

Chapter 14

A Glimpse at Rapidly Rotating Turbulence

We touched on rapidly rotating turbulence in §6.9 where we saw that, after initiating the turbulence, columnar vortices aligned with the rotation axis tend to form on the *fast* timescale of Ω^{-1}. Such vortices dominate the large scales in laboratory experiments of rotating turbulence, though not the small scales if the Reynolds number is large. We now return to this topic, filling in some of the gaps. Throughout, we adopt a rotating frame of reference and take the Rossby number, Ro $= u/\Omega\ell_{\perp}$, to be low, where Ro is defined using the route-mean-square (rms) of the fluctuating velocity and $\ell_{\perp}$ is the integral scale of the turbulence (*i.e.* the size of the large eddies), measured in a plane normal to $\boldsymbol{\Omega}$.

We shall see that there are two distinct theoretical frameworks which have been used to describe the dynamics of rapidly rotating turbulence. The initial formation of the columnar vortices in laboratory experiments is usually well described by linear wave propagation, and indeed the columnar vortices are often just quasi-linear inertial wave packets. This is true even for Rossby numbers as high as Ro ≈ 0.4. However, when the Reynolds number is large, so there exists a wide range of length scales in the turbulence, nonlinear dynamics remain important at the smaller scales, because Ro based on the properties of the small eddies is usually large. Here, weakly nonlinear theory, such as *resonant triad interactions*, can be helpful, at least at scales for which the nonlinearity is not too strong.

14.1 Some Observations of Rapidly Rotating Turbulence

Modern experiments on rapidly rotating turbulence arguably began with Ibbetson & Tritton (1975), who looked at *freely decaying* turbulence in a rotating annulus in which Ro $= O(1)$. In line with most later studies, they observed the rapid growth of columnar vortices aligned with the rotation axis and correctly guessed that this was related to inertial wave propagation. By contrast, Hopfinger et al. (1982) and Dickenson & Long (1983) both used a vertically oscillating grid to *continually force* turbulence in a rotating tank, with Hopfinger et al. noting that long-lived columnar vortices dominated in regions where

The Dynamics of Rotating Fluids. P. A. Davidson, Oxford University Press. © Peter A Davidson (2024).
DOI: 10.1093/9780191994272.003.0014

Ro had fallen below $O(1)$. Moreover, these columnar vortices were observed to be mostly cyclonic, an observation that has since been repeated many times. Dickenson & Long, on the other hand, were interested in how fast the turbulence created by a grid spreads along the rotation axis and away from the grid. They found that, for Ro $> O(1)$, the turbulence spreads by turbulent diffusion at a rate proportional to $\sqrt{t}$, while for Ro $\leq O(1)$, it spreads with a constant velocity, reminiscent of energy transport by waves.

The three experiments mentioned above are all statistically inhomogeneous. Perhaps the first homogeneous rotating turbulence experiment was Wigeland & Nagib (1978), who mounted a rotating honeycomb in a wind tunnel, a technique which was later repeated and refined by Jacquin et al. (1990). Crucially, Jacquin et al. noted that, for Ro $\leq O(1)$, the turbulent integral scale parallel to the rotation axis (*i.e.* the length, in the direction of Ω, of the large eddies) grows at the rate $\ell_\parallel \sim t$. This is again reminiscent of linear wave propagation. It is also consistent with the observation of Dickenson & Long that turbulent energy spreads away from its source at a rate proportional to t.

The results of these early studies were subsequently confirmed and extended in the inhomogeneous experiments of Davidson et al. (2006), and also in the homogeneous experiments of Morize et al. (2005) and Staplehurst et al. (2008), all of which are described below. Perhaps the main point to note, however, is that there were hints from the outset that columnar vortex formation is closely linked to wave propagation when Ro $\lesssim O(1)$. Curiously, though, this seems to have been somewhat overlooked at the time, so that early theoretical studies of rotating turbulence focussed on nonlinear theories, particularly heuristic spectral closure schemes and weakly nonlinear dynamics. Both of these nonlinear theories aimed to emulate the formation of large-scale columnar eddies aligned with Ω, though usually on a *slow* timescale of $\ell_\perp/u$, rather than the observed fast timescale of Ω^{-1}.

14.2 Linear Structure Formation in Inhomogeneous, Freely Decaying Turbulence

The role of linear wave propagation in establishing large-scale columnar vortices is most evident in inhomogeneous experiments, where the columnar vortices are observed to emerge from a cloud of turbulence and propagate into adjacent, quiescent fluid. Consider, for example, the experiment of Davidson et al. (2006), as shown in Figure 14.1. Here turbulence is generated in the upper part of a rotating tank of water by dragging a planar mesh of bars part way

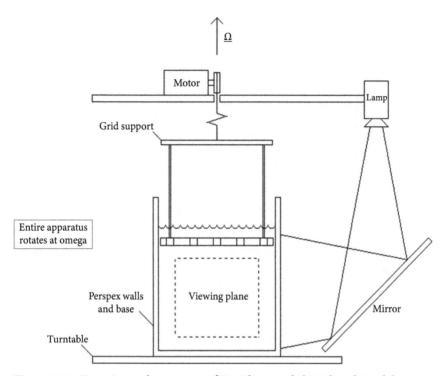

Figure 14.1 Experimental apparatus of Davidson et al. (2006) and Staplehurst et al. (2008).

through the tank, and then removing the mesh. This generates turbulence in which $Ro = u/\Omega\delta \gg 1$, δ being a measure of the size of the large eddies. Since $Ro \gg 1$, no inertial waves are created during the generation of the turbulence. However, as the turbulent kinetic energy decays, Ro declines and eventually we enter a regime in which $Ro \sim O(1)$. In particular, when Ro reaches a value of ≈ 0.4, columnar vortices aligned with the rotation axis are seen to emerge from the turbulent cloud, propagating in the axial direction, as shown schematically in Figure 14.2.

It is instructive to compare Figure 14.2 with the columnar structures shown in the numerical simulation of Figure 6.9(b), which are nothing more than inertial wave packets propagating along the rotation axis. In both cases, the wider vortices are observed to propagate faster than thin ones, which is consistent with the inertial wave packet described by equation (6.28) and shown in Figure 6.8. Given the similarities between Figures 14.2 and 6.9, it seems likely that the columnar structures in this laboratory experiment are also inertial wave packets, and this is readily confirmed. First, note that measurements

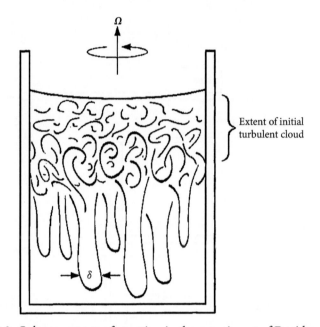

Figure 14.2 Columnar vortex formation in the experiment of Davidson et al. (2006).

show that the columnar vortices elongate at a constant rate, as shown in Figure 14.3(a). This figure shows the axial length, Δz, versus time of the dominant columnar vortex in six experiments which have different rotation rates and different bar sizes, b. In each case, we observe that $\Delta z \sim t$. Second, note that the growth rate is proportional to Ω and to the transverse scale of the vortices, δ, for which the bar size b acts as a proxy. This is evident from Figure 14.3(b), which shows the same data as Figure 14.3(a), but replotted as $\Delta z/b$ versus $2\Omega t$. Thus, we have $\Delta z \sim b\Omega t \sim \delta\Omega t$, which is consistent with Figure 6.7 and confirms that the columnar vortices are indeed quasi-linear inertial wave packets.

As noted in §6.9, one of the interesting features of this experiment is that inertial wave packets can propagate with amplitudes of up to Ro $= \hat{u}/\Omega\lambda \sim 0.4$, where λ is the dominant horizontal wavelength and $\hat{u}$ is the wave amplitude. A similar observation appears in the axisymmetric simulations of Sreenivasan & Davidson (2008), where the transition from quasi-linear wave propagation to the nonlinear suppression of waves starts at Ro ≈ 0.4 for anticyclones and Ro ≈ 1.4 for cyclones. The transition is also surprisingly abrupt, requiring a mere doubling of the Rossby number to suppress inertial waves.

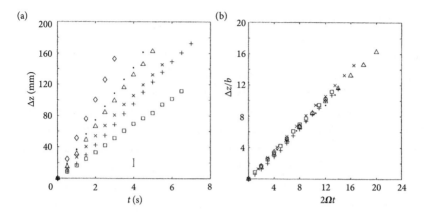

Figure 14.3 The variation of length, Δz, versus time of the dominant columnar vortex in six experiments which have different rotation rates and mesh sizes. (a) Δz versus t shows the columnar vortices growing at a constant rate. (b) $\Delta z/b$ versus $2\Omega t$ confirms that $\Delta z \sim b\Omega t$.

We conclude that the growth of columnar vortices in low-Ro turbulence is driven by linear wave propagation, at least for the geometry shown in Figure 14.2. This, in turn, rests on the remarkable property of inertial waves to spontaneously focus radiated energy onto the rotation axis, as shown in Figures 6.7 and 6.8, and as discussed in §6.2.3.

14.3 Linear Structure Formation in Homogeneous, Freely Decaying Turbulence

Most theories of turbulence are developed within the context of statistically homogeneous turbulence. It is of interest, therefore, to examine columnar vortex formation in rapidly rotating turbulence experiments which are (at least approximately) homogeneous. In particular, it is of interest to see if linear wave dynamics, as opposed to nonlinear interactions, remains the dominant mechanism by which the large eddies elongate along the rotation axis.

We now hit a problem. In statistically homogeneous experiments it is difficult to identify, and then track, individual vortices, unlike the inhomogeneous case shown in Figure 14.2. Consequently, in most homogeneous turbulence experiments, various statistical quantities are measured, such as the two-point velocity correlation

$$Q_{ij}(\mathbf{r}, t) = \langle u_i(\mathbf{x}, t)u_j(\mathbf{x} + \mathbf{r}, t)\rangle,$$

and then inferences made about the structure of the turbulence. (Here < ~ > indicates an ensemble average.) This, in turn, creates a problem for those homogeneous turbulence experiments in which wave propagation plays an important role. The difficulty arises because a great deal of phase information is lost when two-point velocity correlations are formed (see, for example, Davidson, 2015, §8.1.3), and this phase information provides the very basis by which wave packets propagate. The net result is that the statistical correlations normally measured in a homogeneous turbulence experiment tend to be blind, at least to some extent, to wave-packet propagation. Luckily, however, some residual phase information is retained, and this turns out to be sufficient to identify wave packets and so settle the question as to the origin of the columnar vortices in rapidly rotating turbulence. Indeed, the homogeneous experiments, like their inhomogeneous counterparts, suggest that the columnar eddies initially form through linear inertial wave propagation.

14.3.1 What Can Two-Point Velocity Correlations Capture?

Before examining the experimental evidence, perhaps it is worth asking what we might reasonably expect to see in a two-point velocity correlation when the turbulence is being structured by linear wave propagation. To that end, we start with a simple model problem. In §6.4 we used Fourier analysis to show that, if Ro << 1, an initial condition consisting of a single Gaussian eddy,

$$\Gamma = r u_\theta = \Lambda r^2 \exp\left[-(r^2 + z^2)/\delta^2\right], \qquad u_r = u_z = 0, \tag{14.1}$$

disperses to form a pair of columnar vortices described by

$$u_\theta \approx \Lambda\delta \int_0^\infty \kappa^2 e^{-\kappa^2} J_1\left(\frac{2\kappa r}{\delta}\right)\left\{\exp\left[-\left(\frac{z}{\delta} - \frac{\Omega t}{\kappa}\right)^2\right] + \exp\left[-\left(\frac{z}{\delta} + \frac{\Omega t}{\kappa}\right)^2\right]\right\} d\kappa, \tag{14.2}$$

$$u_z \approx \Lambda\delta \int_0^\infty \kappa^2 e^{-\kappa^2} J_0\left(\frac{2\kappa r}{\delta}\right)\left\{-\exp\left[-\left(\frac{z}{\delta} - \frac{\Omega t}{\kappa}\right)^2\right] + \exp\left[-\left(\frac{z}{\delta} + \frac{\Omega t}{\kappa}\right)^2\right]\right\} d\kappa, \tag{14.3}$$

where J_0 and J_1 are the usual Bessel functions, $\kappa = k_r\delta/2$, k_r is the radial wavenumber, and we use cylindrical polar coordinates. The term $\kappa^2 e^{-\kappa^2}$ in the integrands above ensures that these integrals are dominated by $\kappa \sim 1$. Hence, the wave energy is centred at $z \sim \pm\Omega\delta t$, exactly as expected from the equation $c_{gz} = 2\Omega/k_{\text{dom}} \sim \Omega\delta$, where k_{dom} is the dominant transverse wavenumber. Also, because a range of horizontal wavenumbers contribute to these integrals, each of which has a different axial group velocity, the wave energy is spread over an increasingly large area, which grows axially as $\Delta z \sim \Omega\delta t$. Thus, the Gaussian vortex gives rise to two wave packets propagating along the rotation axis, one upwards and one downwards, whose centres are located at $z \sim \pm\Omega\delta t$ and whose axial lengths grow as $\ell_z \sim \Omega\delta t$. (There is an approximation inherent in (14.2) and (14.3), as discussed in §6.4. However, this is shown to be relatively unimportant in Davidson et al., 2006.)

Now consider an initial condition consisting of a sea of such vortices, uniformly but randomly distributed in space. For $t > 0$, and continuing to assume that Ro $<< 1$ and Re $>> 1$, this creates a statistically homogeneous (though anisotropic) velocity field whose statistical properties can be determined using the solution for a single vortex, combined with superposition. The resulting two-point velocity correlations, both the exact correlations and those based on approximations (14.2) and (14.3), can be found in Staplehurst et al. (2008). A comparison of the exact and approximate two-point velocity correlations shows very little difference between the two, with the latter being given by

$$Q_\perp = Q_\perp^S(\mathbf{r}) + \frac{1}{2}\langle\mathbf{u}^2\rangle \int_0^\infty \kappa^3 e^{-\kappa^2} J_0\left(\frac{2\kappa r_\perp}{\delta}\right)\left\{\exp\left[-\left(\frac{r_\parallel}{\delta} - \frac{2\Omega t}{\kappa}\right)^2\right]\right.$$

$$\left. + \exp\left[-\left(\frac{r_\parallel}{\delta} + \frac{2\Omega t}{\kappa}\right)^2\right]\right\} d\kappa, \tag{14.4}$$

$$Q_\parallel = Q_\parallel^S(\mathbf{r}) - \frac{1}{2}\langle\mathbf{u}^2\rangle \int_0^\infty \kappa^3 e^{-\kappa^2} J_0\left(\frac{2\kappa r_\perp}{\delta}\right)\left\{\exp\left[-\left(\frac{r_\parallel}{\delta} - \frac{2\Omega t}{\kappa}\right)^2\right]\right.$$

$$\left. + \exp\left[-\left(\frac{r_\parallel}{\delta} + \frac{2\Omega t}{\kappa}\right)^2\right]\right\} d\kappa, \tag{14.5}$$

where

$$Q_\perp(\mathbf{r}) = \langle u_x(\mathbf{x})u_x(\mathbf{x}+\mathbf{r}) + u_y(\mathbf{x})u_y(\mathbf{x}+\mathbf{r})\rangle, \quad Q_\parallel(\mathbf{r}) = \langle u_z(\mathbf{x})u_z(\mathbf{x}+\mathbf{r})\rangle, \tag{14.6}$$

$r_\parallel = r_z$, $\mathbf{r}_\perp = \mathbf{r}-\mathbf{r}_\parallel$, and $\kappa = k_\perp\delta/2$. The terms $Q_\perp^S(\mathbf{r})$ and $Q_\parallel^S(\mathbf{r})$ are steady contributions to the velocity correlations which are localized around $|\mathbf{r}| < \delta$. It is clear

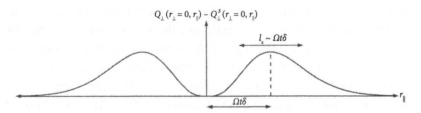

Figure 14.4 Schematic of the time-dependent term in (14.4). Note the area under the curve is conserved.

that the time-dependent parts of these correlations mimic, albeit in a statistical sense, the behaviour of individual vortices, as shown in Figure 14.4. In particular, the axial distance over which the velocities are appreciably correlated grows as $\ell_\parallel \sim \Omega \delta t$, with the peaks in the time-dependent part of the correlations located at $r_\parallel \sim \pm \Omega \delta t$ and the width of the peaks growing as $\Delta r_\parallel \sim \Omega \delta t$. Of course, this is due to inertial waves dispersing along the rotation axis.

Note that, for $\Omega t \gg 1$, (14.4) and (14.5) simplify considerably. Consider (14.4) evaluated at $r_\parallel = 2\Omega \delta t$. Then, for large Ωt, the only significant contribution to the integrand comes from $\kappa = 1 \pm O\left((2\Omega t)^{-1}\right)$, and so (14.4) integrates to give

$$Q_\perp = \frac{\sqrt{\pi}}{2e} \langle \mathbf{u}^2 \rangle J_0 \left(\frac{2r_\perp}{\delta}\right) \frac{1}{2\Omega t}, \quad r_\parallel = 2\Omega \delta t, \quad \Omega t \gg 1. \tag{14.7}$$

Moreover, as noted in Staplehurst et al. (2008), the time-dependent parts of the correlations, $Q_\perp - Q_\perp^s$ and $Q_\parallel - Q_\parallel^s$, are self-similar for large Ωt, provided that $r_\parallel$ is scaled by $2\Omega \delta t$.

Evidently, sufficient phase information is retained by the two-point velocity correlations to capture, in a statistical sense, the axial elongation of the vortices by linear wave propagation. Moreover, it is clear from Figure 6.7 that this behaviour is not merely an artefact of the Gaussian vortex (14.1). Rather, for almost any homogeneous initial condition, we would expect the two-point velocity correlations to become self-similar at large times, except in the region $|\mathbf{r}| < \delta$. It is necessary only to normalize $r_\parallel$ by $2\Omega \delta t$, where δ must now be interpreted as a characteristic transverse scale of the large eddies. This then provides us with a test for columnar vortex formation by linear wave propagation. Note that it is not just the two-point velocity correlations that behave in this way. In the model problem above, the two-point vorticity correlations exhibit similar behaviour.

We conclude this section with a word of warning. It is clear from (14.4) and (14.5) that Q_{ii} is constant, despite the axial growth of the eddies. Moreover, it is shown in Staplehurst et al. (2008) that, under linearized dynamics, $Q_\perp$ has the property

$$\int_0^\infty Q_\perp dr_\parallel = \text{constant}. \tag{14.8}$$

This is important because, for a statistically axisymmetric velocity field, the standard definition of the axial integral scale is

$$\ell_\parallel = \frac{1}{\langle u_\perp^2 \rangle} \int_0^\infty Q_\perp(r_\perp = 0, r_\parallel) dr_\parallel. \tag{14.9}$$

Moreover, $\ell_\parallel$ defined in this way is usually taken to be indicative of the axial length of the vortices. Clearly, definition (14.9) is inappropriate for linear wave propagation, as it fails to capture the axial growth of the wave packets. Of course, this failure of definition (14.9) is because of the loss of phase information in the two-point correlations.

14.3.2 The Experimental Evidence

Let us now consider the experimental evidence of Staplehurst et al. (2008), who generated freely decaying, homogeneous turbulence in the apparatus shown in Figure 14.1. Two-point correlations were obtained from luminosity measurements of light reflected from fine flakes in the water. Figure 14.5(a) shows one such set of two-point correlations as a function of $r_\parallel$, and it can be seen that the axial correlation length, $\ell_\parallel$, increases with time, as expected. The same data is then replotted as a function of $r_\parallel/(t - t_0)$ in Figure 14.5(b), where t_0 is indicative of the time when waves first appear. The collapse of all but the earliest of the curves confirms that, after an initial transient, the axial correlation length grows linearly with time, $\ell_\parallel \sim t$, consistent with linear wave-packet propagation.

Figure 14.6(a) shows the correlations curves (excluding the initial transient) for two experiments performed for different rotation rates, Ω. These curves are plotted as a function of $r_\parallel/(t - t_0)$ and, as expected, the correlation functions for each experiment collapse, but to different curves. Crucially, when replotted against $r_\parallel/2\Omega (t - t_0)$ in Figure 14.6(b), the correlation functions from both experiments collapse to the same curve. It follows that the axial correlation

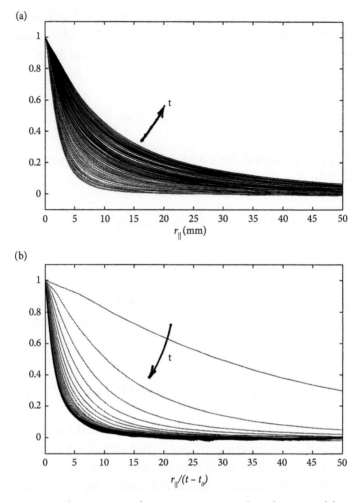

Figure 14.5 Correlation curves for one experiment plotted against: (a) $r_\parallel$; (b) $r_\parallel/(t - t_0)$.

length satisfies $\ell_\parallel \sim \Omega t$, consistent with structure formation by linear wave propagation. Thus, we see that, as with the inhomogeneous experiments, the evidence points to the columnar vortices being linear wave packets.

14.4 Weakly Nonlinear Interactions and Resonant Triad Theory

The initial formation of the large-scale columnar vortices in rapidly rotating turbulence is associated with linear wave propagation because the value of Ro, based on the integral scale, is chosen to be small. However, when Re is large,

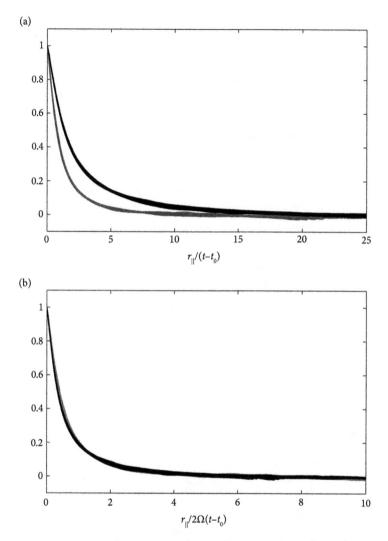

Figure 14.6 Correlation curves for two experiments performed at different rotation rates. These are plotted against: (a) $r_{\parallel}/(t - t_0)$; and (b) $r_{\parallel}/2\Omega (t - t_0)$.

there is always a wide range of length scales in a turbulent flow, and Ro based on the properties of the smallest eddies is invariably large. Thus, the smallest vortices develop through nonlinear processes and there must be some intermediate (or transitional) scale, say λ_{trans}, at which quasi-linear wave propagation gives way to nonlinear dynamics. In this respect, it is interesting to recall that the transition from quasi-linear inertial waves to the nonlinear suppression of wave motion can be surprisingly abrupt (see §14.2), occurring over a narrow range of Ro.

The dynamics of this cross-over range of scales is particularly interesting, because columnar vortices in the form of quasi-linear wave packets, which can grow in length as $\ell_z \sim \lambda_{\text{trans}} \Omega t$, may coexist with smaller, energetically weaker, nonlinear vortices. These smaller eddies then sit in the shadow of the columnar vortices and so, in principle, they can be sheared by the columnar strain field imposed on them, perhaps being wrapped around the larger columnar vortices in a helical fashion. (There is some hint of this kind of helical winding in Figure 14.7.) *If* this were to be a common occurrence (and that is still a matter of debate), then the axial length scale of the columnar wave packets could, in principle, be imprinted on the smaller vortices, leading to a systematic *nonlinear* transfer of kinetic energy towards axially elongated structures.

Irrespective of whether this picture is correct, a nonlinear transfer of kinetic energy towards axially elongated structures has been well documented for a variety of numerical simulations. Consequently, there is some incentive

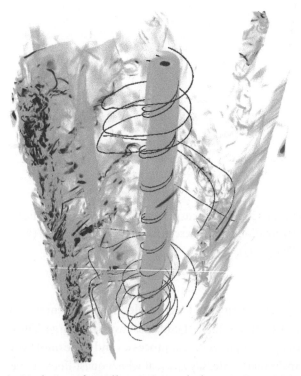

Figure 14.7 A simulation of rapidly rotating turbulence at Ro ~ 0.06, visualized using vorticity. Courtesy of Pablo Mininni and Annick Pouquet.

to characterize the weakly nonlinear dynamics of rotating turbulence, and to determine if there are any systematic statistical trends resulting from the background rotation, possibly including the strain imposed by the columnar vortices on smaller eddies. In this respect, perhaps the most developed weakly nonlinear theory is that of *resonant triad interactions*, which we now describe.

Resonant triad theory has been used to describe a number of weakly non-linear wave-turbulence systems, most notably Rossby-wave turbulence. In the case of weakly nonlinear Rossby waves, two overlapping wave packets with dominant wavevectors $\mathbf{p}$ and $\mathbf{q}$, and corresponding frequencies $\varpi_p = \varpi(\mathbf{p})$ and $\varpi_q = \varpi(\mathbf{q})$, can interact nonlinearly to generate a third wave with a wavevector of $\mathbf{k} = \mathbf{p} \pm \mathbf{q}$, and a frequency of $\varpi = \varpi_p \pm \varpi_q$. The response is weak, however, unless the excited frequency equals the natural frequency of Rossby waves of wavevector $\mathbf{k}$. In short, to get a significant nonlinear interaction, the dispersion relationship must admit solutions of the form

$$\varpi(\mathbf{p} \pm \mathbf{q}) = \varpi(\mathbf{p}) \pm \varpi(\mathbf{q}). \tag{14.10}$$

This resonance condition can indeed be satisfied for a restricted subclass of Rossby waves, and so a significant nonlinear transfer of energy between wave packets can occur, though this happens on the *slow* timescale of ℓ/u (see, for example, Pedlosky, 1979).

A complimentary theory for inertial waves has been developed by Smith & Waleffe (1999), where the dispersion relationship makes it somewhat harder to satisfy (14.10). However, for a small but finite Ro, (14.10) can be relaxed to the requirement that

$$\varpi(\mathbf{p} \pm \mathbf{q}) = \varpi(\mathbf{p}) \pm \varpi(\mathbf{q}) + O(\mathrm{Ro}), \tag{14.11}$$

for which near-resonant triad interactions can be identified. It is natural to ask, therefore, if this theory is consistent with the observation that there is a non-linear transfer of energy to low-frequency, columnar modes. Unfortunately, one problem with triadic interaction theory is that resonant triads possess no arrow of time, in the sense that for every initial condition that pushes energy into columnar modes, there is another that does the opposite. Luckily, there is a general result for conservative resonant triads, due to Hasselmann (1967), which states that a finite-amplitude wave is unstable to (*i.e.* passes energy to) two infinitesimal waves of lower frequency. This suggests, but does not prove, that there is a statistical bias for energy to pass to lower frequencies through triadic interactions.

The nonlinear energy transfers in rotating turbulence have been studied in numerical experiments carried out in so-called *periodic cubes*. This is a popular computational domain used in the field of turbulent simulations, which has the idiosyncratic boundary condition that a wave packet leaving the domain on the right (or top) immediately re-enters on the left (or bottom). In any event, when Ro is small (by virtue of the initial conditions), a nonlinear shift in energy towards columnar structures is indeed observed.

In summary, there are two weakly nonlinear cartoons in which energy accumulates in columnar vortices: (i) columnar wave packets shear smaller, weaker eddies at the transitional scale λ_{trans}; and (ii) a cascade of resonant triads nudges the wave energy to lower frequencies across a range of scales. Indeed, perhaps these two cartoons are flip sides of the same coin. In any event, the relative roles played by these cartoons is somewhat unclear.

14.5 The Cyclone, Anticyclone Asymmetry

We have already mentioned, in §14.1, that the first laboratory experiments showed an asymmetry between cyclonic and anticyclonic columnar vortices, with a dominance of cyclonic vortices. This has since been observed in numerical simulations (see, for example, Bartello et al., 1994) and also in more recent laboratory experiments (see Morize et al., 2005 and Staplehurst et al., 2008). A number of distinct explanations for this asymmetry have been proposed, with different authors favouring different mechanisms.

Perhaps the first explanation was offered by Bartello et al. (1994), who considered axisymmetric columnar vortices for which Ro ~ $O(1)$. They noted that, in an *inertial* frame of reference, the anticyclones were more likely to be unstable than cyclones to Rayleigh's centrifugal instability. The suggestion, therefore, is that anticyclones *do* form, but they are less likely to survive than the cyclones. The attraction of this explanation is its simplicity, though it seems more likely that the anticyclones do not form in the first place, rather than form and then disintegrate.

A related, though distinct, explanation was offered by Sreenivasan & Davidson (2008). They observed that, for the particular initial condition of a Gaussian vortex, the transition from quasi-linear wave dispersion to the nonlinear suppression of inertial waves starts at Ro ≈ 0.4 for anticyclones, but at Ro ≈ 1.4 for cyclones. Hence, in the range 0.4 < Ro < 1.4, cyclonic columnar wave packets spontaneously form while anticyclonic wave packets do not. The suggestion, then, is that for rotating turbulence in which Ro ~ $O(1)$, one might expect to see more cyclones than anticyclones emerge in the form of inertial wave packets. The weakness of this argument, however, is that the observations are restricted to a specific initial condition.

A third explanation has been offered by Morize et al. (2005). They consider the vortex stretching term in the vorticity equation, $\boldsymbol{\omega} \cdot \nabla \mathbf{u}$, and note that, in an inertial frame of reference, this becomes $(\boldsymbol{\omega} + 2\boldsymbol{\Omega}) \cdot \nabla \mathbf{u}$. The suggestion, then, is that cyclonic vorticity is more likely to be amplified by vortex stretching than anticyclonic vorticity. As with the explanation of Bartello et al. (1994), the attraction of this mechanism is its simplicity. Note that there is a significant difference between this explanation and that of Sreenivasan & Davidson (2008), in that it suggests columnar vortex formation is a nonlinear process, whereas Sreenivasan & Davidson (2008) assume that it is wave-like.

There is still some discussion as to which of these mechanisms is the most plausible. However, perhaps it worth noting that more than one mechanism may be present in any given experimental configuration. Moreover, the dominant mechanism may well depend on the value of Ro, and on whether the turbulence is continually forced or is freely decaying.

***

This concludes our brief glimpse at rotating turbulence. As with turbulence in general, there is still much that is poorly understood. In particular, the relative roles played by linear wave packets and nonlinear dynamics in structuring the turbulence remain an open question. Part of the difficulty is that different groups have different views on what the important questions are, and what constitutes the canonical geometry. For some, the canonical object is turbulence in a periodic cube, which can be realized only in a computer, while for others it is a laboratory experiment of one form or another. Some think the initial structuring of the large scales is important, which emphasizes the role of waves, while for others it is the long-term properties of the smaller scales, which emphasizes nonlinearly. In any event, this is a debate that is likely to run for some time.

References

Bartello, P., Metais, O., & Lesieur, M., 1994, Coherent structures in rotating three-dimensional turbulence, *J. Fluid Mech.*, **237**, 1–29.

Davidson, P.A., 2015, *Turbulence*, 2nd Ed., Oxford University Press.

Davidson, P.A., Staplehurst, P.J., & Dalziel, S.B., 2006, On the evolution of eddies in a rapidly rotating system, *J. Fluid Mech.*, **557**, 135–44.

Dickenson, S.C., & Long, R.R., 1983, Oscillating-grid turbulence including effects of rotation, *J. Fluid Mech.*, **126**, 315–33.

Hasselmann, K., 1967, A criterion for nonlinear wave stability, *J. Fluid Mech.*, **30**(4), 737.

Hopfinger, E.J., Browand, F.K., & Gagne, Y., 1982, Turbulence and waves in a rotating tank, *J. Fluid Mech.*, **125**, 505–34.

Ibbetson, A., & Tritton, D.J., 1975, Experiments on turbulence in a rotating fluid, *J. Fluid Mech.*, **68**(4), 639–72.

Jacquin, L., Leuchter, O., Cambon, C., & Mathieu, J., 1990, Homogenous turbulence in the presence of rotation, *J. Fluid Mech.*, **220**, 1–52.

Morize, C., Moisy, F., & Rabaud, M., 2005, Decaying grid-generated turbulence in a rotating tank, *Phys. Fluids*, **17**, 095105.

Pedlosky, J., 1979, *Geophysical Fluid Dynamics*, 2nd Ed., Springer-Verlag.

Smith, L.M., & Waleffe, F., 1999, Transfer of energy to two-dimensional large scales in forced, rotating three-dimensional turbulence, *Phys. Fluids*, **11**(6), 1608–22.

Sreenivasan, B., & Davidson, P.A., 2008, On the formation of cyclones and anticyclones in a rotating fluid, *Phys. Fluids*, **20**(8), 085104.

Staplehurst, P.J., Davidson, P.A., & Dalziel, S.B., 2008, Structure formation in homogeneous freely decaying rotating turbulence, *J. Fluid Mech.*, **598**, 81–105.

Wigeland, R.A., & Nagib, H.M., 1978, Grid generated turbulence with and without rotation about the streamwise direction, *IIT Fluid & Heat Transfer Rep.*, R 78–1.

PART III
ILLUSTRATIVE EXAMPLES
OF ROTATING FLOWS IN NATURE

Chapter 15

Tornadoes, Dust Devils, and Tidal Vortices

We now discuss a range of naturally occurring rotating flows. The discussion throughout is necessarily brief, intended as a steppingstone to further study. We start, in this chapter, with a largely *qualitative* description of tornadoes, dust devils, and tidal vortices.

15.1 Tornadoes and Waterspouts

15.1.1 Observational Evidence as to the Nature of Tornadoes

Most tornadoes occur in North America, numbering around 10^3 per year. They generate some of the highest natural wind speeds on earth, reaching values as large as ~100 m/s. Waterspouts form over water and closely resemble tornadoes.

Tornadoes may be 1–2 km high, while their radius of maximum wind varies from 20 m to roughly 400 m, the mean radius being around 150 m. The majority of tornadoes exhibit maximum wind speeds in the range of 40–90 m/s. They are recognizable from their low-pressure condensation funnels (see Figure 15.1), which may also be filled with dust and other debris. Typically, tornadoes last for around 10 minutes, but a small proportion survives for over an hour. Examples of extreme tornadoes in the USA are listed in Table 15.1.

Detailed measurements of the flow in and around tornadoes are difficult to obtain because of their short lives and capricious nature. Early estimates were based on tracking the movement of dust and debris in movies, though that has now given way to the use of mobile Doppler radar. In addition to the horizontal swirling motion, there is a radial, Ekman-like flow near the ground which spirals inwards towards the base of the tornado, as indicated in Figure 15.2. Moreover, there is an intense updraft within the tornado itself, the updraft being particularly intense near the ground. Curiously, Doppler

The Dynamics of Rotating Fluids. P. A. Davidson, Oxford University Press. © Peter A Davidson (2024).
DOI: 10.1093/9780191994272.003.0015

Figure 15.1 A tornado in Oklahoma in 1999. Source: NOAA.

Table 15.1. Examples of extreme tornadoes in the USA prior to 2020.
Data from NOAA

Highest recorded wind speed	3 May 1999	Oklahoma	135 m/s
Largest recorded tornado	31 May 2013	El Reno	Width of damage path was 4 km
Most deadly recorded tornado	18 March 1925	Missouri, Illinois, and Indiana	Lasted more than 3 hours, 695 deaths
Most costly on record	22 May 2011	Missouri	Cost of $2.8 billion

radar measurements often indicate a down-flow near the axis, as shown in Figure 15.2(b). The same downdraft is frequently observed in laboratory experiments.

Although tornadoes are too small to feel directly the Coriolis force associated with the earth's rotation, their sense of rotation is almost always cyclonic. This is because most tornadoes form below supercell thunderstorms, which *are* influenced by the Coriolis force, albeit indirectly. Supercells are large storm systems that stretch from the ground to the tropopause and are characterized by a deep, persistent, vertical vortex within their interior, called a *mesocyclone*. An example of a supercell is shown in Figure 15.3. The vertical vorticity within the mesocyclone is thought to originate from the wind shear near the ground, whose horizontal vorticity gets caught up in powerful updrafts which tilt the vorticity towards the vertical as it is pulled up into the supercell. It is important to realize, however, that most supercells do not generate tornadoes, and those which do form tornadoes need not have the most intense mesocyclones. Indeed, the precise conditions under which tornadoes form are still

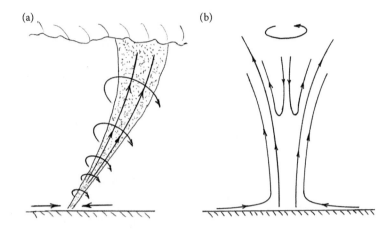

Figure 15.2 (a) A schematic of the three-dimensional flow generated by a tornado. (b) A schematic of the secondary flow in the vertical plane showing flow reversal.

Figure 15.3 A supercell thunderstorm in 2004 in New Mexico. Source: NOAA.

poorly understood. Of course, this makes predicting the appearance of tornadoes difficult. In any event, the sense of rotation in a mesocyclone is almost invariably cyclonic, and this is why tornadoes are predominantly cyclonic.

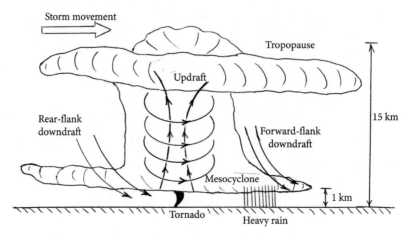

Figure 15.4 A schematic of the flow in a supercell thunderstorm.

Figure 15.4 is a schematic of the flow within a supercell. The mesocyclone may be 2–10 km in diameter, and the tornado forms immediately below it. In addition to the mesocyclone, there is a rear-flank downdraft which drives air towards the tornado, and a forward-flank downdraft which is cooled by rain. These flows, along with their temperature differences, are thought to be important for tornado formation (see, e.g., Rotunno, 2013).

15.1.2 The Ekman-Like Boundary Layer Beneath a Rankine Vortex

The strong updraft within a tornado is striking, with the swirling and axial velocities of similar magnitudes near the ground. The origin of this updraft may be understood by considering the laminar, axisymmetric, boundary-layer analysis of Burggraf et al. (1971).

Consider a thin vortex tube of radius r_0 whose centreline sits on the z-axis and is normal to, and coaxial with, a disc of radius a. It is assumed that $a \gg r_0$, and that the disc is located on the plane $z = 0$, as shown in Figure 15.5. If we adopt cylindrical polar coordinates centred on the vortex tube and disc, then away from the disc the velocity is azimuthal and a function of r only. Outside the vortex tube, Stokes' theorem tells us that this azimuthal velocity takes the form $u_\theta = \Phi/2\pi r$, where Φ is the net flux of vorticity along the vortex tube. Near the axis, on the other hand, u_θ will increase linearly with r, so the overall velocity profile remote from the disc will resemble that of a Rankine vortex.

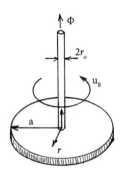

Figure 15.5 A thin vortex tube is normal to, and coaxial with, a disc of radius a.

Near the disc a boundary layer must form, across which the azimuthal velocity drops to zero. Let us focus on the region outside the thin vortex tube, where the velocity external to the boundary layer is $u_\theta = \Phi/2\pi r$. Given Bödewadt's solution for the boundary layer beneath the external flow $u_\theta = \Omega r$ (see §5.1.1 or Bödewadt (1940)), it is natural to look for a similar boundary-layer structure, within which the fluid spirals radially inward. As we shall see, such a solution does indeed exist, at least away from the outer edge of the disc. However, this solution represents only one part of a composite boundary layer, that comprises a thin, viscous, Bödewadt-like solution adjacent to the disc, which is capped by a thicker, almost inviscid, outer layer. The function of the inner layer is to meet the no-slip condition, and that of the outer layer is to carry the bulk of the radial mass flux.

Let us consider the inner, viscous layer first. The essential features of this layer can be established using a simple scaling analysis. The first point to note is that there is a radial inflow near the boundary, driven by a radial pressure gradient. That is, outside the boundary layer there is a radial force balance between $\partial p/\partial r$ and $\rho u_\theta^2/r$, resulting in a low pressure on the axis. This radial pressure gradient is imposed on the fluid within the inner boundary layer where the centrifugal force is smaller, and the resulting imbalance between $\partial p/\partial r$ and $\rho u_\theta^2/r$ drives the fluid radially inward. Now, this radial pressure gradient is resisted by the radial viscous force $\rho \nu \nabla^2 u_r$. It follows that there is a force balance of the form

$$\rho \nu \left|\nabla^2 u_r\right| \sim \rho \nu u_r/\delta_{in}^2 \sim \partial p/\partial r \sim \rho u_\theta^2/r,$$

where δ_{in} is the thickness of the inner boundary layer. Now suppose that, as in Bödewadt's solution, the radial and azimuthal velocities are of similar magnitudes. Then

$$\nu/\delta_{in}^2 \sim u_\theta/r \sim \Phi/2\pi r^2,$$

or equivalently,

$$\delta_{in}(r) \sim r\sqrt{2\pi\nu/\Phi}. \tag{15.1}$$

Given this estimate of the boundary-layer thickness, and given an external velocity of $u_\theta = \Phi/2\pi r$, it is natural to look for a laminar, axisymmetric solution of the form

$$u_r = \frac{\Phi}{2\pi r}f(z/\delta_{in}), \quad u_\theta = \frac{\Phi}{2\pi r}g(z/\delta_{in}), \quad u_z = \frac{\sqrt{\nu\Phi/2\pi}}{r}h(z/\delta_{in}), \tag{15.2}$$

where $\delta_{in}(r) = r\sqrt{2\pi\nu/\Phi}$. Continuity then requires that f and h are related by

$$\eta f'(\eta) = h'(\eta), \quad \eta = z/\delta_{in}, \tag{15.3}$$

while the radial and azimuthal components of the Navier–Stokes equation yield two more equations relating f, g, and h:

$$hf'(\eta) - f\frac{d}{d\eta}(\eta f) + (1 - g^2) = f''(\eta) + \frac{2\pi\nu}{\Phi}\left[\frac{1}{\eta}\frac{d}{d\eta}(\eta^3 f'(\eta))\right], \tag{15.4}$$

$$(h - \eta f)g'(\eta) = g''(\eta) + \frac{2\pi\nu}{\Phi}\left[\frac{1}{\eta}\frac{d}{d\eta}(\eta^3 g'(\eta))\right]. \tag{15.5}$$

For a large Reynolds number, $\Phi/\nu \gg 1$, these simplify to

$$hf'(\eta) - f\frac{d}{d\eta}(\eta f) + (1 - g^2) = f''(\eta), \tag{15.6}$$

$$(h - \eta f)g'(\eta) = g''(\eta), \tag{15.7}$$

provided that $r \gg z$, i.e. provided we avoid the axis.

Finally, we introduce the Stokes streamfunction, Ψ, defined by (2.9). Following Burggraf et al. (1971), we express Ψ in terms of an auxiliary function, $\psi_0(\eta)$:

$$\Psi(r, z) = -r\sqrt{\nu\Phi/2\pi}\,\psi_0(\eta). \tag{15.8}$$

It is readily confirmed that (15.8) ensures $f = \psi'_0(\eta)$ and $h = \eta\psi'_0(\eta) - \psi_0$, and so (15.6) and (15.7) can be rewritten in terms of $\psi_0(\eta)$ and g, according to

$$\psi_0'''(\eta) + \psi_0\psi_0''(\eta) + \psi_0'^2 + (g^2 - 1) = 0, \tag{15.9}$$

and

$$g''(\eta) + \psi_0 g'(\eta) = 0. \tag{15.10}$$

Actually, Burggraf et al. (1971) show that we have $u_\theta \ll u_r$ within the inner layer, and so g can be dropped from (15.9). This then integrates, subject to $\psi_0'(0) = \psi_0(0) = 0$, to give

$$2\psi_0'(\eta) + \psi_0^2 = \eta^2 - 2\alpha\eta, \tag{15.11}$$

for some constant α. Now $f = \psi_0'(\eta)$ remains finite and of order unity for large η, and so (15.11) requires $\psi_0^2(\eta \to \infty) \sim \eta^2 - 2\alpha\eta$. Thus, we find $\psi_0 \sim \alpha - \eta$, and hence $\psi_0'(\eta) = -1$, for $\eta \to \infty$. This added constraint is sufficient to determine α (Burggraf et al., 1971). The function ψ_0 then follows from integrating (15.11), and g from (15.10).

The key point, however, is that the inward radial mass flux within the inner boundary layer is

$$\dot{m}_{\text{in}}(r)/\rho = 2\pi r \int_0^{\delta_{\text{in}}} |u_r| dz = \Phi\delta_{\text{in}} \int_0^1 |f| \, d\eta \sim r\sqrt{2\pi\nu\Phi}. \tag{15.12}$$

This starts out as $\dot{m}_{\text{in}}/\rho \sim a\sqrt{2\pi\nu\Phi}$ towards the outer edge of the disc, but approaches zero as we get close to the vortex tube, and so the inner boundary layer continually *detrains* mass.

It turns out that all of this mass flux gradually transfers to an outer boundary layer, whose depth scales as $\delta_{\text{outer}} \sim a\sqrt{2\pi\nu/\Phi} \sim \delta_{\text{in}}(a)$. Thus, as we approach the vortex tube, the outer boundary layer becomes much thicker than the inner one, and so it sustains considerably less shear stress. It also carries nearly all of the radial mass flux, *i.e.* $\dot{m}_{\text{outer}}/\rho \sim a\sqrt{2\pi\nu\Phi}$, as it approaches the central vortex tube. This mass is then redirected upward into the vortex, causing an eruption of boundary-layer fluid into the vortex core.

The characteristic axial and azimuthal velocities at the base of the vortex tube are therefore

$$u_z \sim \frac{\dot{m}_{\text{outer}}/\rho}{\pi r_0^2} \sim \frac{\Phi a\sqrt{2\pi\nu/\Phi}}{\pi r_0^2} \sim \frac{\Phi\delta_{\text{outer}}}{\pi r_0^2}, \quad u_\theta \sim \frac{\Phi}{2\pi r_0}. \tag{15.13}$$

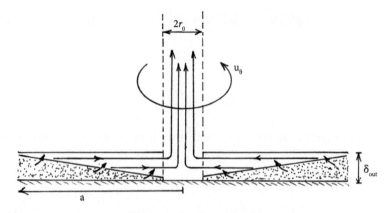

Figure 15.6 A schematic of the composite boundary layer beneath a Rankine vortex. The inner layer, which transfers all of its mass flux to the outer layer, is shown speckled.

Now it is observed that $u_z \sim u_\theta$ at the base of the vortex tube (Phillips & Khoo, 1987), and so we conclude from (15.13) that $\delta_{\text{outer}} \sim r_0$. The radius of the vortex tube is then determined through the relationship $r_0 \sim a\sqrt{2\pi\nu/\Phi}$. The situation is as shown in Figure 15.6. Of course, as with Bödewadt's solution applied to a *finite* disc, there is an outer annulus within which the asymptotic solution (15.2) does not apply, as the boundary layer takes time to build up. In this sense, Figure 15.6 is misleading.

15.1.3 Implications for Tornadoes

The predictions of Burggraf et al. (1971) have been confirmed experimentally by Phillips & Khoo (1987). Although this is a laminar, axisymmetric flow, it is nevertheless highly suggestive as to the origin of the updraft in a tornado. The primary weakness of the comparison is, of course, that there is no obvious analogue of the parameter a in a tornado. Perhaps the best we can do is loosely associate a with some characteristic *domain of influence* of the tornado, a_{doi}, in the sense that, for $r < a_{\text{doi}}$, the velocity induced by the tornado exceeds the characteristic ground-level wind speed associated with the supercell, say u_{sc}. Now, in the analysis above we considered a and Φ as being independently prescribed, and then r_0 determined by the relationship $r_0 \sim a\sqrt{2\pi\nu/\Phi}$. In a tornado, however, a_{doi} and Φ are not independent, but rather related by $u_{\text{sc}} \sim \Phi/2\pi a_{\text{doi}}$. Still, we might expect r_0 and a_{doi} to be related by $r_0 \sim a_{\mathrm{doi}}\sqrt{2\pi\nu_t/\Phi}$, where ν has been replaced by an eddy viscosity, ν_t. Introducing the speed $u_{\max} = \Phi/2\pi r_0$, this suggests that a_{doi} satisfies

$$\frac{a_{\mathrm{doi}}}{r_0} \sim \frac{u_{\max}}{u_{\mathrm{sc}}} \sim \sqrt{u_{\max}r_0/\nu_t}. \qquad (15.14)$$

Whether or not (15.14) is substantially correct, we might anticipate that, for $r < a_{\mathrm{doi}}$, the boundary layer beneath certain classes of tornadoes is reminiscent of Figure 15.6.

15.1.4 Waterspouts

Waterspouts (see Figure 15.7) are separated into two classes: *tornadic* and *fair weather*. Tornadic waterspouts are simply tornadoes that form over water. They are associated with a supercell mesocyclone and have properties which are similar to those described above for tornadoes. Fair-weather waterspouts, on the other hand, are weaker and more common, with wind speeds less than 30 m/s. They form in coastal waters where they are associated with convective cumulus cloud. Several hundred form each year off the coast of Florida.

Studies of waterspouts are largely confined to observational data, the most notable being Golden (1971, 1974). As with tornadoes, much of a mature waterspout is made visible by a low-pressure condensation funnel. Above the sea surface, high wind speeds whip up the surface of the water giving rise to a *spray vortex*, out of which rises an annular *spray sheath*, as shown in Figure 15.8(a). The spray sheath has a calm central eye which is surrounded by

Figure 15.7 A waterspout off the coast of Florida. Courtesy of J. Golden, NOAA.

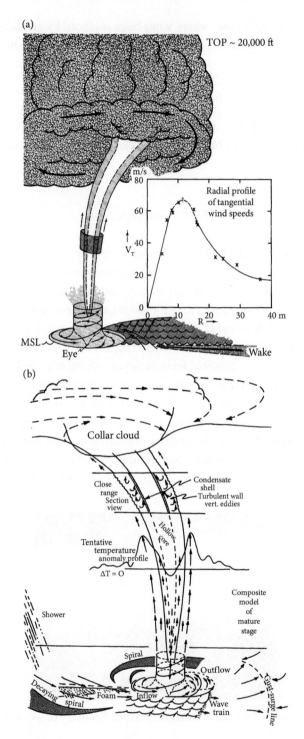

Figure 15.8 Schematics of a waterspout. (a) Reproduced from Golden (1971). (b) Reproduced from Golden (1974).

strong tangential winds. The maximum tangential velocity occurs just outside of the eye and particle tracking within the spray vortex suggests a Rankine-like vortex distribution (see Figure 15.8a). The core of a waterspout is observed to contain subsiding air, as shown in Figure 15.8(b), similar to that detected in tornadoes.

Mature waterspouts generally survive for around 2–20 minutes. Some waterspouts propagate along the water surface, with the base moving faster than the top. In such cases, the vortex soon becomes highly contorted and dissipates. Quite how the vortex first forms is unclear, although it has been suggested that the source of vorticity may be convective stretching and tilting upward of boundary-layer vorticity at the sea surface.

15.1.5 The Evidence for Vortex Breakdown in Tornadoes and Waterspouts

According to Chapter 13, the prerequisites for vortex breakdown are a slender vortex tube, an intense axial vorticity, an order-one axial velocity, and a positive axial pressure gradient. Given the observation that $u_z \sim u_\theta$, and that u_z decreases with height, both tornadoes and waterspouts would seem to be prime candidates for vortex breakdown.

The anecdotal evidence for vortex breakdown in waterspouts is reasonably strong and goes back to 1648. (See Antonescu, 2017 and Lugt, 1989, for the historical evidence of vortex breakdown in waterspouts.) For example, Figure 15.9 shows two of Michaud's 1789 drawings of waterspouts, one of which shows a possible spiral breakdown, and the other a sudden expansion in the condensation funnel.

The observational evidence of vortex breakdown within tornadoes is, by contrast, a little more tenuous, as it relies on an interpretation of the images of condensates, dust, or debris within a highly unsteady flow. Nevertheless, some movies of tornadoes show flows which are suggestive of vortex breakdown. Also, the analysis of Burggraf & Foster (1977) of vortex breakdown in tornado-like vortices yields results consistent with the observations.

Laboratory experiments in vortex chambers also provide evidence of vortex breakdown in 'mini tornadoes' (see Maxworthy, 1982). For example, the laboratory experiments of Phillips (1985) display evidence of both axisymmetric and spiral breakdown, as shown in Figure 15.10. Of course, the

Figure 15.9 Michaud's 1789 drawings of waterspouts near Nice showing a possible spiral breakdown and a sudden expansion in the condensation funnel. Reproduced from Michaud (1801).

helical flow within a vortex chamber is a long way from a real tornado, being laminar or else only mildly turbulent. Nevertheless, it is tempting to extrapolate these observations to real atmospheric vortices, and many choose to do so.

(a)   (b)

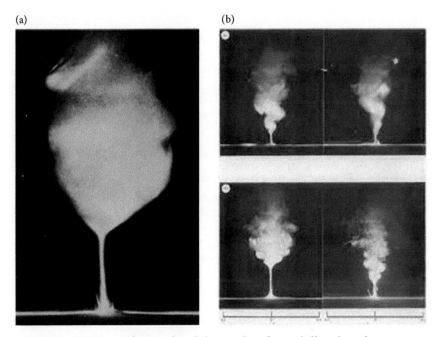

Figure 15.10 Images of vortex breakdown taken from Phillips (1985).
(a) Axisymmetric breakdown. (b) Helical breakdown.

15.2 Dust Devils

Dust devils are columnar vortices not unlike tornadoes, although they are
smaller and weaker, with maximum wind speeds in the range 10–30 m/s.
Terrestrial dust devils are typically around 1–10 m wide and a few tens of
metres tall, but on occasions they can grow to be tens of metres across and
over 1 km high. Lifetimes are measured in minutes, rarely exceeding 30 minutes. Examples of dust devils are shown in Figures 15.11 and 15.12, the latter
being a nineteenth-century woodcut. Although documented by the Romans,
the first detailed study of dust devils is by Baddeley (1860), an army surgeon.
Subsequent scientific interest remained modest, but flourished following the
discovery of dust devils on Mars.

Dust devils form in deserts on hot days under a clear sky and with light
winds. Evidently, their generation mechanism is entirely different to that of
tornadoes. When the radiation from the sun is intense, radiative heating of the
ground can produce an unstable thermal stratification, and hence convection.

(a)   (b)

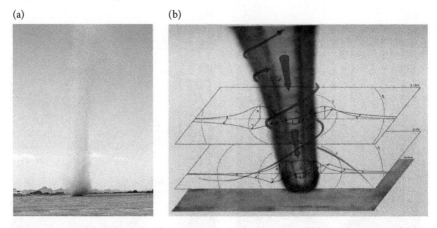

Figure 15.11 (a) A dust devil in Arizona. Source: NASA (b) A schematic of the lower part of a dust devil showing a downdraft near the axis. Reproduced from Sinclair (1966).

Figure 15.12 Dust devils illustrated in a French woodcut from the nineteenth century 'Trombes de Sable dans la Steppe'. Source: Lorenz, R.D. et al., 2016, History and applications of dust devil studies. *Space Sci. Rev.*, 203, 5–37.

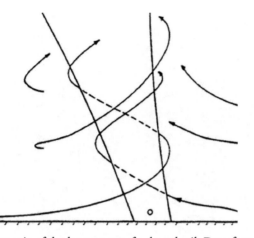

Figure 15.13 Schematic of the lower part of a dust devil. Data from Sinclair (1966).

This provides the energy for dust devils, though not their vorticity. However, if the wind simultaneously creates a shear flow, either in the form of horizontal boundary-layer vorticity, or perhaps inclined vortex sheets resulting from flow around terrain, then the combination of thermal convection and wind shear can give rise to a dust devil (Lugt, 1983). In the case of boundary-layer vorticity, the horizontal vorticity gets caught up in the updraft of the dust devil, which then tilts the vorticity towards the vertical as it is pulled up into the vortex (Maxworthy, 1973).

The flow within a dust devil can be visualized, at least in part, by the dust that gets caught in the updraft. Like tornadoes, there is a radially inward flow along the ground towards the base of a dust devil (see Figure 15.13), as well as a strong updraft in the vortex core. Also, as with waterspouts, a downdraft is typically observed on the axis near the base of the vortex, as shown in Figure 15.11(b). Unlike tornadoes, there is no bias towards cyclonic rotation. (A review of the flow within dust devils may be found in Balme & Greeley, 2006.) Finally, intense electric fields have long been observed in dust devils (Baddeley, 1860), as the smaller dust particles acquire negative charge during collisions.

Dust devils also occur on Mars, being first detected in the late 1970s by Viking 2. Martian dust devils are typically much larger than their terrestrial counterparts, and in November 2016 five such dust devils of heights

Figure 15.14 An illustration of a dust devil on Mars. Courtesy of JPL/MSSS/NASA.

0.5–1.9 km were imaged by the Mars Orbiter Mission. An illustration of a Martian dust devil is shown in Figure 15.14.

15.3 Tidal Vortices

Tidal vortices, sometime called *whirlpools* or *maelstroms*, usually occur in particularly narrow ocean straits which are subject to intense tidal flows. Unlike tornadoes, waterspouts, and dust devils, which have all been reasonably well documented, the observational data for whirlpools is somewhat limited. In any event, they are thought to arise from the interaction of strong tidal currents with: (i) outcrops of rock protruding from the coastline, or (ii) submerged rock formations, or (iii) an unusual seabed topography.

Their dramatic appearance has made tidal vortices popular in fiction, from the Charybdis of Homer's *Odyssey* to Edger Alan Poe's *Descent into the Maelstrom* and Jules Verne's *20,000 Leagues Under the Sea*. The Scandinavian word malstrøm was first applied to the tidal vortices near the Lofoten islands in Norway, which make an appearance in Olaus Magnus' 1539 Carta Marina, labelled as 'horrenda caribdis' on that map (see Figure 15.15). The word was subsequently translated into English as maelstrom by Poe, and the term is now synonymous with a tidal vortex.

The Norwegian coastline is famous for its maelstroms, particularly the *Saltstraumen* near Bodø and the *Moskstraumen* in the Lofoten islands. The former

Figure 15.15 Carta Marina of Olaus Magnus (1539) showing the malstrøm near Lofoten. Source: Wikimedia.

has one of the strongest tidal currents in the world, and when this interacts with coastal outcrops it produces tidal vortices whose diameter can be as large as 10 m (Gjevik, 2009). In the Moskstraumen, on the other hand, tidal currents of up to 10 m/s interact with an uneven seabed to produce vortices in *open sea*. Table 15.2 lists some of the more famous whirlpools.

The Corryvreckan tidal vortex is the third largest in the world (see Figure 15.16) and is located in the Gulf of Corryvreckan, which runs between the islands of Jura and Scarba on the west coast of Scotland. A mannequin, complete with a depth gauge and life jacket, was once thrown into the Corryvreckan. The mannequin disappeared into the whirlpool, and when it resurfaced some distance downstream, the depth gauge read over 200 m. It is said that George Orwell almost drowned in the whirlpool while he was staying on Jura.

Perhaps it is worth describing the Corryvreckan whirlpool in a little more detail, if only to illustrate the complexity of the flow involved. The Gulf of Corryvreckan is 3.2 km long and 1.1 km wide with a maximum depth of around 230 m, as shown in the depth map of Figure 15.17. It is subject to powerful tidal currents with speeds of up to 5 m/s, being classified by the Royal Navy as 'very violent and dangerous'. An intense whirlpool is observed during the east-flowing ebb tide, though not during the west-flowing flood tide. (The flood tide

Table 15.2 Some of the more famous whirlpools

Whirlpool	Location	Description
Saltstraumen	The Saltstraumen strait, 10 km southeast of Bodø, Norway	400 million cubic metres of water flows through this 150 m wide strait in 6 hours. The tide is deflected by coastal outcrops and whirlpools form up to 10 m across.
Moskstraumen	The Lofoten Islands, off the coast of Norway	The Moskstraumen occurs in *open sea* as a result of tidal flow over an uneven seabed. The whirlpools are up to 50 m across.
Skookumchuck	The Skookumchuck narrows at the entrance of the Sechelt Inlet, British Columbia, Canada	A difference in water level of 2 m forces a tidal flow through the Skookumchuck narrows at speeds of up to 8 m/s. This results in standing waves and whirlpools.
Naruto whirlpools	Naruto strait, located between Naruto and Awaji Island, Japan	The tide causes a difference in water level of 1.5 m between the Inland Sea and the Pacific. This forces water through the Naruto strait at speeds of up to 5 m/s, creating vortices up to 20 m in diameter and with rotational speeds of around 3 m/s.
Corryvreckan whirlpool	Gulf of Corryvreckan, located between the islands of Jura and Scarba off the west coast of Scotland	The whirlpool is created by a powerful tidal flow of up to 5 m/s passing around a submerged outcrop which extends from the coast of Scarba. The vortex appears during the east-flowing ebb tide only.

produces mostly steep waves and turbulence.) The location of this whirlpool coincides with a submerged outcrop of rock extending out from the island of Scarba, as indicated by the term 'wall' in Figure 15.17. Evidently, this tidal vortex owes its existence to vortex shedding from the outcrop, although the details are not well understood.

It is curious that the intense whirlpool appears only during the ebb tide, and this suggests that some form of asymmetry in the rock formation is central to the generation of this tidal vortex. A depth map of the region of interest is shown in Figure 15.18. The whirlpool is located slightly downstream of a submerged rock pinnacle, the rock apex being located by the cross AA'–BB' in Figure 15.18. The pinnacle itself sits at the outer edge of a submerged shelf (shown as rust coloured) which extends south from Scarba and into the gulf, extending a distance of around 300–400 m and at a depth around 40 m. There is a sharp cliff-like drop down to the floor of the gulf at the edge of the shelf,

Figure 15.16 The Corryvreckan whirlpool off the island of Scarba. Courtesy of Walter Baxter.

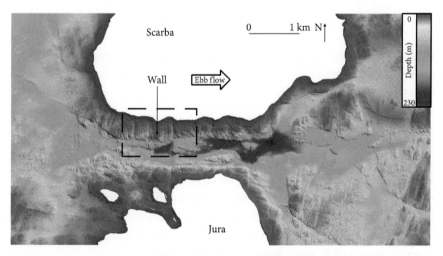

Figure 15.17 A depth map of the gulf of Corryvreckan. Courtesy of Christian Armstrong of the Scottish Association for Marine Science, SAMS.

shown as dark green in the depth map, and a deep basin, shown as blue, southeast of the pinnacle. Since the pinnacle extents upward from the shelf, and sits right at its outer edge, it seems likely that it sheds vorticity whenever there is a strong tidal current.

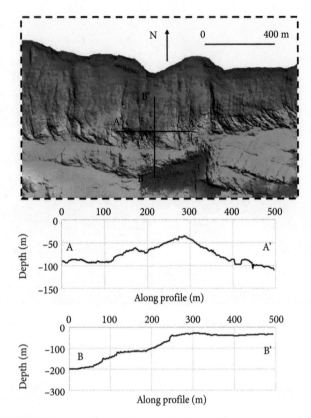

Figure 15.18 Depth map of the submerged outcrop that generates the Corryvreckan whirlpool. See Figure 15.17 for the colour code. Courtesy of Christian Armstrong of SAMS.

The geometrical asymmetry that might favour the ebb tide is a sharp drop in depth just to the east of the pinnacle, shown as dark green and then blue in Figure 15.18. The full significance of this change in depth remains uncertain, but it seems probable that, as the shed vorticity moves into deeper water, it intensifies by vortex stretching, creating an intense tidal vortex. By contrast, the vorticity shed from the pinnacle during the flood tide undergoes no such intensification as it is swept away by the tidal currents.

The Corryvreckan whirlpool constitutes an interesting case study as it illustrates how the formation of whirlpools can often involve a surprisingly complex interaction between tidal currents and coastal terrain.

* * *

That concludes our qualitative discussion of columnar vortices in the atmosphere and oceans. Evidently, much remains poorly understood. In particular, the source of vorticity and the formation mechanisms for these various colum-nar vortices are far from settled. There are many helpful reviews, however, such as Rotunno (2013) on tornadoes, Golden (1974) on waterspouts, and Lorenz et al. (2016) and Balme & Greeley (2006) on dust devils.

References

Antonescu, B., 2017, Tornadoes and Waterspouts in Europe: Depictions from 1555 to 1910. Available from *Blurb*.

Baddeley, P.F.H., 1860, *Whirlwinds and Dust Devils of India*, Bell.

Balme, M., & Greeley, R., 2006, Dust devils on earth and Mars, *Rev. Geophys.*, **44**, 303.

Bödewadt, U.T., 1940, Die Drehstromung uber festem Grunde, *Z. angew Math Mech.*, **20**, 241–53.

Burggraf, O.R., & Foster, M.R., 1977, Continuation or breakdown in tornado-like vortices, *J. Fluid Mech.*, **80**(4), 685–703.

Burggraf, O.R., Stewartson, K., & Belcher, R., 1971, Boundary layer induced by a potential vortex, *Phys. Fluids*, **14**(9), 1821–33.

Gjevik, B., 2009, *Tide along the Coast of Norway and Svalbard*, Farleia Forlag.

Golden, J.H., 1971. Waterspouts and tornadoes over south Florida, *Mon. Weather Rev*, **99**(2), 146–54.

Golden, J.H., 1974, The life cycle of Florida Keys' waterspouts, NOAA Technical Memorandum, ERL NSSL–70.

Lorenz, R.D. et al., 2016, History and applications of dust devil studies, *Space Sci. Rev.*, **203**, 5–37.

Lugt, H.J., 1983, *Vortex Flow in Nature and Technology*, Wiley.

Lugt, H.J., 1989, Vortex breakdown in atmospheric columnar vortices, *American. Met. Soc.*, **70**(12), 1526–37.

Maxworthy, T., 1973, A vorticity source for large-scale dust devils and other comments on naturally occurring columnar vortices, *J. Atmos. Sc.*, **30**(8), 1717–22.

Maxworthy, T., 1982, The laboratory modelling of atmospheric vortices: a critical review. In: *Intense Atmospheric Vortices*. L. Bengtsson & J. Lighthill, eds. Springer, 229–46.

Michaud, M., 1801, Observations sur les trombes de mer vues de Nice en 1789, le 6 Janvier et le 19 Mars, *Memoires de l' Acad. de Turin*, **6**, 3–22.

Phillips, W.R.C., 1985, On vortex boundary layers, *Proc. R. Soc. Lond., A*, **400**, 253–61.

Phillips, W.R.C., & Khoo, B.C., 1987, The boundary layer beneath a Rankine-like vortex, *Proc. R. Soc. Lond., A*, **411**, 177–92.

Rotunno, R., 2013, The fluid dynamics of tornadoes, *Annu. Rev. Fluid Mech.*, **45**, 59–84.

Sinclair, P.C., 1966, A quantitative analysis of the dust devil. PhD thesis, The University of Arizona.

Chapter 16
Tropical Cyclones

16.1 The Observed Properties of Tropical Cyclones

Tropical cyclones are vast, rapidly rotating storms that form over warm tropical seas when the sea surface temperature exceeds ~27°C. Depending on their location, they are also known as *hurricanes or typhoons*, and around 80–90 form each year. Some indication as to their scale is given in Figure 16.1, where the curvature of the earth is evident. Typically, tropical cyclones have radii somewhere in the range 100–1000 km and they extend from the ocean surface up to the tropopause, which has an altitude of $H \sim 15$ km and a temperature of around –75°C. Table 16.1 lists some examples of extreme tropical cyclones.

The energy source for tropical cyclones is the moisture that evaporates from the sea surface to increase the internal energy of the moist air. This moisture then condenses at cooler altitudes, forming clouds and releasing latent heat that warms the air. The precipitation and cloud cover tends to be organized into several curved spiral bands, which are around 10 km wide and extend down from the tropopause. The rotation of the earth also plays an important role in the dynamics of cyclones, as evidenced by the fact that the sense of rotation at sea level is always cyclonic (*i.e.* the azimuthal velocity, u_θ, has the same sign as the background rotation). Thus, tropical cyclones form close to the equator but never *at* the equator, as they require both warm seas and a finite component of the earth's rotation normal to the sea surface. In practice, most tropical cyclones form within a colatitude of 5°–30°, as indicated in Figure 16.2.

We shall adopt cylindrical polar coordinates, (r, θ, z), centred on the cyclone and rotating with the earth, with $z = 0$ at the sea surface. The time-averaged flow in a mature cyclone is more or less axisymmetric and dominated by the azimuthal motion, u_θ. A Bödewadt-like boundary layer forms at the sea surface, and so the air spirals radially inward along that surface, picking up speed through (approximate) angular momentum conservation. The air turns before reaching the centre of the cyclone to spiral vertically upwards towards the tropopause. The flow then spirals radially outward along the tropopause, losing speed to conserve angular momentum. The quiescent region at the centre of the cyclone is known as the *eye*, and it has a radius of $R_{eye} = 10$–40 km.

The Dynamics of Rotating Fluids. P. A. Davidson, Oxford University Press. © Peter A Davidson (2024).
DOI: 10.1093/9780191994272.003.0016

Figure 16.1 Hurricane Isabel as seen from the International Space Station. Courtesy of NASA.

Table 16.1 Documented examples of extreme tropical cyclones. Courtesy of NOAA

Highest sustained wind	2015	Hurricane Patricia	Pacific	96 m/s
Fastest measured gust	1996	Cyclone Olivia	Australia	113 m/s
Largest radius	1979	Typhoon Tip	Pacific	1100 km
Longest lasting	1994	Hurricane John	Pacific	30 days
Highest storm surge	1899	Cyclone Mahina	Australia	~9 m
Largest one-day rainfall	1966	Cyclone Denise	Indian Ocean	180 cm
Most deadly	1970	Cyclone Bhola	Bangladesh	300,000 dead
Most costly (as of 2022)	2005	Hurricane Katrina	New Orleans	~$150 billion

It can be recognized by the lack of cloud cover (see Figure 16.1), and also by the fact that the vertical motion consists of subsiding air, as shown in Figure 16.3. The outer edge of the eye is conical and bounded by the so-called *eyewall*, which is the location of the largest wind speeds within a cyclone, typically in the range $u_\theta \sim$ 30–80 m/s.

Since the θ component of the Coriolis force is $-2\Omega u_r$, Ω being the component of the earth's rotation normal to the sea surface, the motion in a tropical cyclone is cyclonic near the sea surface, where $u_r < 0$. The flow remains cyclonic as it turns up into the tropopause, but because $u_r > 0$ in

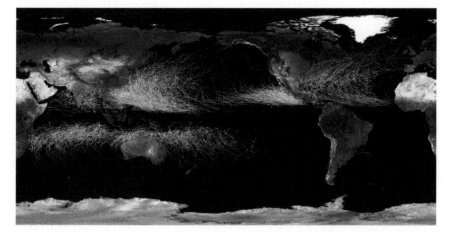

Figure 16.2 The trajectories of tropical cyclones from 1985 to 2005. Courtesy of NASA.

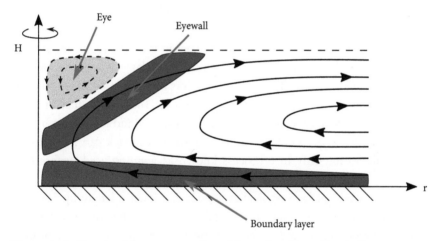

Figure 16.3 Sketch of the eye, eyewall, and boundary layer. Blue indicates $\omega_\theta < 0$.

the tropopause, it eventually becomes anticyclonic at larger radii. The Rossby number based on the vertical scale H, and a local estimate of u_θ, Ro $= |u_\theta|/\Omega H$, is of the order of unity in the bulk of the cyclone, but rises to Ro $\sim$ 40–90 at the eyewall. Thus, the Coriolis force is irrelevant near the eyewall, but dynamically significant elsewhere.

An inflow along the sea surface and an outflow at the tropopause means that the azimuthal vorticity is positive for r significantly greater than R_{eye}, except in the bottom boundary layer where we have $\partial u_r/\partial z < 0$, and hence $\omega_\theta < 0$. Within the eye and eyewall, on the other hand, we have $\omega_\theta < 0$, with $|\omega_\theta|$

particularly intense in the eyewall and the bottom boundary layer, as shown in Figure 16.3. (In Figure 16.3, blue indicates negative ω_θ.) As we shall see, the eyewall acquires its negative azimuthal vorticity from the boundary layer, as ω_θ is stripped off the sea surface and advected upward. That negative vorticity then diffuses across the streamlines from the eyewall to the eye, so that the flow in the eye is mechanically driven, being a passive response to the eyewall. The bottom boundary layer is 1–2 km deep.

16.2 Moist Convection and the Energetics of Tropical Cyclones

Although there is less than 1°C difference in temperature between the sea surface and the adjacent air, the energy source for tropical cyclones is a thermodynamic disequilibrium between the atmosphere and the oceans. In particular, the incoming air near the air–sea interface is under-saturated, and the subsequent evaporation of water transfers enthalpy from the ocean to the moist air above.

The rate of transfer of enthalpy from the ocean is a function of the surface wind speed, *i.e.* the mass flow rate within the boundary layer. If the sea surface were to remain perfectly flat, and there was no generation of sea spray, it would be a linear function of wind speed. However, the wind-induced roughening of the sea surface leads to an enhanced transfer and to a stronger than linear dependence on wind speed. This positive feedback can allow tropical cyclones to intensify in a process known as *wind-induced surface heat exchange*.

The energy budget for a tropical cyclone is discussed at length by Emanuel (1991). The incremental change in specific enthalpy (the enthalpy per unit mass) of dry air is $dh = c_p dT$, where c_p is the usual specific heat. When dealing with moist air we must add to this the enthalpy of the water vapour, which can be estimated as $dH = L_v dm_w$, where L_v is the latent heat of vaporization and m_w the mass of water vapour. The incremental change in specific enthalpy, written for a unit mass of air, then becomes

$$dh_{\mathrm{moist}} = c_p dT + L_v d\left(m_w/m_a\right), \qquad (16.1)$$

where m_w/m_a is the mass of water vapour per unit mass of air and we have neglected the heat capacity of any water droplets dispersed throughout the air/water–vapour mixture.

We now follow a parcel of moist air as it spirals through the bottom boundary layer towards the base of the eyewall, absorbing water vapour in the process. It turns out that the air temperature, T, is roughly constant near the sea surface and so the evaporation of water increases the specific enthalpy of the moist air according to

$$\Delta h_{\text{moist}} = L_v \Delta (m_w/m_a) > 0. \tag{16.2}$$

When this moist air reaches the base of the eyewall, it turns to spiral up towards the tropopause. The steady-flow energy equation applied to a parcel of moist air (including any water droplets) now requires

$$dq = d\left(u^2/2 + gz + h_{\text{moist}}\right) + \mathbf{F} \cdot d\mathbf{r}, \tag{16.3}$$

where dq is the heat transfer to the parcel and $\mathbf{F} \cdot d\mathbf{r}$ is the work done by the fluid against the frictional force $\mathbf{F}$ in moving a distance $d\mathbf{r}$. However, to a first approximation, we may ignore friction and heat transfer in the eyewall, and so (16.3) simplifies to

$$u^2/2 + gz + c_p T + L_v (m_w/m_a) = \text{constant}. \tag{16.4}$$

As the moist air rises up through the eyewall to cooler altitudes, the vapour starts to condense, and so we have

$$d\left(u^2/2 + gz + c_p T\right) = -L_v d (m_w/m_a) > 0. \tag{16.5}$$

Thus, the enthalpy released by condensation goes into increasing the mechanical energy and enthalpy of the dry air. Indeed, the energy cycle of a mature tropical cyclone has been idealized by Emanuel (1991) as a Carnot heat engine which converts heat energy taken from the oceans into mechanical energy. This process is summarized in Figure 16.4.

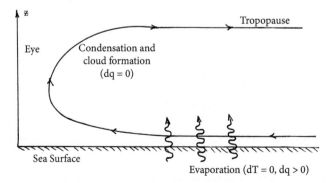

Figure 16.4 The energetics of a tropical cyclone.

16.3 A Toy Model of a 'Dry' Tropical Cyclone

16.3.1 The Model: Its Motivation, Governing Equations, and Weaknesses

We now consider a toy model, or cartoon, of a tropical cyclone which attempts to bypass all of the complexities of fluid turbulence, unsteady evolution, the thermodynamics of moist convection, and a poorly defined air–sea interface involving waves and sea spray. The purpose of this cartoon is to offer a simple, stripped-down system that allows us to focus on the fluid dynamics of cyclones and answer some simple questions, such as why eyes form. This toy model, which was developed by Oruba *et al.* (2017, 2018) and Atkinson *et al.* (2019), is steady and axisymmetric, consisting of flow in a rotating cylindrical container whose base is heated (Figure 16.5). In short, the model replaces unsteady moist convection in an open domain by steady, laminar, dry convection in a closed domain.

Consider the steady motion of a Boussinesq fluid in a shallow, rotating, cylindrical domain of radius R and height H. (See §2.11 for the equations of motion of a Boussinesq fluid.) The lower surface, $z = 0$, and the outer radius, $r = R$, are no-slip boundaries while the upper surface is stress free. The motion is driven by buoyancy with a *fixed* heat flux maintained between the surfaces $z = 0$ and $z = H$, while the outer boundary $r = R$ is thermally insulating. In the absence of motion there is a uniform temperature gradient, $dT_0/dz = -\gamma$, but when in motion the density and temperature fields become

$$\rho = \rho_0(z) + \rho' = \rho_0(z) - \bar{\rho}\beta\vartheta, \quad T = T_0(z) + \vartheta, \qquad (16.6)$$

where ϑ is the perturbation in temperature, $\beta = -(\partial\rho/\partial T)/\bar{\rho}$ the thermal expansion coefficient, and $\bar{\rho}$ a mean density. In a frame of reference that rotates with the boundaries, the momentum and heat equations are

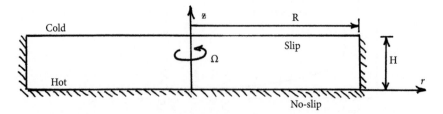

Figure 16.5 Flow domain and boundary conditions in a toy model of a cyclone.

$$\frac{D\mathbf{u}}{Dt} = -\nabla(p/\bar{\rho}) + 2\mathbf{u} \times \mathbf{\Omega} + \nu\nabla^2\mathbf{u} - \beta T\mathbf{g}, \tag{16.7}$$

$$\frac{D\vartheta}{Dt} = \alpha\nabla^2\vartheta + \gamma u_z, \tag{16.8}$$

where $\mathbf{u}$ is the solenoidal velocity measured in the rotating frame, $-\beta T\mathbf{g}$ the buoyancy force, and α the thermal diffusivity.

For axisymmetric motion the azimuthal component of (16.7) becomes an evolution equation for the angular momentum density, $\Gamma = ru_\theta$, in accordance with (2.87). When the Coriolis force is incorporated into (2.87), this becomes

$$\frac{D\Gamma}{Dt} = -2r\Omega u_r + \nu\nabla_*^2\Gamma, \tag{16.9}$$

where ∇_*^2 is the Stokes operator (2.48). The curl of the poloidal components of (16.7), on the other hand, yields an evolution equation for ω_θ/r, in accordance with (2.86). When the Coriolis and buoyancy forces are incorporated into (2.86), we obtain

$$\frac{D}{Dt}\left(\frac{\omega_\theta}{r}\right) = \frac{\partial}{\partial z}\left(\frac{\Gamma^2}{r^4}\right) + \frac{2\Omega}{r^2}\frac{\partial\Gamma}{\partial z} - \frac{\beta g}{r}\frac{\partial\vartheta}{\partial r} + \frac{\nu}{r^2}\nabla_*^2(r\omega_\theta). \tag{16.10}$$

Since Γ and ω_θ between them uniquely determine the instantaneous structure of the flow, the motion is completely determined by the three evolution equations (16.8)→(16.10).

The magnitude of the heat flux is now chosen to ensure that the local Rossby number, $\text{Ro} = |u_\theta|/\Omega H$, is of the order unity at large radius, $r \sim R$, but is much larger than unity near the eyewall, which is typical of a real tropical cyclone. In short, we demand that the Coriolis force is of order unity in the bulk of the cyclone, but negligible near the eyewall. Moreover, the viscosity is chosen so that a suitably defined Reynolds number is large near the eyewall, though not so large that the flow becomes turbulent. This is important because, as we shall see, a moderately large Reynolds number is crucial to eye formation, as it is essential that the boundary-layer vorticity can be stripped off the lower boundary and advected upward to form an eyewall, as indicated in Figure 16.3. (If Re is low-to-moderate, the flow is then too diffusive for the boundary-layer vorticity to be advected upward, and so an eyewall, and hence an eye, cannot form.)

Under these particular conditions, the flow in this idealized model problem resembles that of a real tropical cyclone, including the formation of an

Figure 16.6 A typical solution of the model problem showing ω_θ /r (blue is negative and red positive) overlaid on streamlines. The eyewall and eye are evident, and the boundary layer and eyewall are characterized by a strong negative ω_θ. Courtesy of J. Atkinson.

eye and eyewall within which the azimuthal vorticity is negative, $\omega_\theta < 0$. The fluid then spirals radially inward along the lower boundary and outward near $z = H$, as shown in Figure 16.6. Moreover, the Coriolis force in (16.9) ensures that Γ rises as the fluid spirals inward along the bottom boundary, but falls as it spirals back out along the upper surface. As a result, the flow is cyclonic near the lower boundary and anticyclonic at the upper surface at large radius. In addition, the approximate conservation of angular momentum ensures that particularly high levels of $|u_\theta|$ build up as the fluid spirals in towards the eyewall, with a correspondingly large value of Rossby number within the eyewall.

It is intriguing that an eye and eyewall form in this minimalist model without the need for moist convection. However, it is important to keep in mind that there are essential differences between this stripped-down model and a real tropical cyclone. Perhaps the most important differences are:

   (i) the surface energy transfer in a real tropical cyclone is induced by the winds and not externally imposed;

  (ii) the lower boundary layer in a cyclone is turbulent and occurs over a rough surface;

 (iii) the removal of heat in a real cyclone is distributed throughout the upper regions of the flow in the form of radiative cooling;

 (iv) there are no sidewalls in real cyclones to provide a sink or source of angular momentum;

  (v) real tropical cyclones do not live long enough to achieve a steady angular momentum distribution.

Despite these differences, it seems worthwhile to investigate the fluid mechanics of this simplified model. The hope, of course, is that it might shed some light on the motion in a real cyclone. The nature of eye formation, in particular, is of considerable interest. However, let us start by looking at the global structure of the flow before focussing on the eye.

16.3.2 The Global Structure of the Flow

Let $\Delta T = \gamma H$ be the unperturbed temperature difference between the bottom and top surfaces. We can construct a characteristic velocity scale from ΔT, i.e. $V = \sqrt{gH\beta\Delta T}$, and indeed typical eyewall velocities in this model problem turn out to be of the order of V. We can now form Rossby and Reynolds numbers using V and H, so our key dimensionless groups become

$$\text{Ro} = \frac{V}{\Omega H}, \qquad \text{Re} = \frac{VH}{\nu}, \qquad \text{Pr} = \frac{\nu}{\alpha}. \qquad (16.11)$$

Of course, in rotating convection it is more common to use the Rayleigh and Ekman numbers as control parameters, but these can be expressed in terms of Ro, Re, and Pr according to

$$\text{Ra} = \frac{g\beta\Delta T H^3}{\nu\alpha} = \text{Re}^2\,\text{Pr}, \qquad \text{Ek} = \frac{\nu}{\Omega H^2} = \text{Ro/Re}. \qquad (16.12)$$

The advantage of working with Ro and Re, rather than Ra and Ek, is that they are more directly constrained in this model problem. That is, we require both Ro and Re to be large near the eyewall, the former to mimic what happens in a real cyclone, and the later in order to convect the boundary-layer vorticity, ω_θ, upward to form an eyewall (see below).

By way of an example, the flow for the case of Ro = 45, Re = 450, and Pr = 0.1 (or equivalently, Ek = 0.1 and Ra = 20,000) is shown in Figure 16.7. It is clear that:

(i) an eye and eyewall have formed, with subsidence in the eye;
(ii) the flow is predominantly cyclonic, being anticyclonic only near the upper boundary at large radius;
(iii) a *local* Rossby number, Ro = $(u_\theta)_{\text{max}}/\Omega H$, based on the maximum value of $|u_\theta|$ at a given radius, is around Ro = 30 near the eyewall, but of order unity for larger radii;
(iv) the eye is the warmest part of the cyclone.

Interestingly, all of these features are characteristic of real tropical cyclones.

Oruba et al. (2017) have calculated, for the same flow, the relative magnitudes of the various terms in the azimuthal vorticity equation and shown that the Coriolis and buoyancy forces are negligible in the vicinity of the eye and eyewall. Evidently, the very forces that dominate the global flow pattern in this

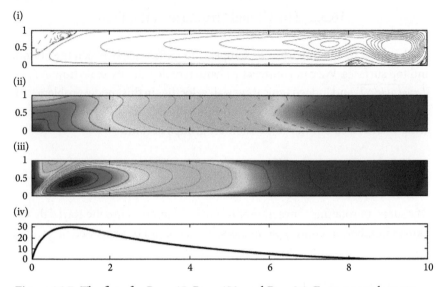

Figure 16.7 The flow for Ro = 45, Re = 450, and Pr = 0.1. From top to bottom: (i) the streamfunction; (ii) the total temperature, $T = T_0(z) + \vartheta$ (red is hot, blue is cold); (iii) the azimuthal velocity (red is large and blue is small or marginally negative, with dashed lines indicating negative values); (iv) the local Rossby number, Ro = $(u_\theta)_{\mathrm{max}}/\Omega H$, based on the maximum value of $|u_\theta|$ at a given radius. Reproduced from Oruba et al. (2017).

model problem are irrelevant in the vicinity of the eyewall, where inertia and shear dominate. We conclude that, while the global flow pattern in a tropical cyclone is established and shaped by the buoyancy and Coriolis forces, the large value of Ro near the eyewall means that the Coriolis force is locally negligible there. It turns out that, in the idealized model problem, the buoyancy force is also negligible near the eyewall.

16.3.3 The Mechanism of Eye Formation

According to §16.3.2, the very forces that establish the global flow pattern in this model problem play no role in the local dynamics of the eye and eyewall, whose behaviour is controlled by the simplified equations

$$\frac{D\Gamma}{Dt} \approx \nu \nabla_*^2 \Gamma, \qquad \frac{D}{Dt}\left(\frac{\omega_\theta}{r}\right) \approx \frac{\partial}{\partial z}\left(\frac{\Gamma^2}{r^4}\right) + \frac{\nu}{r^2}\nabla_*^2(r\omega_\theta). \qquad (16.13)$$

Moreover, we would expect that, outside the lower boundary layer, the viscous forces are negligible, and so Γ is locally constant on a streamline, *i.e.* $\Gamma \approx \Gamma(\Psi)$,

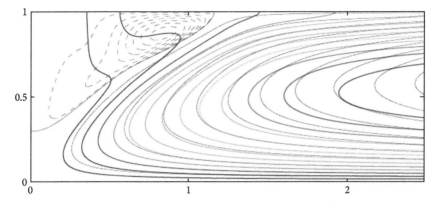

Figure 16.8 The flow near the eye and eyewall for Ro = 45, Re = 450 and Pr = 0.1. Streamlines are grey and the contours of constant Γ coloured. Reproduced from Oruba et al. (2017).

where Ψ is the Stokes stream-function. This is confirmed in Figure 16.8, where the streamlines are shown as grey and the contours of constant Γ are coloured. We conclude that, in the vicinity of the eyewall, but outside the bottom boundary layer, we have

$$\frac{D\Gamma}{Dt} \approx 0, \quad \frac{D}{Dt}\left(\frac{\omega_\theta}{r}\right) \approx \frac{\partial}{\partial z}\left(\frac{\Gamma^2}{r^4}\right). \tag{16.14}$$

Now the eyewall is characterized by an intense negative azimuthal vorticity, as shown in Figure 16.6, and it is natural to ask where this vorticity comes from. It is sometimes suggested that it is generated by the source term $\partial\Gamma^2/\partial z$ in (16.14), which represents a spiralling-up of the poloidal vortex lines by axial gradients in swirl. At first sight this seems plausible, as there are indeed very strong axial gradients of Γ in and around the eyewall. However, it is readily confirmed that the source term $\partial\Gamma^2/\partial z$ is *not* the origin of eyewall vorticity. The point is that the inviscid equations (16.14) can be rewritten as

$$\mathbf{u} \cdot \nabla\left(\omega_\theta/r - \Gamma\Gamma'(\Psi)/r^2\right) = 0, \tag{16.15}$$

as shown in the derivation of (2.90). Evidently, $\omega_\theta/r - \Gamma\Gamma'(\Psi)/r^2$ is conserved along a streamline in a steady, inviscid flow. We now apply (16.15) to a streamline in Figure 16.8 which lies outside the boundary layer. If the streamline enters the domain near the sea surface and exits near the tropopause at the *same radius*, then there is no net change in $\Gamma\Gamma'(\Psi)/r^2$, and so there is no overall change in azimuthal vorticity. We conclude that the negative azimuthal

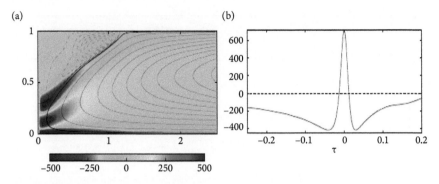

Figure 16.9 (a) Colour map of ω_θ/r in the region of the eyewall, overlaid on streamlines. (b) Variation of ω_θ/r along the streamline that originates in the boundary layer and passes through the centre of the eyewall. (Ro = 45, Re = 450, Pr = 0.1.) Reproduced from Oruba et al. (2017).

vorticity in the eyewall cannot arise from axial gradients in Γ. Rather, it must come directly from the boundary layer at the sea surface.

This becomes clear if we monitor the variation of ω_θ/r along a streamline that originates in the boundary layer and subsequently passes through the centre of the eyewall, as shown in Figure 16.9. As the streamline passes along the boundary layer ω_θ/r becomes progressively more negative, as shown on the left of Figure 16.9(b). There is then a sharp rise and subsequent fall in ω_θ/r as the streamline passes through regions of positive and negative $\partial\Gamma^2/\partial z$. Crucially, however, the rise and subsequent fall in ω_θ/r perfectly cancel, so that the air enters the eyewall with exactly the same value of ω_θ/r that it had on leaving the bottom boundary layer.

In short, the eyewall is a free shear layer that arises from the separation of the bottom boundary layer. The eye is then a passive response to the existence of an eyewall, with the negative azimuthal vorticity in the eye arising from the cross-stream diffusion of vorticity out of the eyewall and into the eye. Thus, while the bulk of the cyclone is thermally driven, the eye is mechanically driven.

Note that this process of eyewall formation relies on the Reynolds number near the eyewall being sufficiently large, so that the boundary-layer vorticity can be advected up into the region of the eyewall. Indeed, as noted in Oruba et al. (2017), when

$$\text{Re}_r = \frac{|u_r|_{\max}H}{\nu} < 40, \tag{16.16}$$

the flow becomes too diffusive to effectively advect the boundary-layer vorticity upward, and so an eyewall (and hence an eye) cannot form. (From now on, a subscript on Re or Ro indicates that Re or Ro is based on the maximum eyewall velocity, as distinct from definition (16.11), which is based on V. The subscript indicates whether $|u_r|_{\max}$ or $(u_\theta)_{\max}$ has been used.) Note also that this process relies on the Rossby number in the eyewall being larger than unity,

$$\mathrm{Ro}_\theta = \frac{(u_\theta)_{\max}}{\Omega H} > 1, \qquad (16.17)$$

otherwise eyewall formation could be suppressed by the Taylor–Proudman theorem. In short, the local values of Re_r and Ro_θ must be sufficiently large in the vicinity of the eyewall.

16.3.4 Scaling Laws for Wind Speed and the Criterion for Eye Formation

We now discuss the scaling laws for velocity in this idealized model problem. These laws are reliant on an overall energy budget for the flow, and since our model problem replaces turbulent moist convection with laminar dry convection, one cannot transcribe any of these scaling estimates directly to a real tropical cyclone (though see the discussion in §16.4). Nevertheless, having established this model problem, it seems natural to take it to its logical conclusion.

An energy equation can be obtained by rewriting (16.7) in the form

$$0 = \mathbf{u} \times (\boldsymbol{\omega} + 2\boldsymbol{\Omega}) - \nabla\left(u^2/2 + p/\bar{\rho}\right) + \nu\nabla^2\mathbf{u} - \beta T\mathbf{g}, \qquad (16.18)$$

and then integrating this once around a closed streamline. The end result is

$$\left|\oint \nu\nabla^2\mathbf{u} \cdot d\mathbf{r}\right| = g\beta \oint T\hat{\mathbf{e}}_z \cdot d\mathbf{r} = g\beta \oint \vartheta\hat{\mathbf{e}}_z \cdot d\mathbf{r}, \qquad (16.19)$$

where we have used the fact that $u^2/2 + p/\bar{\rho}$ is single valued and $T_0(z)\hat{\mathbf{e}}_z$ is conservative. This energy equation can be applied to different streamlines within the flow. Let us focus on a streamline that passes close to the lower boundary. If δ is the thickness of the bottom boundary layer, then (16.19) yields the scaling law

$$\nu \frac{u_r}{\delta^2} R \sim g\beta H\Delta T = V^2, \tag{16.20}$$

where ϑ is assumed to scale on ΔT. To make further progress, we need an estimate of δ.

Now the numerical experiments reported in Oruba et al. (2018) suggest that, provided Ro is less than $\sim 2.5R/H$, δ scales as a conventional Ekman layer, that is, as

$$\delta \sim \sqrt{\nu/\Omega}. \tag{16.21}$$

We might classify this regime as being in strong rotation. However, for Ro greater than $\sim 2.5\,(R/H)$, which we might call moderate-to-weak rotation, δ appears to scale in more or less the usual non-rotating fashion, according to which:

viscous forces $\sim$ stream-wise inertial acceleration.

In the present context, this second scaling translates to

$$\nu \frac{u_r}{\delta^2} \sim \frac{u_r^2}{R}. \tag{16.22}$$

Note that the transition between these two scaling laws for δ, which in practice is not abrupt, can be estimated by substituting $\delta \sim \sqrt{\nu/\Omega}$ into (16.22). This then yields

$$(\mathrm{Ro}_r)_{\mathrm{trans}} \sim R/H, \tag{16.23}$$

which is consistent with the observed break-point of around $(\mathrm{Ro})_{\mathrm{trans}} \sim 2.5\,(R/H)$.

These two estimates for δ can be combined with the energy constraint (16.20) to yield the scaling laws

$$\frac{u_r}{V} \sim \frac{V}{\Omega R} = \frac{H}{R}\mathrm{Ro}, \qquad \text{for } \mathrm{Ro} < 2.5\,(R/H), \tag{16.24}$$

and

$$\frac{u_r}{V} \sim 1, \qquad \text{for } \mathrm{Ro} > 2.5\,(R/H). \tag{16.25}$$

Note that these scaling laws are *independent of the molecular diffusivities*, both the kinematic viscosity and thermal diffusivity. Moreover, both of these estimates are consistent with the numerical experiments reported in Oruba et al. (2018), which tentatively suggest

$$\frac{u_r}{V} \approx 0.15\frac{H}{R}\text{Ro}, \qquad \text{for Ro} < 2.5\,(R/H), \qquad (16.26)$$

and

$$\frac{u_r}{V} \approx 0.37, \qquad \text{for Ro} > 2.5\,(R/H), \qquad (16.27)$$

with the transition between the two at $(\text{Ro}_r)_{\text{trans}} \approx 0.9R/H$.

We now return to the conditions required for eye formation in our model problem. We have already noted that Oruba et al. (2017) observe that eyes form only when

$$\text{Re}_r = \frac{|u_r|_{\text{max}}H}{\nu} > 40,$$

since Re_r must be large enough to advect the boundary-layer vorticity upward to form an eyewall. We have also suggested that eye formation requires

$$\text{Ro}_\theta = \frac{(u_\theta)_{\text{max}}}{\Omega H} \gg 1,$$

to avoid the imposition of the Taylor–Proudman theorem in the vicinity of the eyewall, which could suppress any vertical structuring of the flow.

Oruba et al. (2018) provide a more thorough discussion of this issue, having monitored eye formation in over 150 numerical experiments for various values of Re, Ro, Pr, and H/R. They find that, for the range of parameters included in their study, one prerequisite for eye formation is

$$\text{Re}_r = \frac{|u_r|_{\text{max}}H}{\nu} > 40 \pm 10,$$

with the critical value of Re_r dependent on the exact values of Ro, Pr, and H/R, especially Ro. They also observe that eye formation imposes both upper and lower bounds on Ro. In particular, eyes form only when

$$\text{Ro} = \frac{V}{\Omega H} = 12 \to 35.$$

The lower bound on Ro is to be expected, as noted above. The upper bound arises because it is important that the local value of Ro_θ away from the eye is not much larger than unity, so that the Coriolis force can play a central role in the global dynamics of the cyclone, allowing an Ekman layer to form. This, in turn, means that Ro cannot become too large, at least in this model problem. Finally, Oruba et al. (2018) note that eyes do not form in this toy model when the aspect ratio, H/R, rises above 0.3. This tentatively suggests that tropical cyclones are intrinsically disc-like objects.

16.3.5 Points of Connection to, and Departure from, Real Tropical Cyclones

Perhaps we should close this section by highlighting the similarities and differences between the idealized model problem discussed above and a real tropical cyclone. As already noted, the obvious deficiencies of our simple cartoon are:

   (i) the surface energy transfer in a real cyclone is induced by the winds and not externally imposed, so there is a wind-induced feedback mechanism that allows real tropical cyclones to intensify;
  (ii) there is no latent heat release in the model problem to boost the eyewall convection;
 (iii) the lower boundary layer in a cyclone is turbulent and occurs over a rough surface;
 (iv) the removal of heat in a real cyclone is distributed throughout the upper regions of the flow in the form of radiative cooling;
  (v) there are no sidewalls in a real tropical cyclone;
 (vi) real tropical cyclones do not last long enough to get close to a statistically steady state.

Clearly, our idealized model problem cannot begin to emulate the true energy cycle of a tropical cyclone. Nevertheless, if we limit ourselves to simple fluid dynamics, it does capture a number of important features observed in real cyclones. These include:

   (i) an eye and eyewall form, with subsidence within the eye, provided that the local values of Re and Ro are sufficiently large near the eyewall;
  (ii) the eye is the warmest part of the cyclone (see Figure 16.7) and, unlike the rest of the cyclone, its motion is mechanically driven;

(iii) we observe that $Ro_\theta = (u_\theta)_{max}/\Omega H \sim 1$ in the bulk, yet $Ro_\theta = (u_\theta)_{max}/\Omega H \gg 1$ near the eyewall;

(iv) there is an inward (Ekman-like) swirling flow within the bottom boundary layer;

(v) the motion is cyclonic near the lower boundary but anticyclonic at the upper boundary (at large radii);

(vi) Γ is almost constant on a streamline near the eyewall, *i.e.* $\Gamma \approx \Gamma(\Psi)$;

(vii) the intense azimuthal vorticity within the eyewall, ω_θ, has its origins in the boundary-layer vorticity below;

(viii) thermally driven cyclones with eyes are intrinsically disc-like objects.

So, perhaps there is some merit in this simple, minimalist cartoon, at least when it comes to exposing those mechanisms that are essentially fluid dynamical in origin.

16.4 Estimating the Maximum Wind Speed in Real Tropical Cyclones

It is important to emphasize that moist convection ensures that the energy cycle of a tropical cyclone is very different to that of our toy problem, and this means that the derivation of the velocity scaling laws (16.24) and (16.25) in §16.3.4, which is based on the mechanical energy balance (16.19), does not apply to a real tropical cyclone. That is to say, the mechanical energy input (per unit mass of dry air) in the model problem is of order $g\beta H\Delta T = V^2$, whereas the enthalpy input (per unit mass of moist air) in a real tropical cyclone is, according to (16.2), $\Delta h_{moist} = L_v\Delta(m_w/m_a)_{sea}$, where the subscript 'sea' indicates a change that occurs within the sea-surface boundary layer.

If we wish to adapt the energy analysis of §16.3.4 to estimate the wind speed in a real tropical cyclone, we must allow for moist convection. We also need to estimate the rate of dissipation of mechanical energy in the boundary layer near the sea surface, which is far from trivial because the air flow is highly turbulent and the sea surface rough. One way to account for the effects of moist convection is to follow the thermodynamic argument outlined in Emanuel (1991, 2003). Here the cyclone is assumed to behave like a reversible Carnot engine with an ideal efficiency of $e = (T_{sea} - T_{trop})/T_{sea}$, where T_{sea} is the sea-surface temperature and T_{trop} the temperature of the tropopause. This means that, of all of the heat transferred to a fluid parcel near the sea surface, Δq_{sea}, only a fraction is used to do mechanical work on the environment, *i.e.* $e\Delta q_{sea}$.

The rest of the energy is lost to space through electromagnetic radiation. The mechanical energy balance, (16.19), is then replaced by

$$e\Delta q_{sea} = \oint \mathbf{F} \cdot d\mathbf{r}, \tag{16.28}$$

where $\mathbf{F} \cdot d\mathbf{r}$ is the work done by the fluid against frictional forces in moving a distance $d\mathbf{r}$. Of course, the integral on the right is dominated by friction at the sea surface.

It is tempting to equate Δq_{sea} to the enthalpy transfer $\Delta h_{moist} = L_v \Delta(m_w/m_a)_{sea}$, but this neglects the dissipative heating of the air, and so a better estimate is (Emanuel, 2003)

$$\Delta q_{sea} \approx \Delta h_{moist} + \int_{sea} \mathbf{F} \cdot d\mathbf{r}. \tag{16.29}$$

Substituting for Δq_{sea} in (16.28), and assuming that nearly all the frictional work is performed within the sea-surface boundary layer, we find (Emanuel, 2003),

$$\frac{T_{sea} - T_{trop}}{T_{trop}} L_v \Delta(m_w/m_a)_{sea} \approx \int_{sea} \mathbf{F} \cdot d\mathbf{r}. \tag{16.30}$$

Note that the temperature ratio in (16.30) is *not* the Carnot efficiency. Note also that (16.30) is equivalent to (16.19), but with a thermodynamic energy source replacing $g\beta H\Delta T$.

One is still faced with the task of relating the energy dissipation rate in the sea-surface boundary layer to the surface wind speed, and of estimating Δh_{moist}. The usual approach is to: (i) write the friction integral in (16.30) as $c_d V_{max}^2$, where c_d is a dimensionless drag coefficient and V_{max} the maximum velocity in the eyewall; and (ii) write the enthalpy rise at the sea surface as $\Delta h_{moist} = c_h (h_{sat} - h_{in})$, where h_{in} is the enthalpy of the inflowing air, h_{sat} the enthalpy of saturated air at temperature T_{sea}, and c_h is yet another dimensionless coefficient. This yields (Emanuel, 2003)

$$V_{max}^2 \approx \frac{c_h}{c_d} \frac{T_{sea} - T_{trop}}{T_{trop}} (h_{sat} - h_{in}). \tag{16.31}$$

Of course, the difficulty of estimating the boundary layer dissipation has not gone away, but rather is hidden in the coefficient c_d, which is a poorly constrained function of the state of the sea surface. For simplicity, the ratio c_h/c_d is usually given a value of order one.

$$* * *$$

That concludes our brief excursion into the fluid dynamics of tropical cyclones. They are both powerful and devastating, as noted in Table 16.1. Indeed, tropical cyclones dissipate mechanical energy at a rate of $\sim 10^3$ G Watts, which is roughly equivalent to the electrical power consumption of the entire USA. As with tornadoes, they are extremely complicated objects, and so there is still much that is poorly understood. However, extensive reviews may be found in Emanuel (1991, 2003) and Montgomery & Smith (2017).

References

Atkinson, J.W., Davidson, P.A., & Perry, J.E.D., 2019, Dynamics of a trapped vortex in rotating convection, *Phys. Rev. Fluids*, **4**, 074701.

Emanuel, K.A. 1991, The theory of hurricanes, *Annu. Rev. Fluid Mech.*, **23**, 179–96.

Emanuel, K.A. 2003, Tropical cyclones, *Annu. Rev. Earth Planet. Sci.*, **31**, 75–104.

Montgomery, M.T., & Smith, R.K., 2017, Recent developments in the fluid dynamics of tropical cyclones, *Annu. Rev. Fluid Mech.*, **49**, 541–74.

Oruba, L., Davidson, P.A., & Dormy, E., 2017, Eye formation in rotating convection, *J. Fluid Mech.*, **812**, 890–904.

Oruba, L., Davidson, P.A., & Dormy, E., 2018, Formation of eyes in large-scale cyclonic vortices, *Phys. Rev. Fluids*, **3**, 013502.

Chapter 17

Convective Motion in the Earth's Core and the Geodynamo

We now consider convective motion within the liquid core of the earth, motion which is strongly influenced by the Coriolis force. This buoyant convection is important as it sustains the terrestrial magnetic field against the natural forces of decay through a form of fluid dynamo. Such a dynamo converts mechanical energy into magnetic energy. Indeed, without fluid motion, Ohmic dissipation would have extinguished the earth's magnetic field on a timescale of 10^5 years, whereas the earth's field has been around for at least 10^9 years.

17.1 The Structure of the Planets

17.1.1 The Structure of the Earth, Sources of Motion, and the Geomagnetic Field

As noted in §10.1, the earth is almost spherical with a mean outer radius of 6371 km and a rotation rate of $\Omega = 7.29 \times 10^{-5}$ s^{-1}. It comprises an iron core, a rocky mantle, and a thin crust, as shown in Figure 17.1. The iron core, which contains roughly one third of the earth's total mass, is divided into two parts. The liquid outer core has a radius of $R_C = 3485$ km, while the inner core is solid and has a radius of $R_i = 1215$ km. The maximum temperature in the outer core is at the inner core boundary (ICB) and is around 6000 K. The inner core is solid, despite this high temperature, because of the intense pressure at the centre of the earth, and indeed the inner core is slowly growing by solidification. The temperature difference between the ICB and the mantle drives convection.

The composition of the core can be estimated from seismology, and also from guesses as to the chemical makeup of the iron-rich meteorites that helped form the earth. The inner core is thought to be an alloy of iron and nickel, being

The Dynamics of Rotating Fluids. P. A. Davidson, Oxford University Press. © Peter A Davidson (2024).
DOI: 10.1093/9780191994272.003.0017

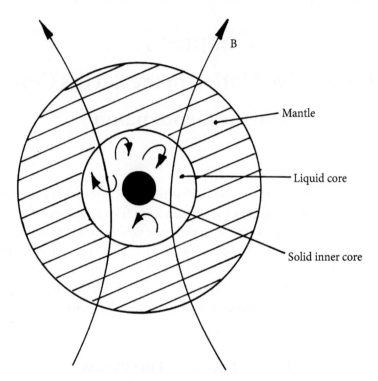

Figure 17.1 The structure of the earth.

around 92% iron. The outer core is also composed largely of iron and nickel, although seismology shows a sharp drop in density, of around 5%, at the ICB. In fact, the outer core is estimated to be around 15% lighter than the inner core, which requires the presence of lighter elements, such as silicon, sulphur, or oxygen.

It is probable that, initially, the core was entirely molten, and that the solid inner core first nucleated around 10^9 years ago. In any event, it is currently growing at a rate of ~1 mm/year, and this slow growth of the inner core contributes to the convection in the outer core. That is, the iron that freezes onto the inner core is free of the light impurities which are dispersed across the outer core. Hence, solidification leaves behind a thin layer of fluid adjacent to the ICB which is rich in the lighter elements. This light fluid then floats out to the mantle, with the resulting *compositional convection* supplementing the more conventional thermal convection. In numerical simulations of the core, this combined buoyancy flux is usually found to be concentrated near the rotation axis, in the form of buoyant upwellings, and also in and around the equator, where it takes the form of plumes which meander radially out along the equatorial plane towards the mantle.

The external magnetic field of the earth is dipolar, with a dipole moment of $m = 7.9 \times 10^{22}$ Am^2, and a magnetic axis more or less aligned with the rotation axis. Now, currents that are confined to a spherical volume V satisfy the relationship

$$\mathbf{m} = \frac{1}{2} \int_V \mathbf{x} \times \mathbf{J} dV = \frac{3}{2\mu} \int_V \mathbf{B} dV, \tag{17.1}$$

where $\mathbf{J}$ is the current density, $\mathbf{B}$ the magnetic field strength, μ the *permeability of free space*, and $\mathbf{m}$ the dipole moment, which is defined by the first equality in (17.1). (Expression 17.1 follows from the Biot–Savart law, as shown in Jackson, 1998.) This allows us to estimate the mean axial magnetic field in the earth's conducting core, V_C, which yields

$$\bar{B}_z = \frac{1}{V_C} \int_{V_C} B_z dV = 3.7 \text{ Gauss.} \tag{17.2}$$

Paleomagnetic records show that the earth's dipole field exhibits random reversals, which typically take around 10^4 years to complete. This is longer than the convective turnover time in the core, of around 300 years, but much faster than the average time between reversals, which is around 10^6 years. Interestingly, the reversal timescale is similar to the time taken for a magnetic field to diffuse through the solid inner core, which is $t_d = R_i^2/(\lambda \pi^2) \sim 6800$ years, where λ is the *magnetic diffusivity* of iron. This has led some to suggest that the timescale for field reversals may be set, at least in part, by the time taken for the magnetic field to diffuse in or out of the solid inner core, though others argue that the inner core is too small to have such a dominant influence.

The underlying reason for field reversals is still uncertain, but the simplest explanation rests on the fact that the core motion is highly chaotic. This randomness could result in transient fluctuations in the thermal convection pattern, and it is conceivable that, on occasions, this might push the core flow into an anti-dynamo configuration, leading to the fluid dynamo temporarily shutting down. If, when the dynamo resumes, the emergent magnetic field has the same polarity to the original one, an excursion is said to have occurred.

However, in those cases where the emergent field has the opposite polarity, we have a reversal.

17.1.2 The Structure of the Other Solar System Planets and Their Magnetic Fields

Some of the physical properties of the solar system planets are documented in Figure A3.1 and Table A3.2 of Appendix 3. These planets are grouped into three distinct categories. Mercury, Venus, the earth, and Mars are known as the *terrestrial planets*, as they all have rocky mantles and liquid iron cores. Moving further out, Jupiter and Saturn are composed of hydrogen and helium and are referred to as the gas giants. The extreme pressure within their interiors gives rise to metallic hydrogen, which is electrically conducting. Finally, Uranus and Neptune are composed mostly of water, ammonia, and methane, with a rocky inner core and an atmosphere rich in hydrogen and helium. These are the ice giants. The mean axial magnetic fields in the conducting cores of these planets, $\bar{B}_z$, can be calculated from their observed dipole moments, $\mathbf{m}$, using (17.1), and these values are tabulated in Table A3.2. This mean field varies from around 0.01 Gauss (10^{-6} Tesla) for Mercury, up to ~20 Gauss (0.002 Tesla) for Jupiter. However, when expressed in the appropriate dimensionless form, the range is narrower, with

$$\frac{\bar{B}_z / \sqrt{\rho \mu}}{\Omega R_C} \sim 0.3 \times 10^{-6} \rightarrow 13 \times 10^{-6}. \tag{17.3}$$

Let us start with the terrestrial planets. Mercury's external magnetic field is dipolar and more or less aligned with the rotation axis, although it is very weak. The reason why Mercury's field is so weak is still debated, although one suggestion is that convection within Mercury's core may be partially suppressed by a stable stratification. Venus and Mars do not have dynamos, despite the fact that their internal structures and sizes are similar to that of the earth. While Venus has no detectable magnetic field, Mars exhibits some remnant magnetism in its southern hemisphere, which suggests that Mars once had an active dynamo. The absence of a fluid dynamo in Venus is probably linked to the fact that the core convection is thought to be weak. This, in turn, is partly because Venus has not yet nucleated a solid inner core, and so there is no compositional buoyancy, and partly because Venus' highly insulating mantle suppresses the heat flux through the core and mantle, thus limiting the thermal power available to drive convection. (There are no tectonic plates on Venus, and this suppresses the heat flux through the mantle.) The extinction of Mars' dynamo has also been attributed to a decline in core convection, possibly because the mantle in Mars may have changed from earth-like to Venus-like.

Both Jupiter and Saturn have dipolar magnetic fields with a magnetic axis more or less aligned with the rotation axis. The metallic hydrogen in their cores has an electrical conductivity which is an order of magnitude smaller than that of liquid iron, but still large enough to maintain a convective dynamo. One important difference between the gas giants and the terrestrial planets, however, is the absence of a rocky mantle to confine the conducting core. Instead, there is a gradual transition from metallic hydrogen at small radii to molecular hydrogen at larger radii. In Jupiter, the metallic hydrogen occupies much of the planet, and so the conducting core is large, while in Saturn the metallic hydrogen is confined to the inner part of the planet, which limits the spatial extent of the fluid dynamo. One striking feature of Saturn's external magnetic field is that it is almost perfectly axisymmetric, with a close alignment of the rotation and magnetic axes.

Uranus and Neptune possess magnetic fields which are quite different to those of the other planets. The external magnetic fields considered so far are all strongly dipolar and roughly axisymmetric. However, the external fields of the ice giants have a much more complicated spatial structure, with only a weak contribution from the axisymmetric dipole component. Moreover, the dipolar contribution to the external magnetic field is more closely aligned with the equator than the poles. It is therefore probable that the origin of the magnetic fields in the ice giants is quite different to those of the other planets.

17.2 Flow in the Liquid Core of the Earth

17.2.1 Dimensionless Groups and Dominant Forces in the Earth's Core

We shall provide a crash course on the electrodynamics of dynamo theory in §17.3. In the interim, it is sufficient to note that a magnetic field evolving in an electrically conducting, incompressible fluid is governed by an equation which is structurally identical to the vorticity equation (2.56), but with $\boldsymbol{\omega}$ and ν replaced by $\mathbf{B}$ and λ, i.e.

$$\frac{\partial \mathbf{B}}{\partial t} = \nabla \times (\mathbf{u} \times \mathbf{B}) + \lambda \nabla^2 \mathbf{B}, \quad \lambda = (\mu \sigma)^{-1}, \tag{17.4}$$

where σ is the electrical conductivity. Moreover, we shall see that the fluid is subject to a *Lorentz force* per unit mass of

$$\mathbf{F} = \mathbf{J} \times \mathbf{B}/\rho = (\nabla \times \mathbf{B}) \times \mathbf{B}/\rho\mu, \tag{17.5}$$

which tells us that $|\mathbf{B}|/\sqrt{\rho\mu}$ has the dimensions of a velocity.

Let us now consider the various dimensionless groups that characterize motion in the core of the earth, starting with the magnetic analogue of the Reynolds number, $R_m = u\ell/\lambda$. A characteristic fluid velocity in the core can be estimated from the rate at which the magnetic field drifts along the core–mantle boundary, and this suggests $|\mathbf{u}| \sim 0.2$ mm/s. Also, estimates of the electrical conductivity of the molten iron in the core suggest a magnetic diffusivity of $\lambda \approx 0.7$ m^2/s. A typical value of the *magnetic Reynolds number* for large-scale motion is then

$$R_m = \frac{|\mathbf{u}|\,(R_C - R_i)}{\lambda} \sim 600. \tag{17.6}$$

Moreover, virtually all of the dissipation of energy in the core is Ohmic, rather than viscous. Hence, an estimate of the minimum length-scale for motion in the core is sometimes obtained by assuming that Rm ~ 1 at the smallest dynamically active scale, just as the small scales in turbulence are characterized by $u\ell_{\min}/\nu \sim 1$. Taking $|\mathbf{u}| \sim 0.2$ mm/s and $u\ell_{\min}/\lambda \sim 1$ then tentatively suggests $\ell_{\min} \sim 4$ km.

Let us now consider the relative magnitude of the forces operating in the earth's core. The Rossby number is very small,

$$\mathrm{Ro} = \frac{|\mathbf{u}|}{\Omega\,(R_C - R_i)} \sim 10^{-6}, \tag{17.7}$$

and so inertia, $\mathbf{u} \cdot \nabla\mathbf{u}$, is totally irrelevant in the core. This is true even at the smallest scale of the motion, and so we are obliged to neglect $\mathbf{u} \cdot \nabla\mathbf{u}$ in any force balance. Note that the low value of Ro also means that inertial waves are likely to be ubiquitous and so, although we must drop $\mathbf{u} \cdot \nabla\mathbf{u}$ in the governing equations, we cannot neglect $\partial\mathbf{u}/\partial t$.

The kinematic viscosity of liquid iron is $\nu \approx 10^{-6}$ m^2/s and so the Ekman and *magnetic Prandtl* numbers are

$$\mathrm{Ek} = \nu/\Omega R_C^2 \sim 10^{-15}, \quad \mathrm{Pr}_m = \nu/\lambda \sim 10^{-6}, \tag{17.8}$$

the small value of Pr_m ensuring that Ohmic dissipation dominates over viscous dissipation. These tiny values of Ek and Pr_m are impossible to achieve in numerical simulations of the core, because of the relatively small length-scales associated with them. Consequently, great care must be exercised when interpreting the results of these simulations. Also, the estimate Ek ~ 10^{-15} demands that the viscous stresses are tiny within the core, and so these stresses are unlikely to play any significant dynamical role. The physical properties of the earth's core, as well as these estimates of the various dimensionless groups, are summarized in Table A3.1 of Appendix 3.

Given that the viscous stresses and $\mathbf{u} \cdot \nabla \mathbf{u}$ are much too small to play any important role, the dominant force balance in the core is between buoyancy, which drives the motion, and the Coriolis and Lorentz forces. Now the mean density in the outer core is 10.9×10^3 kg/m^3, from which we can calculate the magnitude of our final dimensionless group, the *Elsasser number*,

$$\Lambda = \frac{\sigma B^2}{\rho \Omega} = \frac{B^2/\rho\mu}{\lambda \Omega}. \tag{17.9}$$

This parameter is usually taken to be indicative of the ratio of the Lorentz force to the rotational part of the Coriolis force. However, the arguments in support of that statement are a little convoluted and rest on the fact that the core flow is dominated by thin, columnar vortices aligned with the rotation axis and whose length is comparable with R_C and width is some multiple of ℓ_{min} (see Figure 17.2a). To justify this interpretation of Λ, we first perform a Helmholtz decomposition on the Coriolis force (see Appendix 1). This yields

$$2\mathbf{u} \times \mathbf{\Omega} = 2\mathbf{\Omega} \cdot \nabla \mathbf{a} - \nabla \varphi, \tag{17.10}$$

where $\varphi = 2\mathbf{\Omega}\cdot\mathbf{a}$ is a scalar potential and $\mathbf{a}$ is the (solenoidal) vector potential for $\mathbf{u}$, defined by $\nabla \times \mathbf{a} = \mathbf{u}$ and $\nabla \cdot \mathbf{a} = 0$. Evidently, $\mathbf{a}$ is of order $|\mathbf{a}| \sim |\mathbf{u}|\,\ell_{min}$. Pressure then balances $\nabla\varphi$ and the ratio of the azimuthally averaged Lorentz force to the rotational part of $2\mathbf{u}\times\mathbf{\Omega}$ is

$$\frac{\langle \mathbf{J} \times \mathbf{B} \rangle}{\rho\mathbf{\Omega} \cdot \nabla\mathbf{a}} = \frac{\langle (\nabla \times \mathbf{B}) \times \mathbf{B} \rangle}{\rho\mu\mathbf{\Omega} \cdot \nabla\mathbf{a}} \sim \frac{B^2/R_C}{\rho\mu\Omega u\ell_{min}/R_C} \sim \frac{\sigma B^2}{\rho\Omega} \cdot \frac{\lambda}{u\ell_{min}} \sim \Lambda, \tag{17.11}$$

where we have assumed that $u\ell_{min}/\lambda \sim 1$. If we take $B = \bar{B}_z \approx 4$ Gauss then we find $\Lambda \approx 0.2$, which suggests an approximate balance between the Lorentz force and the rotational part of the Coriolis force. Since buoyancy drives the

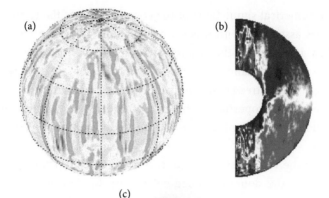

(c)

Figure 17.2 Numerical simulation of a dynamo at $\mathrm{Ra}/(\mathrm{Ra})_{\mathrm{crit}} \approx 40$. (a) Alternating cyclonic–anticyclonic columnar vortices. (b) Azimuthally averaged helicity in a vertical plane, blue for $h < 0$, red for $h > 0$. (c) Un-averaged helicity. Reproduced from Davidson & Ranjan (2023).

motion, it is usually assumed that the rotational parts of the Coriolis, Lorentz, and buoyancy forces are all of the same order of magnitude, with pressure absorbing the irrotational parts of these various forces.

17.2.2 Numerical Simulations of the Core Flow and the Role of Columnar Vortex Pairs

We discussed the onset of convection in a rotating spherical shell in §12.4. In particular, we considered the critical value of the Rayleigh number,

$$\mathrm{Ra} = \frac{g\beta\Delta T d^3}{\nu\alpha}, \qquad d = R_C - R_i,$$

at which convection first sets in. An example of a mildly supercritical convection pattern is shown in Figure 12.9. It consists of columnar convection cells aligned with the rotation axis and arranged around the *tangent cylinder* (an imaginary cylinder that circumscribes the inner core). The columns span the outer core and act as conduits for helical flow driven by Ekman pumping. Moreover, the axial vorticity within these columns alternates between cyclonic, $\boldsymbol{\omega}\cdot\boldsymbol{\Omega} > 0$, and anticyclonic, $\boldsymbol{\omega}\cdot\boldsymbol{\Omega} < 0$.

The columns in a mildly supercritical flow do little other than grind away on the mantle while drifting slowly eastward. However, as the value of Ra in the numerical simulations is increased, the convection pattern becomes progressively more dynamic. The basic building block remains the same: helical, columnar, cyclone–anticyclone pairs aligned with the rotation axis and located outside the tangent cylinder. However, as Ra is increased, the columns become thinner and more dynamic, drifting through the outer core in an apparently random fashion. Also, progressively fewer columns intersect with the mantle, so the source of the intense helicity in the columns is no longer Ekman pumping. Figure 17.2(a) shows an example of a numerical simulation in which $\mathrm{Ra}/(\mathrm{Ra})_{\mathrm{crit}} \approx 40$.

The strength of the convection in the earth's core is controlled by the heat flux through the mantle, which is a poorly constrained quantity. Consequently, it is uncertain just how supercritical convection in the earth's core might be, although a typical estimate is $\mathrm{Ra} \sim 10^5\, \mathrm{Ra}_{\mathrm{crit}}$. The flow is therefore likely to be highly chaotic. The numerical simulations of the core cannot get close to $\mathrm{Ra} \sim 10^5\mathrm{Ra}_{\mathrm{crit}}$, nor to $\mathrm{Ek} \sim 10^{-15}$. Rather, they are overly viscous by a factor of $\sim 10^9$ and significantly underpowered. Nevertheless, many of the simulations yield dynamos that exhibit earth-like features, and so the thin, helical, columnar vortices seen in the simulations may well be representative of the flow in the earth's core.

As we shall see, helicity is central to a planetary dynamo, and indeed dynamo action in the numerical simulations is driven by the helical flow within the thin, columnar vortices. It turns out that, outside the tangent cylinder, the helicity in the numerical dynamos is, on average, negative in the north ($h \approx u_z\omega_z < 0$) and positive in the south ($h \approx u_z\omega_z > 0$), as shown in Figure 17.2(b). Moreover, we shall see that just such an antisymmetric distribution of mean helicity is optimal for a planetary dynamo. The numerical simulations also exhibit vigorous upwellings within the tangent cylinder. However, this does not contribute significantly to the dynamo, which is firmly located *outside* the tangent cylinder.

17.2.3 The Tendency for the Buoyancy Flux to Concentrate Around the Equatorial Plane

As mentioned in §17.1.1, the buoyancy flux in the numerical simulations is usually found to be concentrated inside the tangent cylinder, in the form of buoyant upwellings, and also in and around the equator, where it takes the form of plumes which meander radially out along the equatorial plane towards the mantle (see Figure 17.3a). The temperature distribution is also observed to be biased towards the equator, with much of the hot, buoyant material located close to the equatorial plane, as shown in Figure 17.3(c).

The reason for the high temperatures in and around the equatorial plane is rather subtle. Plots of the azimuthally averaged radial velocity tend to show a jet-like outflow centred on the equatorial plane, which is sometimes called the *equatorial jet*. One such example is shown in Figure 17.3(b), and we shall see in §17.4.6 that this jet-like structure is driven by Lorentz forces. It is tempting to suppose that this equatorial jet is responsible for the high temperatures at the equator, as it carries heat from the inner core to the mantle. However, it turns out that this is not the dominant effect. Rather, the high equatorial temperature is related to the (statistically) antisymmetric distribution of helicity, $h \approx u_z \omega_z$, outside the tangent cylinder (Davidson & Ranjan, 2023). In the north, where the helicity is negative, the streamlines within the columnar vortices consist of left-handed spirals aligned with the axis of the columns. In the south, the spirals are right-handed. Now the axial vorticity in a cyclone or anticyclone does not change sign on passing through the equator, and so the change in the sign of helicity at the equator must arise from a reversal in the direction of the axial velocity. In short, what makes the equator special is that (statistically) the motion there is almost two-dimensional, with little axial velocity.

Now the columnar vortices move in a chaotic manner, and so much of the heat transfer outside the tangent cylinder is driven by turbulent diffusion. Near the equator, where the motion is almost two-dimensional, that diffusion is highly anisotropic, with the turbulent diffusivity being considerably larger in the radial direction than in the axial direction. Thus, the heat that leaves the ICB near the equator diffuses primarily out along the equatorial plane, with little turbulent diffusion to the north or south. In short, the high equatorial temperatures are a hydrodynamic, rather than an electrodynamic, phenomenon.

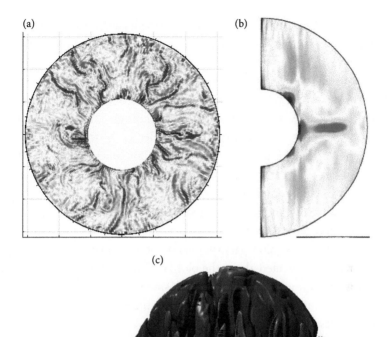

Figure 17.3 (a) Axial vorticity in the equatorial plane showing buoyant plumes floating outward (Ra/(Ra)$_{crit}$ ≈ 40). (b) Azimuthally averaged radial velocity in a vertical plane. Reproduced from Sakuraba & Roberts (2009). (c) Temperature on a spherical surface (Ra/(Ra)$_{crit}$ ≈ 40).

17.2.4 The Hidden Role of Inertial Waves in the North–South Segregation of Helicity

Since the Rossby number is very small, the earth's core is a wave-bearing system, and any disturbance will trigger inertial waves, possibly modified by the ambient magnetic field. The buoyant material floating out along the equatorial plane is important in this respect, as it represents a disturbance that almost certainly triggers waves. Moreover, as shown in Figure 6.9, upward travelling inertial waves carry negative helicity and downward travelling waves positive helicity, consistent with the distribution of h in Figure 17.2(b).

Perhaps it is worth taking a moment to remind ourselves of the dispersion pattern generated by a buoyant blob drifting through a rapidly rotating fluid. We discussed this in §6.5.1, where we considered an isolated blob of buoyant material of scale δ which sits in a rotating, Boussinesq fluid with $\mathbf{\Omega} = \Omega\hat{\mathbf{e}}_z$, $\mathbf{g} = -g\hat{\mathbf{e}}_x$ and $u/\Omega\delta \ll 1$. Such a configuration is reminiscent of a buoyant blob drifting out along the equatorial plane in the earth's core, with the x and y axes locally aligned with r and θ, as shown in Figure 17.4(b). (We shall adopt cylindrical polar coordinates (r, θ, z) when discussing the earth's core.) In §6.5.1 we showed that the dispersion pattern consists of a cyclone–anticyclone pair of columnar vortices located above the blob, matched to a cyclone–anticyclone pair below, with the anticyclones located at smaller θ, and the cyclones at larger θ. Also, the flow diverges from the r–θ plane in the anticyclones, and converges in the cyclones, as shown in Figure 17.4(b).

Figure 17.4(a) shows such a dispersion pattern, where x (or r) points out of the page, y (or θ) points to the right, and the buoyant blob is shown black. Blue indicates $u_z < 0$ while red represents $u_z > 0$, with the cyclones appearing on the right. Also shown in Figure 17.4(a) is a dynamo simulation with its cyclone–anticyclone pairs, and the similarity is striking. This tentatively suggests that the alternating cyclones–anticyclones observed in the dynamo simulations may be inertial wave packets triggered by the equatorial buoyancy flux. Finally, Figure 17.5 shows inertial wave packets (coloured by helicity) dispersing from a layer of buoyant blobs which are slowly drifting across the x–y plane. This shows a segregated helicity pattern not unlike that seen outside the tangent cylinder in Figure 17.2(b). Once again, the similarity is striking.

Davidson (2016) and Davidson & Ranjan (2018a, b) have explored the possibility that the cyclone–anticyclone pairs observed in the dynamo simulations are indeed inertial wave packets triggered by buoyant blobs near the equator, and they argue that this is in fact the case. Certainly, inertial waves (probably modified by the ambient magnetic field) are ubiquitous in the simulations, as demonstrated by the space–time Fourier transform of $\partial u_z/\partial t$ shown in Figure 17.6. The power spectrum is dominated by frequencies in the range $-2\Omega < \bar{\omega} < 2\Omega$, which is precisely the range occupied by inertial waves. Also, inertial waves propagating away from the equator provide a simple, robust

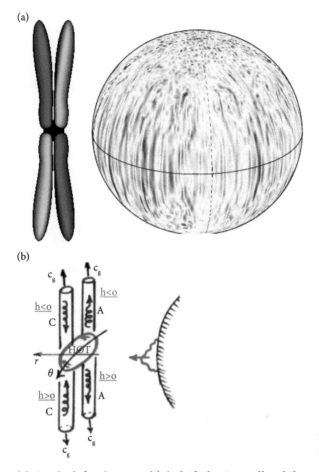

Figure 17.4 (a) On the left, a buoyant blob drifts horizontally while emitting inertial waves at Ro = 0.01. $u_z < 0$ is shown as blue and $u_z > 0$ as red. On the right, a dynamo simulation showing alternating cyclones–anticyclones. Reproduced from Davidson & Ranjan (2015). (b) Cartoon of inertial waves dispersing from a blob at the equator. C stands for cyclone, A for anticyclone.

explanation for the north–south segregation of helicity which is so evident in Figure 17.2(b).

Interestingly, the lack of a clear segregation in Figure 17.2(c), which is not azimuthally averaged, tells us that the story is a little more complicated, and that complication arises from the reflection of the wave packets at the mantle. That is to say, an important difference between Figures 17.5 and 17.2 is that there are no boundaries in Figure 17.5, whereas wave packets reflect

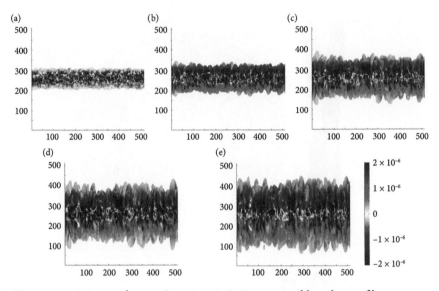

Figure 17.5 Wave packets at Ωt = 2, 4, 6, 8, 10 generated by a layer of buoyant blobs slowly drifting in a Boussinesq fluid. Colour represents helicity. From Davidson & Ranjan (2015).

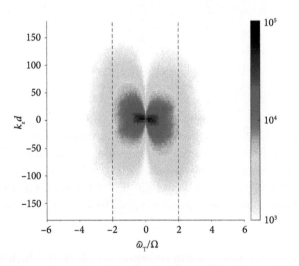

Figure 17.6 The z–t space–time Fourier transform of $\partial u_z/\partial t$ at a given cylindrical radius in a fully developed numerical dynamo. From Davidson & Ranjan (2018a).

off the mantle in Figure 17.2, and such reflections reverse the helicity of the wave packets. It is the tendency for reflected waves to cancel the helicity of the incoming waves that makes Figure 17.2(c) so messy. Crucially, however,

Ohmic dissipation ensures that the reflected waves are invariably weaker than the incoming ones, and so the cancellation is only partial, leaving a statistical bias towards negative helicity in the north and positive helicity in the south, as indicated by Figure 17.2(b).

Of course, there are other possible explanations for the north–south segregation of helicity. For example, in mildly supercritical simulations, such as that shown in Figure 12.9, Ekman pumping can yield just such a segregation pattern. However, in the highly supercritical simulations, very few of the columnar structures intersect with the mantle (see Figure 17.2), and so this is unlikely to explain the segregation shown in Figure 17.2(b). Indeed, numerical simulations employing slip boundary conditions on the mantle, in which there is no Ekman pumping, yield helical dynamos not unlike their no-slip counterparts.

Another common argument relies on dimensional analysis and symmetry. In particular, it is sometimes suggested that the distribution of helicity, which is a pseudo-scalar, must mimic the spatial distribution of the pseudo-scalar $\mathbf{g}\cdot\mathbf{\Omega}$ (which is antisymmetric about the equator). However, this is certainly *not* the case in Figure 17.5, where $\mathbf{g}\cdot\mathbf{\Omega} = 0$ yet h is antisymmetric about the mid-plane, nor in Figure 6.15, where $\mathbf{g}\cdot\mathbf{\Omega} < 0$ at all points and h is again antisymmetric. Evidently, this argument lacks rigour. It also leaves too many questions unanswered. For example, it does not explain why the north–south segregation of helicity is clear only in *statistically averaged* helicity plots. More importantly, it does not explain why the north–south segregation of the azimuthally averaged helicity is observed to collapse in the numerical dynamos when the small-scale Rossby number, $u/\Omega\delta$, rises above 0.4 (see McDermott & Davidson, 2019). In short, it seems that a more mechanistic explanation for the helicity segregation is essential if progress is to be made.

This brings us to what is, perhaps, the strongest argument in favour of the cyclone–anticyclone pairs being inertial wave packets (possibly modified by the magnetic field). It is observed that, in the numerical dynamos, the helical columnar vortices disappear, and the dipole collapses, when the small-scale Rossby number rises above $\mathrm{Ro} = u/\Omega\delta = O(1)$ (see §17.6). But we have met the threshold of $\mathrm{Ro} = u/\Omega\delta = O(1)$ before, and in a completely different context. As noted in §6.9, laboratory experiments of rapidly rotating turbulence show that inertial waves can freely propagate when $\mathrm{Ro} < 0.4$, but are suppressed for values of Ro significantly larger than 0.4. This supports the hypothesis that the columnar structures observed in the dynamo simulations are maintained by inertial wave packets.

17.3 The Simplified Version of Maxwell's Equations Required for Dynamo Theory

In the remainder of this chapter, we shall focus on the mechanisms by which the flow in the earth's core maintains the terrestrial magnetic field. As a prelude to this, we provide an introduction to the reduced form of Maxwell's equations used in magnetohydrodynamics (or MHD for short), and hence used in planetary dynamo theory. Let us start by describing (qualitatively) how the velocity and magnetic fields interact in an electrical conductor.

The mutual interaction of a magnetic field, **B**, and a moving conducting fluid arises for the following reason. Relative movement of the magnetic field and the fluid causes an *electromotive force* (or EMF) to develop, of the order of **u** × **B**. Ohm's law then tells us that this EMF drives a current, whose density is of the order of σ**u** × **B**, where σ is the electrical conductivity of the fluid. This current has two effects. First, the induced current gives rise to a second magnetic field. This adds to the original magnetic field and the change is such that the fluid appears to drag the magnetic field lines along with it. Second, the combined magnetic field interacts with the induced current density, **J**, to give rise to a Lorentz force per unit volume, **J** × **B**. This force acts on the fluid and is such as to oppose the relative motion of the fluid and magnetic field.

These two effects are evident in conventional electrodynamics. Consider the situation shown in Figure 17.7, where a wire loop is pulled through a magnetic field. Current is induced in the wire by the relative movement of the wire and field. This current distorts the magnetic field as shown, so the wire seems to drag the magnetic field with it. The current also interacts with **B** to produce a Lorentz force which opposes the relative motion.

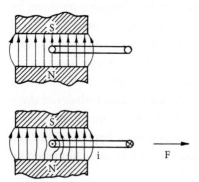

Figure 17.7 The interaction of a magnet and a moving wire loop.

17.3.1 Charge Conservation and the Laws of Ohm, Ampère, and Faraday

Let us now quantify this picture by introducing the laws of Ohm, Ampère, and Faraday. Ohm's equation is an empirical law which relates to the current density, $\mathbf{J}$, in a conducting medium to the electric field, $\mathbf{E}$. For a stationary conductor it states that $\mathbf{J} = \sigma\mathbf{E}$, where σ is the electrical conductivity. We may interpret this as $\mathbf{J}$ being proportional to the Coulomb force, $\mathbf{f} = q\mathbf{E}$, which acts on the free charge carriers, q being their charge. If, however, the conductor is moving with a velocity $\mathbf{u}$ in a magnetic field $\mathbf{B}$ then the free charges experience an additional force of $\mathbf{f} = q\mathbf{u} \times \mathbf{B}$, and so Ohm's law is modified to read,

$$\mathbf{J} = \sigma(\mathbf{E} + \mathbf{u} \times \mathbf{B}). \tag{17.12}$$

The quantity $\mathbf{E} + \mathbf{u} \times \mathbf{B}$, which is the net electromagnetic force per unit charge, is often called the *effective electric field*, since it is the electric field which would be measured in a frame of reference moving with velocity $\mathbf{u}$ (relativistic effects apart). We use the symbol $\mathbf{E}_e$ to represent $\mathbf{E} + \mathbf{u} \times \mathbf{B}$. Thus, the force on a charge q, and Ohm's law, may be written as,

$$\mathbf{f} = q(\mathbf{E} + \mathbf{u} \times \mathbf{B}) = q\mathbf{E}_e, \qquad \mathbf{J} = \sigma(\mathbf{E} + \mathbf{u} \times \mathbf{B}) = \sigma\mathbf{E}_e.$$

Faraday's law, on the other hand, tells us about the EMF which is generated in a conducting medium as a result of either: (i) a time-dependent magnetic field; or else (ii) the motion of the conductor within the magnetic field. In either case, the integral version of Faraday's law states that

$$\oint_C \mathbf{E}_e \cdot d\mathbf{r} = \oint_C (\mathbf{E} + \mathbf{v} \times \mathbf{B}) \cdot d\mathbf{r} = -\frac{d}{dt}\int_S \mathbf{B} \cdot d\mathbf{S}. \tag{17.13}$$

Here C is *any* closed curve composed of line elements $d\mathbf{r}$ and S is any surface which spans the curve C. The velocity $\mathbf{v}$ in (17.13) is the velocity of the line element $d\mathbf{r}$, so that $\mathbf{E}_e$ is the effective electric field as measured in a frame of reference moving with $d\mathbf{r}$. Now the EMF around a closed curve C is defined as $\oint_c \mathbf{E}_e \cdot d\mathbf{r}$, and so Faraday's law tells us that an EMF is induced around the curve C whenever there is a net rate of change of magnetic flux, $\Phi = \int \mathbf{B} \cdot d\mathbf{S}$, through any surface which spans C. Note that C may be stationary, move with the conducting medium, or execute some entirely different motion. It does not matter. For the particular case where C is a *material curve*, C_m, we have $\mathbf{v} = \mathbf{u}$. The laws of Faraday and Ohm then combine to tell us that the induced EMF will drive a current through the conductor according to

$$\frac{1}{\sigma} \oint_{C_m} \mathbf{J} \cdot d\mathbf{r} = -\frac{d}{dt} \int_S \mathbf{B} \cdot d\mathbf{S} = -\frac{d\Phi}{dt}. \tag{17.14}$$

Note that Faraday's law is often written in differential form, as discussed below.

Next, we turn to Ampère's law. This tells us about the magnetic field associated with a given distribution of current, $\mathbf{J}$. In integral form, it may be written as,

$$\oint_C \mathbf{B} \cdot d\mathbf{r} = \mu \int_S \mathbf{J} \cdot d\mathbf{S}, \tag{17.15}$$

where S spans the curve C. Alternatively, Stokes' theorem tells us that this integral equation is equivalent to the differential expression

$$\nabla \times \mathbf{B} = \mu \mathbf{J}. \tag{17.16}$$

Crucially, Ampère's law may be inverted using the Bio–Savart law, which states that, in an unbounded domain,

$$\mathbf{B}(x) = \frac{\mu}{4\pi} \int \frac{\mathbf{J}(x') \times \mathbf{r}}{r^3} dx' \quad , \quad \mathbf{r} = x - x'. \tag{17.17}$$

Finally, we need to introduce charge conservation which, by analogy with (2.6), takes the form

$$\nabla \cdot \mathbf{J} = -\partial \rho_e / \partial t, \tag{17.18}$$

where ρ_e is the charge density. Somewhat disturbingly, this is inconsistent with Ampère's law, whose divergence demands $\nabla \cdot \mathbf{J} = 0$. This led Maxwell to add the so-called *displacement current* to Ampère's law,

$$\nabla \times \mathbf{B} = \mu \left[\mathbf{J} + \varepsilon_0 \frac{\partial \mathbf{E}}{\partial t} \right], \tag{17.19}$$

where ε_0 is the permittivity of free space. This is known as the *Ampère–Maxwell equation* and we shall see shortly that this is fully compatible with charge conservation. Comparing (17.16) with (17.19), we see that Ampère's law is strictly only correct for static (or slowly varying) electric fields, or for cases where the charge density is negligible.

17.3.2 The Reduced Form of Maxwell's Equations Required for Dynamo Theory

We shall now establish the reduced form of Maxwell's equations required to describe a conducting fluid moving through a magnetic field. We start by summarizing the governing equations of electrodynamics. The macroscopic equations of electrodynamics, applicable to materials which are neither polarized nor magnetized, are:

- Ohm's law: $\quad\quad\quad\quad\quad\quad\quad\quad\quad\quad \mathbf{J} = \sigma(\mathbf{E} + \mathbf{u} \times \mathbf{B})$ (17.20)

- Charge conservation: $\quad\quad\quad\quad\quad\quad \nabla \cdot \mathbf{J} = -\partial\rho_e/\partial t$ (17.21)

- Force on a charge moving with velocity $\mathbf{u}$: $\quad \mathbf{f} = q(\mathbf{E} + \mathbf{u} \times \mathbf{B})$ (17.22)

- Gauss' law: $\quad\quad\quad\quad\quad\quad\quad\quad\quad \nabla \cdot \mathbf{E} = \rho_e/\varepsilon_0$ (17.23)

- Solenoidal nature of $\mathbf{B}$: $\quad\quad\quad\quad\quad \nabla \cdot \mathbf{B} = 0$ (17.24)

- Faraday's law in *differential* form: $\quad \nabla \times \mathbf{E} = -\dfrac{\partial \mathbf{B}}{\partial t}$ (17.25)

- Ampère - Maxwell equation: $\quad\quad\quad \nabla \times \mathbf{B} = \mu\left[\mathbf{J} + \varepsilon_0 \dfrac{\partial \mathbf{E}}{\partial t}\right]$ (17.26)

Of course, (17.25) looks very different to (17.13), but we shall establish the equivalence of the differential and integral versions of Faraday's law shortly.

The final four equations in the list above are known collectively as *Maxwell's equations*. Note that not all of these equations are independent. For example, taking the divergence of (17.25) gives $\partial(\nabla \cdot \mathbf{B})/\partial t = 0$, from which, with suitable initial conditions, we can deduce (17.24). Similarly, taking the divergence of (17.26), and invoking Gauss' law, we have

$$\nabla \cdot \mathbf{J} = -\varepsilon_0 \frac{\partial}{\partial t}\nabla \cdot \mathbf{E} = -\frac{\partial \rho_e}{\partial t},$$

which is charge conservation.

It turns out that, in MHD, we can simplify these equations considerably. In particular, ρ_e turns out to be extremely small, to the extent it may be totally neglected. This is because the high electrical conductivity of liquid metals ensures that nearly all of the excess charge within the interior of the fluid is expelled to the boundaries by electrostatic forces, and this occurs on an extremely fast timescale known as the *charge relaxation time*. So, in MHD, we use a reduced set of Maxwell's equations which are equivalent to assuming that the charge density is infinitesimally small. Thus, (17.21) simplifies to $\nabla \cdot \mathbf{J} = 0$, while the second term on the right of (17.26), which Maxwell introduced specifically to allow for the possibility that $\mathbf{J}$ is non-solenoidal, must be omitted. The final simplification comes from the force law (17.22). When this is integrated over a unit volume of conductor, it becomes $\boldsymbol{F} = \rho_e \mathbf{E} + \mathbf{J} \times \mathbf{B}$. Since ρ_e is negligible, this body force simplifies to $\boldsymbol{F} = \mathbf{J} \times \mathbf{B}$. Thus, the reduced set of Maxwell's equations used in MHD are:

- Ohm's law plus the Lorentz force per unit volume:

$$\mathbf{J} = \sigma(\mathbf{E} + \mathbf{u} \times \mathbf{B}), \tag{17.27}$$

$$\boldsymbol{F} = \mathbf{J} \times \mathbf{B}. \tag{17.28}$$

- Faraday's law in differential form plus the solenoidal constraint on $\mathbf{B}$:

$$\nabla \times \mathbf{E} = -\frac{\partial \mathbf{B}}{\partial t}, \tag{17.29}$$

$$\nabla \cdot \mathbf{B} = 0. \tag{17.30}$$

- Ampère's law plus charge conservation:

$$\nabla \times \mathbf{B} = \mu \mathbf{J}, \tag{17.31}$$

$$\nabla \cdot \mathbf{J} = 0. \tag{17.32}$$

To these we must add the Navier–Stokes equation in which $\mathbf{J} \times \mathbf{B}$ appears as a body force per unit volume. Note that Gauss' law is omitted in the MHD approximation, since it merely specifies ρ_e, which is a small quantity whose distribution is of no interest.

Often Faraday's law appears in integral form, and indeed this is how we introduced it in the previous section. The relationship between (17.13) and (17.29) may be established as follows. In §2.9.2 we saw that, if $\mathbf{G}$ is a solenoidal vector field, and S is any open surface which spans the closed material curve C_m, then we have the kinematic relationship

$$\frac{d}{dt} \int_S \mathbf{G} \cdot d\mathbf{S} = \int_S \frac{\partial \mathbf{G}}{\partial t} \cdot d\mathbf{S} - \oint_{C_m} (\mathbf{u} \times \mathbf{G}) \cdot d\mathbf{r} = \int_S \left[\frac{\partial \mathbf{G}}{\partial t} - \nabla \times (\mathbf{u} \times \mathbf{G}) \right] \cdot d\mathbf{S}.$$

$$\tag{17.33}$$

(Here the fluid velocity $\mathbf{u}$ is also the velocity of any point on the material curve C_m.) However, the differential form of Faraday's law tells us that

$$\nabla \times (\mathbf{E} + \mathbf{u} \times \mathbf{B}) = -\left[\frac{\partial \mathbf{B}}{\partial t} - \nabla \times (\mathbf{u} \times \mathbf{B})\right], \tag{17.34}$$

and combining these two expressions with $\mathbf{G} = \mathbf{B}$ yields

$$\frac{d}{dt}\int_S \mathbf{B} \cdot d\mathbf{S} = -\int_S \nabla \times (\mathbf{E} + \mathbf{u} \times \mathbf{B}) \cdot d\mathbf{S} = -\oint_{C_m} (\mathbf{E} + \mathbf{u} \times \mathbf{B}) \cdot d\mathbf{r}.$$

Finally, we introduce the effective electric field $\mathbf{E}_e = \mathbf{E} + \mathbf{u} \times \mathbf{B}$, and define the EMF to be the closed line integral of $\mathbf{E}_e$ around C_m. This yields the integral version of Faraday's law:

$$\text{EMF} = \oint_{C_m} \mathbf{E}_e \cdot d\mathbf{r} = -\frac{d}{dt}\int_S \mathbf{B} \cdot d\mathbf{S}. \tag{17.35}$$

Actually, (17.33), and hence (17.35), applies to *any* closed curve, C, spanned by the open surface, S. In particular, C need not be a material curve. It is necessary only to ensure that $\mathbf{E}_e$ is evaluated using the velocity $\mathbf{v}$ of the line element $d\mathbf{r}$, $\mathbf{E}_e = \mathbf{E} + \mathbf{v} \times \mathbf{B}$.

17.3.3 An Evolution Equation for the Magnetic Field

Returning now to the differential form of Faraday's law, we may eliminate $\mathbf{E}$ using Ohm's law and then $\mathbf{J}$ using Ampère's law. This yields an evolution equation for $\mathbf{B}$ which we first introduced in §17.2.1,

$$\frac{\partial \mathbf{B}}{\partial t} = \nabla \times (\mathbf{u} \times \mathbf{B}) + \lambda \nabla^2 \mathbf{B}. \tag{17.36}$$

Here λ is the magnetic diffusivity $(\mu\sigma)^{-1}$, from which we can construct the magnetic Reynolds number $R_m = u\ell/\lambda$. It seems that, for a given $\mathbf{u}$, vorticity and magnetic fields evolve in identical ways, except that they have different diffusivities. This suggests that, for a perfectly conducting fluid, there should exist the MHD analogues of Helmholtz's two laws. This turns out to be true. For a perfect conductor ($\lambda = 0$), we shall see that:

Theorem 1 : The **B** – lines are 'frozen' into the fluid,   (17.37)

Theorem 2 : $\dfrac{d}{dt}\displaystyle\int_S \mathbf{B} \cdot d\mathbf{S} = 0,$ (17.38)

where S is any open surface which spans a closed material curve, C_m. Theorem 1, which is called *Alfvén's theorem*, is illustrated in Figure 17.8.

Theorem 2, which is the analogue Helmholtz's second law, or equivalently Kelvin's theorem, follows directly from Faraday's law in integral form applied to a closed material curve, C_m:

$$\oint_{C_m} \mathbf{E}_e \cdot d\mathbf{r} = \frac{1}{\sigma}\oint_{C_m} \mathbf{J} \cdot d\mathbf{r} = -\frac{d}{dt}\int_S \mathbf{B} \cdot d\mathbf{S}. \qquad (17.39)$$

If we let $\sigma \to \infty$, while demanding that $|\mathbf{J}|$ stays finite, the line integral on the left tends to zero, which yields (17.38). The first theorem then follows from the second if we consider Figure 17.9. This shows a magnetic flux tube sitting in a perfectly conducting fluid. Since **B** is solenoidal, the flux of **B** along the tube, Φ, is constant. Now consider a material curve C_m which at some initial instant, $t = 0$, encircles the flux tube. The flux enclosed by C_m is equal to Φ at $t = 0$, and from Theorem 2 this is true at all subsequent times. That is, as C_m is swept around by the flow it always encircles the same magnetic flux. This is also true of each and every material curve which encloses the tube at $t = 0$. The only

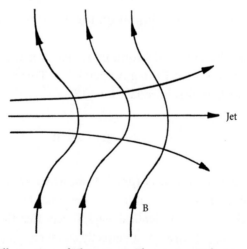

Figure 17.8 An illustration of Theorem 1. Flow across B-lines causes the field to bow out.

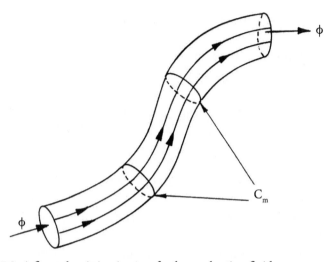

Figure 17.9 A flux tube sitting in a perfectly conducting fluid.

way this can be realized for all possible velocity fields is if the flux tube, like the material curve C_m, is frozen into the fluid, so that they execute the same motion. If we now let the flux tube have a vanishingly small cross-section, we recover Theorem 1.

17.3.4 An Energy Equation

We now derive an energy equation for MHD. As usual, we restrict the discussion to incompressible fluids. If we take the dot product of $\mathbf{u}$ with the Navier–Stokes equation, and ignore buoyancy, we obtain

$$\frac{D}{Dt}\left(\frac{\rho \mathbf{u}^2}{2}\right) = \frac{\partial}{\partial x_j}\left(u_i \tau_{ij}\right) - 2\rho\nu S_{ij}S_{ij} + (\mathbf{J}\times\mathbf{B})\cdot\mathbf{u}, \qquad (17.40)$$

where S_{ij} is the rate of strain tensor, τ_{ij} is Cauchy's stress tensor, $2\rho\nu S_{ij}S_{ij}$ is the viscous dissipation rate, and $(\mathbf{J}\times\mathbf{B})\cdot\mathbf{u}$ is the rate of generation of kinetic energy by the Lorentz force. Next, we note that the dot product of $\mathbf{B}$ with Faraday's law yields

$$\frac{\partial}{\partial t}\left(\frac{\mathbf{B}^2}{2}\right) = -\mathbf{B}\cdot\nabla\times\mathbf{E} = -\nabla\cdot(\mathbf{E}\times\mathbf{B}) - \mathbf{E}\cdot(\mu\mathbf{J}),$$

where we have substituted for $\nabla\times\mathbf{B}$ using Ampère's law. Moreover, Ohm's law gives us

$$(\mathbf{J} \times \mathbf{B}) \cdot \mathbf{u} = -\mathbf{J} \cdot (\mathbf{u} \times \mathbf{B}) = \mathbf{J} \cdot (\mathbf{E} - \mathbf{J}/\sigma),$$

and on substituting for $\mathbf{J} \cdot \mathbf{E}$ we obtain an expression for the rate of change of $\mathbf{B}^2/2\mu$, the magnetic energy density,

$$\frac{\partial}{\partial t}\left(\frac{\mathbf{B}^2}{2\mu}\right) = -(\mathbf{J} \times \mathbf{B}) \cdot \mathbf{u} - \nabla \cdot \left[\frac{\mathbf{E} \times \mathbf{B}}{\mu}\right] - \frac{\mathbf{J}^2}{\sigma}. \tag{17.41}$$

The first term on the right, which appears with opposite signs in (17.40) and (17.41), is the rate of exchange of energy between the $\mathbf{B}$ and $\mathbf{u}$ fields, while the last term is the Ohmic dissipation. It follows that the divergence must represent a flux of magnetic energy, $(\mathbf{E} \times \mathbf{B})/\mu$ being the energy flux per unit area (called the *Poynting vector*).

Finally, we add (17.40) and (17.41), eliminating $(\mathbf{J} \times \mathbf{B}) \cdot \mathbf{u}$ in the process. This yields

$$\frac{\partial}{\partial t}\left(\frac{\rho \mathbf{u}^2}{2} + \frac{\mathbf{B}^2}{2\mu}\right) = \nabla \cdot \left[u_i \tau_{ij} - \frac{\rho \mathbf{u}^2}{2}\mathbf{u} - \frac{\mathbf{E} \times \mathbf{B}}{\mu}\right] - 2\rho\nu S_{ij}S_{ij} - \mathbf{J}^2/\sigma. \tag{17.42}$$

To interpret the terms on the right of (17.42) it is convenient to integrate the equation over a control volume V with bounding surface S. The three divergences then represent the rate of working of the fluid stresses on S, the rate of transport of kinetic energy across S, and the flux of magnetic energy through S. If V extends to include all space the surface integrals vanish, to give

$$\frac{d}{dt}\int_{V_\infty}\left(\frac{\rho \mathbf{u}^2}{2} + \frac{\mathbf{B}^2}{2\mu}\right)dV = -\int_{V_\infty}(\mathbf{J}^2/\sigma)dV - 2\rho\nu\int_{V_\infty}S_{ij}S_{ij}dV. \tag{17.43}$$

Thus, the total electro-mechanical energy declines due to viscous dissipation and Ohmic heating. If buoyancy is retained in the analysis, it appears as a source on the right of (17.43).

17.3.5 Maxwell's Stresses, Faraday's Tension, and Alfvén's Waves

Finally, we turn to the phenomenon of Alfvén waves, which is a distinctive feature of MHD. To understand how these waves arise we need to introduce the idea of *Faraday's tension*. Faraday pictured magnetic field lines as if they were elastic bands held in tension. To appreciate where this idea comes from we need to think about the form of the Lorentz force. From Ampére's law, this may be rewritten as

$$\mathbf{J} \times \mathbf{B} = \mathbf{B} \cdot \nabla \left(\mathbf{B}/\mu \right) - \nabla \left(\mathbf{B}^2/2\mu \right).$$

In the absence of free surfaces, the second term on the right is unimportant in an incompressible fluid as it may be simply absorbed into the pressure gradient. Indeed, $\mathbf{B}^2/2\mu$ is referred to as the *magnetic pressure*. So the important term is the first one, which may be written in terms of curvilinear coordinates using (2.3):

$$\mathbf{B} \cdot \nabla \left(\frac{\mathbf{B}}{\mu} \right) = \frac{\partial}{\partial x_j} \left(\frac{B_i B_j}{\mu} \right) = \frac{B}{\mu} \frac{\partial B}{\partial s} \hat{\mathbf{e}}_t - \frac{B^2}{\mu R} \hat{\mathbf{e}}_n. \tag{17.44}$$

Here $B = |\mathbf{B}|$, $\hat{\mathbf{e}}_t$ and $\hat{\mathbf{e}}_n$ are unit vectors tangential and normal to a magnetic field line, R is the local radius of curvature of the field line, and s is a coordinate measured along the field line. Comparing the central term in (17.44) with (2.21), we see that the mechanical effect of the force $\mathbf{B} \cdot \nabla \left(\mathbf{B}/\mu \right)$ is entirely equivalent to an imaginary set of stresses, $B_i B_j/\mu$, acting on the fluid. Hence, $B_i B_j/\mu$ is known as the *Maxwell stress*.

Let us now try to understand where Faraday's notion of tension in the field lines comes from. Consider a thin, isolated magnetic flux tube which is curved, as shown on the left of Figure 17.10. From a consideration of the Maxwell stresses, the effect of the force $\mathbf{B} \cdot \nabla \left(\mathbf{B}/\mu \right)$ acting on the fluid within the flux tube should be entirely equivalent to the effect of a tensile stress of $\sigma(s) = B^2/\mu$ acting along the axis of the tube, or equivalently, a tensile force (Faraday's tension) of $T(s) = \left(B^2/\mu \right) A(s)$, where $A(s)$ is the cross-sectional area of the flux tube. An independent confirmation of this statement may be obtained from a consideration of the terms on the right of (17.44). That is to say, a string of cross-sectional area A, and carrying a tensile force of $T(s) = \sigma(s)A(s)$, experiences a force per unit length of $\delta \mathbf{F} = (dT/ds) \hat{\mathbf{e}}_t - (T/R) \hat{\mathbf{e}}_n$. If we now substitute B^2/μ for $\sigma(s)$, note that $\Phi = BA$ is independent of s in a flux tube, and divide through by $A(s)$ to convert to force per unit volume, we arrive back at (17.44).

Now, we have already seen that, when $\lambda = 0$, the magnetic field lines are frozen into the fluid. We now see that the field lines also behave as if they are in tension. These two features combine to give us Alfvén waves. Consider what happens if we apply an impulsive force to a fluid which is threaded by a uniform magnetic field. If the medium is highly conducting then the **B**-lines are frozen into the fluid, and so the field lines will start to bow out, as shown in Figure 17.11. However, the resulting curvature of the lines creates a back reaction, $B^2/\mu R$, on the fluid. So, as the field lines become more distorted, the Lorentz force increases and eventually the fluid comes to rest. The Faraday tension then reverses the flow and pushes the fluid back to its starting point.

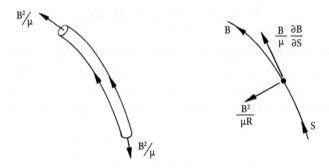

Figure 17.10 $\mathbf{B} \cdot \nabla \, (\mathbf{B}/\mu)$ is equivalent to a tensile stress of B^2/μ in the field lines.

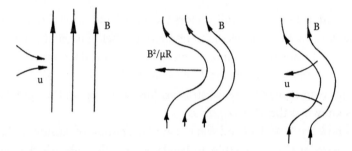

Figure 17.11 B-lines act like elastic bands frozen into the fluid, resulting in Alfvén waves.

However, the inertia of the fluid ensures that it overshoots the neutral point and the entire process now begins in reverse, leading to oscillations.

The properties of these waves are readily established. Suppose we have a uniform magnetic field, $\mathbf{B}_0$, which is perturbed by an infinitesimally small velocity field, $\mathbf{u}$. Let $\mathbf{j}$ and $\mathbf{b}$ be the resulting perturbations in current density and magnetic field. Then, from (17.36),

$$\frac{\partial \mathbf{b}}{\partial t} = \nabla \times (\mathbf{u} \times \mathbf{B}_0) + \lambda \nabla^2 \mathbf{b} \quad , \quad \nabla \times \mathbf{b} = \mu \mathbf{j},$$

which yields

$$\frac{\partial \mathbf{j}}{\partial t} = \frac{1}{\mu} (\mathbf{B}_0 \cdot \nabla) \, \boldsymbol{\omega} + \lambda \nabla^2 \mathbf{j}. \qquad (17.45)$$

Now consider the vorticity equation. Since $\nabla \times (\mathbf{u} \times \boldsymbol{\omega})$ is quadratic in the small quantity $\mathbf{u}$, this simplifies to

$$\frac{\partial \boldsymbol{\omega}}{\partial t} = \frac{1}{\rho} (\mathbf{B}_0 \cdot \nabla) \mathbf{j} + \nu \nabla^2 \boldsymbol{\omega}. \tag{17.46}$$

The next step is to eliminate $\mathbf{j}$ from (17.45) and (17.46). This yields the governing equation for Alfvén waves,

$$\left(\frac{\partial}{\partial t} - \lambda \nabla^2\right)\left(\frac{\partial}{\partial t} - \nu \nabla^2\right) \boldsymbol{\omega} = \frac{1}{\rho\mu}(\mathbf{B}_0 \cdot \nabla)^2 \boldsymbol{\omega}. \tag{17.47}$$

In liquid metals we have $\lambda \gg \nu$, and so we may neglect the viscous term in (17.47). It is readily confirmed that the resulting equation supports plane-wave solutions of the form $\boldsymbol{\omega} \sim \hat{\boldsymbol{\omega}} \exp\left[j(\mathbf{k} \cdot \mathbf{x} - \bar{\omega}t)\right]$, and when λ is small, the corresponding dispersion relationship is

$$\bar{\omega} = \pm v_a k_{//} - \left(\lambda k^2/2\right) j. \tag{17.48}$$

Here $k_{//}$ is the component of $\mathbf{k}$ in the direction of $\mathbf{B}_0$ and $\mathbf{v}_a$ is the *Alfvén velocity* $\mathbf{B}_0/(\rho\mu)^{1/2}$. This represents the propagation of a weakly damped, transverse wave, with a group velocity equal to $\pm\mathbf{B}_0/(\rho\mu)^{1/2}$. So, as we might have expected, these waves carry energy in the $\pm\mathbf{B}_0$ directions, the magnetic field lines acting like plucked strings.

17.4 An Introduction to Geodynamo Theory

We now introduce some basic ideas in planetary dynamo theory, starting with three kinematic results.

17.4.1 Some Kinematics: The Need for a Large R_m, the Ω-effect, and Cowling's Theorem

We adopt a frame of reference rotating with the earth and use cylindrical polar coordinates centred on the core. When using such coordinates, it is natural to divide the magnetic field into its azimuthal, $\mathbf{B}_\theta = (0, B_\theta, 0)$, and poloidal, $\mathbf{B}_p = (B_r, 0, B_z)$, components. For axisymmetric fields, $\mathbf{B}_\theta$ and $\mathbf{B}_p$ are individually solenoidal and the current that supports the poloidal field is azimuthal, while $\mathbf{B}_\theta$ is associated with a poloidal current distribution, as shown in Figure 17.12. Note that Ampère's law, $\oint \mathbf{B}_\theta \cdot d\mathbf{r} = \mu \int \mathbf{J}_p \cdot d\mathbf{S}$, ensures that $\mathbf{B}_\theta$ cannot extend past the core–mantle boundary if $\mathbf{B}$ is axisymmetric.

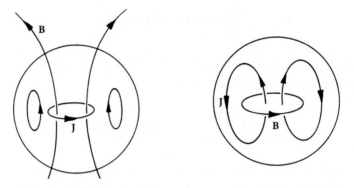

Figure 17.12 Poloidal and azimuthal components of an axisymmetric magnetic field and the currents that support them. Note that, since **J** is restricted to the conducting core, so is $\mathbf{B}_\theta$.

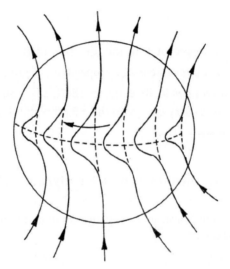

Figure 17.13 Differential rotation sweeps out an azimuthal field from the dipole field.

There are three simple kinematic ideas which are central to planetary dynamo theory:

(i) There is a lower limit on the magnetic Reynolds number needed for dynamo action.

(ii) A combination of angular momentum conservation plus natural convection will tend to cause departures from rigid-body rotation. Such differential rotation, when strong enough, will spiral out an azimuthal field from the dipole field. This is called the Ω-*effect* and it is illustrated in Figure 17.13. However, the Faraday tension in the dipole field resists being sheared in the θ-direction and so tends to suppress the Ω-effect.

(iii) *Cowling's theorem* states that an axisymmetric dynamo is not possible. This was the first of a number of anti-dynamo theorems, all of which assert that a dynamo will fail if the flow and magnetic field are constrained to be overly symmetric.

Let us start with the lower limit on R_m. Integrating (17.41) over all space, we obtain

$$\frac{d}{dt} \int_{V_\infty} (\mathbf{B}^2/2\mu) dV = - \int_{V_C} [\mathbf{u} \cdot (\mathbf{J} \times \mathbf{B})] \, dV - \int_{V_C} (\mathbf{J}^2/\sigma) \, dV. \tag{17.49}$$

The first integral on the right is the rate of working of the Lorentz force and, in a dynamo, it represents an exchange of energy from the velocity field to the magnetic field. The second integral is the Ohmic dissipation, which we label D. In order to maintain a dynamo we need

$$\dot{W} = - \int_{V_C} [\mathbf{u} \cdot (\mathbf{J} \times \mathbf{B})] \, dV \geq \int_{V_C} (\mathbf{J}^2/\sigma) \, dV = D, \tag{17.50}$$

where $\dot{W}$ is the rate of conversion of kinetic energy into magnetic energy.

The ratio of $\dot{W}$ to D is of the order of $\sigma u B/J$, or equivalently $R_m = u\ell/\lambda$, for some suitably defined u and ℓ. It seems that the magnetic Reynolds number must be $O(1)$, or larger, to achieve a dynamo and indeed we may show that there is a minimum value of R_m below which a dynamo is not possible. We can identify such a minimum by placing bounds on $\dot{W}$ and D. For example, the Schwarz inequality yields

$$\mu^2 \dot{W}^2 \leq u_{max}^2 \left[\int_{V_C} (\nabla \times \mathbf{B}) \times \mathbf{B} dV \right]^2 \leq u_{max}^2 \int_{V_\infty} \mathbf{B}^2 dV \int_{V_C} (\nabla \times \mathbf{B})^2 dV, \tag{17.51}$$

where u_{max} is the maximum velocity in the core. In addition, for current confined to a sphere, the calculus of variations yields

$$\int_{V_\infty} \mathbf{B}^2 dV \le \frac{R_C^2}{\pi^2} \int_{V_C} (\nabla \times \mathbf{B})^2 dV, \tag{17.52}$$

which combines with (17.51) to give us

$$|\dot{W}| \le \frac{u_{\max} R_C}{\pi \lambda} D. \tag{17.53}$$

A necessary condition for a dynamo is then

$$R_m = u_{\max} R_C / \lambda \ge \pi. \tag{17.54}$$

Tighter bounds exist, and in practice it is found that dynamo action in a sphere is not possible unless R_m exceeds a few multiples of 10. In the numerical dynamos, a magnetic Reynolds number in excess of ~100 is usually required for a self-sustaining dynamo.

Next we consider the role played by differential rotation within the core. The Ω-effect operates as follows. In an inertial frame of reference, fluid parcels tend to conserve their angular momentum as they move, pressure, and Lorentz forces apart. So we might expect $(\mathbf{x} \times \mathbf{u})_z = r u_\theta$ to be roughly conserved by a parcel in an inertial frame. Thus, as buoyant fluid moves out to greater r, driven by the convection, u_θ will tend to fall as r increases. We might expect, therefore, the fluid near the inner core to spin slightly faster than that near the mantle. In a frame of reference rotating with the mantle, the fluid surrounding the inner core will then have a positive value of u_θ. If this is indeed the case, then an east–west field will be swept out from the dipole field. In addition, since the solid inner core is coupled to the surrounding fluid by Maxwell stresses, we might expect the inner core to also spin slightly faster than the mantle, being dragged by the surrounding fluid.

However, this picture neglects the back reaction of the dipole field on the fluid, whose Faraday tension will resist being sheared in the θ-direction and so tend to suppress differential rotation. So, there are tentative grounds for believing that differential rotation will accompany convection, but only when the dipole field is relatively weak. A strong dipole field, on the other hand, will suppress differential rotation, and indeed just such a suppression is usually observed outside the tangent cylinder in the numerical simulations, where there is little or no Ω-effect. (See, for example, Davidson & Ranjan, 2023.)

The consequences of differential rotation acting on the dipole field are most easily seen when we have axial symmetry, which we now assume. Let us split

B and u into azimuthal and poloidal components. The θ component of the evolution equation (17.36) is then

$$\frac{\partial \mathbf{B}_\theta}{\partial t} = \nabla \times \left[\mathbf{u}_p \times \mathbf{B}_\theta\right] + \nabla \times \left[\mathbf{u}_\theta \times \mathbf{B}_p\right] + \lambda \nabla^2 \mathbf{B}_\theta, \qquad (17.55)$$

where

$$\nabla \times \left[\mathbf{u}_p \times \mathbf{B}_\theta\right] = -r\mathbf{u}_p \cdot \nabla \left(B_\theta/r\right) \hat{\mathbf{e}}_\theta,$$

and

$$\nabla \times \left[\mathbf{u}_\theta \times \mathbf{B}_p\right] = r\mathbf{B}_p \cdot \nabla \left(u_\theta/r\right) \hat{\mathbf{e}}_\theta.$$

On substituting for the convective terms in (17.55) we obtain, for an axisymmetric field,

$$\frac{D}{Dt}\left(\frac{B_\theta}{r}\right) = \mathbf{B}_p \cdot \nabla \left(\frac{u_\theta}{r}\right) + \lambda r^{-2} \nabla_*^2 \left(rB_\theta\right). \qquad (17.56)$$

The interaction of the axial field, B_z, with differential rotation, $\partial \left(u_\theta/r\right)/\partial z$, is now clear. If the dipole field points to the north, and u_θ/r is a maximum near the inner core, then $\mathbf{B}_p \cdot \nabla \left(u_\theta/r\right)$ will be negative in the north and positive in the south. Differential rotation then generates an azimuthal field which is antisymmetric about the equator and negative in the north (see Figure 17.14).

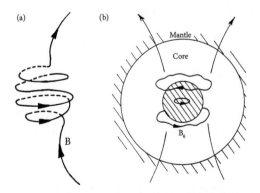

Figure 17.14 (a) The generation of an east–west field by differential rotation. (b) B_θ generated by the Ω-effect is antisymmetric about the equator and negative in the north.

However, the existence of significant differential rotation in the earth's core, and hence the status of B_θ, has proved to be controversial. The point is that the dipole field resists being sheared, and so tends to suppress differential rotation. Indeed, seismic studies which seek to detect differential rotation of the inner core are either inconclusive or negative. Moreover, in the numerical simulations of the geodynamo, the Ω-effect plays little or no role in the dynamo, with dynamo action located outside the tangent cylinder and any differential rotation largely restricted to within the tangent cylinder. (We shall discuss the simulations in §17.4.6.)

Finally, let us turn to Cowling's theorem (Cowling, 1934). This states that an axisymmetric dynamo is not possible. There are several ways in which the theorem may be understood. Perhaps the simplest is *Cowling's neutral-point argument*. Suppose that we have a steady, axisymmetric dynamo in which **B** and **u** are poloidal, and **J** is azimuthal. Since the dynamo is steady, Faraday's law requires $\nabla \times \mathbf{E} = 0$ and Ohm's law reduces to $\mathbf{J} = \sigma(-\nabla V + \mathbf{u} \times \mathbf{B})$. The potential V is governed by the divergence of Ohm's law,

$$\nabla^2 V = \nabla \cdot (\mathbf{u} \times \mathbf{B}) = \mathbf{B} \cdot \boldsymbol{\omega} - \mu \mathbf{u} \cdot \mathbf{J},$$

and since **J** and $\boldsymbol{\omega}$ are both azimuthal, we find $V = 0$. It follows that, for this steady configuration, $\mathbf{J} = \sigma \mathbf{u} \times \mathbf{B}$. Now an axisymmetric poloidal field always possesses at least one *neutral ring*, N, where $\mathbf{B} = 0$ and the **B**-lines are locally closed in the r–z plane (see Figure 17.15). Clearly $\mathbf{J} = 0$ at the neutral ring, yet Ampère's law, $\oint \mathbf{B} \cdot d\mathbf{r} = \mu \int \mathbf{J} \cdot d\mathbf{S}$, applied to a field line surrounding N demands that **J** is non-zero close to N. This ultimately leads to a contradiction, as we now show. Let ℓ be the length of the field line surrounding N and $\bar{B}$ be defined by $\oint \mathbf{B} \cdot d\mathbf{r} = \bar{B}\ell$. Then $\mathbf{J} = \sigma \mathbf{u} \times \mathbf{B}$ combines with Ampère's law to give

$$\oint \mathbf{B} \cdot d\mathbf{r} = \lambda^{-1} \int (\mathbf{u} \times \mathbf{B}) \cdot d\mathbf{S} \sim \frac{u\ell}{\lambda} \bar{B}\ell,$$

which requires $u\ell/\lambda \sim 1$. Clearly, this cannot be true in the limit of the field line in Ampère's law shrinking onto the point N, *i.e.* $\ell \to 0$. This is the anticipated contradiction.

An alternative proof of Cowling's theorem proceeds as follows. We allow for both poloidal and azimuthal components of the velocity and magnetic fields, which may be steady or unsteady, but which are axisymmetric:

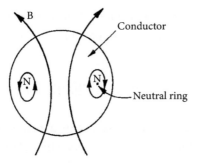

Figure 17.15 The neutral ring, N, in an axisymmetric poloidal field.

$$\mathbf{B}\,(r,\ z,\ t) = \mathbf{B}_P + \mathbf{B}_\theta, \qquad \mathbf{u}\,(r,\ z,\ t) = \mathbf{u}_P + \mathbf{u}_\theta.$$

The induction equation, (17.36), may also be divided into poloidal and azimuthal parts. The azimuthal part is simply (17.56), while the poloidal contribution is evidently

$$\frac{\partial \mathbf{B}_p}{\partial t} = \nabla \times \left[\mathbf{u}_p \times \mathbf{B}_p\right] + \lambda \nabla^2 \mathbf{B}_p.$$

Since $\mathbf{B}_P$ is solenoidal, we can introduce a vector potential, A_θ, defined by

$$\mathbf{B}_P = \nabla \times A_\theta = \nabla \times \left[(\chi/r)\ \hat{\mathbf{e}}_\theta\right],$$

where χ is the Stokes streamfunction for $\mathbf{B}_p$. We now uncurl the poloidal part of the induction equation to give

$$\frac{\partial A_\theta}{\partial t} = \mathbf{u}_p \times \mathbf{B}_p + \lambda \nabla^2 A_\theta,$$

which in turn yields

$$\frac{D\chi}{Dt} = \lambda \nabla_*^2 \chi, \tag{17.57}$$

where ∇_*^2 is the Stokes operator (2.48). The absence of any source term in (17.57) tells us that χ will eventually diffuse to zero, at which point $\mathbf{B}_P$ vanishes, as does the Ω-effect in (17.56). However, in the absence of an Ω-effect, (17.56) is also an advection-diffusion equation with no source term. Thus, as

$\mathbf{B}_P$ decays, so does $\mathbf{B}_\theta$, and the magnetic field is eventually extinguished. In short, an axisymmetric dynamo is not possible.

17.4.2 Dynamics at Last: The Taylor Constraint

Cowling's theorem was the first of several anti-dynamo theorems that greatly constrain the type of dynamo that can operate in a planetary core. A second important constraint was pointed out some 30 years later by J. B. Taylor. This relates to the angular momentum balance for a cylindrical control volume which is aligned with the rotation axis.

Our starting point is the angular momentum equation written for an inviscid, low-Ro flow. Since the buoyancy force is radial, it plays no role in this balance and we find

$$\frac{\partial}{\partial t}\left(\rho\mathbf{x}\times\mathbf{u}\right) = 2\rho\mathbf{x}\times(\mathbf{u}\times\boldsymbol{\Omega}) + \nabla\times(p\mathbf{x}) + \mathbf{x}\times(\mathbf{J}\times\mathbf{B}).$$

The z component of this may be rewritten in (r, θ, z) coordinates using the identity

$$\nabla\cdot\left[(x^2 - z^2)\,\Omega\mathbf{u}\right] = 2\Omega(\mathbf{x}\cdot\mathbf{u} - zu_z) = -[2\mathbf{x}\times(\mathbf{u}\times\boldsymbol{\Omega})]_z,$$

to give

$$\frac{\partial}{\partial t}(\rho\mathbf{x}\times\mathbf{u})_z = -\nabla\cdot\left[\rho r^2\Omega\mathbf{u}\right] + [\nabla\times(p\mathbf{x})]_z + r(\mathbf{J}\times\mathbf{B})_\theta.$$

We now integrate this over a cylindrical control volume V of radius R which spans the core, is aligned with $\boldsymbol{\Omega}$, and is centred on the core of the earth. The pressure term drops out as it cannot exert an axial torque on the surface of the control volume, while continuity kills the Coriolis term. The end result is the simple angular momentum balance

$$\frac{d}{dt}\int_V \rho(\mathbf{x}\times\mathbf{u})_z dV = \int_V r(\mathbf{J}\times\mathbf{B})_\theta dV, \tag{17.58}$$

which we rewrite in terms of Maxwell stresses using Ampère's law,

$$\frac{d}{dt} \int_V r u_\theta dV = \frac{1}{\rho\mu} \int_V \left[\frac{1}{r} \frac{\partial}{\partial r} \left(r^2 B_r B_\theta \right) + \frac{\partial}{\partial z} \left(r B_\theta B_z \right) \right] dV.$$

When converted to a surface integral, the Lorentz torque has two contributions, one from the Maxwell stress $B_r B_\theta / \mu$ acting on the cylindrical surface $r = R$, and one from the stresses $B_r B_\theta / \mu$ and $B_\theta B_z / \mu$ acting at the mantle. Now Ampère's law demands that $B_\theta = 0$ at the mantle if the field is axisymmetric, and it follows that $B_\theta \approx 0$ if the field is close to axisymmetric. We shall see later that the terrestrial magnetic field is indeed close to axisymmetric, and so we might tentatively ignore the Maxwell stresses at the mantle. The end result is (Taylor, 1963)

$$\frac{d}{dt} \int_V r u_\theta dV = \frac{1}{\rho\mu} \int_{r=R} \left[r B_r B_\theta \right] dS. \tag{17.59}$$

This is a powerful constraint, because it demands that any steady dynamo must satisfy

$$\int_{r=R} \left[r B_r B_\theta \right] dS = 0$$

on *all* cylindrical surfaces, *i.e.* the Maxwell stresses must balance on all cylindrical annuli. This is known as *Taylor's constraint*, and a field which satisfies this is called a *Taylor state*.

In practice, the core flow is highly chaotic and unsteady, and so a Taylor state is never achieved. Nevertheless, we shall see that the dominant contribution to **B** is approximately axisymmetric and quasi steady (except during global reversals), and so it seems plausible that the core fluctuates chaotically about a Taylor state. One way to picture this behaviour is to think of the core as divided up into many concentric cylindrical annuli, with the radial magnetic field threading through these annuli, providing a magnetic spring which tends to oppose any differential rotation between adjacent annuli. Of course, the net Lorentz torque arising from a closed system of currents interacting with its 'self' magnetic field is always zero. So, in a global sense, the Taylor constraint is automatically satisfied. Thus, if one cylindrical annulus is subject to a net positive torque, another must experience a negative torque. Now consider an annulus which is subject to a net positive torque. It will experience an azimuthal acceleration in accordance with (17.59), and so shear the radial magnetic field which links it to the two adjacent annuli. The resulting Maxwell stresses will act to oppose the azimuthal acceleration of the annulus, while passing its excess angular momentum onto the adjacent annuli. If we now allow for inertia, oscillations will develop at the Alfvén frequency.

Thus, it seems likely that torsional oscillations develop between adjacent annuli, with the radial magnetic field acting as a magnetic spring, linking the annuli. Such oscillations are governed by (17.59) and damped by Ohmic dissipation. In short, the core flow fluctuates stochastically about a Taylor state, subject to damped torsional oscillations.

17.4.3 More Kinematics: Integral Equations for the Axial and Azimuthal Magnetic Fields

We now return to kinematics to deduce an evolution equation for the mean axial magnetic field in the core. This provides some hints as to how to circumvent Cowling's theorem. We start with (17.1),

$$\mathbf{m} = \frac{1}{2} \int_{V_C} \mathbf{x} \times \mathbf{J} dV = \frac{3}{2\mu} \int_{V_C} \mathbf{B} dV, \tag{17.60}$$

which can be rewritten as

$$\frac{3}{2\mu} \int_{V_C} \mathbf{B} dV = \frac{\sigma}{2} \int_{V_C} \mathbf{x} \times (\mathbf{E} + \mathbf{u} \times \mathbf{B}) \, dV. \tag{17.61}$$

We now transform this using

$$\mathbf{x} \times \mathbf{E} = \nabla \times \left[(\tfrac{1}{2}\mathbf{x}^2)\mathbf{E} \right] - (\tfrac{1}{2}\mathbf{x}^2)\nabla \times \mathbf{E} = \nabla \times \left[(\tfrac{1}{2}\mathbf{x}^2)\mathbf{E} \right] + \frac{1}{2}\mathbf{x}^2 \frac{\partial \mathbf{B}}{\partial t},$$

where we have combined a vector identity with Faraday's law. This yields

$$\frac{3}{2\mu} \int_{V_C} \mathbf{B} dV = \frac{\sigma}{2} \int_{V_C} \mathbf{x} \times (\mathbf{u} \times \mathbf{B}) \, dV - \frac{\sigma}{4} \int_{V_C} \left(R_C^2 - \mathbf{x}^2 \right) \frac{\partial \mathbf{B}}{\partial t} dV, \tag{17.62}$$

which may be rearranged to give an integral evolution equation for the magnetic field:

$$\frac{d}{dt} \int_{V_C} \left(R_C^2 - \mathbf{x}^2 \right) \mathbf{B} dV = 2 \int_{V_C} \mathbf{x} \times (\mathbf{u} \times \mathbf{B}) \, dV - 6\lambda \int_{V_C} \mathbf{B} dV. \tag{17.63}$$

The axial component of (17.63) is evidently,

$$\frac{d}{dt} \int_{V_C} \left(R_C^2 - \mathbf{x}^2 \right) B_z dV = 2 \int_{V_C} r(\mathbf{u} \times \mathbf{B})_\theta dV - 6\lambda \int_{V_C} B_z dV. \tag{17.64}$$

This tells us that the axial magnetic field is maintained by the volume integral of $r(\mathbf{u} \times \mathbf{B})_\theta$. In the steady state, where $\nabla \times \mathbf{E} = 0$ and $\mathbf{J} = \sigma(-\nabla V + \mathbf{u} \times \mathbf{B})$, (17.64) reduces to

$$3\lambda \int_{V_C} B_z dV = \int_{V_C} r(\mathbf{u} \times \mathbf{B})_\theta dV = \int_{V_C} r \left(J_\theta/\sigma \right) dV. \tag{17.65}$$

This conforms to expectation since, by virtue of Ampère's law, an axial magnetic field requires azimuthal currents to support it, and these are maintained by the azimuthal EMF, $(\mathbf{u} \times \mathbf{B})_\theta$.

Now the middle integral in (17.65) is necessarily zero when the velocity and magnetic fields are both axisymmetric. This can be seen as follows. If $\mathbf{B}$ is axisymmetric, we can write $\mathbf{B} = \mathbf{B}_p + \mathbf{B}_\theta$ where $\mathbf{B}_p$ is solenoidal and so can be expressed in terms of a vector potential, i.e. $\mathbf{B}_p = \nabla \times A_\theta$. We then have $r(\mathbf{u} \times \mathbf{B})_\theta = -\mathbf{u}_p \cdot \nabla (rA_\theta)$, and hence

$$\int_{V_C} r(\mathbf{u} \times \mathbf{B})_\theta dV = - \int_{V_C} \mathbf{u}_p \cdot \nabla (rA_\theta) \, dV = - \oint_{S_C} (rA_\theta) \mathbf{u}_p \cdot d\mathbf{S} = 0, \tag{17.66}$$

where we have used the fact that $\nabla \cdot \mathbf{u}_p = 0$ in an axisymmetric velocity field. Evidently, sustained dynamo action is not possible if $\mathbf{u}$ and $\mathbf{B}$ are both axisymmetric. Of course, this is just Cowling's theorem.

Let us now remove the restriction of axial symmetry and imagine a flow (and magnetic field) that consists of an axisymmetric component plus a fluctuating part which is associated with the columnar vortices shown in Figure 17.2(a). We assume that the scale δ of the local fluctuations (the width of the columnar eddies) is much less than R_C and introduce the azimuthal average

$\langle \mathbf{B} \rangle = \langle B_r \rangle \, \hat{\mathbf{e}}_r + \langle B_\theta \rangle \, \hat{\mathbf{e}}_\theta + \langle B_z \rangle \, \hat{\mathbf{e}}_z$. We can then write $\mathbf{B} = \langle \mathbf{B} \rangle + \mathbf{b}$ and $\mathbf{u} = \langle \mathbf{u} \rangle + \mathbf{v}$, where $\langle \mathbf{B} \rangle$ and $\langle \mathbf{u} \rangle$ are axisymmetric. Since $\mathbf{b}$ and $\mathbf{v}$ have zero azimuthal means, the cross terms $\mathbf{v} \times \langle \mathbf{B} \rangle$ and $\langle \mathbf{u} \rangle \times \mathbf{b}$ integrate to zero and so (17.64) yields

$$\frac{d}{dt} \int_{V_C} \left(R_C^2 - x^2 \right) \langle B_z \rangle \, dV = 2 \int_{V_C} r \left\langle (\mathbf{v} \times \mathbf{b})_\theta \right\rangle dV - 6\lambda \int_{V_C} \langle B_z \rangle \, dV. \qquad (17.67)$$

Evidently, the axial field is maintained by an azimuthal EMF generated by non-axisymmetric fluctuations. This provides the key to circumventing Cowling's theorem. As we shall see, the EMF $(\mathbf{v} \times \mathbf{b})_\theta$ arises from the interaction of thin, columnar, helical vortices with the mean east–west field, in which the helical flow locally *lifts and twists* the azimuthal magnetic field to generate a *local* EMF on the scale of the columnar vortex. Moreover, if the helicity is predominantly of one sign in the north, and of another sign in the south, say left-handed spirals in the north and right-handed spirals in the south, as shown in Figure 17.2(b), then the sense of twist, and hence the sign of the locally induced EMF, is more or less uniform in each hemisphere. In such a case the local EMFs, $(\mathbf{v} \times \mathbf{b})_\theta$, are additive in each hemisphere, giving rise to global azimuthal currents which support the dipole field. This suggests that an important part of the dynamo story is the local EMF induced by thin, helical disturbances lifting and twisting the background magnetic field.

There is, however, a second important consequence of (17.67). It tells us that, provided the small-scale EMFs are reasonably additive in each hemisphere, then

$$\langle (\mathbf{v} \times \mathbf{b})_\theta \rangle \sim \frac{\lambda}{u R_C} u \langle B_z \rangle \ll u \langle B_z \rangle,$$

since $u R_C / \lambda \gg 1$. So, the perturbed field $\mathbf{b}$ is much weaker than the mean dipole field $\langle B_z \rangle$.

Let us now turn to the azimuthal component of $\mathbf{B}$. We start by combining Faraday's law with the azimuthally averaged version of Ohm's law. This gives

$$\langle \mathbf{E} \rangle = - \langle \partial \mathbf{A} / \partial t \rangle - \nabla \Phi = \langle \mathbf{J} \rangle / \sigma - \langle \mathbf{u} \times \mathbf{B} \rangle,$$

where $\mathbf{A}$ is the (solenoidal) vector potential for $\mathbf{B}$ and Φ is a scalar potential. We now integrate the poloidal component of this equation in a clockwise sense around a closed loop in the r–z plane. If the loop C encloses a torus of volume V, then

$$\frac{d}{dt}\oint_C \langle \mathbf{A}_p \rangle \cdot d\mathbf{r} = \frac{d}{dt}\int_V \frac{\langle B_\theta \rangle}{2\pi r}dV = \oint_C \left[\langle (\mathbf{u} \times \mathbf{B})_p \rangle - \langle \mathbf{J}_p \rangle / \sigma \right] \cdot d\mathbf{r}.$$

Writing $\mathbf{B} = \langle \mathbf{B} \rangle + \mathbf{b}$ and $\mathbf{u} = \langle \mathbf{u} \rangle + \mathbf{v}$, this yields

$$\frac{d}{dt}\int_V \frac{\langle B_\theta \rangle}{2\pi r}dV = \oint_C (\langle \mathbf{u} \rangle \times \langle \mathbf{B} \rangle)_p \cdot d\mathbf{r} + \oint_C \left[\langle (\mathbf{v} \times \mathbf{b})_p \rangle - \langle \mathbf{J}_p \rangle / \sigma \right] \cdot d\mathbf{r},$$

or

$$\frac{d}{dt}\int_V \frac{\langle B_\theta \rangle}{2\pi r}dV = (\Omega - \text{effect}) + \oint_C \left[\langle (\mathbf{v} \times \mathbf{b})_p \rangle - \langle \mathbf{J}_p \rangle / \sigma \right] \cdot d\mathbf{r}, \qquad (17.68)$$

where the term 'Ω-effect' represents the line integral of $\langle \mathbf{u} \rangle \times \langle \mathbf{B} \rangle$. (The Ω-effect in (17.68) is the contribution that arises from integrating the non-diffusive terms on the right of (17.55) over V, or equivalently, the contribution that arises from the non-diffusive terms in (17.56).)

Note the appearance of the EMF $\langle \mathbf{v} \times \mathbf{b} \rangle_p$ in (17.68). We shall see shortly that this arises from the interaction of thin, helical vortices with the mean radial field $\langle B_r \rangle$, in which the helical flow lifts and twists the radial magnetic field to generate a local EMF.

17.4.4 Parker's Lift-and-Twist Mechanism and the Alpha Effect

In 1955 Gene Parker introduced the lift-and-twist mechanism into dynamo theory (Parker 1955), which is now known as the α-effect. This involves small-scale, helical disturbances punching through the mean east–west field. These disturbances lift and twist the ambient magnetic field lines, creating small, twisted flux loops, as shown in Figure 17.16.

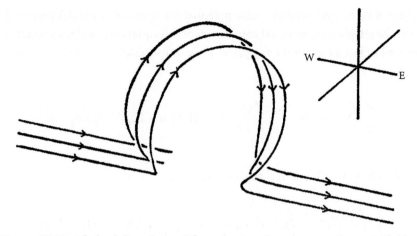

Figure 17.16 A helical disturbance lifts and twists the east–west magnetic field lines.

This lift-and-twist mechanism drives a local EMF which, as we shall see, is often aligned with the ambient magnetic field, being parallel to the background field if the helicity of the disturbance is negative, and antiparallel if it is positive. Thus, as suggested earlier, if the helicity of the disturbances is predominantly of one sign in the north, and the opposite sign in the south, these local EMFs combine in each hemisphere, like a bank of batteries wired in series, to drive large-scale azimuthal currents in the core. These large-scale azimuthal currents then give rise to a dipole field. Similarly, small-scale helical disturbances punching through the radial component of the mean poloidal field will induce local EMFs aligned with B_r, and these can combine to drive a large-scale poloidal current in the core.

We now calculate the spatially averaged EMF generated by a random, but statistically homogeneous, sea of columnar vortices spiralling through the mean east–west field. We follow the arguments in Davidson (2014). We shall assume that the horizontal scale of the vortices is sufficiently small, say $\delta \sim 10$ km, that we may take the background magnetic field to be locally uniform and reinterpret the average $< \sim >$ as a horizontal spatial average. We focus on the simplest case where the magnetic Reynolds number is of order unity at the scale of a columnar vortex, $u\delta/\lambda \sim 1$, so that we can use the so-called low-R_m approximation. In brief, this allows us to adopt a pseudo-static approximation in which $\nabla \times \mathbf{E} = -\partial\mathbf{b}/\partial t$ is replaced by $\nabla \times \mathbf{E} \approx 0$ on the grounds that $\mathbf{b}$ is small, and so Ohm's law simplifies to $\mathbf{j} = \sigma(-\nabla V + \mathbf{u} \times \langle\mathbf{B}\rangle)$. We adopt local Cartesian coordinates, with z aligned with the rotation axis and the background magnetic field lying in the x–y plane. Finally, we assume that the axial gradients in velocity are much weaker than transverse gradients and that the flow has maximal helicity, $h = \mathbf{v} \cdot \boldsymbol{\omega} = \pm \delta^{-1} \mathbf{v}^2$, where the sign of the helicity determines whether the $+$ or $-$ sign is used. Our velocity field then takes the form

$$\mathbf{v} = \mathbf{v}(x, y), \quad \boldsymbol{\omega} = \nabla \times \mathbf{v} = \pm \delta^{-1} \mathbf{v}, \tag{17.69}$$

where δ is a constant.

We now calculate the EMF induced by this helical flow. The calculation uses the fact that statistical homogeneity ensures that spatial gradients of averaged quantities are zero. For example, it is readily confirmed that homogeneity combined with (17.69) requires

$$\langle v_z^2 \rangle = \langle v_x^2 + v_y^2 \rangle, \quad \langle v_x v_z \rangle = \langle v_y v_z \rangle = 0. \tag{17.70}$$

Now, in the low-R_m approximation, the induced magnetic field is governed by

$$\mathbf{j}/\sigma = -\nabla V + \mathbf{v} \times \langle \mathbf{B} \rangle = \lambda \nabla \times \mathbf{b}, \tag{17.71}$$

whose curl is evidently,

$$\langle \mathbf{B} \rangle \cdot \nabla \mathbf{v} = \lambda \nabla \times \nabla \times \mathbf{b}. \tag{17.72}$$

Introducing a solenoidal vector potential for $\mathbf{v}$, i.e. $\mathbf{v} = \nabla \times \mathbf{a}$ with $\nabla \cdot \mathbf{a} = 0$, and noting that (17.69) requires $\mathbf{a} = \pm \delta \mathbf{v}$, (17.72) uncurls to give

$$\langle \mathbf{B} \rangle \cdot \nabla \mathbf{a} = \pm \delta \langle \mathbf{B} \rangle \cdot \nabla \mathbf{v} = \lambda \nabla \times \mathbf{b} = \mathbf{j}/\sigma. \tag{17.73}$$

Comparing (17.72) with (17.73) we see that $\mathbf{j}$ and $\mathbf{b}$ are also helical, with

$$\mathbf{b} = \mathbf{b}(x, y), \quad \mu \mathbf{j} = \nabla \times \mathbf{b} = \pm \delta^{-1} \mathbf{b}. \tag{17.74}$$

Moreover, (17.72) yields the simple result

$$\mathbf{b} = \frac{\delta^2}{\lambda} \langle \mathbf{B} \rangle \cdot \nabla \mathbf{v}. \tag{17.75}$$

The spatially averaged EMF associated with the columnar vortices is now readily found. To focus thoughts, suppose that $\langle \mathbf{B} \rangle$ points in the x direction, say $\langle \mathbf{B} \rangle = B_0 \hat{\mathbf{e}}_x$, so that

$$\mathbf{v} \times \mathbf{b} = \frac{\delta^2 B_0}{\lambda} \mathbf{v} \times \frac{\partial \mathbf{v}}{\partial x}. \tag{17.76}$$

Then (17.69) allows us to rewrite the cross product in (17.76) as

$$\mathbf{v} \times \frac{\partial \mathbf{v}}{\partial x} = \mp 2\delta^{-1} \left(v_y^2, -v_x v_y, 0 \right) + \frac{\partial}{\partial x} \left(-v_y v_z, v_x v_z, v_x v_y \right) + \frac{\partial}{\partial y} \left(0, 0, v_y^2 \right), \tag{17.77}$$

and the terms involving spatial gradients disappear on averaging. The averaged EMF is now

$$\langle \mathbf{v} \times \mathbf{b} \rangle = \mp \frac{2\delta B_0}{\lambda} \left\langle \left(v_y^2, -v_x v_y, 0 \right) \right\rangle. \tag{17.78}$$

Finally, for simplicity, we shall assume that $\langle v_x v_y \rangle = 0$ and $\langle v_x^2 \rangle = \langle v_y^2 \rangle$, as would be the case for disturbances which are isotropic in the x–y plane. Invoking (17.70), we now obtain

$$\langle \mathbf{v} \times \mathbf{b} \rangle = \mp \frac{\delta}{2\lambda} \langle \mathbf{v}^2 \rangle \langle \mathbf{B} \rangle = -\frac{\delta^2}{2\lambda} \langle h \rangle \langle \mathbf{B} \rangle. \tag{17.79}$$

This turns out to be a good approximation to the EMF induced by a sea of low-R_m inertial waves propagating along the rotation axis (see Davidson & Ranjan, 2015). However, Figure 17.2(c) shows that there is a great deal of cancellation of helicity of opposite signs when evaluating $\langle h \rangle$ in the numerical dynamos, and so a better low-R_m estimate becomes

$$\langle \mathbf{v} \times \mathbf{b} \rangle \sim \mp \varepsilon \, (\delta/\lambda) \langle \mathbf{v}^2 \rangle \langle \mathbf{B} \rangle \sim -(\delta^2/\lambda) \langle h \rangle \langle \mathbf{B} \rangle, \quad \varepsilon << 1. \tag{17.80}$$

In any event, if $u\delta/\lambda \leq O(1)$ (which may or may not be true in the earth's core), $\langle \mathbf{v} \times \mathbf{b} \rangle$ is aligned with $\langle \mathbf{B} \rangle$ and takes the opposite sign to that of $\langle h \rangle$ (see Figure 17.17). It is usual to rewrite (17.80) as $\langle \mathbf{v} \times \mathbf{b} \rangle = \alpha \langle \mathbf{B} \rangle$, where $\alpha = -\delta^2 \langle h \rangle / 2\lambda$, which is the origin of the phrase 'α-effect'. Note that α, like h, is a pseudo-scalar. Note also that (17.80) may be applied to either B_θ or B_r, and so we may write $\langle \mathbf{v} \times \mathbf{b} \rangle = \alpha \langle \mathbf{B}_\perp \rangle$, where $\mathbf{B}_\perp = \mathbf{B} - \mathbf{B}_z$.

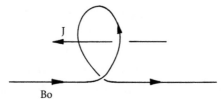

Figure 17.17 A disturbance with positive helicity generates an EMF antiparallel to **B**.

Let us now return to a global picture and to azimuthal averages. Since **b** and **v** have zero azimuthal means, the azimuthally averaged induction equation, (17.36), now becomes

$$\frac{\partial \langle \mathbf{B} \rangle}{\partial t} = \nabla \times [\langle \mathbf{u} \rangle \times \langle \mathbf{B} \rangle] + \nabla \times [\alpha \langle \mathbf{B}_\perp \rangle] + \lambda \nabla^2 \langle \mathbf{B} \rangle. \tag{17.81}$$

This is the governing equation of so-called *mean-field electrodynamics*. It is axisymmetric and so its poloidal part is best rewritten in terms χ, the Stokes streamfunction for $\mathbf{B}_p$. This allows us to incorporate the α-effect into (17.56) and (17.57), which now become

$$\frac{D}{Dt}\left(\frac{B_\theta}{r}\right) = \mathbf{B}_p \cdot \nabla \left(\frac{u_\theta}{r}\right) + \frac{1}{r}\frac{\partial}{\partial z}(\alpha B_r) + \lambda r^{-2}\nabla_*^2 (rB_\theta), \tag{17.82}$$

$$\frac{D\chi}{Dt} = \alpha r B_\theta + \lambda \nabla_*^2 \chi, \tag{17.83}$$

where the angled brackets have been omitted for clarity. Note that αB_θ appears as a source of azimuthal current, and hence poloidal field, while αB_r is a source of poloidal current, and hence east–west field. It is readily confirmed that these axisymmetric equations support a dynamo, provided that $|\alpha| R_C/\lambda$ is sufficiently large.

Evidently, Parker's introduction of the lift-and-twist mechanism was a major step forward in circumventing Cowling's theorem. As Cowling himself put it: 'clearly his [Parker's] suggestion deserves a good deal of attention' (Cowling, 1957).

17.4.5 Two Classes of Planetary Dynamos

There are two popular cartoons of the geodynamo based on mean-field electrodynamics. In α-Ω dynamos, differential rotation converts the dipole into an east–west field. To complete the cycle $\mathbf{B}_p \to \mathbf{B}_\theta \to \mathbf{B}_p$, Parker's α-effect is then used to convert $\mathbf{B}_\theta$ back into a dipole field. By way of contrast, in α^2 dynamos, any differential rotation is incidental to dynamo action. Instead, helical columnar vortices generate $\mathbf{B}_\theta$ from the radial magnetic field via an α-effect, and then convert $\mathbf{B}_\theta$ back into a poloidal field using a second α-effect. Note that, in either model, we require that the helicity is of uniform sign (at least statistically) in each hemisphere, so that the local EMFs are additive. We now consider these two cartoons in more detail, taking the dipole to point to the north.

In an α-Ω dynamo, the Ω-effect sweeps out an azimuthal field that is negative in the north and positive in the south, as shown in Figure 17.14. Indeed, such an effect was evident in the some of the early numerical dynamos, such as that shown in Figure 17.18. However, in order to support the dipole field in an α-Ω dynamo, we require azimuthal currents that are positive in both hemispheres. Given that $\langle \mathbf{v} \times \mathbf{b} \rangle_\theta \sim -\langle h \rangle \langle \mathbf{B} \rangle_\theta$, and the signs of B_θ in Figure 17.14, such a dynamo requires a helicity density that is positive in the north and negative in the south. But the majority of recent dynamo simulations yield a helicity distribution of the opposite sign, at least outside the tangent cylinder where the α-effect resides (see Figure 17.2). In addition, they show only a weak Ω-effect outside the tangent cylinder (Schaeffer et al., 2017). So, these dynamos are not of the α-Ω type, but rather α^2.

An example of an α^2 dynamo is shown in Figure 17.19 where, in line with most of the recent numerical simulations, the helicity outside the tangent cylinder is negative in the north and positive in the south. Here an α-effect operating on the radial field, $\langle \mathbf{v} \times \mathbf{b} \rangle_r \sim -\langle h \rangle \langle \mathbf{B} \rangle_r$, drives a positive radial current in the north and south, and since $\langle \mathbf{J}_p \rangle$ is solenoidal, this returns near the equator, establishing a quadrupole structure for $\langle \mathbf{J}_p \rangle$. Ampère's law then tells us that these currents give rise to an azimuthal magnetic field which is positive in the north and negative in the south (precisely the opposite to that expected from the Ω-effect). A second α-effect then operates on the azimuthal magnetic field, $\langle \mathbf{v} \times \mathbf{b} \rangle_\theta \sim -\langle h \rangle \langle \mathbf{B} \rangle_\theta$, which yields azimuthal currents that are positive in both the north and the south, exactly as required to support the original dipole field. Note that we still have a self-consistent dynamo if $\mathbf{B}$ and $\mathbf{J}$ are reversed, but h kept the same.

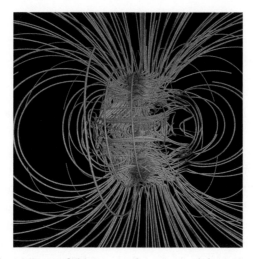

Figure 17.18 The magnetic field in an early numerical dynamo. Courtesy of G. Glatzmaier.

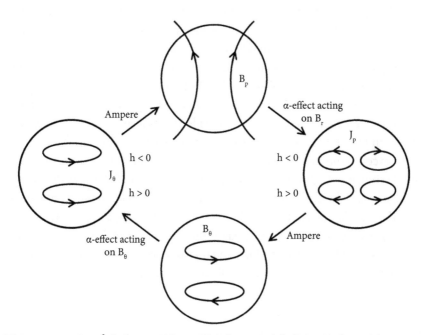

Figure 17.19 An α^2 dynamo with positive (negative) helicity in the south (north).

Note that this kind of α^2 dynamo relies crucially on a helicity distribution which is statistically antisymmetric about the equator, and so it is important to understand what maintains this helicity distribution and why it is spatially segregated in the way that it is. This brings us back to the discussion in §17.2.4, where we suggested that the cyclone–anticyclone pairs are, in fact, inertial wave packets launched in the equatorial regions by the local buoyancy flux. Certainly, this would explain the observed helicity segregation.

17.4.6 The Numerical Simulations of the Geodynamo and the Equatorial Jet

We shall now confirm that the dynamo cycle shown in Figure 17.19 is indeed observed in the numerical dynamos, at least in some zero-order sense. Consider Figure 17.20, which displays the azimuthally averaged results of three distinct numerical dynamos (S1, S2, and S3), whose properties are detailed in Table 17.1. The panels all consist of north–south slices, with the simulations S1, S2, and S3 arranged from left-to-right. Figure 17.20(a) shows B_θ (as a colour map) and the $\mathbf{J}_p$-lines, while Figure 17.20(b) shows J_θ (as a colour map) and the $\mathbf{B}_p$-lines. Figure 17.20 confirms that, outside the tangent cylinder, $\mathbf{J}_p$ has the same quadrupole structure as shown in Figure 17.19. It also shows that (outside the tangent cylinder) B_θ is predominantly positive in the north and negative in the south, consistent with Figure 17.19 and inconsistent with azimuthal field generation through the Ω-effect.

In all three simulations, there is a dominance of negative helicity in the north and positive helicity in the south (outside the tangent cylinder), exactly as shown in Figure 17.2(b), which is in fact simulation S1. The same simulations show excess temperatures in the equatorial regions and also an equatorial jet, just as shown in Figure 17.3, although the jets are not as sharp as that shown in Figure 17.3(b). Interestingly, when the magnetic field is removed from S1 and the simulation continued, the equatorial jet disappears, but the excess equatorial temperature remains (Davidson & Ranjan, 2023). This supports the suggestion of §17.2.3 that the high equatorial temperature is a hydrodynamic phenomenon, whereas the equatorial jet is electrodynamic in nature, driven by Lorentz forces.

Table 17.1 The properties of three numerical dynamos. Pr = 1 in all cases (The full details of the simulations may be found in Ranjan et al., 2020)

Case	Ra	Ra/(Ra)$_{\text{crit}}$	Ek	Pr$_m$	Ro	Re	R_m	Λ
S1	1.2×10^8	42.4	3×10^{-5}	2.5	0.01	361	902	27
S2	3×10^8	28.4	1×10^{-5}	2.5	0.003	328	820	32
S3	1.0×10^9	21.6	3×10^{-6}	0.5	0.0015	494	247	2

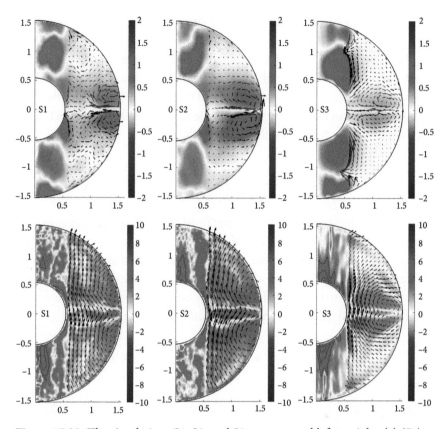

Figure 17.20 The simulations S1, S2, and S3 are arranged left-to-right. (a) $\langle B_\theta \rangle$ as a colour map and $\langle \mathbf{J}_p \rangle$ as black arrows. (b) $\langle J_\theta \rangle$ as a colour map and $\langle \mathbf{B}_p \rangle$ as black arrows. Only the B_p field lines outside the tangent cylinder are shown. From Davidson & Ranjan (2023).

The key to understanding the origin of the equatorial jet is to note that, according to Figure 17.19, an α^2 dynamo requires a negative radial current in the equatorial regions. The leading-order azimuthal force balance near the equator is then between the Coriolis and Lorentz forces, there being no buoyancy in the azimuthal direction. Since $\langle \mathbf{B} \rangle$ is axial at the equator, this gives

$$2\rho\Omega \langle u_r \rangle = \langle (\mathbf{J} \times \mathbf{B})_\theta \rangle = - \langle J_r B_z \rangle. \tag{17.84}$$

So the jet is driven by the negative radial current near the equator (Sakuraba & Roberts, 2009).

17.5 Scaling Laws for the Geodynamo

We shall now pull together the results of §17.4.3, where we derived evolution equations for the integrated magnetic field, with those of §17.4.4→§17.4.6, which show the numerical dynamos to be of the α^2 type. We shall see that this leads to scaling laws which relate the dipole field strength and characteristic fluctuating velocity to the heat flux through the core.

17.5.1 Force and Energy Balances in an α^2 Dynamo

Let us summarize the results we have obtained so far for the geodynamo. The numerical simulations suggest that the dynamo is of the α^2 type, driven by cyclone–anticyclone pairs located outside the tangent cylinder. These columnar vortices are highly helical and it is possible that they are inertial wave packets triggered in the equatorial regions by buoyant anomalies. We have also seen that the axial field is governed by the integral equation

$$\frac{d}{dt} \int_{V_C} \left(R_C^2 - \mathbf{x}^2\right) \langle B_z \rangle \, dV = 2 \int_{V_C} r \langle (\mathbf{v} \times \mathbf{b})_\theta \rangle \, dV - 6\lambda \int_{V_C} \langle B_z \rangle \, dV,$$

which suggests

$$\int_{V_C} r \langle (\mathbf{v} \times \mathbf{b})_\theta \rangle \, dV \approx 3\lambda \int_{V_C} \langle B_z \rangle \, dV, \qquad (17.85)$$

where $< \sim >$ is an azimuthal average. A similar expression can be deduced from (17.68) for the mean east–west field. This gives, for a statistically steady dynamo with little Ω-effect,

$$\oint_C \langle (\mathbf{v} \times \mathbf{b})_p \rangle \cdot d\mathbf{r} \approx \lambda \oint_C \mu \langle \mathbf{J}_p \rangle \cdot d\mathbf{r} = \lambda \oint_C \langle \nabla \times \mathbf{B}_\theta \rangle \cdot d\mathbf{r}. \qquad (17.86)$$

Provided that the EMFs are reasonably additive in each hemisphere, these yield

$$\langle \mathbf{v} \times \mathbf{b} \rangle_\theta \sim \frac{\lambda}{R_C} \langle B_z \rangle, \qquad \langle \mathbf{v} \times \mathbf{b} \rangle_r \sim \frac{\lambda}{R_C} |\langle B_\theta \rangle| . \qquad (17.87)$$

We note in passing that (17.87) combines with the low-R_m EMF estimate (17.80) to give

$$\varepsilon \sim (u\delta/\lambda)^{-1}(uR_c/\lambda)^{-1} \ll 1.$$

We shall now perform an energy balance, which requires an estimate of the Ohmic dissipation per unit mass. Our starting point is the energy equation (17.49) which, for a statistically steady dynamo, requires

$$\int_{V_C} (J^2/\sigma)\, dV = -\int_{V_C} [\mathbf{u} \cdot (\mathbf{J} \times \mathbf{B})]\, dV.$$

We shall assume that the small-scale magnetic field is helical, $\mu\mathbf{j} = \nabla \times \mathbf{b} = \pm\delta^{-1}\mathbf{b}$, where, as in §17.4.4, the upper sign applies in the south (where the mean helicity is positive), while the lower sign holds in the north (where the mean helicity is negative). We shall see shortly that $|\mathbf{J}|$ is dominated by the small-scale current $\mathbf{j}$, while $|\mathbf{B}|$ is dominated by the large-scale field $\langle\mathbf{B}\rangle$, so we have

$$\mathbf{j}^2/\sigma \sim -\mathbf{u} \cdot (\mathbf{j} \times \langle\mathbf{B}\rangle) = -\langle\mathbf{B}\rangle \cdot (\mathbf{u} \times \mathbf{j}) = \mp(\mu\delta)^{-1}\langle\mathbf{B}\rangle \cdot (\mathbf{u} \times \mathbf{b}),$$

with an azimuthally averaged Ohmic dissipation rate of

$$\langle\mathbf{j}^2\rangle/\sigma \sim \mp(\mu\delta)^{-1}\langle\mathbf{v} \times \mathbf{b}\rangle \cdot \langle\mathbf{B}\rangle. \tag{17.88}$$

Given our north–south convention for $\mp$, Figure 17.19 confirms that all of the contributions on the right of (17.88) are indeed positive. Crucially, combining (17.88) with (17.87), and noting that viscous dissipation is negligible in the geodynamo, we obtain the energy balance

$$\frac{\langle\mathbf{j}^2\rangle}{\rho\sigma} \sim \frac{\lambda}{R_C\delta}\frac{\langle\mathbf{B}\rangle^2}{\rho\mu} \sim P, \tag{17.89}$$

where P is the mean rate of working of the buoyancy force per unit mass. We note in passing that, since $\mu\langle\mathbf{J}\rangle \sim \langle\mathbf{B}\rangle/R_C$ while $\mu|\mathbf{j}| \sim |\mathbf{b}|/\delta$, (17.89) yields

$$|\mathbf{j}|/|\langle\mathbf{J}\rangle| \sim \sqrt{R_C/\delta}, \qquad |\mathbf{b}|/|\langle\mathbf{B}\rangle| \sim \sqrt{\delta/R_C}, \tag{17.90}$$

which confirms that $|\mathbf{j}| \gg |\langle\mathbf{J}\rangle|$ and $|\mathbf{b}| \ll |\langle\mathbf{B}\rangle|$, as claimed above.

Our next step is to estimate P. If T' is the temperature fluctuation about the purely conductive state, and β the expansion coefficient, then $P = -\beta \langle T'\mathbf{u}\rangle \cdot \mathbf{g}$. This can be written in terms of the convective heat flux per unit area, $\dot{\mathbf{q}} = \rho c_p \langle T'\mathbf{u}\rangle$, as $P = \left(g\beta/\rho c_p\right)\dot{q}_R$, where c_p is the specific heat and $\dot{q}_R$ is the radial (in a spherical polar sense) component of $\dot{\mathbf{q}}$. Estimates of the flux $\dot{Q} = 4\pi R_C^2 \dot{q}_R$, and hence P, are available for several of the solar system planets. In the case of the terrestrial planets, for example, virtually all of the thermal resistance lies in the mantle, and so we may think of P as prescribed and independent of the motion in the core. In the gas giants, on the other hand, P may be estimated directly from their luminosity. Either way, tentative estimates of P are available for the earth ($P \approx 3{\times}10^{-13}$ m^{2}s^{-3}, Christensen, 2010) and for Jupiter and Saturn ($P \sim 10^{-10}$ m^{2}s^{-3}, Read et al, 2015). In any event, given P, we may estimate $|\mathbf{B}|$ using (17.89) in the form

$$\frac{\lambda}{\delta}\frac{\langle \mathbf{B}\rangle^2}{\rho\mu} \sim PR_C. \tag{17.91}$$

Now, the rotational part of the Coriolis force is of the same order of magnitude as the buoyancy force, the larger irrotational part being balanced by pressure. So, (17.10) yields

$$2\boldsymbol{\Omega} \cdot \nabla\mathbf{a} \sim \beta T'g \sim P/u, \qquad |\mathbf{a}| \sim |\mathbf{u}|\,\delta, \tag{17.92}$$

where $\mathbf{a}$ is the solenoidal vector potential for $\mathbf{u}$. On the assumption that axial gradients scale as R_C, this yields the estimate $\Omega\delta u^2 \sim PR_C$, which combines with (17.91) to give

$$\frac{\lambda}{\delta}\frac{\langle \mathbf{B}\rangle^2}{\rho\mu} \sim \Omega\delta u^2 \sim PR_C, \tag{17.93}$$

(Davidson, 2016). This triple balance provides a basis for developing scaling laws for the geodynamo. That is, if P is prescribed and δ (somehow) estimated, then estimates of $|\mathbf{B}|_{\mathrm{rms}}$ and $|\mathbf{u}|_{\mathrm{rms}}$ can be found. For example, if we assume $u\delta/\lambda \sim 1$, then (17.93) yields $\Lambda \sim 1$.

Note, however, that the estimate of the dissipation in (17.88) is somewhat simplistic, and so subject to some uncertainty. Indeed, alternative putative scaling laws exist, such as

$$\langle|\mathbf{B}|\rangle^3/\left(\rho\mu\right)^{3/2} \sim \Omega\delta u^2 \sim PR_C, \tag{17.94}$$

(see Davidson, 2013). In practice, it is difficult to distinguish between (17.93) and (17.94) in the numerical dynamos, as the low-Ek dipolar simulations tend to have u, $|\mathbf{B}|/\sqrt{\rho\mu}$ and λ/δ all of similar magnitudes, which blurs the distinction between (17.93) and (17.94).

17.5.2 Scaling Laws for the Rossby and Lehnert Numbers in the Numerical Dynamos

In most of the numerical dynamos (but *not* the planets) the scale of the columnar eddies, δ, is set by a balance between the viscous and Coriolis forces, $2\mathbf{\Omega} \cdot \nabla \mathbf{a} \sim \nu\nabla^2\mathbf{u}$, which yields $\delta \sim (\text{Ek})^{1/3} R_C$. So, for the *numerical dynamos*, our scaling law for u, (17.93), becomes

$$\text{Ro} = \frac{u_{\text{rms}}}{\Omega R_C} \sim \Pi_P^{1/2}\text{Ek}^{-1/6}, \qquad \Pi_p = \frac{P}{\Omega^3 R_C^2}. \qquad (17.95)$$

Scaling (17.95) for the global Rossby number has been proposed by several authors and found to be a good match to numerical dynamo data sets. For example, Figure 17.21 compares (17.95) with a suite of numerical dynamos tabulated in Christensen & Aubert (2006). When discussing the simulations it is usual to replace Π_P and Ek by the related parameters $\text{Ra}_Q = \Pi_P/a(1-a)^2 = 6.76\Pi_P$ and $E = \text{Ek}/(1-a)^2 = 2.36\text{Ek}$, where a is the ratio of the inner and outer core radii. We follow that convention in Figure 17.21. The comparison is evidently favourable, with the best-fit pre-factor in (17.95) being 1.1.

By way of contrast, several different scalings for $|\mathbf{B}|_{\text{rms}}$ have been proposed over the years and the matter is still a little uncertain. If we combine (17.93) with the viscous scaling $\delta \sim (\text{Ek})^{1/3} R_C$, then we expect $|\mathbf{B}|_{\text{rms}}$ in the numerical dynamos to scale as

$$\frac{B_{\text{rms}}/\sqrt{\rho\mu}}{\Omega R_C} \sim \Pi_P^{1/2}\text{Pr}_m^{1/2}\text{Ek}^{-1/3}. \qquad (17.96)$$

The dimensionless group on the left of (17.96) is called the *Lehnert number*, Le. Figure 17.22 compares (17.96) with the same suite of numerical dynamos. One complication here is that there is significant viscous dissipation in the numerical dynamos, yet (17.93) is based on an energy balance in which all of the dissipation is assumed to be Ohmic, the viscous dissipation in the core of the earth being negligible. To compensate for this, Le in Figure 17.22 has been

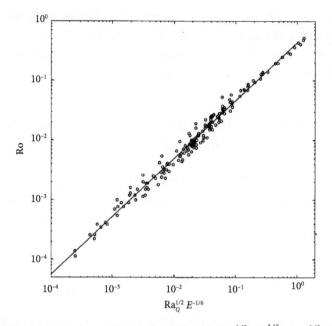

Figure 17.21 A comparison of the scaling law Ro $\sim \Pi_P^{1/2}\text{Ek}^{-1/6} \sim \text{Ra}_Q^{1/2}E^{-1/6}$ with the numerical data set of Christensen & Aubert (2006). The straight line has a slope of unity.

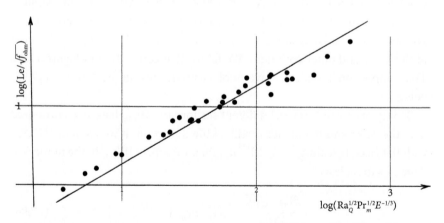

Figure 17.22 A comparison of the scaling law $\text{Le}/\sqrt{f_{\text{ohm}}} \sim \Pi_P^{1/2}\text{Pr}_m^{1/2}\text{Ek}^{-1/3} \sim \text{Ra}_Q^{1/2}\text{Pr}_m^{1/2}E^{-1/3}$ with the same data set as in Figure 17.21. Only data for which $f_{\text{ohm}} > 0.5$ has been used.

replaced by $\text{Le}/\sqrt{f_{\text{ohm}}}$, where f_{ohm} is the ratio of the Joule dissipation to the total dissipation. Also, only data for which $f_{\text{ohm}} > 0.5$ is shown. The comparison with (17.96) is plausible, but not as convincing as Figure 17.21.

Of course, a favourable comparison of (17.93) with the numerical dynamos does not guarantee that it applies also to the geodynamo. Moreover, we encounter a problem when applying (17.93) to the earth's core, in that it is not obvious how δ is determined, with the suggestion $u\delta/\lambda \sim 1$ being somewhat speculative. Still, it is noticeable that those numerical dynamos which are more earth-like, in the sense that they have particularly low viscosities, *e.g.* $f_{\text{ohm}} > 0.9$, and are strongly dipolar, *e.g.* $f_{\text{dip}} > 0.9$ (see below), do tend to have $u\delta/\lambda \sim 1$.

17.6 The Collapse of the Dipole Field in the Numerical Dynamos for $u/\Omega\delta > 0.4$

We now turn to the observation that inertial waves cannot propagate when $u/\Omega\delta$ exceeds ~ 0.4, as discussed in §6.9. If (magnetically modified) inertial waves are indeed central to maintaining the α-effect, then $u/\Omega\delta \approx 0.4$ should signal the collapse of the dipole, heralding either a disorganized multipolar dynamo or else no dynamo at all. Combining $\text{Ro} \sim \Pi_P^{1/2}\text{Ek}^{-1/6}$ with the viscous scaling $\delta/R_C \sim \text{Ek}^{1/3}$ yields $u/\Omega\delta \sim \Pi_P^{1/2}\text{Ek}^{-1/2}$, and in what follows we shall take the pre-factor in this estimate to be equal to unity, *i.e.* $u/\Omega\delta = \sqrt{P/\nu\Omega^2}$. Combining this with the transition criterion $u/\Omega\delta \approx 0.4$, and replacing Π_P and Ek by $\text{Ra}_Q = 6.76\Pi_P$ and $E = 2.36Ek$, we conclude that the transition from a dipolar to a multipolar dynamo in the numerical simulations (but *not* the geodynamo) should occur at $2\text{Ra}_Q/E \approx 1$. Figure 17.23 shows f_{dip} plotted as a function of $2\text{Ra}_Q/E$, where f_{dip} is the strength of the dipole divided by the field contained in the harmonics of degree 1 to 12. It is conventional to regard dipolar cases as corresponding to $f_{\text{dip}} > 0.5$, and multipolar dynamos as characterized by $f_{\text{dip}} < 0.35$.

Evidently, the criterion $2\text{Ra}_Q/E \approx 1$ does a reasonable job of distinguishing between dipolar and non-dipolar dynamos. This lends support to the hypothesis that the cyclone–anticyclone vortex pairs observed in the simulations are indeed magnetically modified inertial wave packets. (Note, however, that the threshold $u/\Omega\delta \approx 0.4$ applies only to *pure* inertial waves, so the actual threshold may vary somewhat with the strength of the ambient magnetic field.) We also note in passing that the earth lies in the wrong quadrant in Figure 17.23, which reminds us that the scaling $\delta/R_C \sim \text{Ek}^{1/3}$ is an artefact of the excessive viscosity in the numerical dynamos, and does not apply to the planets.

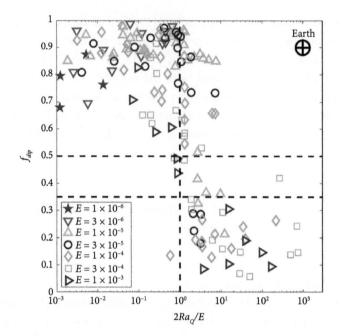

Figure 17.23 f_{dip} versus $2Ra_Q/E$ for the same data set as Figure 17.21. From Davidson (2016).

17.7 Some Comments on Reversals of the Geodynamo

We close this chapter by discussing possible mechanisms by which the terrestrial field can reverse. Our starting point is to note that the mechanism of dipole collapse described in the preceding section is extremely unlikely to occur in the earth, as inconceivably narrow columns ($\delta < 100$ m) would be required in order to reduce the small-scale Rossby number, $u/\Omega\delta$, to a value of order one. There are, however, other ways by which the EMF $\langle(\mathbf{v} \times \mathbf{b})_\theta\rangle$ could be reduced sufficiently to kill off the dipole, thus heralding a reversal.

Recall that $m_z = (3/2\mu)\int B_z dV$ is the axial dipole moment, while $t_d = R_C^2/(\lambda\pi^2)$ is the magnetic diffusion time for the slowest decaying mode (the so-called *free-decay dipole mode*). Then (17.67),

$$\frac{d}{dt}\int_{V_C} \frac{1}{2}\left(R_C^2 - \mathbf{x}^2\right)\langle B_z\rangle\, dV + 3\lambda\int_{V_C} \langle B_z\rangle\, dV = \int_{V_C} r\langle(\mathbf{v} \times \mathbf{b})_\theta\rangle\, dV,$$

which is exact, can be conveniently rewritten as

$$\frac{d}{dt}(S(t)m_z) + \frac{m_z}{t_d} = \frac{\pi^2}{2\mu R_C^2} \int_{V_C} r\langle (\mathbf{v} \times \mathbf{b})_\theta \rangle \, dV. \qquad (17.97)$$

Here $S = 1$ if $\mathbf{B}_p$ is the free-decay dipole mode, but $S < 1$ if the axial field is displaced towards the mantle. Of course, (17.97) can be integrated to give $m_z(t)$ only if both $S(t)$ and $\int r\langle (\mathbf{v} \times \mathbf{b})_\theta \rangle \, dV$ can be somehow estimated. However, (17.80) tentatively suggests

$$\langle (\mathbf{v} \times \mathbf{b})_\theta \rangle \approx -\frac{\delta^2}{\lambda} \langle h \rangle \langle B_\theta \rangle, \qquad (17.98)$$

and so taking $S \approx 1$ gives us the approximate evolution equation

$$\frac{dm_z}{dt} + \frac{m_z}{t_d} \approx \frac{\pi^2}{2\mu R_C^2} \int_{V_C} r\langle (\mathbf{v} \times \mathbf{b})_\theta \rangle \, dV \approx -\frac{\sigma\pi^2\delta^2}{2R_C^2} \int_{V_C} r\langle h \rangle \langle B_\theta \rangle \, dV. \qquad (17.99)$$

Clearly, a reversal in dipole moment requires $\int r\langle (\mathbf{v} \times \mathbf{b})_\theta \rangle \, dV$ to change sign. Now, paleomagnetic records show that the reversal timescale is around $t_r \sim 10^4$ years, which is faster than the diffusion time of $t_d \sim 5 \times 10^4$ years. It follows that the left-hand side of (17.99) is dominated by m_z/t_d before a reversal, but by dm_z/dt during a reversal. Consequently, the value of the integral on the right of (17.99) must change from order m_z/t_d to order $-m_z/t_r$ as a reversal gets underway. In short, a reversal cannot be regarded as a passive response to the temporary removal of the lift-and-twist EMF, as represented by $\langle (\mathbf{v} \times \mathbf{b})_\theta \rangle$. Rather, the integral on the right of (17.99) must switch signs on a timescale somewhat faster than t_r, actively driving down the dipole.

Since B_θ appears in (17.99), it seems appropriate to find an evolution equation for the east–west field which is analogous to (17.97). Suppose we apply (17.68) to one of the azimuthal flux patches shown near the equator in Figure 17.20(a), either the positive flux patch to the north, or the negative patch to the south. This yields a second (exact) result

$$\frac{d}{dt} \int_A \langle B_\theta \rangle \, dA = (\Omega - \text{effect}) + \oint_C \left[\langle (\mathbf{v} \times \mathbf{b})_p \rangle - \langle \mathbf{J}_p \rangle/\sigma \right] \cdot d\mathbf{r},$$

where A is the area of the flux patch enclosed by the curve C in the r–z plane. This can be rewritten, using $\nabla \times \langle \mathbf{B}_\theta \rangle = \mu \langle \mathbf{J}_p \rangle$, in the approximate form

$$\frac{d}{dt} \int_A \langle B_\theta \rangle \, dA + \frac{\lambda \pi^2}{\ell^2} \int_A \langle B_\theta \rangle \, dA \approx (\Omega - \text{effect}) + \oint_C \langle (\mathbf{v} \times \mathbf{b})_p \rangle \cdot d\mathbf{r}, \quad (17.100)$$

$$\langle (\mathbf{v} \times \mathbf{b})_p \rangle \approx -\frac{\delta^2}{\lambda} \langle h \rangle \langle \mathbf{B}_r \rangle, \quad (17.101)$$

where $\ell(t)$ is of the order of the height of the flux patch. In the numerical dynamos, there is normally little or no Ω-effect outside the tangent cylinder, and so it is tempting to drop that term from (17.100). The equations above are then analogous to (17.97) and (17.98), the main difference being that the diffusion time for $\langle B_\theta \rangle$ is an order of magnitude faster than t_d, since $R_C^2 \sim 10\ell^2$. To obtain reversals, however, we may need to retain the Ω-effect.

These equations look promising, but the sparsity of the observational data ensures that building on them rapidly pulls us into the realms of speculation. So, let us speculate! Perhaps the first thing to note is that, as pointed out in §17.4.5, an α^2-dynamo with negative $\langle h \rangle$ in the north and positive $\langle h \rangle$ in the south (the usual distribution) can operate equally with the dipole pointing to the north or to the south. That is to say, if we reverse both $\mathbf{B}$ and $\mathbf{J}$, but keep the averaged helicity unchanged, we still have a self-consistent dynamo. So, the sign of the dipole is arbitrary and (in a numerical dynamo) determined by the initial conditions. Perhaps the first question to ask, then, is can we find a way to reverse the lift-and-twist EMF in (17.99) and so drive the dipole down to zero, effectively creating a new initial condition. Such an event is called an *excursion* and it is a necessary precursor to a reversal. In what follows, we focus on excursions, leaving aside the more difficult question of how a dynamo regrows from a weak, incoherent seed field.

Inspection of (17.99)→(17.101) tentatively suggests that there are at least two ways in which excursions in m_z might be triggered: (i) the transient growth of an Ω-effect, or (ii) stochastic variations in the intensity of $\langle h \rangle$. We consider these one at a time. Recall that the dipole field resists being sheared and this is why the Ω-effect is weak outside the tangent cylinder in the numerical dynamos. Now suppose that a temporary fall in m_z allows an Ω-effect to develop. According to Figure 17.14, when the dipole points to the north, the Ω-effect generates a negative B_θ in the north and a positive B_θ in the south, which is the opposite to that associated with the α-effect (see Figure 17.19). So, if the Ω-effect is sufficiently strong, it can reverse the sign of B_θ in the equatorial flux patches, and hence reverse the sign of the lift-and-twist EMF in (17.99). This, in turn, will further reduce the dipole field, conceivably pushing m_z down to zero. If, when the dipole recovers, the resurgent dipole is of the opposite polarity, we have a reversal.

One difficulty with this argument is that the initial fall in m_z needs to be explained, which brings us to our second source of fluctuations in m_z. Another way to trigger a change in the lift-and-twist EMF is to modify the azimuthally averaged helicity. This, in turn, can be achieved through a redistribution of the buoyancy field. For example, one could conceive of turbulence pushing hot polar fluid out of the tangent cylinder near the mantle, which would trigger inertial waves propagating inward from the mantle towards the equator, hence reducing $\langle h \rangle$. Of course, this is just one of several possibilities, but the key underlying point is that, in a statistically steady dynamo, the helicity distribution is highly chaotic and so there is a great deal of cancellation of helicity when constructing an azimuthal average (see Figure 17.2). So, it seems likely that modest changes in $h(\mathbf{x}, t)$ resulting from a change to the buoyancy field could significantly alter $\langle h \rangle$.

In any event, suppose that, for whatever reason, changes in the buoyancy field cause the magnitude of $\langle h \rangle$ to fall. Then the EMF in (17.99) will also fall, causing m_z to drift downward on the diffusive timescale of $t_d = R_C^2/(\lambda \pi^2) \sim 5 \times 10^4$ years. If $\langle h \rangle$, and hence m_z, becomes small enough, then an Ω-effect can develop. This will reverse B_θ in the equatorial flux patches on the *faster* timescale of $\ell^2/(\lambda \pi^2) \sim 5 \times 10^3$ years, causing a reversal of the EMF in (17.99). This reversed EMF will, in turn, further reduce m_z, and as m_z approaches zero, the Ω-effect weakens. At this point we have an excursion, but not a reversal, with a weak dipole and a reversed east–west field. It seems probable that the dipole then oscillates about zero with a small amplitude and a period of $\sim \ell^2/(\lambda \pi^2)$. Finally, when the buoyancy field, and hence $\langle h \rangle$, returns to its original distribution, the dipole presumably recovers its pre-excursion amplitude, possibly with a reversed polarity.

Of course, this is all rather speculative. Indeed, given the sparsity of the observational data, we have very little hard evidence as to how excursions and reversals actually occur.

* * *

This concludes our brief excursion into the earth's core. The subject is still somewhat in a state of flux and much remains uncertain. In particular, the scaling law (17.96), as well as our discussion of reversals, must be regarded with caution. There are, however, several useful review papers, such as Jones (2011), Christensen (2011), and Roberts & King (2013).

References

Christensen, U.R., 2010, Dynamo scaling laws and application to the planets, *Space Sci. Rev.*, **152**, 565–90.

Christensen, U.R., 2011, Geodynamo models: tools for understanding properties of Earth's magnetic field, *Phys. of Earth and Planetary Interiors*, **187**, 157–69.

Christensen, U.R., & Aubert, J. 2006, Scaling properties of convection-driven dynamos in rotating spherical shells and applications to planetary magnetic fields, *Geophys. J. Int.*, **166**, 97–114.

Cowling, T.G., 1934, The magnetic field of sunspots, *Mon. Not. Roy. Astro. Soc.*, **94**, 39–48.

Cowling, T.G., 1957, *Magnetohydrodynamics*, Interscience Publishers.

Davidson, P.A., 2013, Scaling laws for planetary dynamos, *Geophys. J Int.*, **195**, 67–74.

Davidson, P.A., 2014, The dynamics and scaling laws of planetary dynamos driven by inertial waves, *Geophys. J Int.*, **198**(3), 1832–47.

Davidson, P.A., 2016, Dynamos driven by helical waves: scaling laws for numerical dynamos and for the planets, *Geophys. J. Int.*, 207(2), 680–90.

Davidson, P.A., & Ranjan, A., 2015, Planetary dynamos driven by helical waves: Part 2, *Geophys. J. Int.*, **202**, 1646–62.

Davidson, P.A., & Ranjan, A., 2018a, Are planetary dynamos driven by helical waves? *J. Plasma Phys.*, **84**, 735840304.

Davidson, P.A., & Ranjan, A., 2018b, On the spatial segregation of helicity by inertial waves in dynamo simulations and planetary cores, *J. Fluid Mech.*, **852**, 268–87.

Davidson, P.A., & Ranjan, A., 2023, The connection between the equatorial temperature bias and north-south helicity segregation in simulations of the geodynamo, *Geophys. J. Int.*, **233**, 2254–68.

Jackson, J.D., 1998, *Classical Electrodynamics*, 3rd Ed., Wiley.

Jones, C.A., 2011, Planetary magnetic fields and fluid dynamos, *Ann. Rev. Fluid Mech.*, **43**, 583–614.

McDermott, B.R., & Davidson, P.A., 2019, A physical conjecture for the dipolar-multipolar dynamo transition, *J. Fluid Mech.*, **874**, 995–1020.

Parker, E.N., 1955, Hydromagnetic dynamo models, *Astrophys. J.*, **122**, 293–314.

Ranjan, A., Davidson, P.A., Christensen, U.R., & Wicht, J., 2020, The generation and segregation of helicity in geodynamo simulations, *Geophys. J. Int.*, **221**, 741–57.

Read, P.L. et al., 2015, Global energy budgets and 'Trenberth diagrams' for the climates of terrestrial and gas giant planets, *Q. J. R. Meteorol. Soc.*, **142**, 703–20.

Roberts, P., & King, E., 2013, On the genesis of the Earth's magnetism, *Rep. Prog. Phys.*, **76**, tag 096801.

Sakuraba, A., & Roberts, P.H., 2009, Generation of a strong magnetic field using uniform heat flux at the surface of the core, *Nature Geosci.*, **2**, 802–5.

Schaeffer, N., Jault, D., Nataf, H.-C., & Fournier, A., 2017, Turbulent geodynamo simulations: a leap towards Earth's core, *Geophys. J. Int.*, **211**, 1–29.

Taylor, J.B., 1963, The magnetohydrodynamics of a rotating fluid and the earth's dynamo problem, *Proc. Royal Soc. London, Ser. A*, **274**, 274–83.

Chapter 18

Zonal (East–West) Winds and Rossby-Wave Turbulence

One of the most striking features of Jupiter and Saturn are the intense east–west winds (called *zonal winds*) that dominate the surface weather layer (see Figure 1.4). Similar zonal flows are observed in the oceans on earth, though they tend to meander somewhat. There has been a long history of trying to explain this robust phenomenon, and, as we shall see, that effort is still ongoing. Traditional models assumed the flow is quasi-geostrophic and shallow water (QGSW), being subject to the β-effect (a latitudinal variation in background rotation). In the terminology of §9.4, the flow is said to sit on the β-plane. As discussed in Chapter 9, QGSW flow on the β-plane can support Rossby waves, and since the zonal winds on the gas giants are observed to be extremely turbulent, an explanation for these winds was sought in the interplay between Rossby waves and turbulence. That interplay is still thought to be central to the phenomenon.

18.1 A Return to the β-plane: Turbulence Versus Rossby Waves

Let us return to QGSW flow on the β-plane, as discussed in §9.4. Recall that the β-plane is tangent to the surface of a planet, as shown in Figure 18.1, with x pointing to the east and y to the north. The z component of the background rotation is then allowed to vary with latitude, according to $2\Omega = 2\Omega_0 + \beta y$, where $\beta > 0$.

When friction is ignored, flow on the β-plane is governed by the QGSW equation

$$\frac{Dq}{Dt} = \frac{D}{Dt}\left[\omega + \frac{\psi}{R_d^2} + \beta y + \frac{2\Omega_0 h_b}{h_0}\right] = 0, \tag{18.1}$$

where q is the potential vorticity (PV), ψ the streamfunction for $\mathbf{u}(x, y)$, h_0 the nominal depth of the layer, $\omega = -\nabla^2\psi$ the vorticity, $R_d = \sqrt{gh_0}/2\Omega_0$ the Rossby deformation radius, and h_b the undulation of the lower boundary, as

The Dynamics of Rotating Fluids. P. A. Davidson, Oxford University Press. © Peter A Davidson (2024).
DOI: 10.1093/9780191994272.003.0018

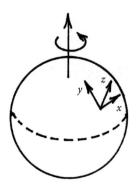

Figure 18.1 The β-plane.

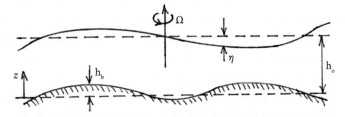

Figure 18.2 Shallow-water flow on the β-plane.

shown in Figure 18.2. With the exception of §18.4, we take the lower boundary to be horizontal, and so our governing equation simplifies to

$$\frac{D}{Dt}\left[\omega + \frac{\psi}{R_d^2}\right] = -\beta u_y. \tag{18.2}$$

To focus thoughts, let us temporarily consider the case where R_d is much greater than any characteristic scale of the flow, which is called the *rigid-lid approximation* because free-surface perturbations are negligible in this regime. Our governing equation is now

$$\frac{\partial \omega}{\partial t} + \mathbf{u} \cdot \nabla \omega = -\beta u_y + D, \tag{18.3}$$

where we have introduced some kind of large-scale friction, as represented by D. When $|\mathbf{u} \cdot \nabla \omega| \gg |\beta u_y|$ we may ignore the β term in (18.3), which then describes conventional two-dimensional motion, including two-dimensional turbulence. Conversely, if $|\beta u_y| \gg |\mathbf{u} \cdot \nabla \omega|$, we can drop the nonlinear convective term and we obtain the governing equation for Rossby waves on the β-plane. An estimate of the ratio of βu_y to the convective term is

$$\frac{\beta u_y}{\mathbf{u} \cdot \nabla \omega} \sim \frac{\beta \ell}{\tilde{\omega}} \sim \frac{\beta \ell^2}{u}, \tag{18.4}$$

where u is a characteristic velocity, say the route-mean-square (rms) velocity, $\tilde{\omega}$ is the rms vorticity, and ℓ some characteristic scale of the motion.

It is conventional to introduce the length scales $L_{\mathrm{Rh}} = \sqrt{u/\beta}$ and $L_\beta = \tilde{\omega}/\beta$, which are known as the *Rhines scale* and the *β scale*, respectively. Expression (18.4) then becomes

$$\frac{\beta u_y}{\mathbf{u} \cdot \nabla \omega} \sim \frac{\ell}{L_\beta} \sim \frac{\ell^2}{L_{\mathrm{Rh}}^2}, \tag{18.5}$$

and we conclude that we will observe Rossby waves whenever $\ell \gg L_{\mathrm{Rh}}$, or else $\ell \gg L_\beta$, and conventional two-dimensional motion, say two-dimensional turbulence, when $\ell \ll L_{\mathrm{Rh}}$, or $\ell \ll L_\beta$. Moreover, if $\ell \sim L_{\mathrm{Rh}}$, we expect waves and turbulence to coexist.

Now, it is a property of conventional two-dimensional turbulence that the size of the largest eddies, called the *integral scale* of the turbulence, continually grows until the domain size is reached. This is known as the *inverse energy cascade* and it is true of both freely decaying and forced-dissipative turbulence in two dimensions (see, for example, Davidson, 2015). Suppose, therefore, that we create some two-dimensional turbulence on the β-plane whose initial integral scale, ℓ, is significantly smaller than the Rhines scale, $\ell \ll L_{\mathrm{Rh}}$. Then ℓ will grow and, after some time, we will reach the point where $\ell \sim L_{\mathrm{Rh}}$ and Rossby waves start to be generated by the turbulence. Eventually, *Rossby-wave turbulence* develops, where turbulent eddies and Rossby waves coexist over a range of scales, with the Rossby waves partially inhibiting the inverse energy cascade. A curious phenomenon is now observed, in which persistent east–west jets of alternating sign spontaneously develop, jets that are immersed in turbulence. These are reminiscent of the zonal winds on Jupiter.

One of the central questions in β-plane turbulence is why zonal jets are so ubiquitous. The traditional answer, now largely abandoned, was that weakly nonlinear interactions between Rossby waves, called *resonant triad interactions* (see §14.4), can systematically nudge energy towards 'zonal modes', in which the velocity is predominantly in the east–west direction. However, resonant triad interactions are inefficient at transporting energy between wavevectors, requiring multiple interactions compounded over a prolonged period of time. Moreover, this interpretation of events has received relatively little support, either from observations or numerical experiments. Consequently, alternative explanations have been sought for the spontaneous appearance of zonal flows, as we now discuss.

18.2 Zonal Flows and Jets on the β-plane

18.2.1 A Cartoon of the Early Stages of Development of Zonal Flow: The Vallis Dumbbell

One rationalization for the formation of zonal flows relies on the partial suppression of the inverse energy cascade by Rossby waves, a mechanism that seems likely to characterize the early stages of development of zonal flow. Perhaps the hallmark of this cartoon is the *assumption* of a rapid transition (in spectral space) from conventional two-dimensional turbulence to quasi-linear wave propagation, so that the flow can be classified as being either turbulent or wave-like. Our description follows that of Vallis & Multrud (1993) and the starting point is to re-examine the wave-turbulence crossover scale, embracing the anisotropic nature of Rossby waves. For simplicity, we continue to take $R_d \to \infty$.

We would expect turbulence to excite Rossby waves whenever the wave frequency, $\hat{\omega}$, is matched to the inverse timescale of the large eddies, $\tilde{\omega}$. Equation (9.72), $\hat{\omega} = -\beta k_x/k^2$, then suggests that the crossover from turbulence to waves occurs when

$$|k_x|/k^2 = \hat{\omega}/\beta \sim \tilde{\omega}/\beta = L_\beta, \tag{18.6}$$

where $k = |\mathbf{k}|$. The same estimate can be obtained from the force balance in (18.3). First we note that the kinematic relationship $\partial \omega/\partial x = \nabla^2 u_y$ yields $k_x \omega \sim k^2 u_y$, and hence $\beta u_y \sim (\beta k_x/k^2)\omega$. Given that $(\mathbf{u} \cdot \nabla)\omega \sim \tilde{\omega}^2$, the wave-turbulence crossover criterion $(\mathbf{u} \cdot \nabla)\omega \sim \beta u_y$ now requires $\tilde{\omega}^2 \sim (\beta |k_x|/k^2)\tilde{\omega}$, which brings us back to (18.6).

Now, if this transition in behaviour is relatively abrupt, we might expect relatively few waves for $|k_x|/k^2 < L_\beta$, but a dominance of waves for $|k_x|/k^2 > L_\beta$. Introducing the angle θ defined by $k_x = k \cos \theta$, and allowing $\hat{\omega}$ in (9.72) to be of either sign, (18.6) becomes

$$(k_x, k_y) L_\beta = \pm \left(\cos^2 \theta, \sin \theta \cos \theta \right). \tag{18.7}$$

Figure 18.3 shows this spectral crossover boundary. It is dumbbell-like, with the right-hand side of (18.7) consisting of two circles of radius 1/2 centred

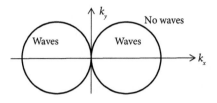

Figure 18.3 The turbulence-wave crossover boundary in the k_x–k_y plane.

on $(\pm 1/2, 0)$. Within the circles the exchange of energy between wavevectors is inhibited, because such a transfer requires resonant triad interactions, and these are inefficient at shifting energy between wavevectors.

Now, suppose we have freely decaying turbulence on the β-plane in which the initial integral scale, ℓ, is much smaller than $L_\beta = \tilde{\omega}/\beta$. The β-effect is then initially weak and we have conventional two-dimensional turbulence in which the integral scale grows while $\tilde{\omega}$, and hence L_β, declines. The energy then migrates across the k_x–k_y plane and eventually we approach the condition $\ell \sim L_\beta$, as shown by the numerical simulation illustrated in Figure 18.4. As the energy approaches the wave-turbulence crossover it starts to pile up at the boundary, because energy transfer between wavevectors is suppressed in regions where Rossby waves dominate. However, the most important feature of Figure 18.4 is the way in which the energy migrates *along* the crossover boundary towards the k_y axis. That is to say, two-dimensional turbulence always tries to increase its integral scale, ℓ, and on the β-plane it does that by migrating towards the k_y axis, which is not blocked by the wave-dominated dumbbell. In short, anisotropy builds up as the large eddies are preferentially elongated in the east–west direction.

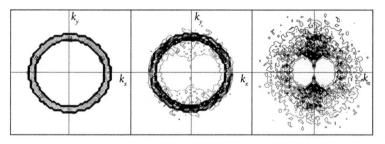

Figure 18.4 Numerical simulation of turbulence on the β-plane showing the migration of the energy across the k_x–k_y plane. Reproduced from Vallis & Multrud (1993).

Note that this is more of a rationalization than an explanation for the spontaneous emergence of zonal flow, since we have not given any explanation for the growth of the integral scale in conventional 2D turbulence. In fact, there are competing explanations for this growth, which vary depending on whether the turbulence is freely decaying or forced.

18.2.2 Strengthening the Zonal Jets: The Jet-Sharpening Model of Dritschel and McIntyre

Once zonal jets start to form, perhaps in the way described above, a second mechanism comes into play that reinforces the jets. This is sometimes referred to as *jet-sharpening* and it is reviewed in Dritschel & McIntyre (2008), (but see also Dritschel & Scott, 2011). Once again, we start with the case where $R_d \to \infty$. The key point is that

$$\frac{Dq}{Dt} = \frac{D}{Dt}\left[\omega + \beta y\right] = D \tag{18.8}$$

has the intriguing property that zonal flows are self-reinforcing through the process of *potential vorticity mixing*, as we now discuss.

Consider the situation where the flow is initially everywhere zero, so the PV is simply $q = \beta y$. Now suppose that the PV becomes well mixed in the band $y_1 < y < y_2$, perhaps because of the passage of a turbulent zonal jet that mixes q in accordance with $Dq/Dt = D$. In the extreme case of perfect mixing a step will form in the PV profile, as shown on the left of Figure 18.5, with $q = \beta(y_1 + y_2)/2$ in the strip $y_1 < y < y_2$. The equivalent vorticity profile is given by $\omega = q - \beta y$, which yields

$$\omega = -\frac{\partial u_x}{\partial y} = \frac{1}{2}\beta(y_1 + y_2 - 2y), \quad y_1 < y < y_2. \tag{18.9}$$

This integrates to give

$$u_x = -\frac{1}{2}\beta(y - y_1)(y_2 - y), \quad y_1 < y < y_2, \tag{18.10}$$

with $u_x = 0$ outside the strip, as shown on the right of Figure 18.5.

In summary, PV mixing in the band $y_1 < y < y_2$, perhaps initiated by the passage of a turbulent jet, gives rise to a step in the PV profile, a negative vorticity gradient, $d\omega/dy = -\beta$, and a westward zonal jet. This process is known

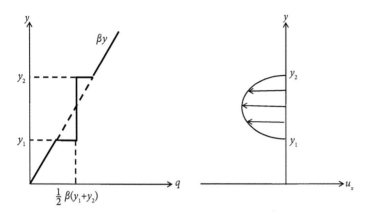

Figure 18.5 Jet formation through PV mixing in the strip $y_1 < y < y_2$, with $R_d \rightarrow \infty$.

as *jet-sharpening* because it tends to sharpen the edges of the jet that did the original mixing. Note that there is a positive feedback mechanism here, in which mixing induces a jet that causes more mixing.

Let us now relax the assumption that $R_d \rightarrow \infty$, and explore the more general consequences of PV mixing in the strip $y_1 < y < y_2$. As before, we have

$$q - \beta y = \frac{1}{2}\beta\left(y_1 + y_2 - 2y\right) \tag{18.11}$$

within the strip and $q - \beta y = 0$ outside the strip. Since $q - \beta y = \omega + \psi/R_d^2$, this yields

$$\frac{d^2\psi}{dy^2} - \frac{\psi}{R_d^2} = \frac{1}{2}\beta\left(2y - (y_1 + y_2)\right), \quad y_1 < y < y_2, \tag{18.12}$$

and

$$\frac{d^2\psi}{dy^2} - \frac{\psi}{R_d^2} = 0, \quad y < y_1, \ y > y_2. \tag{18.13}$$

These equations are readily integrated, subject to the requirements that the solution is continuous at y_1 and y_2 and ψ falls to zero at large $|y|$. After a little algebra, we find

$$\frac{u_x}{\frac{1}{2}\beta R_d^2} = \left[\frac{b}{R_d} + 1\right]\exp\left(-\frac{y - y_1}{R_d}\right) + \left[\frac{b}{R_d} - 1\right]\exp\left(-\frac{y - y_2}{R_d}\right), \quad y > y_2, \tag{18.14}$$

$$\frac{u_x}{\frac{1}{2}\beta R_d^2} = \left[\frac{b}{R_d} + 1\right]\exp\left(-\frac{y - y_1}{R_d}\right) + \left[\frac{b}{R_d} + 1\right]\exp\left(\frac{y - y_2}{R_d}\right) - 2, \quad y_1 < y < y_2,$$

(18.15)

$$\frac{u_x}{\frac{1}{2}\beta R_d^2} = \left[\frac{b}{R_d} - 1\right]\exp\left(\frac{y - y_1}{R_d}\right) + \left[\frac{b}{R_d} + 1\right]\exp\left(\frac{y - y_2}{R_d}\right), \quad y < y_1, \quad (18.16)$$

where $y_2 - y_1 = 2b$. It is readily confirmed that these expressions revert to the westward jet of (18.10) for $R_d \gg b$. Conversely, the solution for $R_d \ll b$ is

$$\frac{u_x}{\frac{1}{2}\beta R_d b} = \exp\left(-|y - y_1|/R_d\right) + \exp\left(-|y - y_2|/R_d\right), \quad (18.17)$$

which represents two narrow *eastward* jets of characteristic width R_d, centred on y_1 and y_2. The intermediate case of $b = R_d$ is shown in Figure 18.6, which evidently consists of a broad westward jet across most of the strip $y_1 < y < y_2$, flanked by sharper eastward jets centred on y_1 and y_2.

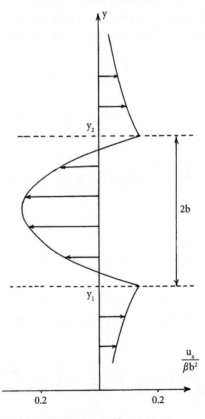

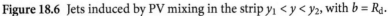

Figure 18.6 Jets induced by PV mixing in the strip $y_1 < y < y_2$, with $b = R_d$.

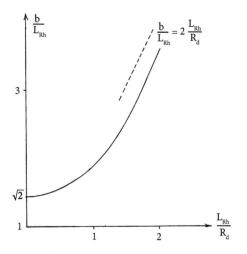

Figure 18.7 The variation of b/L_{Rh} with the ratio L_{Rh}/R_d.

Dritschel & McIntyre (2008) introduce a Rhines length scale based on the difference between the maximum eastward and westward velocities, $\Delta u = u_x(y = y_2) - u_x(y = y_1 + b)$, which is independent of the frame of reference of the observer. Using (18.15), it is readily confirmed that this length scale takes the form

$$L_{Rh} = \sqrt{\Delta u/\beta} = \frac{1}{\sqrt{2}} R_d \sqrt{1 + \chi} (1 - e^{-\chi}),\qquad(18.18)$$

where $\chi = b/R_d$. From this, one can determine how b/L_{Rh} varies with the ratio L_{Rh}/R_d, and the result is shown in Figure 18.7. For large deformation radius, $R_d \gg L_{Rh}$, we find that the separation of the eastward jets scales as $b \sim L_{Rh}$, as has been traditionally assumed. However, for $R_d \ll L_{Rh}$, the predicted scaling changes to $b \approx 2L_{Rh}^2/R_d$, although in practice the jets tend to be unstable for $R_d < L_{Rh}/2$, and so this prediction is difficult to verify.

18.2.3 Some Observations from the Numerical Experiments

There have been numerous numerical experiments of β-plane turbulence. Typically, these are forced-dissipative simulations involving some kind of artificial small-scale forcing. Often, they start with zero motion, so that $q = \beta y$ at $t = 0$, and then the forcing produces small-scale, two-dimensional turbulence. This then rearranges the PV field while the integral scale of the turbulence increases. After some time, the PV field is often seen to take the form a staircase, consisting of multiple steps not unlike that shown in Figure 18.5, with each step associated with zonal jets. Rossby waves then concentrate around the boundaries of the steps, where there are large PV gradients that support such waves.

One such example is shown in Figure 18.8, where $R_d \sim L_{Rh}$. The left-hand panel shows the staircase distribution of q that develops after some time, while the right-hand panel shows the associated zonal jets. In this particular case, the jet spacing is around $2b \approx 2\pi R_d$.

Perhaps we should note that this kind of spontaneous layering of an advected scalar due to small-scale stirring is seen in other turbulent, wave-bearing systems. For example, the density field in a stably stratified fluid tends to form a staircase if the fluid is subject to small-scale stirring. (See, for example, Dritschel & McIntyre, 2008.) The feedback mechanism here is that mixing pushes the density gradients to the edges of the bands, which facilitates more mixing within the bands and concentrates the gravity waves at the interfaces between steps. These strongly stratified interfaces then act as *barriers* that partition the turbulence either side of them.

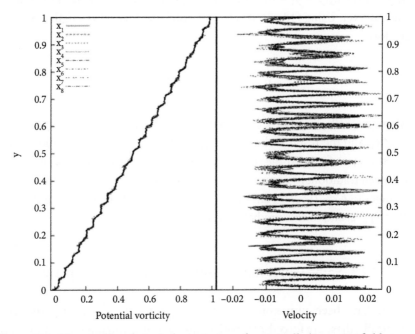

Figure 18.8 The small-scale turbulent stirring of an initially linear PV field results in the spontaneous appearance of a PV staircase (left-hand panel). Each step in the staircase is associated with zonal jets (right-hand panel). Data from Marcus & Shetty (2011).

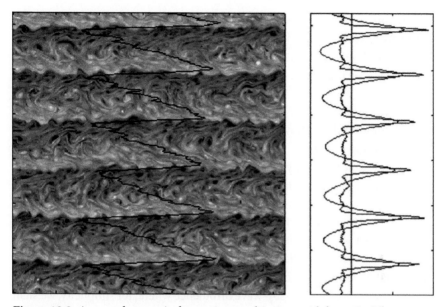

Figure 18.9 A second numerical experiment, this time with large R_d. The left-hand panel shows a snapshot of the vorticity field, along with the zonally averaged vorticity profile (solid line). The right-hand panel shows the zonally averaged distribution of u_x (thick line) and $d\omega/dy$ (thin line). Adapted from Danilov & Gurarie (2004).

A second example is shown in Figure 18.9, where in this case $R_d \to \infty$. The left-hand panel shows the instantaneous vorticity field, along with the zonally averaged vorticity profile (solid line). The banding is clear, as is the negative gradient of ω within each band, consistent with $d\omega/dy = -\beta$ (although the gradient is less than β). The right-hand panel shows the zonally averaged distributions of u_x (thick line) and $d\omega/dy$ (thin line). Note that what one regards as westward or eastward within the u_x distribution depends on the observer's frame of reference. In any event, it is clear from Figures 18.6 and 18.9 that there is an asymmetry in the shape of the westward and eastward jets.

18.3 Energy Spectra in β-plane Turbulence

The spectral properties of Rossby-wave turbulence were first set out in a seminal paper by Rhines (1975). We shall touch on those properties below, but first we need to pause to introduce the *energy spectrum*, $E(k)$, which is central to so much of homogeneous turbulence theory.

The formal definition and properties of $E(k)$ may be found in books on turbulence, such as Davidson, 2015. However, for the present purposes, all we need to know is that it is a function of wavenumber $k = |\mathbf{k}|$ and has the following three important properties:

(i) $E(k) > 0$;
(ii) for a random array of Gaussian-like eddies of fixed size s, $E(k)$ peaks around $k \sim \pi/s$;
(iii) the kinetic energy density of the turbulence is $\frac{1}{2}\langle\mathbf{u}^2\rangle = \int_0^\infty E(k)dk$, where $\langle\sim\rangle$ represents a volume average.

In the light of these properties, it is customary to interpret $E(k)dk$ as the contribution to $\frac{1}{2}\langle\mathbf{u}^2\rangle$ from eddies with wavenumbers in the range $k \to k + dk$, where $k \sim \pi/s$. Moreover, the characteristic kinetic energy of eddies of scale π/k is usually taken to be $u_k^2 \sim kE(k)$.

Let us now return to Rossby-wave turbulence where, for simplicity, we shall take R_d to be large enough for its influence to be ignored. Rhines suggested that the *large scales* in fully developed β-plane turbulence are characterized by a scale-by-scale balance between the wave frequency, $\tilde{\omega} = \beta\,|k_x|/k^2$, and the inverse of the eddy turn-over time, $u_k/s \sim ku_k$. Ignoring anisotropy in the dispersion relationship, this yields $ku_k \sim \beta/k$, or equivalently $u_k^2 \sim \beta^2 k^{-4}$. So Rhines tentatively suggested that the large scales in Rossby-wave turbulence have an energy spectrum of the form

$$E(k) = C_{\mathrm{Rh}}\beta^2 k^{-5}, \tag{18.19}$$

where C_{Rh} is a dimensionless constant. The inverse wavenumber, k^{-1}, now acts like a scale-dependent Rhines length, *i.e.* $k^{-1} \sim \sqrt{u_k/\beta}$.

The hypothesis that there is a scale-by-scale balance between eddy turn-over time and wave period has more recently been proposed for a variety of turbulent, wave-bearing systems, such as turbulence in a stratified fluid and magnetohydrodynamic (MHD) turbulence. It is often called the *critical balance hypothesis* and there is considerable experimental support for the idea. Certainly, spectrum (18.19) has been reported in several studies of β-plane turbulence, at least for the zonally averaged spectrum, which is characterized by $k_x \to 0$. That is to say, the large scales in β-plane turbulence are observed to exhibit the spectrum

$$E_{\mathrm{zon}}(k_y, k_x = 0) \sim \beta^2 k_y^{-5}, \tag{18.20}$$

although the exponent -5 is often difficult to determine unambiguously because of sharp peaks in the spectrum at the harmonics of π/b, associated with the jet periodicity. (It turns out that these sharp peaks have an underlying spectrum of $E_{zon} \sim k^{-4}$.)

Notice that this spectrum applies only at large scales. Suppose, for example, that we have some small-scale stirring which maintains statistically steady turbulence. If the scale of the stirring is much smaller than the Rhines scale, this will produce more or less conventional two-dimensional turbulence (at least at small scales), with its inverse energy cascade. Such turbulence has a Kolmogorov spectrum of $E(k) \sim \Pi^{2/3} k^{-5/3}$, where Π is the mean energy injection rate per unit mass, or equivalently, the flux of energy to larger scales. As energy cascades to larger scales (smaller k) it eventually reaches a scale where $u_k^2 \sim \beta^2 k^{-4}$, at which point the turbulence excites waves. The inverse energy cascade is then partially suppressed by the waves, because resonant triad interactions are inefficient at cascading energy between wavenumbers, and so a larger kinetic energy, u_k^2, is required to maintain the same energy flux, Π. That is to say, the spectrum must now steepen. In fact, $E(k)$ is observed to switch to the steeper Rhines spectrum, as shown in Figure 18.10.

Note that the crossover scale is given by the requirement that $\Pi^{2/3} k^{-5/3} \sim \beta^2 k^{-5}$, which yields $k_\Pi^{-1} \sim L_\Pi = \left(\Pi/\beta^3\right)^{1/5}$. The wavenumber at the peak in the spectrum, on the other hand, is fixed by

$$\langle \mathbf{u}^2 \rangle = u_{rms}^2 \sim \int_{k_{peak}}^{k_\Pi} E dk \sim \beta^2 k_{peak}^{-4}, \qquad (18.21)$$

which gives us $k_{peak}^{-1} \sim L_{Rh} = \sqrt{u_{rms}/\beta}$. The observation that a distinctive PV staircase emerges only if $L_{Rh} > 10 L_\Pi$ (see, Scott & Dritschel, 2012), translates to a requirement that there is at least one decade of the Rhines spectrum.

The Rhines spectrum can be thought of as a kind equilibrium spectrum, in the following sense. Suppose that, at some instant, and at some particular scale, the turbulence deviates from (18.19). For example, we may have $u_k^2 > \beta^2 k^{-4}$, in which case nonlinearity will be stronger than the β effect, the waves at that scale will be suppressed, and the flux of energy to larger scales will be enhanced. The energy at that scale, u_k^2, will then relax back down to the equilibrium spectrum, $u_k^2 \sim \beta^2 k^{-4}$, through the enhanced energy flux to larger scales. Conversely, if we temporarily have $u_k^2 < \beta^2 k^{-4}$ at some scale k, then nonlinearity will be weakened, waves become more dominant, and the flux of energy to large scales suppressed. The kinetic energy at scale k will then grow, because

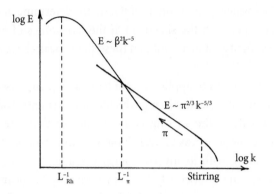

Figure 18.10 The transition from a Kolmogorov spectrum at small scales to a Rhines spectrum at large scales in β-plane turbulence.

energy is passed less rapidly to larger scales. Once again, the equilibrium spectrum, $u_k^2 \sim \beta^2 k^{-4}$, is recovered. Although this is a rather heuristic argument, it does at least provide some rational for the hypothesis of critical balance in Rossby-wave turbulence.

Actually, the situation is a little more complicated than suggested above. At scales smaller than the crossover scale, $L_\Pi = \left(\Pi/\beta^3\right)^{1/5}$, one can still detect hints of a subdominant zonally averaged Rhines spectrum, *i.e.* $E_{\text{zon}}(k_y, k_x = 0) \sim \beta^2 k_y^{-5}$. Conversely, at scales larger than L_Π, the Rhines spectrum $E_{\text{zon}} \sim \beta^2 k_y^{-5}$ is supplemented by a weaker Kolmogorov spectrum, $E \sim \Pi^{2/3} k^{-5/3}$. In short, there is not a sharp transition in spectral properties, but rather a gradual overlap, as indicated by the crossed lines in Figure 18.10.

18.4 Jupiter's Zonal Flows and the Need for Deeper Models

The QGSW equation (18.1), and the system it describes (Figure 18.2), is highly idealized and it is natural to ask how well it describes real phenomena, such as the zonal flows observed in the *weather layer* (or cloud layer) on Jupiter or Saturn. Certainly, a superficial comparison of the QGSW simulation of Figure 18.9 with Jupiter's weather layer (Figure 18.11) is encouraging. Here the zonal bands of alternating east–west flow, which can exceed speeds of 100 m/s, are around $10^4 \rightarrow 2 \times 10^4$ km across. This might be compared to the depth of the weather layer, which is ~70 km, or to a Rossby deformation radius at mid-latitudes, of around 2.5×10^3 km.

There are, however, several challenges when it comes to applying the QGSW equation to the weather layer on Jupiter, the most problematic being that

(a)   (b)

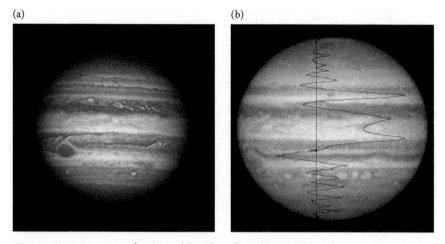

Figure 18.11 Images of Jupiter. (a) Taken from the Hubble telescope, shows zonal bands and the great red spot. (b) Shows the mean zonal velocity as a function of latitude. Courtesy of NASA and ESA/Hubble.

there is *no rigid lower boundary* to that layer. Rather, Jupiter is continuously stratified and the weather layer (which, like the troposphere on earth, contains moist convection, clouds, and thunderstorms), sits on top of a denser and much deeper *dry-convective zone*. Traditionally, QGSW models were applied to the weather layer only, ignoring the convective zone below. However, measurements from the Juno mission suggest that the zonal jets (but not the turbulent vortices) evident in the cloud layer extend down into the dry-convective zone, possibly as far as 3000 km. This has led to the suggestion that the origin of the jets lies deep within the dry-convective zone. The situation is as shown schematically in Figure 18.12.

The Juno data suggests that a two-layer model, or perhaps even a fully three-dimensional analysis (see, for example, Jones & Kuzanyan, 2009), is required to truly capture the zonal jets. However, so-called 1.5-layer models retain the simplicity of the one-layer QGSW equation while embracing the coupling between layers. These take the form

$$\frac{Dq}{Dt} = \frac{D}{Dt}\left[\omega + \frac{\psi - \bar{\psi}_{\text{deep}}}{R_d^2} + \beta y\right] = F + D, \tag{18.22}$$

where D represents some form of dissipation and F is the small-scale forcing, designed to mimic the action of moist convection in Jupiter's weather layer. The new term, $\bar{\psi}_{\text{deep}}(y)$, couples the two layers and represents the geostrophic streamfunction for the zonally symmetric, steady, zonal flow in

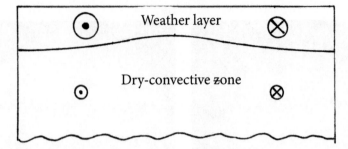

Figure 18.12 Schematic of the weather (or cloud) layer overlying the dry-convective zone. Zonal jets of opposite sign are shown in the cloud layer. These jets extend down into the dry-convective zone below. Adapted from Thomson & McIntyre (2016).

the dry-convective layer, $\bar{u}_{x,\text{deep}} = d\bar{\psi}_{\text{deep}}/dy$. In 1.5-layer models, $\bar{u}_{x,\text{deep}}(y)$ is *prescribed*, typically taken to be sinusoidal, perhaps with an offset. (Note that any offset in $\bar{u}_{x,\text{deep}}$ simply alters the effective value of β.)

Of course, this introduces two additional model parameters into the problem, the amplitude and wavelength of $\bar{u}_{x,\text{deep}}(y)$, which must be somehow estimated. Such models are then predictive if there is an independent means of estimating $\bar{u}_{x,\text{deep}}(y)$. Typical of these 1.5-layer studies is Thomson & McIntyre (2016), who obtain reasonably realistic results when the model parameters are suitably chosen and the forcing strong enough.

It is probably fair to say that the modelling of the zonal jets in the gas giants is still in its infancy, with little agreement as to what constitutes the minimal predictive model. Of course, this reflects the complexity of the processes involved, as well as the uncertainties associated with the role of the dry-convective zone.

* * *

That concludes our brief introduction to zonal flows and turbulence on the β-plane. There is still much which is uncertain, even when it comes to interpreting the results of the idealized, single-layer, QGSW simulations. For example, opinion differs as to the best answer to the simplest question of all: why are zonal jets ubiquitous in β-plane turbulence? The situation is even more uncertain when it comes to applying these idealized models to real flows. Questions abound, such as: why are the zonal jets on Jupiter straight while the zonal jets on earth meander? Readers seeking more detail could do worse than consult the extensive compilation of reviews in Galperin & Read (2019).

References

Danilov, S., & Gurarie, D., 2004, Scaling, spectra and zonal jets in β-plane turbulence, *Phys. Fluids*, **16**(7), 2592–603.

Davidson, P.A., 2015, *Turbulence, An Introduction for Scientists and Engineers*, 2nd Ed., Oxford University Press.

Dritschel, D.G., & McIntyre, M.E., 2008, Multiple jets as PV staircases: the Phillips effect and the resilience of eddy-transport barriers, *Amer. Meteor. Soc.*, **65**, 855–74.

Dritschel, D.G., & Scott, R.K., 2011, Jet sharpening by turbulent mixing, *Phil. Trans. A*, **369**, 754–70.

Galperin, B., & Read, P.L., 2019, *Zonal Jets: Phenomenology, Genesis and Physics*, Cambridge University Press.

Jones, C.A., & Kuzanyan, K.M., 2009, Compressible convection in the deep atmospheres of giant planets, *Icarus*, **204**, 227–38.

Marcus, P.S., & Shetty, S., 2011, Jupiter's zonal winds: are they bands of homogenized potential vorticity organized as a monotonic staircase? *Phil. Trans. A*, **369**, 771–95.

Rhines, P.B., 1975, Waves and turbulence on a β-plane, *J. Fluid Mech.*, **69**, 417–43

Scott, R.K., & Dritschel, D.G., 2012, The structure of zonal jets in geostrophic turbulence, *J. Fluid Mech.*, **711**, 576–98.

Thomson, S.I., & McIntyre, M.E., 2016, Jupiter's unearthly jets: a new turbulent model exhibiting statistical steadiness without large-scale dissipation, *J. Atmospheric Sciences*, **73**(3), 1119–41.

Vallis, G.K., & Multrud, M.E., 1993, Generation of mean flows and jets on the beta plane and over topography, *J. Phys. Oceanogr.*, **23**, 1346–62.

Chapter 19
Accretion Discs in Astrophysics

In this final chapter we consider those vast swirling discs of gas that surround young and dying stars, allowing mass to accrete onto the star. This is a particularly active area of research. For example, protoplanetary discs, which also contain dust and ice, have been much studied in recent years as they establish the environment for planet formation. Let us begin with a brief discussion of why accretion discs form in the first place.

19.1 Accretion Discs and Why They Form

The gravitational collapse of a protostellar cloud onto a young protostar does not occur in a spherically symmetric manner. Rather, a thin, rotating disc of gas first emerges from the cloud. This is centred on the star and has an axis aligned with the star's rotation axis. Mass then falls directly onto the disc, after which it spirals radially inward to be deposited on the surface of the young star. This is shown schematically in Figure 19.1. Accretion discs also form around black holes, and around the more massive star in a binary star system, allowing mass to transfer from the least massive to the more massive star.

As mass spirals through the disc, pulled in by gravity, it must somehow shed its angular momentum. To some extent, this can be achieved magnetically, provided there is an external magnetic field which threads through the disc that is capable of carrying off some of the angular momentum. However, it is thought that the dominant mechanism is the turbulent diffusion of angular momentum out towards the edge of the disc, as indicated in Figure 19.2. Indeed, as we shall see in §19.2, a simple model of accretion discs has the rate of accretion of mass onto the star as proportional to the turbulent diffusivity for momentum within the disc. Consequently, there has been a considerable focus on understanding how turbulence is generated and maintained within accretion discs.

For hot discs with high electrical conductivities (say in a binary star system or around a black hole), it is widely believed that the mechanism that

The Dynamics of Rotating Fluids. P. A. Davidson, Oxford University Press. © Peter A Davidson (2024).
DOI: 10.1093/9780191994272.003.0019

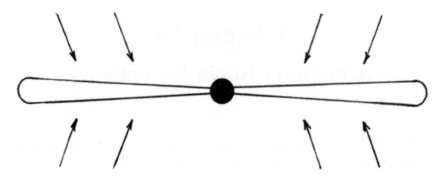

Figure 19.1 Mass is accreted onto a protostar by first falling onto an accretion disc.

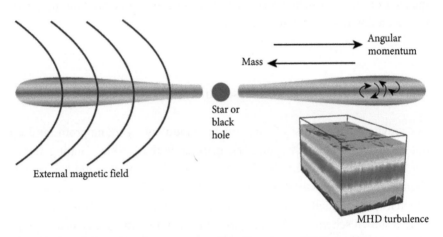

Figure 19.2 Schematic of an accretion disc. Courtesy of Phil. Armitage.

triggers and maintains the turbulence is the classical Chandrasekhar–Velikhov instability, often referred to within the accretion disc literature as the *magneto-rotational instability* (or MRI). In this instability a weak magnetic field which threads through the disc destabilizes the motion, even though the flow is notionally stable by Rayleigh's centrifugal criterion (see §19.5). However, the origin of the turbulence in the cooler protoplanetary discs is still a topic of debate, with MRI, if it occurs at all, confined to the inner, more ionized, parts of the disc.

The reason why discs form in the first place is angular momentum conservation. Let us go back to the case of a protostellar cloud collapsing under gravity onto a nascent protostar. Suppose that the cloud rotates about the z-axis and consider the early stages of collapse, when the gas is in freefall and before a disc is established. As the gas in the cloud falls inward towards the protostar, conserving angular momentum in the process, rotational effects become increasingly dominant. A distinction soon arises between mass that approaches the star along the rotation axis and that which tries to spin inward along the equatorial plane. In particular, the flow inward along the equatorial plane is inhibited because conservation of angular momentum demands that the gas spins faster, increasing its kinetic energy, yet there is only a finite reservoir of potential energy that can feed this growth of kinetic energy. Of course, mass approaching along the rotation axis is subject to no such constraint, and so we conclude that the equatorial collapse of the cloud is suppressed, possibly even halted, while motion directed parallel to the rotation axis is unimpeded. After some time, then, the gas flowing parallel to the z-axis dominates the collapse of the cloud, and so a thin disc of relatively dense gas gradually builds up in the equatorial plane.

A somewhat artificial model problem provides some qualitative insights into the initial stages of collapse of a diffuse, spherical cloud in freefall. The model ignores pressure gradients and shear stresses, something we cannot do when considering dynamics within an accretion disc. It also ignores the effects of any ambient magnetic field and (artificially) assumes that the gravitational field is dominated by the central star, so that the gravitational potential energy per unit mass is $-GM/|\mathbf{x}|$, where M is the mass of the star and G is the universal gravitational constant. (This last assumption is artificial because, during the early stages of collapse, the mass of the cloud dominates.) Finally, we assume that the motion is axisymmetric, and so we adopt cylindrical polar coordinates, (r, θ, z), centred on the star. The materially conserved specific angular momentum is then $\Gamma = r u_\theta$.

We now consider gas spiralling in along the equatorial plane, whose (conserved) mechanical energy per unit mass is

$$ e = \frac{\Gamma^2}{2r^2} + \frac{u_r^2}{2} - \frac{GM}{r}, $$

u_r being the radial velocity. This yields

$$ u_r^2 = \frac{2GM}{r} - \frac{\Gamma^2}{r^2} + 2e_\infty \approx \frac{2GM}{r} - \frac{\Gamma^2}{r^2}, \tag{19.1} $$

where e_∞ is the mechanical energy (per unit mass) far from the star, assumed small. Thus, as mass spirals inward, potential energy is released at a rate of r^{-1}, while the azimuthal contribution to the kinetic energy grows as r^{-2}. Clearly, the release of potential energy cannot continue to feed the growth in kinetic energy for all r, and so the radial inflow along the equatorial plane ceases at around $r \approx \Gamma^2/2GM$. By contrast, the gas flowing in along the z-axis is subject to no such constraint and can fall directly onto the star. Consequently, accretion from a rotating cloud cannot remain spherically symmetric, and a disc of relatively dense gas gradually builds up in the equatorial plane.

Put another way, the cloud collapses in such a way as to minimize the inevitable fall in its moment of inertia, thus minimizing its rise in kinetic energy.

19.2 The Angular Momentum Budget and Energy Dissipation in a Steady Disc

We now consider a minimal model of a fully developed accretion disc. The main difference to the freefall discussion in §19.1 is that we must embrace the role of turbulent stresses in diffusing angular momentum across the disc and in converting mechanical energy into heat. However, we continue to assume the flow is statistically axisymmetric and ignore the effects of any magnetic field that threads through the disc. As above, we adopt cylindrical polar coordinates, (r, θ, z), centred on the star and aligned with the axis of rotation. Also, the central object is typically much more massive than the disc itself and so dominates the gravitational field, whose potential is then $-GM/|\mathbf{x}|$, where M is the mass of the star. Finally, we assume the flow is steady-on-average and we handle the turbulence in a naive way by simply replacing ν in the laminar equation of motion with an *eddy viscosity*, $\nu_t(\mathbf{x})$.

Let $\Omega(r)$ be the angular velocity distribution across the disc. Since the radial velocity is typically much less than the azimuthal velocity, and the disc is assumed to be too thin to support any significant radial pressure gradient, the leading-order radial force balance is between $\rho u_\theta^2/r$ and gravity, *i.e.*

$$\Omega^2 r = \frac{GM}{r^2}. \tag{19.2}$$

This leads to the Keplerian orbit $\Omega(r) = \sqrt{GM/r^3}$, and to a specific angular momentum distribution of

$$\Gamma(r) = r^2 \Omega(r) = \sqrt{GMr}. \tag{19.3}$$

Moreover, since the gravitational potential energy within the disc is $-GM/r$, the mechanical energy per unit mass is

$$e = \frac{1}{2}\Omega^2 r^2 - \frac{GM}{r} = -\frac{GM}{2r}. \tag{19.4}$$

Note that (19.3) and (19.4) tell us that, unlike in §19.1, the specific angular momentum and mechanical energy are not materially conserved. Rather, turbulence acts to diffuse Γ across the disc and to convert mechanical energy into heat.

We focus first on the turbulent diffusion of angular momentum. Since the flow is steady-on-average, the azimuthal equation of motion is

$$\rho\frac{D\Gamma}{Dt} = \rho(\mathbf{u}\cdot\nabla)\Gamma = \frac{1}{r}\frac{\partial}{\partial r}\left(r^2\tau_{r\theta}\right), \tag{19.5}$$

where $\tau_{r\theta}$ is the time-averaged shear stress, which consists of a (negligible) viscous stress and a Reynolds stress associated with the turbulent fluctuations. In the eddy viscosity approximation, $\tau_{r\theta}$ is evaluated by replacing v in the laminar estimate of $\tau_{r\theta}$ by $v_t(r, z)$:

$$\tau_{r\theta} = \rho v_t r\frac{d\Omega}{dr}. \tag{19.6}$$

Now, the steady-on-average mass conservation equation is $\nabla\cdot\left(\rho\mathbf{u}\right) = 0$, and so (19.5) and (19.6) combine to give

$$\nabla\cdot\left(\rho\Gamma\mathbf{u}\right) = \frac{1}{r}\frac{\partial}{\partial r}\left[\rho v_t r^3\frac{d\Omega}{dr}\right] = \nabla\cdot\left[\rho v_t r^2\frac{d\Omega}{dr}\,\hat{\mathbf{e}}_r\right], \tag{19.7}$$

which is our evolution equation for Γ.

Next, we integrate (19.7) over an annular portion of the disc, of radial extent δr. This yields

$$\oint \rho\Gamma\mathbf{u}\cdot d\mathbf{S} = -\frac{d}{dr}\left(\dot{m}\Gamma\right)\delta r = 2\pi\frac{d}{dr}\left[r^3\frac{d\Omega}{dr}\int_{-\infty}^{\infty}\rho v_t dz\right]\delta r,$$

where $\dot{m}$ is the radial mass flow rate through the disc, defined as *positive inward*. We conclude that Γ is governed by

$$\frac{d}{dr}(\dot{m}\Gamma) = -2\pi\frac{d}{dr}\left[r^3\frac{d\Omega}{dr}\int_{-\infty}^{\infty}\rho v_t dz\right], \qquad (19.8)$$

which integrates to give

$$\dot{m}(\Gamma - \Gamma_0) = -2\pi r^3\frac{d\Omega}{dr}\int_{-\infty}^{\infty}\rho v_t dz, \qquad (19.9)$$

for some constant Γ_0.

Note that (19.9) is not restricted to a Keplerian orbit, and so it applies equally to the bulk of the disc, which *is* Keplerian, and to the inner, transitional region where the disc adapts to the surface of the star, which is *not* Keplerian. Note also that Γ_0 is usually taken to be the value of $\Gamma(r)$ at the point where the disc (almost) touches the star, $\Gamma_0 = \Gamma(r_{in})$, because it is observed that, usually, $d\Omega/dr = 0$ at the inner radius, r_{in}. (The star usually rotates more slowly than the inner part of the disc, so Ω is a maximum at r_{in}.) Finally, a Keplerian orbit requires $r^3 d\Omega/dr = -(3/2)\Gamma$, and so have, within the bulk of the disc,

$$\dot{m}[1 - \Gamma_0/\Gamma] = \dot{m}\left[1 - \sqrt{r_{in}/r}\right] = 3\pi\int_{-\infty}^{\infty}\rho v_t dz, \qquad (19.10)$$

which clearly demonstrates that the rate of accretion of mass is controlled by the turbulent diffusivity for momentum within the disc. Of course, to make progress, we need to estimate the eddy viscosity in (19.10). We postpone our discussion of this until §19.4.

Let us now consider the role of turbulence in converting mechanical energy into heat. Equation (19.4) tells us that, as mass spirals radially inward, half of the gravitational potential energy is converted into kinetic energy. Since the disc is assumed steady, the other half must be converted into heat and subsequently radiated away at the surfaces of the disc. If we ignore the redistribution of mechanical energy and heat within the disc by turbulent fluctuations, then the energy budget for an annular portion of the disc is

$$2\left[\sigma T_s^4\right] 2\pi r(\delta r) = \frac{d}{dr}(\dot{m}e)\,\delta r, \tag{19.11}$$

where σ is the Stefan–Boltzmann constant, T_s the surface temperature of the disc, and σT_s^4 the radiated energy per unit area and per unit time. (In writing down (19.11), we have assumed that the disc may be treated as an optically thick black body.) On substituting for e using (19.4), our energy balance yields

$$\sigma T_s^4 = \frac{\dot{m}GM}{8\pi r^3}. \tag{19.12}$$

If the disc has a large outer radius ($r_{\text{out}} \gg r_{\text{in}}$), the luminosity of the disc is then

$$L_{\text{disc}} = 2\int_{r_{\text{in}}}^{\infty}\left[\sigma T_s^4\right] 2\pi r dr = \frac{\dot{m}GM}{2r_{\text{in}}}, \tag{19.13}$$

as demanded by (19.4) and (19.11). Note that measuring L_{disc} allows $\dot{m}$ to be estimated.

In fact, the energy budget (19.11) is too naive, as it does not allow for the redistribution of mechanical energy across the disc by turbulent Reynolds stresses, *i.e.* by the divergence $\nabla \cdot (\tau_{ij}u_i)$ in (2.34). In an accretion disc, this divergence takes the form

$$\nabla \cdot (\tau_{r\theta}u_\theta \hat{\mathbf{e}}_r) = \frac{1}{r}\frac{\partial}{\partial r}\left(\rho v_t r \frac{d\Omega}{dr}\Gamma\right), \tag{19.14}$$

and when the corresponding redistribution of energy is accounted for, we obtain the modified energy budget

$$2\left[\sigma T_s^4\right] 2\pi r = \frac{d}{dr}(\dot{m}e) + 2\pi \frac{d}{dr}\left(r\frac{d\Omega}{dr}\Gamma\int_{-\infty}^{\infty}\rho v_t dz\right), \tag{19.15}$$

as discussed in §19.4. When combined with (19.10), (19.15) yields, after a little algebra,

$$\sigma T_s^4 = \frac{\dot{m}GM}{8\pi r^3}3\left(1 - \frac{\Gamma_0}{\Gamma}\right) = \frac{\dot{m}GM}{8\pi r^3}3\left(1 - \sqrt{r_{\text{in}}/r}\right), \tag{19.16}$$

which also integrates to give $L_{\text{disc}} = \dot{m}GM/2r_{\text{in}}$ (provided that $r_{\text{out}} \gg r_{\text{in}}$).

19.3 A Glimpse at Protoplanetary Discs and Discs in Binary Star Systems

We shall provide a more detailed model of accretion discs, usually called the α *model*, in §19.4. However, prior to that, perhaps it is worth describing, albeit briefly, the physical properties of these discs. We focus on protoplanetary discs and discs in binary star systems.

Protoplanetary discs form around young stars and they are much studied as they provide an environment favourable to planet formation. (The hypothesis that the planets in our solar system condensed out of a gaseous disc dates back to Kant and Laplace.) Such discs are largely composed of cool ($10 \text{ K} < T < 10^3$ K) hydrogen gas plus some dust. They have aspect ratios (thickness to radius) less than 0.1, orbital periods ranging from $T_{\text{in}} \sim 10$ days $\rightarrow T_{\text{out}} \sim 10^3$ years, and an outer radius of the order of $r_{\text{out}} \sim 100$ AU. (1 AU is the distance from the earth to the sun.) Their mass is rarely greater than one tenth of the mass of the central star, and they have accretion rates of around $\dot{m} \sim 10^{-8} \rightarrow 10^{-6} M_\odot/\text{year}$. (Here $M_\odot$ is the mass of our sun.) As already mentioned, the primary function of these discs is to facilitate accretion by removing excess angular momentum from the inflowing gas through the action of turbulent diffusion.

Protoplanetary discs are cool and hence only weakly ionized, so that magnetic fields have only a modest influence on the accretion dynamics. The disc ionization is probably high enough for MRI to take root and generate the turbulence needed for accretion in just two regions: (i) the hotter, inner parts of the disc, say $r < 1$ AU; and (ii) within thin surface layers at larger radii, which are ionized by X-rays emitted by the protostar. Consequently, it is likely that there are extensive *dead zones* in protoplanetary discs where the gas is overwhelmingly neutral and MRI cannot operate (see Armitage, 2011). So, the primary mechanism of angular momentum transport across such discs is far from settled. Indeed, despite their importance for planet formation, their low ionization level and complex chemistry means that the accretion dynamics of protoplanetary discs is less well understood than that of other disc systems. We shall return to protoplanetary discs in §19.6.

Many stars partner to form binary systems, in which two adjacent stars rotate about their combined centre of mass. Consider such a system in which the stars have masses M_1 and M_2 and are located at distances r_1 and r_2 from their combined centre of mass. To fix thoughts, we shall take $M_1 > M_2$, and hence $r_1 < r_2$. It is convenient to move into a frame of reference which rotates with the stars and whose origin is located at their combined centre of mass, so that the stars are stationary in the chosen frame. We then take both stars to lie on the x-axis, located at $x = -r_1$ and $x = r_2$, as shown in Figure 19.3.

The surfaces of constant *effective potential* (the potential energy modified to allow for background rotation) are shown in Figure 19.3 for a typical binary system. Close to M_1 and M_2 the equipotential surfaces are spherical and centred on the two masses, while further out the surfaces are dumbbell shaped and enclose both stars. A figure-of-eight marks the transition from one geometry to the other, and the saddle point L_1 is known as the *inner Lagrangian point*. (A test mass placed at L_1 is unstable to perturbations in the x-direction.) The two halves of the figure-of-eight passing through L_1 are known as the *Roche lobes* of M_1 and M_2, named after the French mathematician Edouard Roche.

Now, a star in equilibrium is bounded by an equipotential surface. It follows that stars much smaller that $r_1 + r_2$ are approximately spherical and when both stars satisfy this condition we have a so-called *detached binary*. By way of contrast, in a *semidetached binary*, one of the stars swells to fill its Roche lobe. Mass can then escape through L_1 to fall onto the companion star, causing an accretion disc to form around the more compact star (see Figure 19.4). It turns out that the more massive star, M_1, has the larger of the two Roche lobes (see Figure 19.3), and so it is the less massive star, M_2, that is most likely to fill its

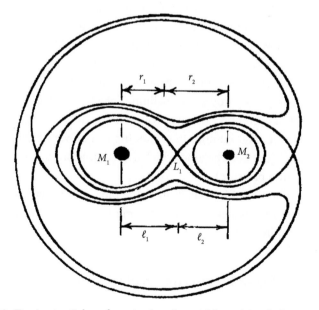

Figure 19.3 Equipotential surfaces in the plane of the orbit of a binary star system.

Figure 19.4 Schematic of an accretion disc in a binary system. The star on the right fills its Roche lobe and transfers mass to a neutron star on the left. Image courtesy of Marc van der Sluys.

Roche lobe. In summary, then, in a typical semidetached binary, mass flows through L_1 from the less massive to the more massive star.

The more massive a star, the faster it evolves. Consequently, in a semidetached binary, the more massive star is typically the compact and dense remnants of a dying star, *e.g.* a white dwarf or a neutron star, while the less massive star is typically middle-aged (a main-sequence star). Hence, in a semidetached binary in which the more massive star is compact, mass flows through L_1 from M_2 to M_1, and an accretion disc is established around M_1 in the orbital plane of the binary system (see Figure 19.4).

Such discs are composed of relatively dense and hot (10^3K $< T < 10^7$K) plasma, consisting of hydrogen and helium. They have aspect ratios less than 0.03, and typical accretion rates are around $\dot{m} \sim 10^{-9} \rightarrow 10^{-8} M_\odot$/year. For a neutron star the total luminosity of the disc can be very high, close to the Eddington limit (the limit beyond which accretion is not possible because of the high radiation pressure).

With that all too brief introduction to the physical properties of discs, we now return to our main theme, which is the minimal mathematical description of accretion within discs.

19.4 A Non-Magnetic Description of Discs and the α Disc Model

We now describe the standard model of accretion discs, which is sometimes called the α model (for reasons that will become obvious). As before, we adopt cylindrical polar coordinates, (r, θ, z), centred on the star and aligned with the rotation axis, and take the disc to be statistically axisymmetric. We continue to neglect magnetic fields and radial pressure gradients, except close to the central star where pressure gradients and magnetic fields cannot normally be ignored. We also take $|u_r| \ll \Omega r$, and assume that the gravitational field is dominated by the mass of the star, M. All of this allows us to adopt a Keplerian orbit, with $\Gamma(r) = \sqrt{GMr}$. We work with time-averaged quantities, averaged over the fast timescale of the turbulence, but allow the statistical properties of the disc, though not M, to evolve on a slow timescale. Finally, we use variables which have been averaged across the thickness of the disc, according to the rule

$$\langle f \rangle = \frac{1}{\Sigma} \int_{-\infty}^{\infty} f\rho\, dz, \quad \Sigma = \int_{-\infty}^{\infty} \rho\, dz, \tag{19.17}$$

where $\langle \sim \rangle$ indicates an average, $f(r, z, t)$ is any property of the disc, and $\Sigma(r, t)$ is known as the *surface density*. Thus, for example, the accretion rate is

$$\dot{m} = -2\pi r \int_{-\infty}^{\infty} \rho u_r\, dz = -2\pi r \langle u_r \rangle \Sigma, \tag{19.18}$$

while the continuity equation

$$\frac{\partial \rho}{\partial t} + \nabla \cdot (\rho \mathbf{u}) = 0, \tag{19.19}$$

integrates through the thickness of the disc to give

$$2\pi r \frac{\partial \Sigma}{\partial t} = \frac{\partial \dot{m}}{\partial r}. \tag{19.20}$$

Let us now consider the azimuthal equation of motion. Since $\Gamma(r)$ is independent of t and z, this takes the form

$$\rho \frac{D\Gamma}{Dt} = \rho u_r \frac{d\Gamma}{dr} = \frac{1}{r} \frac{\partial}{\partial r} \left(r^2 \tau_{r\theta} \right) + r \frac{\partial \tau_{z\theta}}{\partial z}, \tag{19.21}$$

where $\tau_{r\theta}$ and $\tau_{z\theta}$ are turbulent Reynolds stresses. This integrates through the disc to give

$$\dot{m} \frac{d\Gamma}{dr} = -2\pi \frac{\partial}{\partial r} \left[r^2 \int_{-\infty}^{\infty} \tau_{r\theta}\, dz \right]. \tag{19.22}$$

We now model the turbulence using an eddy viscosity, v_t, just as we did in §19.2. So, the Reynolds stress $\tau_{z\theta}$ is given by (19.6) and our azimuthal equation of motion becomes

$$\dot{m}\frac{d\Gamma}{dr} = -2\pi\frac{\partial}{\partial r}\left[r^3\frac{d\Omega}{dr}\int_{-\infty}^{\infty}\rho v_t dz\right] = -2\pi\frac{\partial}{\partial r}\left[r^3\frac{d\Omega}{dr}\langle v_t\rangle\Sigma\right]. \qquad (19.23)$$

For a statistically steady disc, where $\dot{m}$ is independent of r, this is nothing more than (19.8). Noting that $r^3 d\Omega/dr = -(3/2)\,\Gamma$ in a Keplerian disc, (19.23) simplifies to

$$\dot{m}\frac{d\Gamma}{dr} = 3\pi\frac{\partial}{\partial r}\left[\Gamma\langle v_t\rangle\Sigma\right], \qquad (19.24)$$

or equivalently, since $\Gamma \sim \sqrt{r}$,

$$\dot{m} = 6\pi\,r^{1/2}\frac{\partial}{\partial r}\left[r^{1/2}\langle v_t\rangle\Sigma\right]. \qquad (19.25)$$

Finally, eliminating $\dot{m}$ from (19.20) and (19.25) gives us a (diffusion-like) evolution equation for an evolving disc,

$$\frac{\partial\Sigma}{\partial t} = \frac{3}{r}\frac{\partial}{\partial r}\left[r^{1/2}\frac{\partial}{\partial r}\left(r^{1/2}\langle v_t\rangle\Sigma\right)\right], \qquad (19.26)$$

which is readily solved in terms of Bessel functions if $\langle v_t\rangle$ is uniform and constant.

Statistically steady discs, where $\dot{m}$ is independent of r, are an important special case, and so we focus on steady discs from now on. Here (19.23) integrates to give

$$\dot{m}(\Gamma - \Gamma_0) = -2\pi r^3\frac{d\Omega}{dr}\int_{-\infty}^{\infty}\rho v_t dz, \qquad (19.27)$$

which is (19.9). As mentioned in §19.2, it is common to take the constant of integration, Γ_0, to be the value of $\Gamma(r)$ near the surface of the star since, often, it is observed that Ω is a maximum there, *i.e.* the star rotates more slowly than the inner disc. We then have

$$\dot{m}\left[\Gamma - \Gamma(r_{in})\right] = -2\pi r^3 \frac{d\Omega}{dr} \int_{-\infty}^{\infty} \rho v_t dz, \tag{19.28}$$

which, for the Keplerian parts of the disc (away from the central star), simplifies to

$$\dot{m}\left[\Gamma - \Gamma(r_{in})\right] = 3\pi \Gamma \int_{-\infty}^{\infty} \rho v_t dz. \tag{19.29}$$

We have arrived back at (19.10). Finally, noting that $\Gamma \sim \sqrt{r}$, we obtain

$$\dot{m}\left[1 - \sqrt{r_{in}/r}\right] = 3\pi \langle v_t \rangle \Sigma, \tag{19.30}$$

which satisfies (19.25). Clearly, the accretion rate is controlled by the eddy viscosity.

We now turn to the rate of loss of mechanical energy associated with the Reynolds stresses. The energy dissipation rate per unit volume is $\tau_{ij}S_{ij}$ (see (2.34)), which becomes

$$\tau_{r\theta}\, r\frac{d\Omega}{dr} = \rho v_t \left[r\frac{d\Omega}{dr}\right]^2 \tag{19.31}$$

in a disc. So the rate of generation of heat, integrated through the disc thickness, is

$$\left[r\frac{d\Omega}{dr}\right]^2 \int_{-\infty}^{\infty} \rho v_t dz = \left[r\frac{d\Omega}{dr}\right]^2 \langle v_t \rangle \Sigma. \tag{19.32}$$

For a steady disc, we may use (19.30) to substitute for $\langle v_t \rangle \Sigma$ in (19.32), and assuming that all of the heat is immediately radiated away as black-body radiation, we obtain

$$2\left(\sigma T_s^4\right) = \frac{\dot{m}}{3\pi}\left[r\frac{d\Omega}{dr}\right]^2 \left(1 - \sqrt{r_{in}/r}\right), \tag{19.33}$$

where T_s is the surface temperature of the disc. It is readily confirmed that this is equivalent to (19.15). Specifically, the two terms on the right of (19.15) sum to give the rate of dissipation of mechanical energy appearing on the right of (19.33). For a Keplerian disc, (19.33) simplifies to

$$\sigma T_s^4 = \frac{3\dot{m}GM}{8\pi r^3}\left(1 - \sqrt{r_{\text{in}}/r}\right), \tag{19.34}$$

which brings us back to (19.16).

Let us now turn to the vertical structure of the disc. Because all of the motion lies in the plane of the disc, there is a balance between the axial pressure gradient and gravity,

$$\frac{\partial p}{\partial z} = -\frac{GM\rho\,z}{r^3} = -\rho\Omega^2 z. \tag{19.35}$$

If H is the *half-width* of the disc, this requires that $p/H \sim \rho\Omega^2 H$, or equivalently $\sqrt{p/\rho} \sim \Omega H$. Introducing the isothermal sound speed, $c_s = \sqrt{p/\rho}$, we obtain a convenient measure of the disc thickness, $H \sim c_s/\Omega$. We also note that (19.35) requires

$$\frac{\partial p}{\partial r} \sim \rho\,\Omega^2 r(H/r)^2 \ll \rho\,\Omega^2 r, \tag{19.36}$$

which justifies the neglect of radial pressure gradients, except close to the central star.

Finally, in order to have a predictive model, we need to specify the eddy viscosity, $\langle \nu_t \rangle$. The standard model follows Shakura & Sunyaev (1973), who propose $\langle \nu_t \rangle = \alpha c_s H$, which is known as the *alpha-viscosity prescription*. Here α is a dimensionless constant of order unity and c_s is based on the mid-plane pressure and temperature. This is a type of *mixing-length* model, in which the size of the turbulent eddies is assumed to scale on H and the turbulent velocity fluctuations to scale on c_s. If we ignore the contribution to p from the radiation pressure, we can write $c_s = \sqrt{R_g T_{\text{mid}}}$, where R_g is a gas constant and T_{mid} the mid-plane temperature. Moreover, since $H \sim c_s/\Omega$, we conclude that

$$\langle \nu_t \rangle = \alpha c_s H \sim \alpha c_s^2/\Omega = \alpha R_g T_{\text{mid}}/\Omega. \tag{19.37}$$

We are now in a position to obtain scaling laws for H, T_{mid}, and the mid-plane density, ρ_0, in a steady disc. First, we note that (19.30) gives us, for $r \gg r_{\text{in}}$,

$$\dot{m} = 3\pi \langle \nu_t \rangle \, \Sigma = 3\pi \left(\alpha\sqrt{R_g T_{\text{mid}}}\,H\right)\Sigma \sim \alpha\rho_0\sqrt{R_g T_{\text{mid}}}\,H^2. \tag{19.38}$$

This combines with $H \sim c_s/\Omega$ to provide the estimates

$$H^2 \sim \frac{\dot{m}}{\alpha \rho_0 \sqrt{R_g T_{\text{mid}}}} \sim \frac{R_g T_{\text{mid}}}{GM/r^3}, \qquad r \gg r_{\text{in}}. \tag{19.39}$$

Second, we relate the disc mid-plane and surface temperatures through a radiative transport model. In particular, we write

$$T_{\text{mid}}^4/T_s^4 = \tau \sim \kappa\Sigma,$$

where τ is called the *integrated optical depth* and κ is known as the *Rosseland mean opacity* (see, for example, Frank et al., 2002, or else Armitage, 2011). The opacity, κ, is often estimated using *Kramers' law*, $\kappa = \kappa_0 \rho_0 T_{\text{mid}}^{-7/2}$, where κ_0 is a constant, although this applies only within a certain temperature range (see below). In any event, equation (19.34) now reads

$$\sigma T_{\text{mid}}^4 \sim \tau \frac{\dot{m}GM}{r^3} \sim \kappa\Sigma \frac{\dot{m}GM}{r^3}, \qquad r \gg r_{\text{in}}. \tag{19.40}$$

Crucially, (19.39) and (19.40) combine to yield the *α-disc scaling laws* for T_{mid}, H, and ρ_0.

These are shown in Table 19.1, expressed in three different ways. Note that the scalings in the bottom row are restricted to Kramers' law, while the other two rows are quite general.

A number of cases now arise. If, at a given radius, the disc is isothermal (*i.e.* $\tau = 1$), which is rarely a good approximation, then the scaling laws for T_{mid}, H, and ρ_0 in the first row are the most convenient. On the other hand, if the opacity is uniform (*i.e.* κ independent of r), then the scaling laws in the second row are the most useful. Uniform κ tends to be appropriate for a very hot

Table 19.1 The α-disc scaling laws written in terms of: τ (row 1), κ (row 2), and κ_0 (row 3)

T_{mid}	H	ρ_0
$\dfrac{\tau^{1/4}\dot{m}^{1/4}}{\sigma^{1/4}}\left[\dfrac{GM}{r^3}\right]^{1/4}$	$\dfrac{\tau^{1/8}R_g^{1/2}\dot{m}^{1/8}}{\sigma^{1/8}}\left[\dfrac{GM}{r^3}\right]^{-3/8}$	$\dfrac{\sigma^{3/8}\dot{m}^{5/8}}{\tau^{3/8}R_g^{3/2}\alpha}\left[\dfrac{GM}{r^3}\right]^{5/8}$
$\dfrac{\kappa^{1/5}\dot{m}^{2/5}}{\sigma^{1/5}R_g^{1/5}\alpha^{1/5}}\left[\dfrac{GM}{r^3}\right]^{3/10}$	$\dfrac{\kappa^{1/10}R_g^{2/5}\dot{m}^{1/5}}{\sigma^{1/10}\alpha^{1/10}}\left[\dfrac{GM}{r^3}\right]^{-7/20}$	$\dfrac{\sigma^{3/10}\dot{m}^{2/5}}{\kappa^{3/10}R_g^{6/5}\alpha^{7/10}}\left[\dfrac{GM}{r^3}\right]^{11/20}$
$\dfrac{\kappa_0^{1/10}\dot{m}^{3/10}}{\sigma^{1/10}R_g^{1/4}\alpha^{1/5}}\left[\dfrac{GM}{r^3}\right]^{1/4}$	$\dfrac{\kappa_0^{1/20}R_g^{3/8}\dot{m}^{3/20}}{\sigma^{1/20}\alpha^{1/10}}\left[\dfrac{GM}{r^3}\right]^{-3/8}$	$\dfrac{\sigma^{3/20}\dot{m}^{11/20}}{\kappa_0^{3/20}R_g^{9/8}\alpha^{7/10}}\left[\dfrac{GM}{r^3}\right]^{5/8}$

plasma, say the inner regions of the disc surrounding a neutron star or white dwarf. For lower temperatures, but temperatures in excess of 10^4 K, Kramers' law is often a good approximation. In this case, the bottom row is the relevant one. Note that Kramers' law does not apply in protoplanetary discs, where the opacity is provided by dust and ice grains. Note also that Kramers' law has τ independent of r, i.e. $\tau \sim \dot{m}^{1/5}\alpha^{-4/5}$, and so the radial structures of T_{mid}, H, and ρ_0 are the same for a vertically isothermal disc ($\tau = 1$) and for one that obeys Kramers' law.

This completes our introduction to the α model. It is a minimal model, ignoring as it does many important factors, such as external radiation from the central star, and the magnetic field which threads through the disc, exerting torques on the plasma and carrying off angular momentum. The magnetic field can be particularly strong near the central star, often channelling plasma up along the field lines and into the polar regions of the star. The α model also suffers from a naive description of the turbulence.

19.5 Turbulence I: The Chandrasekhar–Velikhov Instability in Hot Discs

It is clear from (19.30) that the turbulent diffusion of angular momentum out to the edge of the disc is an essential part of accretion dynamics. It is important, therefore, to identify the mechanisms by which turbulence is triggered and maintained in a disc. This is all the more important because Rayleigh's discriminant, $\Phi(r)$, is positive in a Keplerian disc,

$$\Phi(r) = \frac{1}{r^3}\frac{d\Gamma^2}{dr} = \frac{GM}{r^3} = \Omega^2, \tag{19.41}$$

and so accretion discs are notionally stable to the centrifugal instability. It was Bulbus and Hawley (see, for example, Bulbus & Hawley, 1998) who first noticed that a weak magnetic field can destabilize an otherwise stable disc, provided the disc is sufficiently ionized (which possibly rules out protoplanetary discs). It turns out that the instability identified by Bulbus & Hawley is essentially the classical Chandrasekhar–Velikhov instability, described in Chandrasekhar (1960, 1961), so perhaps it is worth taking a moment to describe that instability. For simplicity, we will consider an incompressible, inviscid, perfectly conducting fluid and restrict ourselves to axisymmetric perturbations.

Consider the steady, cylindrical base state $\mathbf{u}_0 = (0, r\Omega(r), 0)$ and $\mathbf{B}_0 = (0, 0, B_0)$, described in cylindrical polar coordinates, (r, θ, z), where B_0 is uniform. Chandrasekhar showed that the presence of a magnetic field modifies Rayleigh's stability criterion in such a way that a sufficient (but *not* necessary) condition for stability becomes

$$r\frac{d\Omega^2}{dr} > 0. \tag{19.42}$$

Indeed, (19.42) continues to hold even in the limit of a vanishingly weak magnetic field (although the growth rate goes to zero along with B_0). This is a more stringent test than Rayleigh's strictly hydrodynamic criterion, $\Phi(r) > 0$, and indeed a Keplerian disc fails to satisfy (19.42). Chandrasekhar also showed that any adverse gradient in Ω can always be stabilized provided B_0 is made sufficiently large. The suggestion, then, is that the magnetic field which threads through a hot disc can destabilize it, especially if that field is weak.

The (linearized) instability mechanism is readily identified. We write $\mathbf{u} = \mathbf{u}_0 + \mathbf{v}$, where $\mathbf{v}$ is the perturbation in velocity, and let $\mathbf{b}$ and $\mathbf{j}$ be the perturbations in the magnetic field and current density, assumed small. The azimuthal equation of motion,

$$\rho\frac{D}{Dt}(ru_\theta) = r(\mathbf{j} \times \mathbf{B}_0)_\theta, \tag{19.43}$$

then linearizes to give

$$\rho\left(r\frac{\partial v_\theta}{\partial t} + v_r\frac{d\Gamma_0}{dr}\right) = -rj_rB_0 = \frac{r}{\mu}\frac{\partial b_\theta}{\partial z}B_0, \tag{19.44}$$

where $\Gamma_0 = r^2\Omega(r)$. Moreover, the azimuthal magnetic field is governed by (17.56),

$$\frac{D}{Dt}\left(\frac{B_\theta}{r}\right) = (\mathbf{B}_p \cdot \nabla)\left(\frac{u_\theta}{r}\right), \tag{19.45}$$

from which we find

$$\frac{\partial b_\theta}{\partial t} = r\frac{d\Omega}{dr}b_r + B_0\frac{\partial v_\theta}{\partial z}. \tag{19.46}$$

Finally, the perturbation in the poloidal magnetic field is governed by (17.4), which yields

$$\frac{\partial b_r}{\partial t} = B_0 \frac{\partial v_r}{\partial z}. \tag{19.47}$$

It is readily confirmed that, in the weak-field limit of $B_0 \rightarrow 0$, we can drop the time derivative in (19.44), because marginally unstable disturbances are non-oscillatory and grow very slowly, and also the axial derivative in (19.46), because it is pre-multiplied by B_0. (See, for example, Davidson, 2013, §7.3.7.) Our linearized equations then simplify to the weak-field, quasi-static equations

$$v_r \frac{1}{r} \frac{d\Gamma_0}{dr} = \frac{B_0}{\rho\mu} \frac{\partial b_\theta}{\partial z}, \tag{19.48}$$

$$\frac{\partial b_\theta}{\partial t} = r \frac{d\Omega}{dr} b_r, \tag{19.49}$$

$$\frac{\partial b_r}{\partial t} = B_0 \frac{\partial v_r}{\partial z}. \tag{19.50}$$

These combine to yield

$$\left[\Phi(r) \frac{\partial^2}{\partial t^2} - r \frac{d\Omega^2}{dr} \frac{B_0^2}{\rho\mu} \frac{\partial^2}{\partial z^2} \right] v_r = 0, \quad \Phi(r) = \frac{1}{r^3} \frac{d\Gamma_0^2}{dr}, \tag{19.51}$$

so the flow is unstable to *weak fields* when $\Phi(r) > 0$ and $d\Omega^2/dr < 0$, with growth rates of the order of $kB_0/(\rho\mu)^{1/2}$, where k is an axial wavenumber. Since $rd\Omega^2/dr = \Phi - (2\Omega)^2$, there are many flows that are stable by Rayleigh's criterion, yet unstable to (19.51).

The instability mechanism is also clear from equations (19.48)→(19.50). Suppose we have an axisymmetric radial jet which perturbs the axial magnetic field, as shown in the bottom disc in Figure 19.5. The outward radial movement, $v_r > 0$, sweeps out a radial field, b_r, from B_0, as described by (19.50). The differential rotation, $d\Omega/dr$, then spirals out b_r to create an azimuthal field, b_θ, in accordance with (19.49). However, the magnetic field lines tend to resist this shearing and set up an opposing force, $-j_r B_0 \hat{e}_\theta$, and hence an opposing torque, $\mu^{-1} r B_0 (\partial b_\theta / \partial z)$, which appears on the right of (19.48), a torque which is balanced by the inertial force on the left. Crucially, the direction of this torque, and hence the direction of the balancing inertial force, depends on the sign of $d\Omega^2/dr$.

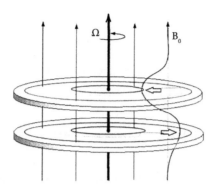

Figure 19.5 The mechanism of MRI. A radial flow shears the background field, $\mathbf{B}_0$.

Consider the case where $d\Omega^2/dr < 0$. Then the field which has been displaced radially outward (the bottom disc in Figure 19.5) is sheared in the negative θ direction and so the field lines, which act like elastic bands in tension, exert an opposing force on the fluid in the *positive* θ direction. From (19.48), we see that the inertial force, which balances this Lorentz force, requires a positive radial velocity. (We assume here that $\Phi > 0$, and so $d\Gamma_0/dr > 0$.) The initial radial movement is then reinforced, signalling an instability. Similarly, if the initial radial movement is inward (the top disc in Figure 19.5), then the azimuthal shearing of the deformed magnetic field lines is such that the Lorentz force is now in the *negative* θ direction, and the matching inertial force requires a negative radial velocity. Once again, the initial radial movement is reinforced, signalling an instability. We conclude, therefore, that the flow is unstable whenever $d\Omega^2/dr < 0$. Similar arguments show the flow to be stable when $d\Omega^2/dr > 0$, consistent with (19.42).

When the unperturbed magnetic field contains an azimuthal component, $\mathbf{B}_0 = (0, B_\theta(r), B_0)$ in (r, θ, z) coordinates, the situation is a little more complicated. However, an energy argument may be used to show that a sufficient (but not necessary) condition for axisymmetric stability is

$$r\frac{d\Omega^2}{dr} > \frac{1}{r^3}\frac{d}{dr}\frac{(rB_\theta)^2}{\rho\mu}, \tag{19.52}$$

provided that $B_0 \neq 0$. (See, for example, Davidson, 2013, §7.3.5.) When $B_0 = 0$, it turns out that a necessary *and* sufficient condition for stability to axisymmetric disturbances is

$$\Phi(r) > r \frac{d}{dr} \frac{(B_\theta/r)^2}{\rho\mu}, \tag{19.53}$$

while the sufficient (but not necessary) conditions for stability to three-dimensional disturbances are (see Davidson, 2013),

$$\Phi(r) > \frac{1}{r^3} \frac{d}{dr} \frac{(rB_\theta)^2}{\rho\mu}, \quad (\Omega r)^2 < \frac{B_\theta^2}{\rho\mu}. \tag{19.54}$$

Returning to the case where $B_\theta = 0$, stronger results can be obtained (for arbitrary B_0) by considering the short wavelength limit. Here it may be shown that the necessary and sufficient conditions for instability are

$$r \frac{d\Omega^2}{dr} < 0, \quad k^2 \frac{B_0^2}{\rho\mu} = k^2 v_a^2 < r \left| \frac{d\Omega^2}{dr} \right|, \tag{19.55}$$

where v_a is the Alfvén velocity and $k = |\mathbf{k}|$ (Balbus & Hawley, 1998). This generalizes the weak-field criterion and is consistent with Chandrasekhar's observation that the instability is suppressed if B_0 is large enough. In a thin disc $k \approx k_z$, and so a Keplerian disc is unstable when $v_a k_z < \sqrt{3}\,\Omega$, with the fastest growth rate at $v_a k_z = \sqrt{15}\Omega/4$. Also, in a disc of width $2H$ we have $k_z \geq \pi/2H$ (corresponding to a half wavelength), and so instability requires

$$v_a < \frac{2\sqrt{3}}{\pi} \Omega H \approx \frac{2\sqrt{3}}{\pi} c_s. \tag{19.56}$$

However, in practice, a finite magnetic diffusivity, λ, sets a lower limit on the value of v_a that can accommodate an instability. Indeed, order-of-magnitude arguments suggests that instability requires $v_a H/\lambda > O(1)$, and so the range of unstable v_a is limited to

$$O(\lambda/H) < v_a < \left(2\sqrt{3}/\pi\right) \Omega H \sim c_s. \tag{19.57}$$

MRI is a robust source of turbulence in hot discs because it may be triggered by a wide range of magnetic field strengths and because the instability is a local one, insensitive to remote boundary conditions. Typically, rapid exponential growth soon gives way to fully developed turbulence in which the mean shear, $d\Omega/dr$, continually feeds energy into the turbulence, with α as high as $\alpha \sim 0.1$. However, the lower limit in (19.57) means that MRI is difficult to achieve in the cooler parts of protoplanetary discs.

19.6 Turbulence II: The Origins of Turbulence in Protoplanetary Discs

The more ionized parts of a protoplanetary disc are the hot, inner regions close to the central star, say $r < 1$ AU, and the thin surface layers which are subjected to external radiation from the star. It is conceivable that MRI can operate in such regions, but it is far from certain that such a limited source of turbulence could sustain the turbulent diffusion of angular momentum across the entire disc. There is, therefore, a need to identify alternative sources of turbulence.

One promising candidate is the instability associated with self-gravitation of the disc, as reviewed in Kratter & Lodato (2016). A somewhat idealized criterion for such an instability can be obtained from a linear stability analysis of a razor-thin, non-magnetic, Keplerian disc where self-gravity of the disc supplements the gravity of the central star. Treating the properties of the disc, such as Σ and c_s, as locally uniform, and looking for axisymmetric perturbations of the form $\exp(kr - \varpi t)$, a *local* dispersion relationship, valid for $kr \gg 1$, can be found:

$$\varpi^2 = \Omega^2 + (c_s k)^2 - 2\pi G\Sigma |k|. \tag{19.58}$$

The first term on the right, if taken by itself, represents the dispersion relationship for a generalized inertial wave associated with the Keplerian background rotation. The second is simply the dispersion relationship for acoustic waves. This is stabilizing, because radial pressure gradients resist the local accumulation of mass within the disc driven by self-gravitation. The final term represents the destabilizing influence of self-gravitation.

For a given $G\Sigma$, the minimum value of ϖ^2 in (19.58), considered as a function of k, is

$$\varpi^2_{min} = \Omega^2 - (\pi G\Sigma/c_s)^2, \tag{19.59}$$

at $k_{min} = \pi G\Sigma/c_s^2$. We conclude that a gravitational instability will occur in regions where

$$Q(r) = \frac{c_s \Omega}{\pi G\Sigma} < 1, \tag{19.60}$$

with $Q(r)$ is known as the *Toomre parameter*. The critical wavenumber can also be expressed in terms of Q, as

$$k_{crit}H = (\Omega H/c_s) \, Q^{-1} \approx Q^{-1}, \tag{19.61}$$

where we have used $H \approx c_s/\Omega$. Given that $\Omega^2 = GM/r^3$, this instability criterion can be rewritten as

$$Q(r) = \frac{c_s M}{\pi r^3 \Omega \Sigma} \approx \frac{M}{(\pi r^2)\Sigma} \frac{H}{r} < 1. \tag{19.62}$$

Finally, it is useful to introduce the disc mass,

$$M_{disc}(r) = \int_{r_{in}}^{r} 2\pi r \Sigma \, dr \approx 2(\pi r^2)\Sigma, \quad r \gg r_{in}, \tag{19.63}$$

where we have used the fact that, at least approximately, $\Sigma \sim r^{-1}$ in a protoplanetary disc. This allows us to rewrite the instability criterion in the physically more intuitive form

$$\frac{M_{disc}(r)}{M} > 2\frac{H}{r}. \tag{19.64}$$

This is the most useful form of the instability criterion, as it brings out the importance of the ratio M_{disc}/M. Note, however, that there are many idealizations inherent in the analysis above, particularly the localized nature of the dispersion relationship (19.58), the approximation $\Sigma \sim r^{-1}$, and the assumed extreme thinness of the disc. (A finite thickness helps stabilize the disc, typically reducing Q_{crit} from 1 to 0.68.) Consequently, the pre-factor of 2 on the right of (19.64) is unreliable and should be treated merely as indicative of a number of order unity. Note also that the formal restriction $kr \gg 1$, combined with the estimate $k_{crit}H \approx Q^{-1} \sim 1$, tells us that the analysis can be justified only for large radii, $r \gg H$.

In any event, (19.64) tells us that a gravitational instability is most likely to occur at large radius, and in the more massive (*i.e.* younger) discs, say discs where $M_{disc} \gtrsim 0.1M$. Numerical simulations suggest that, once triggered, the instability can lead to either fully developed turbulence, as in the case of MRI, or to a fragmentation of the disc, which could be relevant to the formation of brown dwarfs or massive planets. Self-gravitation is thought to be a good candidate for sustained turbulence when $\dot{m} > 10^{-7} M_\odot$/year (Armitage, 2011), although the turbulence does not have the classical broad-band structure, but is restricted to scales of the order of H. This turbulent flow typically develops trailing spiral arms which help transport angular momentum out to the rim of the disc, with values of α as high as $\alpha \sim 0.1$ in the outer parts of the disc.

Purely hydrodynamic instabilities have also been suggested as a source of turbulence, such as those driven by shear, or by an unstable radial entropy gradient. However, their potential for angular momentum transport is poor, with α predicted to be very low. Of these various hydrodynamic mechanisms, perhaps the most promising is a finite-amplitude, baroclinic instability which feeds off an unstable radial entropy gradient. These various instabilities (MRI, self-gravitation, entropy-driven), and their potential for angular momentum transport in protoplanetary discs, are all reviewed in Armitage (2011).

* * *

That concludes our brief introduction to astrophysical accretion discs. Readers seeking more background could do worse than consult Frank et al. (2002) for the classical picture of accretion, Armitage (2011) for a glimpse into the difficult topic of protoplanetary discs, Bulbus & Hawley (1998) for an overview of MRI, and Kratter & Lodato (2016) for a review of self-gravitational instability in discs.

References

Armitage, P.J., 2011, Dynamics of protoplanetary discs, *Ann. Rev. Astron. Astrophys.*, **49**, 195–236.

Bulbus, S.A., & Hawley, J.F., 1998, Instability, turbulence and enhanced transport in accretion discs, *Rev. Modern Phys.*, **70**(1), 1–53.

Chandrasekhar, S., 1960, The stability of non-dissipative Couette flow in hydromagnetics, *Proc. Nat. Acad. Sci.*, **46**, 253–57.

Chandrasekhar, S., 1961, *Hydrodynamic and Hydromagnetic Stability*, Oxford University Press.

Davidson, P.A., 2013, *Turbulence in Rotating, Stratified and Electrically Conducting Fluids*, Cambridge University Press.

Frank, J., King, A., & Raine, D., 2002, *Accretion Power in Astrophysics*, 3rd Ed., Cambridge University Press.

Kratter, K.M., & Lodato, G., 2016, Gravitational instabilities in circumstellar discs, *Ann. Rev. Astron. Astrophys.*, **54**, 271–311.

Shakura, N.I., & Sunyaev, R.A., 1973, Black holes in binary systems: observational appearance, *Astron. Astrophys.*, **24**, 337–55.

Vector Identities and Theorems

Grad, div and curl in Cartesian coordinates:

$$\nabla\varphi = \frac{\partial\varphi}{\partial x}\mathbf{i} + \frac{\partial\varphi}{\partial y}\mathbf{j} + \frac{\partial\varphi}{\partial z}\mathbf{k}$$

$$\nabla\cdot\mathbf{A} = \frac{\partial A_x}{\partial x} + \frac{\partial A_y}{\partial y} + \frac{\partial A_z}{\partial z}$$

$$\nabla\times\mathbf{A} = \left[\frac{\partial A_z}{\partial y} - \frac{\partial A_y}{\partial z}\right]\mathbf{i} + \left[\frac{\partial A_x}{\partial z} - \frac{\partial A_z}{\partial x}\right]\mathbf{j} + \left[\frac{\partial A_y}{\partial x} - \frac{\partial A_x}{\partial y}\right]\mathbf{k}$$

Grad, div and curl in cylindrical polar coordinates (r, θ, z)**:**

$$\nabla\varphi = \frac{\partial\varphi}{\partial r}\hat{\mathbf{e}}_r + \frac{1}{r}\frac{\partial\varphi}{\partial\theta}\hat{\mathbf{e}}_\theta + \frac{\partial\varphi}{\partial z}\hat{\mathbf{e}}_z$$

$$\nabla\cdot\mathbf{A} = \frac{1}{r}\frac{\partial}{\partial r}(rA_r) + \frac{1}{r}\frac{\partial A_\theta}{\partial\theta} + \frac{\partial A_z}{\partial z}$$

$$\nabla\times\mathbf{A} = \left[\frac{1}{r}\frac{\partial A_z}{\partial\theta} - \frac{\partial A_\theta}{\partial z}\right]\hat{\mathbf{e}}_r + \left[\frac{\partial A_r}{\partial z} - \frac{\partial A_z}{\partial r}\right]\hat{\mathbf{e}}_\theta + \left[\frac{1}{r}\frac{\partial}{\partial r}(rA_\theta) - \frac{1}{r}\frac{\partial A_r}{\partial\theta}\right]\hat{\mathbf{e}}_z$$

$$\nabla^2\varphi = \frac{1}{r}\frac{\partial}{\partial r}\left(r\frac{\partial\varphi}{\partial r}\right) + \frac{1}{r^2}\frac{\partial^2\varphi}{\partial\theta^2} + \frac{\partial^2\varphi}{\partial z^2}$$

$$\nabla^2\mathbf{A} = \left[\nabla^2 A_r - \frac{1}{r^2}A_r - \frac{2}{r^2}\frac{\partial A_\theta}{\partial\theta}\right]\hat{\mathbf{e}}_r + \left[\nabla^2 A_\theta - \frac{1}{r^2}A_\theta + \frac{2}{r^2}\frac{\partial A_r}{\partial\theta}\right]\hat{\mathbf{e}}_\theta + \left(\nabla^2 A_z\right)\hat{\mathbf{e}}_z$$

Vector identities:

$$\nabla(\varphi\psi) = \varphi\nabla\psi + \psi\nabla\varphi$$

$$\nabla(\mathbf{A}\cdot\mathbf{B}) = (\mathbf{A}\cdot\nabla)\mathbf{B} + (\mathbf{B}\cdot\nabla)\mathbf{A} + \mathbf{A}\times(\nabla\times\mathbf{B}) + \mathbf{B}\times(\nabla\times\mathbf{A})$$

$$\nabla\cdot(\varphi\mathbf{A}) = \varphi\nabla\cdot\mathbf{A} + \mathbf{A}\cdot\nabla\varphi$$

$$\nabla\cdot(\mathbf{A}\times\mathbf{B}) = \mathbf{B}\cdot(\nabla\times\mathbf{A}) - \mathbf{A}\cdot(\nabla\times\mathbf{B})$$

$$\nabla\cdot(\nabla\times\mathbf{A}) = 0$$

$$\nabla\times(\varphi\mathbf{A}) = \varphi(\nabla\times\mathbf{A}) + (\nabla\varphi)\times\mathbf{A}$$

$$\nabla\times(\mathbf{A}\times\mathbf{B}) = \mathbf{A}(\nabla\cdot\mathbf{B}) - \mathbf{B}(\nabla\cdot\mathbf{A}) + (\mathbf{B}\cdot\nabla)\mathbf{A} - (\mathbf{A}\cdot\nabla)\mathbf{B}$$

$$\nabla\times(\nabla\varphi) = 0$$

$$\nabla^2\mathbf{A} = \nabla(\nabla\cdot\mathbf{A}) - \nabla\times(\nabla\times\mathbf{A})$$

Integral theorems:

$$\int_V \nabla \cdot \mathbf{A} dV = \oint_S \mathbf{A} \cdot d\mathbf{S}, \quad \int_V \nabla \varphi dV = \oint_S \varphi \, d\mathbf{S}$$

$$\int_V \nabla \times \mathbf{A} dV = -\oint_S \mathbf{A} \times d\mathbf{S}$$

$$\int_S \nabla \times \mathbf{A} \cdot d\mathbf{S} = \oint_C \mathbf{A} \cdot d\mathbf{l}$$

$$\int_S \nabla \varphi \times d\mathbf{S} = -\oint_C \varphi d\mathbf{l}$$

Helmholtz's decomposition:
A vector field $\mathbf{B}$ may be expressed as the sum of an irrotational and a solenoidal vector field. The irrotational field may be written in terms of a scalar potential, φ, and the solenoidal field in terms of a vector potential, $\mathbf{A}$, such that

$$\mathbf{B} = -\nabla \varphi + \nabla \times \mathbf{A}, \quad \nabla \cdot \mathbf{A} = 0.$$

These two potentials are solutions of the Poisson equations

$$\nabla^2 \varphi = -\nabla \cdot \mathbf{B}, \qquad \nabla^2 \mathbf{A} = -\nabla \times \mathbf{B},$$

which may be inverted in an infinite domain using the Green's function solution below.

Green's inversion of Poisson's equation:
Consider the Poisson equation

$$\nabla^2 \varphi = -\delta(\mathbf{x}), \quad \varphi_\infty = 0,$$

where $\delta(\mathbf{x})$ is a unit delta function located at the origin. The solution of this Poisson equation is

$$\varphi(\mathbf{x}) = \frac{1}{4\pi |\mathbf{x}|},$$

in the sense that: (i) $\nabla^2 \varphi = 0$ for $|\mathbf{x}| \neq 0$; and (ii) integrating over the sphere $|\mathbf{x}| \leq R$ gives

$$\int_V \nabla^2 \varphi d\mathbf{x} = \oint_S \nabla \varphi \cdot d\mathbf{S} = -\oint_S \frac{1}{4\pi |\mathbf{x}|^2} dS = -1 = -\int \delta(\mathbf{x}) d\mathbf{x},$$

as demanded by the Poisson equation. If the source is a delta function of strength s' located at $\mathbf{x}'$, then our solution clearly generalises to $\varphi(\mathbf{x}) = s'/4\pi |\mathbf{x} - \mathbf{x}'|$. Now consider the case

$$\nabla^2 \varphi = -s(\mathbf{x}), \quad \varphi_\infty = 0,$$

where the source, $s(\mathbf{x})$, is any localised function of position. Superposition now yields

$$\varphi(\mathbf{x}) = \frac{1}{4\pi} \int \frac{s(\mathbf{x'})}{|\mathbf{x} - \mathbf{x'}|} \, d\mathbf{x'},$$

where $\mathbf{x'}$ is a dummy variable that samples the source at different locations and $d\mathbf{x'}$ is a volume element. This is known as the Green's function solution of Poisson's equation.

Navier-Stokes Equation in Cylindrical Polar Coordinates

The viscous contribution to the stress tensor:

$$\tau_{rr} = 2\rho\nu\frac{\partial u_r}{\partial r}, \qquad \tau_{\theta\theta} = 2\rho\nu\left[\frac{1}{r}\frac{\partial u_\theta}{\partial \theta} + \frac{u_r}{r}\right], \qquad \tau_{zz} = 2\rho\nu\frac{\partial u_z}{\partial z}$$

$$\tau_{r\theta} = \rho\nu\left[r\frac{\partial}{\partial r}\left(\frac{u_\theta}{r}\right) + \frac{1}{r}\frac{\partial u_r}{\partial \theta}\right], \qquad \tau_{\theta z} = \rho\nu\left[\frac{1}{r}\frac{\partial u_z}{\partial \theta} + \frac{\partial u_\theta}{\partial z}\right], \qquad \tau_{zr} = \rho\nu\left[\frac{\partial u_r}{\partial z} + \frac{\partial u_z}{\partial r}\right]$$

Navier–Stokes equation:

$$\frac{\partial u_r}{\partial t} + \left[(\mathbf{u}\cdot\nabla)u_r - \frac{u_\theta^2}{r}\right] = -\frac{1}{\rho}\frac{\partial p}{\partial r} + \nu\left[\nabla^2 u_r - \frac{u_r}{r^2} - \frac{2}{r^2}\frac{\partial u_\theta}{\partial \theta}\right]$$

$$\frac{\partial u_\theta}{\partial t} + \left[(\mathbf{u}\cdot\nabla)u_\theta + \frac{u_r u_\theta}{r}\right] = -\frac{1}{\rho r}\frac{\partial p}{\partial \theta} + \nu\left[\nabla^2 u_\theta - \frac{u_\theta}{r^2} + \frac{2}{r^2}\frac{\partial u_r}{\partial \theta}\right]$$

$$\frac{\partial u_z}{\partial t} + (\mathbf{u}\cdot\nabla)u_z = -\frac{1}{\rho}\frac{\partial p}{\partial z} + \nu\nabla^2 u_z$$

Navier-Stokes equation in terms of stresses:

$$\frac{\partial u_r}{\partial t} + \left[(\mathbf{u}\cdot\nabla)u_r - \frac{u_\theta^2}{r}\right] = -\frac{1}{\rho}\frac{\partial p}{\partial r} + \frac{1}{\rho}\left[\frac{1}{r}\frac{\partial}{\partial r}(r\tau_{rr}) + \frac{1}{r}\frac{\partial}{\partial \theta}(\tau_{r\theta}) + \frac{\partial \tau_{rz}}{\partial z} - \frac{\tau_{\theta\theta}}{r}\right]$$

$$\frac{\partial u_\theta}{\partial t} + \left[(\mathbf{u}\cdot\nabla)u_\theta + \frac{u_r u_\theta}{r}\right] = -\frac{1}{\rho r}\frac{\partial p}{\partial \theta} + \frac{1}{\rho}\left[\frac{1}{r^2}\frac{\partial}{\partial r}(r^2\tau_{\theta r}) + \frac{1}{r}\frac{\partial}{\partial \theta}(\tau_{\theta\theta}) + \frac{\partial \tau_{\theta z}}{\partial z}\right]$$

$$\frac{\partial u_z}{\partial t} + (\mathbf{u}\cdot\nabla)u_z = -\frac{1}{\rho}\frac{\partial p}{\partial z} + \frac{1}{\rho}\left[\frac{1}{r}\frac{\partial}{\partial r}(r\tau_{zr}) + \frac{1}{r}\frac{\partial}{\partial \theta}(\tau_{z\theta}) + \frac{\partial \tau_{zz}}{\partial z}\right]$$

Geophysical Data

The Earth's Interior:

Table A3.1 Estimated properties of the core of the earth and various dimensionless groups.

Equatorial radius	R_{equ}	6378 km
Polar radius	R_{pole}	6357 km
Ellipticity	$(R_{equ} - R_{pole})/R_{equ}$	1/299
Liquid core radius	R_C	3485 km
Ellipticity of liquid core	e	1/400
Inner core radius	R_i	1215 km
Angular velocity	Ω	$7.29 \times 10^{-5}\,\text{s}^{-1}$
Mean density of fluid in outer core	ρ	$10.9 \times 10^3\,\text{kg/m}^3$
Magnetic diffusivity of liquid core	λ	$0.7\,\text{m}^2/\text{s}$
Mean axial field in the conducting core	$\bar{B}_z$	3.7 Gauss
Characteristic velocity of fluid in core	$|\mathbf{u}|$	$\sim 0.2\,\text{mm/s}$
Minimum length-scale for motion	ℓ_{min}	$\sim 4\,\text{km}$?
Magnetic Reynolds number	$R_m = \frac{|\mathbf{u}|(R_C-R_i)}{\lambda}$	$O(600)$
Rossby number	$\text{Ro} = \frac{|\mathbf{u}|}{\Omega(R_C-R_i)}$	$O(10^{-6})$
Magnetic Prandtl number	$\text{Pr}_m = \nu/\lambda$	$O(10^{-6})$
Ekman number	$\text{Ek} = \frac{\nu}{\Omega R_C^2}$	$O(10^{-15})$
Elsasser number	$\Lambda = \frac{\sigma B^2}{\rho \Omega}$	$O(1)$

The Planets:

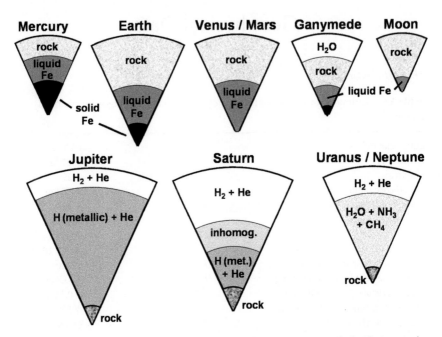

Figure A3.1 Approximate structure of the planets. (Figure courtesy of Uli Christensen)

Table A3.2 Properties of the planets.

Planet	Rotation period (days)	Mean radius of planet (10^3 km)	Radius, R_C, of conducting core (10^3 km)	Dipole moment (10^{22} Am^2)	Mean axial field in core, $\bar{B}_z$ (Gauss)	Dipole angle	$\dfrac{\bar{B}_z/\sqrt{\rho\mu}}{\Omega R_C} \times 10^6$
Mercury	58.6	2.44	1.8	0.004	0.014	5°	5.5
Venus	243	6.05	3.2	0	-	-	-
Earth	1	6.37	3.48	7.9	3.7	11°	13
Mars	1.03	3.39	1.8	0	-	-	-
Jupiter	0.413	69.9	55	150,000	18	9.6°	5.2
Saturn	0.440	58.2	29	4600	3.8	<0.1°	2.3
Uranus	0.718	25.3	~18	390	~1.3	59°	~1.0
Neptune	0.671	24.6	~20	200	~0.5	47°	~0.3

The Physical Properties of Common Fluids

Table A4.1 The physical properties of some common liquids at 20°C

Liquid	Density $\rho\,(\mathrm{kg\,m^{-3}})$	Thermal diffusivity $\alpha\,(\mathrm{m^2 s^{-1}})$	Kinematic viscosity $\nu\,(\mathrm{m^2 s^{-1}})$	Prandtl number Pr
Glycerol	1260	9.4×10^{-8}	1.2×10^{-3}	1.3×10^4
Mercury	13,500	4.6×10^{-6}	1.2×10^{-7}	0.026
Olive oil	920	9.4×10^{-8}	9.1×10^{-5}	970
Engine oil	888	8.7×10^{-8}	9.0×10^{-4}	1.0×10^4
Seawater	1025	1.5×10^{-7}	1.1×10^{-6}	7.3

Table A4.2 The physical properties of gasses at atmospheric pressure and 300°K

Gas	Density $\rho\,(\mathrm{kg\,m^{-3}})$	Thermal diffusivity $\alpha\,(\mathrm{m^2 s^{-1}})$	Kinematic viscosity $\nu\,(\mathrm{m^2 s^{-1}})$	Prandtl number Pr
Carbon dioxide	1.80	1.06×10^{-5}	8.32×10^{-6}	0.78
Hydrogen	0.0819	1.55×10^{-4}	1.10×10^{-4}	0.71
Nitrogen	1.14	2.20×10^{-5}	1.56×10^{-5}	0.71
Oxygen	1.30	2.24×10^{-5}	1.59×10^{-5}	0.71
Helium	0.163	1.76×10^{-4}	1.23×10^{-4}	0.70

Table A4.3 The physical properties of water

Temperature °C	Density $\rho\,(\mathrm{kg\,m^{-3}})$	Thermal diffusivity $\alpha\,(\mathrm{m^2 s^{-1}})$	Kinematic viscosity $\nu\,(\mathrm{m^2 s^{-1}})$	Prandtl number Pr
10	999.7	1.38×10^{-7}	1.30×10^{-6}	9.4
20	998.2	1.42×10^{-7}	1.00×10^{-6}	7.1
30	995.7	1.46×10^{-7}	0.802×10^{-6}	5.5
40	992.3	1.52×10^{-7}	0.659×10^{-6}	4.3

Table A4.4 The physical properties of air at atmospheric pressure

Temperature °K	Density ρ (kg m^{-3})	Thermal diffusivity α (m^{2}s^{-1})	Kinematic viscosity ν (m^{2}s^{-1})	Prandtl number Pr
250	1.41	1.57×10^{-5}	1.13×10^{-5}	0.72
300	1.18	2.22×10^{-5}	1.57×10^{-5}	0.71
350	0.998	2.98×10^{-5}	2.08×10^{-5}	0.70
400	0.883	3.76×10^{-5}	2.59×10^{-5}	0.69

Table A4.5 The U.S. standard atmosphere

z km	T °C	p kPa	ρ kg/m^3
0	15.0	101.3	1.225
1	8.5	89.9	1.112
2	2.0	79.5	1.007
3	−4.5	70.1	0.909
4	−11.0	61.7	0.819
5	−17.5	54.0	0.736
6	−24.0	47.2	0.660
8	−37.0	35.6	0.526
10	−49.9	26.5	0.413
12	−56.5	19.3	0.311
14	−56.5	14.1	0.226
16	−56.5	10.3	0.165

Table A4.6 The physical properties of liquid metals

Metal	Melting point °C	Reference temp. °C	Density ρ 10^3 kg/m^3	Viscosity ν 10^{-6} m^2/s	Electrical conductivity 10^6 Ω^{-1}m^{-1}	Thermal conductivity Wm^{-1}C^{-1}
Titanium	1685	1700	4.1	1.3	0.58	-
Steel (0.2% C)	1495	1600	7.0	0.88	0.71	26
Iron	1535	1600	7.0	0.80	0.72	41
Nickel	1454	1500	7.9	0.62	1.2	-
Copper	1083	1100	7.9	0.51	4.8	160
Aluminium	660	700	2.4	0.60	4.1	95
Magnesium	650	700	1.6	0.80	3.6	81
Tin	232	280	6.9	0.28	2.1	31
Sodium	98	100	0.92	0.68	10	89
Potassium	64	70	0.82	0.58	7.0	52
Galium	30	70	6.1	0.31	3.8	30
Mercury	−38	30	13.5	0.12	1.0	8.0

Subject Index